CRIME SCENE PHOTOGRAPHY

Visit the *Crime Scene Photography* website at:

http://books.elsevier.com/companions/9780123693839

The *Crime Scene Photography* website contains hundreds of color images from the book, as well as another 500 images arranged as PowerPoint™ slideshows that demonstrate some of the key concepts covered in the text. These additional images are meant to aid students in their own practical experience by showing examples of successes and failures in crime scene photography.

CRIME SCENE PHOTOGRAPHY

Edward M. Robinson

With a Foreword by

Gerald B. Richards

FBI Special Agent (Retired);
Former Chief, Special Photographic Unit,
FBI Laboratory

ELSEVIER

AMSTERDAM • BOSTON • HEIDELBERG • LONDON
NEW YORK • OXFORD • PARIS • SAN DIEGO
SAN FRANCISCO • SINGAPORE • SYDNEY • TOKYO

Academic Press is an imprint of Elsevier

Acquisitions Editor:	Jennifer Soucy
Assistant Editor:	Kelly Weaver
Marketing Manager:	Diane Jones
Project Manager:	Jeff Freeland
Cover Designer:	Alisa Andreola
Compositor:	SPi Technologies India Pvt. Ltd.
Cover Printer:	Phoenix Color Corp.
Text Printer/Binder:	RR Donnelly

Academic Press is an imprint of Elsevier
30 Corporate Drive, Suite 400, Burlington, MA 01803, USA
525 B Street, Suite 1900, San Diego, California 92101-4495, USA
84 Theobald's Road, London WC1X 8RR, UK

This book is printed on acid-free paper. ∞

Library of Congress Cataloging-in-Publication Data
Robinson, Edward M.
  Crime scene photography/Edward M. Robinson.
    p. cm.
  Includes bibliographical references and index.
  ISBN-13: 978-0-12-369383-9(alk. paper)
  ISBN-10: 0-12-369383-7(alk. paper)
  1. Legal photography. I.Title.

  TR822.R63 2007
  779'.936325–dc22   2006051910

British Library Cataloguing-in-Publication Data
A catalogue record for this book is available from the British Library.

ISBN-13: 978-0-12-369383-9
ISBN-10: 0-12-369383-7

For information on all Academic Press publications visit our
Web site at www.books.elsevier.com

Printed in the United States of America
07 08 09 10 11 12   8 7 6 5 4 3 2 1

This is dedicated to my wife, Sue.
Thank you for holding down the fort
while I was off writing this book,
and for always being there.

TABLE OF CONTENTS

FOREWORD   xi
ACKNOWLEDGMENTS   xiii
INTRODUCTION   xvii

CHAPTER 1   COMPOSITION AND CARDINAL RULES   1
            Learning Objectives   1
            Use-Once (Or, Use One Time) Camera Versus
               Professional Camera System   1
            Composition and Cardinal Rules   4
            Cardinal Rules of Crime Scene Photography   25
            Chapter Summary   25
            Discussion Questions   27
            Practical Excercises   27
            Further Reading   28

CHAPTER 2   BASIC EXPOSURE (NON-FLASH) CONCEPTS   29
            Learning Objectives   29
            Key Terms   29
            The Proper Exposure Triangle   30
            Shutter Speed as Motion Control   61
            Reciprocal Exposures   67
            The Reflective Light Meter   72
            "Normal" and "Non-Normal" Scenes   76
            Tools for Determining "Proper" Exposures with
               Tricky Scenes   85
            Bracketing   92
            The F/16 Sunny Day Rule   97
            Causes for Complete Rolls of Film with Exposure Errors 102
            Common Filters   103
            The Eye Cup Cover   113
            Summary   115
            Discussion Questions   117
            Exercises   117
            Further Reading   118
            Additional Reading   118

CHAPTER 3   FOCUS, DOF, AND LENSES   **119**
Learning Objectives   119
Key Terms   120
Focus   120
Depth of Field   157
Lenses   168
Summary   212
Discussion Questions   212
Exercises (All Non-Flash Shots)   213
Further Reading   213

CHAPTER 4   ELECTRONIC FLASH   **215**
Learning Objectives   215
Key Terms   216
Guide Numbers   216
Flash Sync Speeds   218
Set the Flash for the Film Used   224
Manual Flash Mode   224
The Inverse Square Law   245
Automatic and Dedicated Flash Exposure Modes   253
Built-In Flash Units   261
Fill-In Flash   261
Oblique Light, Both Flash and Non-Flash (Flashlight)   269
Bounce Flash   295
Painting with Light   302
Summary   318
Discussion Questions   318
Exercises   319
Further Reading   320

CHAPTER 5   CRIME SCENE PHOTOGRAPHY   **321**
Learning Objectives   321
Key Terms   321
Photo Documentation Forms   321
Overall Photographs   331
Midrange Photographs   344
Close-Up Photographs   352
The Photographic Documentation of Bodies
   and Wounds   357
Summary   380
Discussion Questions   381
Exercises   381
Further Reading   382

CHAPTER 6   ULTRAVIOLET, INFRARED, AND FLUORESCENCE   **383**
Learning Objectives   383
Key Terms   383
The Electromagnetic Spectrum (EMS)   384
Ultraviolet Light (UV)   391

Infrared Light (IR) on the Electromagnetic Spectrum   398
Visible Light Fluorescence   406
Summary   417
Discussion Questions   417
Exercises   418
Further Reading   419

CHAPTER 7   **PHOTOGRAMMETRY**   421
Learning Objectives   421
Key Terms   421
Introduction to Photogrammetry   421
Perspective Grid Photogrammetry   425
Perspective Disc Photogrammetry   441
Natural Grid Photogrammetry   446
Reverse Projection Photogrammetry   454
Rhino Photogrammetry   456
Summary   462
Discussion Questions   462
Exercises   463
Further Reading   464

CHAPTER 8   **DIGITAL IMAGING**   465
Contributed by David "Ski" Witzke
Learning Objectives   465
Key Terms   465
In The Beginning   466
Expose Yourself to Digital Imaging Concepts: Bits,
   Bytes, Pixels, and Dots   478
Take a Picture...It Lasts Longer. Or, Does It?   491
Image S&M...Storage and Management of Your
   Digital Images   498
Summary   511
Further Reading   511

CHAPTER 9   **SPECIAL PHOTOGRAPHY SITUATIONS**   513
Learning Objectives   513
Key Terms   514
Accident Photography   514
Surveillance Photography   533
Aerial Photography   546
Underwater Photography   553
Summary   565
Discussion Questions   565
Exercises   566
Further Reading   567

CHAPTER 10 **LEGAL ISSUES RELATED TO PHOTOGRAPHS AND**
DIGITAL IMAGES **569**
Learning Objectives 569
Key Terms 569
Criteria of Photographs and Digital Images as Evidence 569
The Purpose of Crime Scene Photographs 582
Photographs of Suspects and Evidence 583
Legal Implications of Digital Imaging 585
International Association for Identification (IAI)
Resolution 97-9 597
Case Law Citations Relevant to Film Photos and
Digital Images 598
Summary 624
Discussion Questions 625
Further Reading 626

APPENDIX **TIPS, TRICKS, AND MACGYVERS** **627**
Evidence on Vertical Walls 627
Coaxial Lighting 629
Mirrors 631
Distortion Correction, Redux 632

GLOSSARY **635**
INDEX **655**

FOREWORD

Since the introduction of photography in 1839 by Louis Daguerre, this art and science has been vigorously applied to law enforcement and the justice system of every civilized nation. From its humble beginnings as a means of recording criminal faces, in the form of "mug shots," to the 21st century wherein almost all major cases depend heavily on photographic imaging, photography has become the silent archivist and witness in untold thousands of civil and criminal matters. We only have to look at the millions of images produced each year in recording accidents, crime scenes, and evidence and for demonstrative exhibits to understand why the forensic photographer is truly the "master of all photographic trades." The overall field of forensic photography covers a wide variety of photographic and imaging disciplines, each having a multitude of techniques, equipment, and levels of experience. When we really think of "forensic photography" (photography as it applies to legal matters), it covers the entire gambit of imaging know-how, from the portrait (mug shot) photographs to aerial, macro, micro, spectral, studio, field, underwater, and many other facets of the photographic profession. In addition, the modern forensic photographer must be part chemist, physicist, color expert, mechanic, lighting technician, computer geek, artist, communicator, and investigator, in order to be successful.

Crime Scene Photography is an extremely well-organized book. At the beginning of each of the ten chapters are the Learning Objectives and Key Terms that will be used for the remainder of the text. Within the text are tips and Rules of Thumb that highlight important aspects of the chapter. The chapters then conclude with a succinct summary, discussion questions, exercises, and a comprehensive list of further readings. What I find to be the most compelling attribute of this book is its abundance of well thought-out, high-quality photographs that complement the text and lessons being presented. The photographs not only show the reader how to do it, but just as important, how not to do it. After all, what is a text on photography without good photographs?

This literary work outlines the basic foundation of forensic picture-taking in substantial detail, making an ideal tutorial for a person just entering the field. It covers the basic crime scene scenario by providing important details to guide the student and professional through the process in a step-by-step manner. Not only is the photographic aspect addressed, but the handling and care of evidence, as well as maintenance of the crime scene, are integrated into the process.

What I perhaps appreciate most about *Crime Scene Photography* is the in-depth explanation of subjects that seem to confound and confuse even many experienced practitioners, such as diffraction in close-up photography (Chapter 5), inverse square law (Chapter 2), and ultraviolet/infrared imaging (Chapter 6). Good explanations of these subjects are difficult to find. Another important topic that is non-existent in other forensic photography books is "photogrammetry," discussed in Chapter 7. The photogrammetry techniques discussed here are basic single-image graphical methods that can be implemented with little previous experience or equipment, and can provide a reasonable degree of accuracy if the underlying principles are followed. This chapter, along with the chapter covering lenses (Chapter 3), further emphasizes important basic concepts regarding angle of view and perspective. Having a chapter devoted to digital imaging (Chapter 8), the technology of the future, is perhaps one of the most valuable aspects of *Crime Scene Photography*. It is a wealth of succinct information that the forensic photographer will deal with on a daily basis.

Not since Charles C. Scott's 1942 epic text, *Photographic Evidence,* has there been a book this detailed and well written. Ted Robinson has taken a complex subject, strewn with technical jargon, detailed processes, both old and cutting-edge technology, and woven it into a comprehensive, authoritative work that serves well both the student and seasoned professional. Having taught this subject at the college level for many years, it most surely would be the text book I would select for my students. And if an experienced practitioner were to acquire only one book on forensic photography for their bookshelf this year, *Crime Scene Photography* should be their choice.

Gerald B. Richards, FBI Special Agent (Retired)
Former Chief, Special Photographic Unit, FBI Laboratory

ACKNOWLEDGMENTS

This text had its origin in 1989, when I decided it was necessary to write a "how-to" pamphlet on police photography for the Arlington County Police Department, Virginia. They gave me the time to write the *Basic Police Photography for Police Agents Handbook,* so my list of acknowledgements must begin with the ACPD. At that time, I had no idea those 59 pages would lead to this text.

Within the last year, as this book began to materialize, I realized I would need more images than I currently had at the time. I returned to the ACPD and requested images from three of my past working buddies: Marc Hackett, Lisa Haring, and Keith Ahn. They came through for me again, and their images make this a much better book.

With one of the intended target readers of this text being students in an educational environment, it made sense to me to utilize the resources I had available. Since I was currently teaching a course on forensic photography at GWU, I decided to use the local talent: my students. I provided them with early drafts of this book, and charged them with forcing me to make the concepts easier for them to understand. I told them if it wasn't currently written in a way that made concepts clear, point it out, and make me explain it a different way. This was risky: asking for criticism! But the students were up to the task. My suggestion to would-be authors is to avoid asking for criticism unless you have very thick skin.

In addition to asking students to "proof-read" my early chapter drafts, I asked them to cross reference my text with other photography books in my library. Some were asked to create the PowerPoint™ slides of extra images, now located on the companion website, which accompanies this text. Some students provided images for the text to exemplify concepts or principals needing a visual aid. Some wrote discussion questions. These now appear in the main text, and some appear in the Instructor's Manual.

These students were: Kim Atkinson, Naila Bhatri, Jackie Britton, Tom Bush, Heather Calloway, Brian Carroll, Michelle Cobb, Cliff Cook, Ryan Costello, Daniel

Cowen, Kim Criss, Jessica Dee, Ryan Derstine, Atree Desai, Martin Eaves, Melinda Filman, Adam Garver, Matt Graves, Sylvia Greenman, Valerie Hart, Jason Keller, Chris Kinling, Katalin Korossy, Christina Marney, Nick Oliver, Cara McBrayer, Elizabeth Neuendorf, Devon Pierce, Brigid Reilly, Regan Scott, Denise Sediq, Benjamin Serinsky, Carolyn Sikorski, Pat Stewart, Lauren Stignitto, Rachel Stowens, Claudia Thomas, Martha Ward, and Bianca Whitlock.

Three students worked researching case cites for the last chapter on legal issues, and spent many nights in our law school library: Christina Verren, Joelle Duval, and Alissa Ehr. I had thought that those cases you all found would be a distinct benefit to this text, and I was correct.

These students have made this a better book that it would have been without their help. If this text seems to make sense at times, it is because they have forced me to make it better.

With over 600 images printed in the book, and another 500 images contained on the companion website, I'm afraid I may have omitted attribution for a shot or two. If any of my previous students notice that I have failed to properly credit one of their images, I hope they will get in touch with me so that I can have my omission corrected when the book reprints. I also hope they will accept my apology.

I've been fortunate to receive many fine images from several sources, and some require special thanks. Among those are: Jeff Miller, Fairfax County (Virginia) Police Department ID Unit, who helped supply the images related to the DC Sniper case; Nancy Olds, USSS photographer, for many excellent underwater images; Criminalist Rebecca Shaw, Arapahoe County (Colorado) Sheriff's Office for a myriad of crime scene photos; and Senior Criminalist Thomas W. Adair, Westminster Police Department, Colorado, especially for the UAV images.

The chapters on photogrammetry and UV, IR, and fluorescence contain some very technical information. I specifically asked Gerald Richards, a friend of many years, whose wealth of knowledge I respect enormously, to keep me honest. Jerry suggested many critical improvements, and caught several downright errors. That was exactly why I asked for his help in the first place. Thanks, Jerry!

I can't thank David Witzke enough for writing the chapter on digital imaging. When the prospect of actually writing this book became a reality, I knew this chapter would be essential to the success of the entire book. Although I feel very comfortable with most things digital, I knew one person who still amazes me with the depth of his digital knowledge. There was no search for assistance in writing the chapter on digital imaging. I went straight to "Ski." My only fear was that his hectic schedule of traveling the country teaching basic and advanced digital imaging courses would make it impossible for him to commit to writing this chapter. Fortunately, he immediately agreed to write that chapter. My worries were over. Thanks, Ski.

Academic Press automatically requires a review of an anonymous technical editor whenever any of their books is being prepared for publication. Far from being a rubber stamp, approving entire chapters with glowing accolades, this editor showed no mercy. With equal enthusiasm as some of my previous students, this technical editor delighted in deflating the ego this author sometimes felt was justified, having almost completed a huge task. Rather than digging in and erecting defenses, I took his/her constructive criticism for what it was: an opportunity to make the book better than it would have been otherwise. Many changes later, I do believe this book emerges from the dust and debris of a work-in-progress much better for the experience. Thank you. I hope you agree this book has profited from your suggestions.

Finally, I don't believe this book would have been completed but for the constant encouragement and prodding from my two Academic Press contacts, Jennifer Soucy, Acquisitions Editor, and Kelly Weaver, Assistant Editor. When I was excited, they let me be; when I was down, they brought be back up. What a team! Thank you both for being there through the entire process.

If there are still errors in this book, they are entirely mine. I obviously chose to disregard the advice of many good people. But, give me the chance, and I'll correct them.

INTRODUCTION

I began my career in law enforcement in 1971 with the Arlington County (Virginia) Police Department (ACPD). My first camera as a crime scene photographer was a 4″×5″ Speed Graphic, which is what many people consider the classic photojournalist's camera.

Figure I.1
Speed Graphic Camera.
Courtesy of Les Newcomer.

PACEMAKER SPEED GRAPHIC WITH GRAFLITE

This camera took excellent photographs, but was not known for being able to take many photographs quickly. Next, our department began using a medium format Yashica-Mat TLR (twin lens reflex). We eventually evolved into 35 mm photography when we obtained the Pentax K-1000, known for its tank-like sturdiness. Even though it had no bells or whistles, the K-1000 could get the job done.

In 1989, I was appointed as the training coordinator for our department's Basic and Advanced Evidence Collection Schools. The Basic School was to prepare all candidates for promotion for the position of Police Agent, at that time our department's name for sworn crime scene technicians. The examination process included a written exam and a mock crime scene practical exercise. The Advanced School was for those successfully promoted, who would soon begin processing crime scenes. As the training coordinator, I was asked to develop a list of acceptable references to be used as study guides for the various topics included in the examination process.

When I reviewed the various books on police and crime scene photography that then existed, I concluded none would suffice for our needs. Existing texts often read as laundry lists of what items should be photographed, but did not provide information on how to do the photography. Frustrated at the lack of an adequate reference, I asked if I could write a photography manual that would satisfy our needs. That request was granted, and I soon wrote the 59-page *Arlington County Police Department Basic Police Photography for Police Agents Handbook.*

Figure I.2
1990 ACPD Photography Handbook.

Since then, I have been expanding on that original handbook as I continued to teach crime scene photography within my department. I also eventually taught both the basic and advanced 40-hour photography courses at our local Northern Virginia Criminal Justice Academy, serving multiple law enforcement agencies in the area.

In 1996, after 25 years with the ACPD, I retired, and began a second career as a civilian supervisor with the Baltimore County Police Department's Forensic Service Section, supervising a shift of predominantly civilian Forensic Evidence Technicians. I was also responsible for the training of all new hires within that unit. I was then responsible for teaching civilian crime scene search personnel as I had been teaching police officers and deputy sheriffs. The camera then used by Baltimore County Forensic Evidence Technicians was the Nikon 8008, which did have many bells and whistles. It also had a dedicated flash unit. Life was good! The crime scene photography textbook situation, however, had not changed. None was better than the material I had been acquiring over the years.

Deriving more personal satisfaction and enjoyment from my training function rather than my supervisory duties, I began looking for a full-time teaching position. In 1999, I was hired by National University in San Diego, California to teach courses for their Master of Forensic Science degree. Those duties included teaching a course on forensic photography. In 2000, I was hired by The George Washington University's Department of Forensic Science to create the Crime Scene Investigation concentration for their Master of Forensic Science degree program. I also began teaching their forensic photography course.

I continually search the literature in the field, and available textbooks on general photography, and I have yet to find one I believe superior to the material I have been accumulating. I have been providing my students with my "book-in-progress" as their required reading material for several years now. It is now at a stage I believe is worthy to be printed as a text book.

I believe it is a worthy for several reasons. The theories and techniques contained within its pages enabled me to be successful as I worked crime scenes as a police officer. When other police officers and deputy sheriffs used the techniques I passed on to them, they also have been successful. When civilian evidence technicians utilized my "tricks-of-the-trade" they too were successful. And now, being an Assistant Professor at the George Washington University, I continue to teach a varied student body. My current students include local law enforcement personnel, civilian students continuing their education right after finishing their undergraduate degrees, federal law enforcement officers and special agents, and military (NCIS, AFOSI, Army CID) personnel. I continually receive notes and emails of appreciation from our Alumni, indicating my photography course has helped them in their current careers.

This textbook, therefore, is designed for two principal photography student types. One group is the **student in an academic setting**. Students are expected to acquire a solid grasp of the theories and concepts of their various courses. This is the **K**nowledge point of the "KSAs" frequently required by employers. Students in an academic setting are normally expected to also learn the aspects of photography that have their foundation in physics and optics. But, to make students competitive in the real world when applying for their first jobs, it is also important to have them acquire the **S**kills and **A**bilities which are the remainder of the sought after triad of "KSAs." Current students will have to capture successful images at their future jobs.

A second student type is the **current practitioner** working within various law enforcement agencies. At the police academy, the basic crime scene photography course serves as the first formal photography course for new crime scene photographers. The advanced crime scene photography course is aimed at crime scene investigators who have been in the field a number of years. After a quick review of the basics, to get everyone on the same page, concepts like UV, IR, and fluorescence photography are covered, along with photogrammetry, aerial, and underwater photography. I am often told by these students that they just want to learn how to take a successful picture. Abstract theories are less important to them, they say. Techniques to get the job done are what they look for. These students will hopefully learn the basics of crime scene photography and improve their skill set by reading this text, and eventually become recognized within their agencies as the go-to person when quality images are required.

This marriage of theory and practice is a fundamental aspect of this textbook. I believe that many crime scene photographers not only want to know how to produce successful images, but they also want to understand the basic concepts of photography. The more the educated crime scene photographer understands these theories, the better they will be at explaining the results of their work in a courtroom situation, or while doing in-house training of new crime scene photographers; and, the better they are at these specific aspects of their job, the more valued they will be as an employee. Also, having a firm grasp of the theories and concepts of crime scene photography, as difficulties are encountered at various scenes, trained crime scene photographers will be able to pull from their photographic knowledge to solve the problem.

It should be mentioned that, many times, both crime scene photography student types have never taken a previous formal photography course. They may have read some basic photography books on their own, but they have frequently never been exposed to any structured course purported to cover all the basics of general photography, and certainly not crime scene photography concepts. On campus and at the police academy, my lesson plans could not presume my students had taken either high school or community college photography courses. This text also presumes no previous formal photography experience. This will be beneficial to the true novice to photography. Those with prior training or experience in the field will be able to skim quickly past many sections of this text.

What separates this text from other books on law enforcement photography? As mentioned above, this text is designed to teach the "how to" aspects of photography at major crime and accident scenes. Recommendations are made for specific exposure combinations at every stage of typical crime scene photography. Just as important, the issue of composition is thoroughly examined, and recommendations are made for effective and proper composition. Various focusing techniques are also included in this text. Most non-law enforcement photographers just turn the focus ring until their primary subject comes into focus. This text also explains how to focus on small and large areas, not just individual items. The terms "hyperfocal focusing," "zone focusing," and focusing by the "rule of thirds" are all covered in this text.

I believe this is the first law enforcement photography text to thoroughly cover the issue of diffraction, and how it impacts our "critical comparison" photography, sometimes also called "examination quality" photographs. A quick examination of general photography books will quickly find references to diffraction, and how high-end professional photographers consciously chose exposure variables with diffraction specifically in mind. The recognition of diffraction having a definite impact on our crime scene photography is finally covered here, so when the "sharpest" images are required for our most critical images, the reader will know how to select the optimal camera variables.

Many law enforcement agencies are currently transitioning from film cameras to digital imaging. Some agencies have already made the conversion to digital. The critical aspects of this conversion are covered in this text. Sure, digital cameras rely on much of the same photographic principals as film cameras. But there are definite differences also.

And, since one of the main purposes of crime scene photography is to produce images worthy of being evidence in criminal courts, a comprehensive list of court cases where film and digital imaging have been critical is included. One of the best ways to know for sure your images will survive challenges in court is to be aware of the previous court challenges, where film and digital images have survived attacks.

The inverse square law is thoroughly explained in this text. Many other photography books mention it in passing, but I believe this text covers it more thoroughly than any other book on crime scene photography. The use of electronic flash is dependent on the inverse square law. And, if you ever wondered where those strange f/stop numbers came from, the inverse square law is the answer.

This author is not so arrogant as to believe that every aspect of crime scene and accident scene photography is covered. Many times, it is possible to acquire a great image using slightly different techniques. I know there must be many "variations on a theme" out there. I don't want to suggest the techniques taught in this text are the only way to produce quality photographs. It is not "my way or no way." The techniques taught here, however, do work, and will enable you to capture the images necessary to get the job done. If you were taught a different technique which works for you, great! Continue using it. But, I'm a firm believer that the more tools you have on your Bat-Belt, the better you'll be at problem solving in the field. Consider the techniques offered in this text as alternate tools to attacking photography problems. If you haven't already acquired your "routine" image capture techniques, those offered by this text will work for you too.

I am fully aware of the MacGyver attitude prevalent in the law enforcement community. With duct tape, bubble gum, a bit of string, and a paper clip, MacGyver could do almost anything. Many crime scene photographers also pride themselves on solving problems as they are encountered, sometimes with very creative solutions. I predict the Appendix on "Tips, Tricks, and MacGyvers" will become one of the most valuable parts of this text. I invite all readers to send in their "tips" and solutions to problems they have come up with. Should this text be fortunate enough to go to a second edition, this Appendix should grow dramatically. The best part of coming up with a solution to a problem is being able to share it with others, so more "bad guys" are put away with our talents. I invite each reader to send in a "tip, trick, or MacGyver" that works for them. I will include them, with attribution, in the Appendix of the second edition.

Academic Press has thankfully agreed to print many, many images within this text. It is a book on photography, after all. As the number of images continued to escalate, it became apparent that colored images could not be used while keeping the price of the text manageable. A happy compromise was reached when it was agreed to provide all the images within the text in color on a supplemental website. In addition to the images in the text, used to demonstrate the concepts covered, the website also includes more than 500 additional images. The more images the student can look over, the quicker the reader's photographic eye can be developed. Instructors can also use these additional images to show successes and failures, so students do not have to learn from their own trials and errors.

As the chapters were organized for this text, I maintained my teaching methodology when teaching a semester long course on crime scene photography. I know how satisfying it is to depress that shutter button, and hear the camera making its normal noises. But, if the new student to photography begins taking photos too soon, enormous quantities of film can be wasted. It is essential to learn the basics before being given the first practical exercise. For this reason, the first chapters cover composition, exposure, focus and depth of field issues, and various flash techniques. How can you take that first indoor flash photo without knowing the basics? If the itch is overwhelming, however, or if you already have a command of the basics, jump ahead to the chapter which interests you the most. Then, at your leisure, return to the first four chapters to review the basics. Even experienced crime scene photographers should find some useful information there.

Many students, both on campus and at the police academy, have asked for some indication of the key points to remember. There are enormous amounts of information contained in this text, and it can be overwhelming to the beginning student. Over the years, I've come up with a variety of Cardinal Rules and Rules of Thumb to help satisfy these requests. These will not necessarily be found in other photography books. They are my attempts to highlight photography concepts I feel are most important. I use them as training tools to emphasize key concepts. Hopefully, the reader will also find them helpful.

I wish I knew an Irish prayer related to crime scene photography, so I could end this Introduction with it. A camera can be just as powerful a tool as the gun on the hips of sworn crime scene photographers. Since we use the camera at crime scenes more often than sworn officers use their guns, it only makes sense to use it well. Walk into any camera shop and you will overhear potential customers being told the program exposure mode (dummy mode) will guarantee they obtain proper exposures "every time." The well-trained crime scene photographer smiles and shakes their head when hearing such comments. We know the power of a good photograph. Use your photography tools wisely!

CHAPTER 1

COMPOSITION AND CARDINAL RULES

LEARNING OBJECTIVES

On completion of this chapter, you will be able to . . .

1. Explain how a professional photographer can use a simple point-and-shoot camera more effectively than a novice photographer can use a sophisticated camera system.
2. Explain the three cardinal rules of good photography.
3. Explain how the same subject can be composed differently in various images.
4. Explain why "fill the frame" has two aspects, both a positive and a negative connotation.
5. Explain why good composition partly depends on the point of view of the photographer, and what this point of view is.

As a quick search into the meaning of the word "photography" will reveal, it is derived from two Greek words, "phos" (light) and "graphia" (writing or drawing). Together they mean either "writing with light" or "drawing with light." Photography involves the creation of an image using light.

USE-ONCE (OR, USE ONE TIME) CAMERA VERSUS PROFESSIONAL CAMERA SYSTEM

A substantial amount of sophisticated equipment can be used to create some photographic images; however, it is just as possible to create many fine images with very limited equipment. Many years ago, a professional photographer issued a challenge to a nonprofessional photographer. The professional photographer stated he could take better photographs with an inexpensive, small use-once point-and-shoot camera than the nonprofessional could take with all the camera equipment the professional carried in his three large camera bags. The response was a very confused look on the face of the nonprofessional. How could access to all the most up-to-date and expensive camera equipment, which surely included all the latest bells and whistles professional photographers are fond of, not ensure a superior photograph? The professional photographer suggested the nonprofessional think about the challenge for a while, and then an explanation would be provided.

Even though this professional photographer was not a crime scene photographer, the reader should consider the challenge as well. This effort provides an excellent introduction to the concepts used to create quality photographs in general, and high-quality crime scene photographs in particular.

What aspect of the small inexpensive use-once point-and-shoot camera can the professional photographer use so effectively that the results will be noticeably better than what a nonprofessional photographer can produce with the best equipment available?

Examine an example of such a camera the professional said he could use so well.

Figure 1.1 shows four sides of a camera that will contain the answer.

Side "A":
1. Fun Saver 35 (indicating 35-mm film)
2. A lens designated as 35 mm (indicating the lens focal length)
3. f-11 (indicating the fixed aperture)
4. The front viewfinder

Side "B":
1. Use "in bright or partial sunlight."
2. "Get close…" but "4 ft. min."
3. "Center the subject…."
4. The rear viewfinder.
5. The film advance wheel.

Side "C":
1. 27 Exposures
2. Gold 400 film (indicating color rather than black and white film)
3. Shutter button
4. Pictures remaining window

Side "D":
1. ISO 400/27° (indicating the film's sensitivity to light)
2. 27 Exp. 24 × 36 mm (24 mm × 36 mm is the negative size with 35-mm film)
3. Miscellaneous small print

(a)

(b)

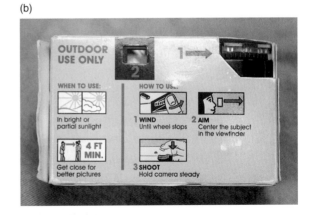

(c)

(d)

Figure 1.1

(a) Front of camera. (b) Back of camera. (c) Top of camera. (d) Bottom of camera.

The answer to the puzzle will be a camera variable some can use well and others will use much less efficiently. If it is an aspect of the camera that cannot be used "differently," it is not the solution to the puzzle.

Examine all the camera parts and information just listed, and the answer becomes more or less clear. It is the viewfinder!

COMPOSITION AND CARDINAL RULES

What is it about the viewfinder that the professional photographer will use so well? Years ago, the professional photographer provided the following answer. He said that professional photographers are much more aware of their compositions, and they take great pains to ensure that both the primary subject and the surrounding setting are contributing to the success of images. He further explained that the snapshot shooter usually is just concerned that the primary subject is in the field of view and then happily (and naïvely) presses the shutter button.

Explaining further, the professional indicated decisions are to be made regarding both the primary subject and the setting or area around the primary subject. The first decision is determining just what the primary subject should be. The professional knows that an image best tells its story when it is limited to just one idea with one primary subject. If more than a single subject is included in an image, the viewer may become confused about the intent of the photographer.

It needs to be emphasized that a "primary subject" does not necessarily mean a single object. The primary subject may, of course, be a single object, such as a gun used in a crime. A close-up photograph of the gun will include just the gun in the field of view. However, the primary subject may also be a small area including just two aspects of the crime scene. For example, a midrange photograph is one composed to show just one item of evidence in relation to a fixed feature of the scene. In this case, the "primary subject" is the item of evidence, the fixed feature of the scene, and the distance between them. Other aspects of the scene, then, become extraneous to the purpose of a midrange photograph and are best eliminated from the field of view. The primary subject may even be a larger area, with many items within its perimeter. For example, one of the purposes of an overall photograph is to show the crime scene in relation to the immediate area around it, which helps acclimate the viewer to the general surroundings around a crime scene. The crime did not occur in a void. In this case the photographer should select one/some aspect(s) of the surroundings that will help the viewer of the overall photograph grasp this relationship. This can be an easily recognized building, for example, or a nearby intersection. If a crime occurred in an alley of a particular business, showing the alley and a recognizable side of the business in one photograph will suffice. If a fatal accident occurred near a particular intersection, photographs from the intersection toward the vehicles in their final position or from the vehicles toward the intersection may be all that is needed. What is important to understand and remember is that once the "primary subject" has been defined in the mind of the photographer, the photographer should then become very aware of everything that is not his or her "primary subject." To produce the most successful image, the photographer should somehow maximize the view of the primary subject while at the same time minimize or exclude entirely all areas and items that are not part of the primary subject as formed in the mind of the photographer.

The primary subject can also be viewed from different distances and from different positions. These differences will make a huge impact on the final image. The photographer should have a particular effect in mind and then compose the primary subject in the viewfinder so the desired effect is maximized. The professional may circle around the primary subject, be it an individual item of evidence or the entire crime scene, until the primary subject is best viewed and composed. This composition will usually satisfy two goals at the same time: (1) it will be the composition that shows the primary subject most effectively, and (2) it will be the viewpoint that excludes as many irrelevant items and areas surrounding the primary subject as possible.

Composition is just one part of creating a great photograph. It may be stated that deliberate, purposeful composition is required for any successful crime scene photograph. Some stunning candid images may exist in the history of photography, but crime scene photography should not depend on such serendipity. The word "composition" suggests intentional choosing, not happenstance.

Some believe composition is the most important and hardest part of crime scene photography, because the other aspects of photography can usually be mastered by most with little difficulty. Although many may believe this, a large amount of this text addresses those other aspects of good crime scene photography. Subtle variations of each aspect of photography need to be learned. What are the other concerns related to photography? Usually, before depressing the shutter button, the photographer must do at least three separate things.

1. The scene or primary subject must be composed.
2. The proper exposure must be determined.
3. The camera must be focused.

Good composition is a skill that can be learned, but it is more art than mechanical expertise. Learning to juggle film speeds, shutter speeds, and f/stops to arrive at a proper exposure seems more like a mathematical problem to many, which suggests there is always one, and just one, proper exposure for any particular subject. Even this text will declare the most effective exposure is sometimes an "incorrect" exposure. For example, if trying to document a witness's ability to recognize a suspect from a particular vantage point at 10 PM, it will be necessary to take a photograph that duplicates the lighting conditions at 10 PM, which might be thought of otherwise as an underexposed image.

Finally, how can focusing be a difficult subject? Just look through the viewfinder and rotate the focus ring until your subject becomes clear! This text will teach other focusing techniques. At times, the use of these new focusing techniques will be the only way to achieve an acceptable image.

These ideas suggest that good photography involves both mechanical skills and an artistic component. Photographic lenses, as variations of imaging optics, like telescopes and microscopes, can be considered a scientific tool. Viewed as dealing with an artist's creation, photography can be considered an art. Just as it is relatively easy to play chess with a nephew or niece, chess can be played on many different levels. Most of us easily learn how the different chess pieces can be moved. Becoming very good at chess, however, requires more than simply understanding how individual chess pieces move. Many can learn how to balance shutter speeds and f/stops to determine a proper exposure. Developing a photographer's eye and becoming very good at photography is not as easy. Two views of the same crime scene may call for the same exposure, but just one of the views may obviously be the better composition.

This chapter introduces material that will be found throughout the text. This material is provided in this chapter, early in the text, because of its importance. Unfortunately, there has not been an appropriate emphasis on these concepts in other texts related to crime scene photography. Perhaps, instead of calling them "Cardinal Rules," they might best be thought of as "tips for optimizing the quality of crime scene photographs." The Cardinal Rules can also be considered key crime scene photography concepts. Here they are:

1. Fill the frame
2. Maximize depth of field
3. Keep the film plane parallel

All the Cardinal Rules will not necessarily apply to other types of photography. Taking photographs of loved ones or pets will not require strict adherence to all of these recommendations. Having an understanding of their meaning, however, will often make your other photography much better.

Adherence to these suggestions will often be the criterion that separates the average crime scene photographer from the truly excellent crime scene photographer. The reader may ask how they might improve the current quality of their crime scene photography. The answer will frequently be, "Follow these Cardinal Rules."

CARDINAL RULE #1: FILL THE FRAME

If something is important enough to photograph, fill the frame with it. Follow this practice whether it is a single item of evidence or an entire crime scene. Too many times the primary subject can get lost in its background. If background elements are not considered a part of the "primary subject," try to minimize or eliminate them.

Get Closer to the Primary Subject

Ask yourself, "Can the primary subject, or the scene of interest, be made bigger by getting closer to it?"

1. Put as many silver halide crystals (using a film camera) as possible over the primary subject.
2. Put as many pixels (using a digital camera) as possible over the primary subject.
3. Zoom in, or get closer, to eliminate unwanted or unneeded areas/objects around the primary subject.

Figures 1.2 and 1.3 are intended to be close-up images of a pistol. Notice in Figure 1.2 that there is a lot of carpet included in the image, when the purpose of the photo was just to photograph the gun. Figure 1.3 succeeds in two important ways. The gun is larger, so it is easier to see; in addition, there is less irrelevant carpet in the image. Why waste image area on irrelevant subject matter? Doing so only detracts from the main purpose of the image. It is also a signal to the viewer of the photograph that the photographer is probably naïve about the importance of using the work space of the photograph to its best advantage.

Other times it will be important to show a gun in relation to the victim or a fixed feature of the crime scene. This is called a midrange photograph. In that case, what is the primary subject matter of this new photograph? It is the gun and the fixed feature of the scene (or, the victim's hand, for example) and the distance between the two. The question then becomes, how can you make the composition, including both the gun and the fixed feature of the scene, as large as possible, while at the same time eliminating as much irrelevant surrounding clutter as possible from the composition? The midrange photograph has one job to do. Let it do just that job. Overall photographs show a larger area around the

Figure 1.2
Pistol not filling-the-frame.

Figure 1.3
Pistol filling-the-frame.

gun and a fixed feature of the scene. They are also necessary photographs in the full documentation of the crime scene and the evidence within the scene; however, different photographs serve different purposes. The midrange photograph has a clear purpose of its own. Once you have that one purpose formed in your mind, compose the shot accordingly.

The purpose of each of the images in Figures 1.4 and 1.5 was to show the gun in relation to a fixed feature of the scene, the corner of the wall. Which image does that job better? Which composition has the primary subject matter (both the gun and a wall corner) larger and easier to see? Which composition excludes nonessential detail? Which photograph indicates the photographer had a clear purpose and knew how to eliminate irrelevant subject matter? Clearly, Figure 1.4 succeeds in all those areas.

4. If the subject is longer than it is wide, and most items are, compose it with the appropriate horizontal or vertical camera viewpoint. If it is long left-to-right, use a horizontal viewpoint; if it is long from near-to-far, use a vertical viewpoint.

   Figures 1.6 and 1.7 are both close-up photos of a pair of scissors. Is one preferable? Hopefully, you will agree that Figure 1.7 is the better image on two counts. First, the scissors are larger in Figure 1.7, and second, there is less carpet showing in Figure 1.7.

5. Avoid, however, having the primary subject so close to the edges of the photograph that parts are cut off in the print.

Eliminate the Irrelevant

"Is there any part of the composition that can be eliminated to make the overall photo better?"

The Background

1. You will be a better photographer if you hold yourself responsible for what appears in the background. The viewer of your photographs should be able to

Figure 1.4
Proper midrange.

Figure 1.5
Too much included!

Figure 1.6
Camera should be vertical.

Figure 1.7
Proper composition.

presume that if anything appears in the background, it is there because you wanted it there.

2. If you do not like what you see in the background, choose another viewpoint that has a "cleaner" background.

3. At times, you may also consider providing your own backdrop that prevents a view of a distracting background. This topic will be revisited in Chapter 5.

4. Another way of eliminating distracting backgrounds is to tilt the camera down more. In doing so, the photographer may compose the distracting background out of the photo. Take for example the two images that follow. They are supposed to be examples of a midrange photograph. It can be presumed that overall photographs, showing the crime scene in relation to the surrounding area, have already been taken. Therefore, it is not necessary for the midrange photograph to do this job. Midrange photographs have their own specific job to do.

   The intent of both Figures 1.8 and 1.9 is to show the scissors in relation to the gate support. Figure 1.9 contains the same details as Figure 1.8; however, Figure 1.9 excludes distracting and irrelevant background elements. Figure 1.9 also has both the scissors and gate support appear larger in the field of view. Why have them appear smaller than necessary? Including extraneous and irrelevant background details frequently requires the primary subject matter become smaller than necessary.

   How do these images comply with Cardinal Rule #3 (discussed later in this chapter), keep the film plane parallel? In this case, an imaginary line connecting the scissors and the gate base is composed to be parallel with the film plane.

5. Or get closer, to crop out distracting or unwanted elements.

6. Or zoom in, to crop out distracting elements.

The Foreground

1. You will be a better photographer if you hold yourself responsible for what appears in the foreground. The viewer of your photographs should be able to presume that if anything appears in the foreground, it is there because you wanted it there.

Figure 1.8

Too much in the background.

Figure 1.9

Proper composition.

2. If you do not like what you see in the foreground, choose another viewpoint that has a "cleaner" foreground.
3. Another way to avoid distracting foregrounds is to raise the camera up more. This may compose the distracting foreground out of the photo.

Figures 1.10 and 1.11 were taken to show the scissors in their relation to the square sign support. It is another midrange photography situation. Figure 1.10, however, includes a second sign base and an area of dirt in the same image. Should we spend time visually searching the dirt for additional evidence? Is the second sign relevant? Figure 1.11 precludes those questions. When the the camera is raised and pivoted to the left a bit, extraneous elements are excluded from the field-of-view. Figure 1.10 is an example of an average image showing the relationship of the scissors to a fixed feature of the scene. Figure 1.11 is obviously composed by a photographer who is aware that a professional carefully chooses not only what to include in the field-of-view but also what to exclude from it. Obviously, if the area of dirt and the second signpost are important at all, they will be included in some of the overall photographs. Midrange photographs, however, have their own specific job to do. To the extent possible, let each image do its own job.

Figure 1.10

Too much in view.

Figure 1.11

Proper composition.

4. Or get closer to crop out distracting or unwanted elements.
5. Or zoom in, to crop out distracting elements.

Areas to the Left and Right

1. You will be a better photographer if you hold yourself responsible for what appears to the left and to the right of your primary subject. The viewer of your photographs should be able to presume that if anything appears on the left or right side of the primary subject, it is there because you wanted it there.
2. Use the same considerations as with the background and foreground.
3. If you want it included in the photo, include it; if not, exclude it somehow.
4. Remember, if the subject is long, compose it vertically or horizontally to fill the frame with the subject, not the surrounding area.

"Fill the frame with the primary subject" is just the positive aspect of a two-sided coin. The other side of the coin contains a negative aspect, which is "eliminate as much as possible that is not the primary subject." *Include what you want; exclude what is unimportant.* Doing so will quickly separate your images from most crime scene photographs. Back to the preceding midrange photography examples: if just one item of evidence and one fixed feature of the scene are the "subject matter" for a particular photograph, include in your composition just those two items and eliminate as much as possible around them.

This author has worked many crime scenes where there is so much clutter around both the items of evidence and the victims that adhering to the preceding suggestions was frequently impossible. Sometimes it will be impossible to completely follow these Cardinal Rules. That does not diminish their importance. Part of the purpose of this text is to help you begin developing a professional photographer's eye. Most crime scene photography will benefit from knowledge of these Cardinal Rules and their implementation. Incorporating these Cardinal Rules into your photographic method will soon be a factor in your crime scene images being recognized as distinctly superior to the average crime scene photographer's work.

Shadow Control

1. Shadows from sunlight:
    a. As you compose photos outside, avoid having your own shadow, or the shadows of co-workers, in the field of view.
    b. If it appears impossible to avoid having your own shadow appearing in the field of view, reposition your body so that your shadow completely covers the area to be photographed.
    c. Have the scene be completely shadow free or completely covered with shadow. Sometime the difference is subtle as Figures 1.12 and 1.13 demonstrate. At times, the difference is more pronounced, as Figures 1.14 and 1.15 show.

Figure 1.12
Sunlight and shadow.

Figure 1.13
Sun blocked: entirely in shadow.

Figure 1.14
Sunlight and shadow.

Figure 1.15
Sun blocked: entirely in shadow.

The shadowless Figures 1.13 and 1.15 are preferable to the shadows apparent in Figures 1.12 and 1.14. One of the shadows in Figures 1.12 and 1.14 is from the casing itself, and one is from the photographer. Knowing both pairs of images are possible, however, and the effort to create Figures 1.13 and 1.15 are minimal, why not opt for Figures 1.13 and 1.15?

2. Shadows from electronic flash:

a. The use of an electronic flash may create a shadow in your photograph, and you should hold yourself responsible for this shadow, even if you cannot see it when the photograph is taken. You must begin to "pre-visualize" your own flash shadows. Distracting flash shadows can ruin otherwise excellent photographs.

b. Consider altering the position of the flash to minimize or control the shadows within the scene.

In Figure 1.16, the photographer held the flash too far to the left and the flash caught the railing, creating an unnecessary railing shadow in the scene. Figure 1.17 avoids this shadow because the photographer intentionally held the flash directly overhead.

Figure 1.16

Flash shadow in view.

Figure 1.17

Proper flash: shadow not in view.

c. If the flash is removed from the camera's hotshoe and connected to the camera by either a PC cord or a remote flash cord, you may position the flash in a variety of locations. Choose the position that minimizes the resultant flash shadow.

With the flash on a PC or remote flash cord, all the preceding shadows are possible, as well as a variety of shadows created with the flash held diagonally to the .410 shot shell casing in Figure 1.18. Because all are possible, is one preferable? Because the shadows themselves are irrelevant to the intent of the photographer attempting to capture a close-up image of the casing, a small shadow is better than a large shadow. Of the two small shadows in Figures 1.18 and 1.21, is one "better" than the other? It may be argued that Figure 1.18 is "better" (or, less annoying), because sometimes shadows suggest movement or directionality (the image of a speeding car with tail light streaks coming from the rear of the car comes to mind), and a shadow at the base of the casing appears more "normal" than one at the front of the casing.

d. If a "hard" shadow is created by the use of oblique flash, use a bounce card/reflector to add light to the shadow area, thereby making the shadow a "soft" shadow in which details can be seen.

Figure 1.22 was created by use of a piece of white paper just out of the field of view on the right. Some light from the flash bounced off this reflector/bounce

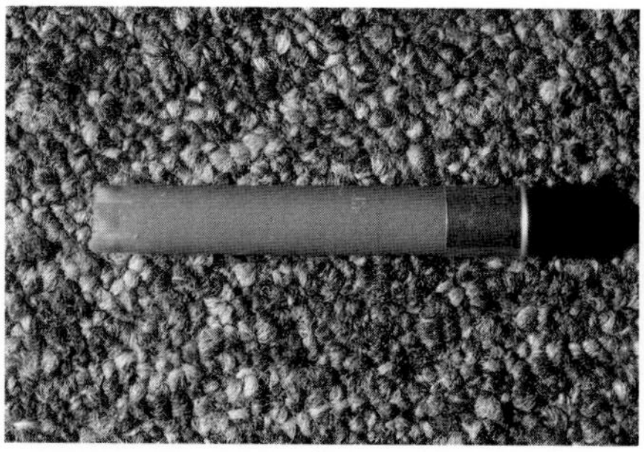

Figure 1.18
Hard shadow on right side.

Figure 1.19
Hard shadow below.

Figure 1.20
Hard shadow on top.

Figure 1.21
Hard shadow on left side.

card and added light to the dark "hard" shadow area. In both Figures 1.22 and 1.23, carpet can be seen in the "soft" shadow areas. This may not seem important in these images of a casing, unless a defense attorney were to question what "evidence" was hidden in the shadow area. Then, being able to see into the shadow area would be a distinct benefit. This ability to create "soft" shadows has very important applications in other situations. At times, the shadow created by an oblique flash falls over other areas of evidence, not just adjacent carpet areas. The most obvious examples are shoe prints or tire tracks in dirt. In both cases, much of the shadow falls within the shoe print or tire track. Obviously, it is critical to be able to see as much of the evidence as possible, and "hiding" any of the evidence within a "hard" shadow is counterproductive. This subject will be revisited in the chapter dealing with oblique electronic flash and oblique flashlight.

Lens Flare

If possible, avoid photographing with the sun in front of you.

1. Lens flare may ruin an otherwise excellent photograph.
2. Objects that are important will be backlit and underexposed.

Figure 1.22
Soft shadow on right side.

Figure 1.23
Reflector causing a soft shadow.

In both of the following images (Figures 1.24 and 1.25), sunlight was able to enter the lens directly, where it created images of lens optical elements and the aperture opening on the film/sensor. This is certainly distracting and detracts from the intent of the photographer. Sometime these flare elements are visible to the photographer while he or she looks through the viewfinder, and sometimes they are not. If you cannot avoid a composition that has the sun in front of you when you press the shutter button, consider using a lens hood, as seen in Figure 1.27. The lens hood in Figure 1.27 extends from the front of the lens, hopefully intersecting the direct light path of the sun into the lens. Light reflected from the scene into the lens is not affected by the lens hood.

Figure 1.28 is a four-picture collage, with the top photographs showing a building with and without lens flare. With the top left image, a lens hood was not used to eliminate the lens flare. The top right image did have the direct sunlight blocked

Figure 1.24
Lens flare.

Figure 1.25
Lens flare.

Figure 1.26
Camera and lens.

Figure 1.27
Camera and lens with hood.

(b)

(a)

(c)

(d)

Figure 1.28
(a) Lens flare. (b) No lens flare. (c) Hand shadow over lens.
(d) Eliminating lens flare.

from coming directly into the lens. The bottom two photographs show the "poor man's lens hood." The effect can be achieved by blocking the sun from coming into the lens with your hand. Alternately, a co-worker can hold up his or her hand or a clipboard to block the sun's rays from coming directly into the camera lens.

To recap the first Cardinal Rule: you will be a better crime scene photographer if you (1) fill the frame with your primary subject, and (2) exclude areas and objects that are not relevant to the success of your image.

CARDINAL RULE #2: MAXIMIZE DEPTH OF FIELD

Depth of field is the variable range, from foreground to background, of what appears to be in focus. The depth of field can be very small, where only one plane or distance from the camera appears to be in focus. It can encompass a small area of the scene, with the foreground and background being noticeably out-of-focus. Or, the depth of field may encompass the entire crime scene seen through the viewfinder.

Figures 1.29 through 1.31 show three views of staggered evidence numbers. Figure 1.29 shows a small depth of field set on the front number. Figure 1.30 shifts the small depth of field to the rear number. Figure 1.31 shows a depth of field

Figure 1.29
Wide aperture: focused in front.

Figure 1.30
Wide aperture: focused at the rear.

Figure 1.31
Small aperture: entire range in focus.

large enough to encompass all the numbers. Were these multiple items of evidence within a crime scene, it would be important to have them all be in focus at the same time. For most crime scene photography, maximizing the depth of field is a critical skill. Having the largest depth of field possible does not usually happen by accident. The photographer must know the camera variables that affect depth of field and then purposefully select the setting of each variable that maximizes the depth of field.

Focusing to maximize the depth of field is one distinguishing feature of crime scene photographers that separates them from other photographers. Portrait photographers, for example, frequently want the background of their images to be out of focus to force the viewer of their images to concentrate on their models. Crime scene photographers, however, should usually strive to ensure everything in their images is in focus.

Why? For several reasons. First, if a substantial amount of the image is out of focus, it may not be a "fair and accurate representation of the scene." When we were at the crime scene, as we looked around it, each part of the scene was in focus. If the photograph of the crime scene does not also show the scene to be in-focus, then it is not as we viewed the scene. Second, a defense attorney may be able to successfully argue that possible exculpatory evidence was located in areas currently appearing to be out of focus. He would remind the court that most photographers know how to minimize the depth of field to intentionally blur selected areas of a scene, suggesting we had done so purposefully. Third, a defense attorney may be able to successfully argue that the photographer was too incompetent to create a properly focused image, and the image should be excluded as being substandard and unprofessional. These ideas and others will be discussed at length in Chapter 10 on the legal issues related to crime scene photography.

Many nonprofessional photographers compose their main subject in the viewfinder and then rotate the focusing ring until their main subject comes into focus. Many readers may now be saying, "Of course, what else would you do?"

Crime scene photographers should be able to optimize their focusing by use of other focusing techniques or other techniques known to provide better depth of field. The following techniques are mentioned here briefly, because they fall under one of the Cardinal Rules this author believes are worth stressing. They will be elaborated on further in the chapters on exposure (Chapter 2) and focusing (Chapter 3).

1. Use the reciprocal exposure that uses the smallest aperture. Multiple shutter speed and aperture combinations result in the same exposure level. The use of small apertures results in a larger depth of field range. For instance, the following exposure combinations are reciprocal exposures, because they will

produce images with the same exposure: f/22 and 1/60; f/16 and 1/125; f/11 and 1/250, and f/8 and 1/500. Although they all produce the exact same exposure, they differ in the resulting depth-of-field ranges. Of the four combinations, the f/22 and 1/60 will result in the best depth of field range. If all four produce good exposures, why not select the one exposure combination that also maximizes the depth-of-field range?

2. Hyperfocal focus when infinity is in the background. Hyperfocal focusing maximizes the depth-of-field range for large outdoor crime scenes. Hyperfocal focusing usually requires the use of a depth-of-field scale (see Figure 1.32), along with the camera's distance scale. Hyperfocal focusing will be explained in detail in Chapter 3.

3. Zone focus when infinity is not in the background. Or, if your lens does not have a depth-of-field scale, focus by the "Rule of Thirds." These two focusing techniques, which result in virtually the same depth-of-field range, maximize the depth of field for smaller scenes. Both zone focusing and focusing by the rule of thirds will be explained in detail in Chapter 3.

4. Prefocus the camera at its closest focusing distance by rotating the focus ring to its shortest focusing distance. This does not require looking through the viewfinder. Then, move the camera, mounted on a tripod, toward the subject (a fingerprint, for instance) until the subject comes into focus. Set the tripod down at that position. It would be possible to arbitrarily place the tripod too close to the evidence, in which case focus will never be achieved. Or, if the tripod were to be placed too far away from the evidence, it would be possible to eventually focus on the evidence, but then the evidence would be smaller in the viewfinder than necessary. Prefocusing and then moving the tripod toward the evidence until it is in focus ensures the evidence is as large as it can be in the viewfinder.

These focusing techniques frequently result in a depth-of-field range that encompasses the entire crime scene or area of interest. What if the depth of field is not adequate to ensure everything will be in focus? If the lens being used has a depth-of-field scale (see Figure 1.32), very precise determinations of the depth-of-field range can be made and known to the photographer before the photograph is taken. If the photographer knows part of the foreground or part of the background will be out of focus, what should be done? The answer should

Figure 1.32
Depth of field scale.

be clear. If you are responsible for not only what appears in your foreground and background but also for whether or not it is in focus, then the same solutions are available. If the foreground or background is distracting to the overall purpose of your image, recompose the scene until the distracting foreground or background is no longer in view. If it is known the foreground or background will be out of focus, recompose the scene until the out-of-focus foreground or background is no longer in view.

This may require raising or lowering the camera's view of the scene. It may require getting closer to the primary subject matter to compose the scene tighter, thereby cropping the scene in the viewfinder. The depth-of-field scale, and the other focusing techniques mentioned previously, will be explained in great detail in the chapter dealing with focusing issues.

CARDINAL RULE #3: KEEP THE FILM PLANE PARALLEL

Which angle of view is best for the primary subject? Often, having the film plane parallel to the main subject is best.

This suggestion is in direct contradiction to many other books in which crime scene photography techniques are taught. In fact, this author was originally taught that diagonal viewpoints were frequently the optimal camera position. Justifications for this departure from the norm will be provided. If you remain unconvinced, or if your agency prescribes a certain protocol be used, then by all means, use the technique required by your agency. If you are not restricted by such protocols, however, these justifications may be convincing.

Overall Photographs

With exterior overall photographs, building facades and walls are best photographed with the film plane parallel to those surfaces.

1. Diagonal views have part of the wall closer to the camera and part of the wall farther from the camera. This makes part of the wall larger and part of it smaller. Often, it is not known during the early crime scene photography which areas are more important than others. Therefore, to the extent possible, it is best to have everything the same size, so there is not an emphasis on one part of the scene by having it larger than another part of the scene. Also, it would be unfortunate if several days after the crime scene photos were taken, it became known that the point of entry/exit was through a door or window at the far end of a wall photographed with a diagonal viewpoint. This would result in the point of entry/exit being very small and difficult to see.

2. Diagonal views have part of the scene closer to the camera and part of the scene farther from the camera. This creates a distance issue between fore-

ground and background. Because it is best to have this entire expanse in focus, it is necessary to manage the depth of field to ensure both the foreground and background are in focus. As the distance between the two becomes greater, it becomes more and more possible that the depth-of-field range may not cover the foreground-to-background distance. To eliminate this concern, compose the scene with the film plane parallel to the façade, wall, or area of primary interest. Having the film plane parallel to the scene will ensure everything is in focus.

3. Diagonal views have part of the scene closer to the camera and part of the scene farther from the camera. When electronic flash is being used to light the scene, the larger the scene the more likely a single flash will be unable to illuminate the entire scene. If the flash is set to properly expose the rear of the scene, the foreground will be overexposed. If the flash is set to properly expose the foreground, the background will be underexposed. These differences of exposure are eliminated when the flash is equidistant to both the left and right side of the scene.

There are some well-known exceptions to this rule.

1. Standing with the film plane parallel to a wall, the lens will be perpendicular to the wall. You will be directly facing the wall you are photographing. If the wall you are facing has a window or mirror, you will be able to see yourself or your flash, as a reflection. It is not recommended that you include yourself and your camera equipment in your own photograph. In this case, it is necessary to compose the scene from a slight diagonal viewpoint, so that you will not be seen in your own photograph.

   Figure 1.33 shows the worst possible situation: the entire wall is composed of windows. The top image of the three shows the reflection of the flash when the photograph was taken. The flash "hot-spot" is in the center of the image. To avoid this, a very slight diagonal view is recommended. This enables the photographer to compose himself or herself, and the flash, out of the image. The two bottom images show the entire wall, but the photographer and flash are no longer in view.

2. If you are using an electronic flash and the wall is highly reflective, maintaining the film plane parallel to the wall will usually result in a bright "hot spot" on the wall. Here again, a slight diagonal viewpoint to the wall will eliminate the flash's "hot spot."

3. Buildings will sometimes be so close together that it is impossible to stand far enough away from the wall to maintain a perspective with the film plane parallel to the wall. When there is no alternative, diagonal viewpoints will be necessary.

Figure 1.33

(a) Film plane parallel: flash reflection seen. (b) Slight diagonal: no flash reflection. (c) Slight diagonal: no flash reflection.

(a)

(b)

(c)

Figure 1.34 shows two examples of exterior overall photographs. The top views show diagonal viewpoints of a building. The bottom views show the same building photographed with the film plane parallel to the building's façade. Both pairs of images encompass the entire front façade of the building. It is hoped the reader will agree the bottom two images would be of more use to document the entry/exit from any particular door or window, if it were not determined until several days after the photographs were taken.

Midrange Photographs

With midrange photos, try to arrange the subject and the fixed feature of the scene so they are the same, or as close to the same, distance from you as is possible. (Think of an isosceles triangle.) Another way to think of this is to have the film plane parallel to an imaginary line formed by connecting the evidence and the fixed feature of the scene. This eliminates perspective distortion, where one item looks closer to another than it really is.

Figure 1.35 shows three improper viewpoints and one proper viewpoint for showing the distance relationship between a cartridge casing and a silver electric box. The top three photos show the casing and electric box aligned from three linear points of view. A linear point of view aligns the photographer, one item of interest, and a second item of interest in a single line. Each shows a different

Figure 1.34

Building facade: diagonal and film plane parallel views.

(a)

(b)

(c)

(d)

Figure 1.35

(a) Linear viewpoint: standing. (b) Linear viewpoint: kneeling. (c) Linear viewpoint: prone. (d) Film plane parallel; isosceles viewpoint.

perceived distance between the casing and the fixed feature of the scene. They were taken by having the photographer standing at full height, kneeling, and then lying prone. These are examples of perspective distortion. Linear viewpoints should always be avoided because of this perspective distortion effect. Each of the three photographs was taken with the camera at different heights. Is any of the three accurate? No. Raising the camera higher or lowering it changes the perceived distance relationship between the casing and the electric box. If it was the purpose of the photographer to show the viewer, and ultimately this will be a judge and a panel of jurors, the real distance relationship between the casing and electric box, all of the three upper photos fail. The photograph on the bottom, where the film plane is parallel to an imaginary line between the casing and the electric box, does show the proper distance relationship between the two items of interest. There is no perceived distance distortion when viewed from this position.

Close-Up Photographs

Having the film plane parallel to the evidence is critical with close-ups of the subject. Frequently, if the film plane is not parallel to the subject, the close-up cannot be used for comparison purposes.

Figure 1.36 shows two items with incorrect and correct close-up photos. Both of the left images were photographed at an angle. Both of the right images were photographed with the film plane parallel to the evidence. Both of the right images were carefully composed so that the back of the camera, the film plane, was parallel to the evidence. If the shoe of a suspect were recovered, only the lower right image would be suitable for comparison purposes. The lower left image of the shoe print would have to be corrected in the darkroom (film) or in a software program like Photoshop (digital) before it could be directly compared with the suspect's shoe.

The three Cardinal Rules will be referred to constantly throughout this text. Adhering to them will make your photography better than the average crime scene photographer's images. Remember them!

CARDINAL RULES OF CRIME SCENE PHOTOGRAPHY

I. Fill the Frame
II. Maximize Depth of Field
III. Keep the Film Plane Parallel

CHAPTER SUMMARY

Many have considered photography to be grounded on both science and art. Photography is thought to be based, in part, on science, because photography

Figure 1.36

(a) Angular viewpoint. (b) Film plane parallel. (c) Angular viewpoint. (d) Film plane parallel.

depends on the physics of optics and the mechanics of determining a proper exposure. Photography is thought to be based, in part, on art, because photography that is considered to be "well done" is often a result of the creative expression of the photographer. Both are usually necessary to create a successful image. This chapter emphasized the need to develop a photographer's eye and consider the elements of composition.

Several elements of composition were stressed. The photographer must first consider their viewpoint. The same object or area may be best viewed from one direction rather than from another direction. What considerations may affect this choice? One viewpoint may include too many distracting elements, which are easily eliminated from a different viewpoint. One viewpoint may include the photographer's shadow, which should always be avoided if possible. If it is impossible to exclude a partial shadow from being seen from one viewpoint, then totally covering a small area in shadow should be considered.

Once the "primary subject" has been determined, additional elements in the foreground and background and to the left and to the right of the "primary subject" should be eliminated from the field of view. This can sometimes be accomplished best by framing the subject with the camera held either horizontally or vertically, depending on the shape of the subject.

Composition suggests the photographer makes choices to optimally frame the subject matter. Composition should be as carefully considered as either exposure issues or focusing techniques.

DISCUSSION QUESTIONS

1. Briefly explain why photography can be considered to be both a scientific enterprise and an artistic expression of the photographer.
2. Can a properly exposed crime scene photograph still be unsuccessful? Why?
3. How can the photographer's viewpoint affect whether an image correctly depicts the spatial relationship between two objects?
4. How can both ambient and flash shadows be controlled?

PRACTICAL EXERCISES

The following are composition exercises only. As such, issues of exposure, focus, depth of field, and flash techniques are not considered. You may use an SLR set to program exposure mode and set the camera to autofocus. Or, you may use a point-and-shoot film or digital camera. Composition is the only issue in these exercises.

Construct an exterior mock crime scene with several items of evidence included in it.

1. Take an exterior overall image that shows the crime scene in its relationship to the surrounding neighborhood. Attempt to locate a single easily recognizable feature of the surroundings, and include this feature with the crime scene. This may be a particular business, a house, intersection street signs, or a natural feature of the scene.
2. Take one exterior overall photograph, encompassing the entire crime scene, while attempting to eliminate everything not within the crime scene.
3. Take a series of images, capturing two sides of any building. Some larger buildings will require overlapping images if a single image will not capture the entire façade.
4. Select one of the items of evidence, and take a midrange photograph of it with a fixed feature of the scene.
5. Take one close-up of that item of evidence, filling the frame with it.

Compare your images with those in this chapter and with other images on the supplemental website of images, referring particularly to the following image folders: close-ups, composition, exterior overalls, and midranges.

FURTHER READING

Davis, P. (1995). "Photography." 7th Ed. (Chap. 2) McGraw-Hill, Boston, Mass.

O'Brien, M. F., and Sibley, N. (1995). "The Photographic Eye: Learning to See with a Camera." Rev. Ed., Davis Publications, Inc, Worcester, Mass.

Adams, A. (1985). "Elements of Composition. The Camera." (Chap. 1). Little, Brown and Company, Boston, Mass.

Doeffinger, D. (1984). "The Art of Seeing." Eastman Kodak Company, Rochester, New York.

Upton, B. L., and Upton, J. (1989). "Photography." 4th Ed. (Chap. 14). Scott, Foresman and Company, Glenview, Ill.

CHAPTER 2

BASIC EXPOSURE (NON-FLASH) CONCEPTS

LEARNING OBJECTIVES

On completion of this section, you will be able to . . .

1. Explain the four exposure variables and their interrelationships.
2. Explain that shutter speeds are not only exposure controls but also motion controls.
3. Explain the concept of the theory of reciprocity.
4. Explain how a reflective light meter works.
5. Explain what "tools" are available to help determine the proper exposure for a tricky scene.
6. Explain the typically encountered "non-normal" scenes requiring exposure adjustments.
7. Explain the various exposure modes available with different cameras.
8. Explain how to bracket in manual and automatic exposure modes.
9. Explain the f/16 sunny day rule.
10. Explain why entire rolls of film or long sequences of digital images may be uniformly improperly exposed.
11. Explain the basic uses of filters as lens protection, reflection and glare removers, and "sunglasses."

KEY TERMS

Aperture
Aperture priority
  exposure mode
Bracketing
Burning and dodging

Critical comparison
Dark noise
Depth of field
Diaphragm
Dirty snow

DX coding
Exposure
  compensation dial
Exposure latitude
Exposure stops

Exterior overall
  photographs
F/16 sunny day rule
Fill-flash
Focal plane shutter
F/stops
Graininess
18% Gray card
Incident light meter

Interior overall
  photographs
Inverse square law
Manual exposure mode
Neutral density filter
Photo identifiers
Polarizer filter
Program exposure
  mode

Reciprocity failure
Reflective light meters
Shutter
Shutter priority
  exposure mode
Single lens reflex
Theory of reciprocity
UV filter
White balance

THE PROPER EXPOSURE TRIANGLE

EXPOSURE STOPS

Exposure stops: An exposure change from an image with a proper exposure is normally expressed as a change that either halves or doubles the overall lighting from the original exposure. A+1 exposure stop change, therefore, doubles the lighting; a−1 exposure stop change halves the lighting.

A photograph either can be properly exposed, underexposed, or overexposed.

Exposure stops are an exposure change from an image with a proper exposure that is normally expressed as a change that either halves or doubles the overall lighting from the original exposure. A +1 exposure stop change, therefore, doubles the lighting; a −1 exposure stop change halves the lighting.

A proper exposure looks "correctly" exposed, neither too light nor too dark. It seems "normal," as we would expect to see it. Details within the image are neither hidden in dark shadows nor washed out by being overexposed.

When discussing exposures, it is common to express differences in exposures by using the term "stops." An image may be properly exposed, which is usually referred to as being "0" or what is believed to be the optimal exposure. Another image of the same subject can be underexposed precisely by 1 stop (−1). A third image of the same subject can be overexposed precisely by 1 stop (+1). See Figures 2.1–2.3. Images can also be 2 stops underexposed (−2) or 2 stops overexposed (+2), as in the partial shoe print photograph sequence in Figures 2.4 to 2.8.

The concept of an exposure stop needs to be explained further. An exposure change from an image with a proper exposure is normally expressed as a change that either halves or doubles the overall lighting from the original exposure. A +1 image doubles the light of the original; a −1 image halves the light of the original. The camera's exposure system allows the photographer to precisely control the exposure of any image by manipulating several camera variables.

Depth of field: The variable range, from foreground to background, of what appears to be in focus.

Each decision to change exposures usually affects other aspects of the image besides the exposure level. For instance, changes in shutter speeds not only affect the image's exposure, they also affect the photographer's ability to either "freeze" motion or enhance the "blur" of moving subjects. Changes in f/stops not only affect the image's exposure, they also affect the photographer's ability to control the **depth of field** in the photograph. Both of these concepts will be addressed in subsequent sections of this text. At this point, only exposure concepts will be discussed.

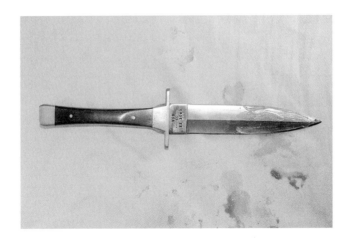

Figure 2.1
+1 (Courtesy of Martin Eaves, GWU MFS student).

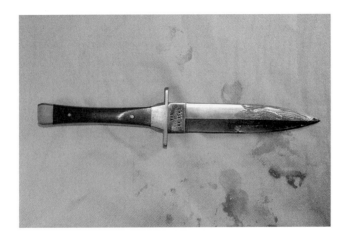

Figure 2.2
Properly exposed, or "0" (Courtesy of Martin Eaves, GWU MFS student).

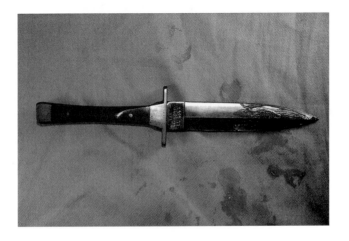

Figure 2.3
−1 (Courtesy of Martin Eaves, GWU MFS student).

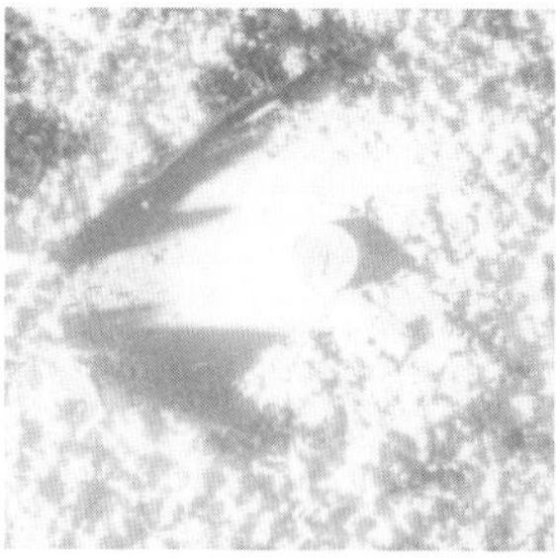

Figure 2.4
Shoe print +2.

Figure 2.5
Shoe print +1.

Figure 2.6
Shoe print normal exposure.

Figure 2.7
Shoeprint −1.

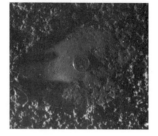

Figure 2.8
Shoeprint −2.

EXPOSURE VARIABLES

Generally, four variables directly affect exposure:

1. Shutter speed
2. Apertures
3. Film speed
4. The lighting of the scene

The four exposure variables can best be understood by relating them to the four sides of a triangle. Yes, a triangle has four sides. The three "normal" sides of the triangle relate to the photographic variables of shutter speeds, f/stops, and film speeds, and the in-"side" of the triangle corresponds to the light at the scene. Any variance in one of these variables results in a change in the exposure of a photograph.

Shutter Speeds

The **shutter** in a **single lens reflex** (SLR) camera is located in the camera body, just in front of the film, where light entering the lens is brought into focus. As such, the shutter in a single lens reflex camera is also called a **focal plane shutter**. In some other cameras, the shutter can be located in the lens itself, rather than the camera body. The single lens reflex camera is considered the optimal camera format to be used at crime scenes by most law enforcement agencies.

The term single lens reflex has two concepts. First, the image viewed by the photographer through the viewfinder and the light striking the film both come from a single lens. Second, to accomplish this dual role, the light coming from the lens is reflected (reflex) by a series of mirrors to the viewfinder. During the actual image capture, a movable mirror, "B" in Figure 2.10, is flipped up to allow the same light to strike the film.

The schematic of an SLR camera body shown in Figure 2.10 has the film located in the back of the camera, indicated by "D" and the shutter indicated by "C." Because the shutter is located near the film plane, where light is brought into focus, it is called a focal plane shutter. Some focal plane shutters open vertically, and some open horizontally.

The film plane is where the camera is designed to focus the light coming in through the lens. When the shutter button is depressed, the shutter opens for a predetermined amount of time, allowing light coming in through the lens to strike the film for different amounts of time. The shutter speeds are normally indicated as whole numbers: 1, 2, 4, 8, 15, 30, 60, 125, 250, 500, 1000. The white numbers in Figure 2.12 are shutter speed indications.

Although the shutter speeds are usually indicated by whole numbers on the shutter speed dial, they are, in fact, fractions. The shutter speeds are fractions of a second. For example, the shutter speed of 15 means that the film will be exposed to light for 1/15th of a second while the picture is being taken; 125 represents 1/125th of a second, 1000 represents 1/1000th of a second, and so on. Modern cameras have shutter speeds as fast as 1/8000th of a second.

Shutter: When the shutter button is depressed, the shutter opens for a predetermined amount of time, allowing light coming in through the lens to strike the film for different amounts of time.

Single lens reflex: The term single lens reflex has two concepts. First, the image viewed by the photographer through the viewfinder and the light striking the film both come from a single lens. Second, to accomplish this dual role, the light coming from the lens is reflected (reflex) by a series of mirrors to the viewfinder. During the actual image capture, a movable mirror is flipped up to allow the same light to strike the film.

Focal plane shutter: Because the shutter of a single lens reflex camera is located near the film plane, where light is brought into focus, it is called a focal plane shutter.

Figure 2.9

The proper exposure triangle.

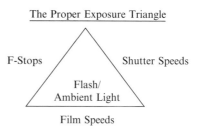

Figure 2.10

Schematic of a 35-mm camera.
(a) Lens elements;
(b) mirror-down; (c) shutter;
(d) film; (e) mirror-up;
(f) lens; (g) pentaprism;
(h) viewfinder. (Courtesy of Jeff
Robinson, scamper.com).

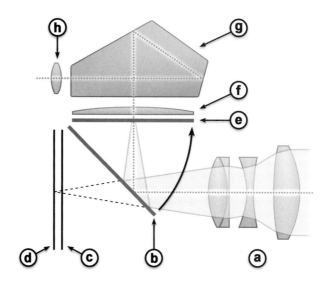

Figure 2.11

A focal plane shutter that
opens vertically.

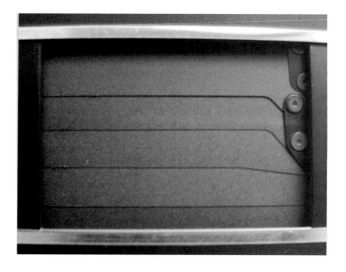

Figure 2.12
Shutter speed dial and threaded shutter button.

Figure 2.13
A bulb on shutter release cable.

Shutter speeds are also longer than 1 second. To distinguish these shutter speeds from those that are just fractions of a second, shutter speeds that are full seconds are usually indicated by the symbol ("). For example, the shutter speed 2" means 2 seconds, whereas the shutter speed indicated by 2 means $\frac{1}{2}$ second; 15" means 15 seconds; 15 means 1/15th of a second. Sometimes, full second shutter speeds are indicated by a different color rather than a ("). In Figure 2.12, the "2" next to the "B" actually stands for a 2-second shutter speed.

Figure 2.12 has a "B" just beyond the 2-second shutter speed, which stands for "bulb" and is the shutter speed setting used when longer shutter speeds than those available on the shutter speed dial are required. When the "bulb" shutter speed has been selected, the shutter will remain open for as long as the shutter button is depressed, allowing very long exposure times. Keeping the shutter button depressed with a finger over long exposure times, however, can result in blurred images from camera shake, which is possible even if the camera is mounted on a tripod for additional camera stability. When the bulb shutterspeed is selected, it is best to use a remote shutter release cable, which is screwed into the threaded shutter button; when the plunger is depressed, the shutter is activated. Long ago, shutter release cables had a rubber squeeze bulb on one end instead of a plunger. When the rubber bulb was squeezed, the shutter would trip.

The term "bulb" still remains, although the rubber bulb itself is not used much.

Any vibration associated with depressing the plunger on the remote shutter release cable is dampened by the flexible cable and does not result in camera shake. Figures 2.14 to 2.16 show the old-style mechanical remote shutter release cable on a film camera and a modern electronic shutter release cable on a digital camera.

Choose any number on the following continuum (1, 2, 4, 8, 15, 30, 60, 125, 250, 500, 1000, 2000), and you will notice that every preceding number seems half as large, and each succeeding number seems twice as large. In reality, it is just

Figure 2.14

Remote shutter release cable.

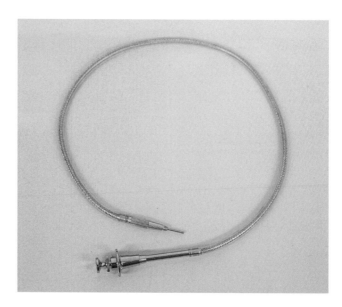

Figure 2.15

Remote shutter release cable screwed into the shutter button.

Figure 2.16

Electronic shutter release cable.

the opposite. For example, beginning with 60 (really 1/60th of a second), 30 (really 1/30th of a second) is twice as long (sometimes referred to as "slower"); 125 (really 1/125th of a second) is half as long (sometimes referred to as "faster"). Every change along the normal shutter speed continuum either halves or doubles the time the shutter remains open, giving the photographer precise control over the time that light is permitted to strike the film.

Many cameras, both film and digital, give the photographer the option of using shutter speeds that are in smaller increments of the full shutter speeds given previously. For instance, if the camera has shutter speed increments in 1/2 stops, the shutter speed between 1/60 and 1/125 would be 1/90. If the camera has shutter speed increments in 1/3 stops, the shutter speeds between 1/60 and 1/125 would be 1/80 and 1/100. Knowing your camera's capabilities is important, because on different cameras a change from one shutter speed to the next shutter speed may be either a full 1-stop change, a 1/2-stop change, or a 1/3-stop change.

Increasing the length of time the shutter is allowed to remain open will cause succeeding images to be progressively lighter and lighter, tending toward overexposure. Decreasing the length of time the shutter is allowed to remain open will cause succeeding images to be progressively darker and darker, tending toward underexposure. Returning to the previous exposures of a knife, shutter speed examples can now be assigned to each of the three images.

If a shutter speed of 1/60 led to a proper exposure, the shutter speed of 1/125 would let in half the light, and the resulting image would be darker. The shutter speed of 1/30 would let in twice the amount of light, and the resulting image would be lighter.

It will eventually be seen that all four of the proper exposure variables are in the same half or double increments, enabling the photographer to increase or decrease the exposure of the film by altering different camera variables.

Changing shutter speeds with a digital camera works precisely the same as with a film camera. The same concept of halves and doubles applies.

F/Stops

F/stops are sometimes referred to as the **aperture** of a lens or the **diaphragm** opening of the lens. The three terms are sometimes used interchangeably. To be more precise, each term will be explained.

The diaphragm of the lens is a set of blades, forming a circular opening, which can be opened to let more light through the lens or closed to restrict the light entering the camera.

The aperture is the size of the resulting diaphragm opening. Apertures are said to be "wide" or "large" or "opened up" when the diaphragm produces a large opening. Apertures are said to be "narrow" or "small" or "closed down" when the diaphragm produces a smaller opening.

F/stop: The term f/stop is actually a fraction relating the size of the aperture opening to the lens currently being used.

Aperture: The aperture is the size of the diaphragm opening.

Diaphragm: The diaphragm of the lens is a set of blades, forming a circular opening, which can be opened to let more light through the lens or closed to restrict the light entering the camera.

Figure 2.17
1/30 (Courtesy of Martin Eaves, GWU MFS student).

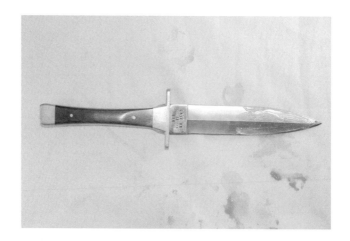

Figure 2.18
1/160 (Courtesy of Martin Eaves, GWU MFS student).

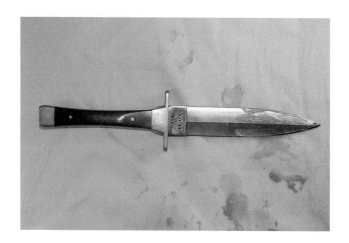

Figure 2.19
1/125 (Courtesy of Martin Eaves, GWU MFS student).

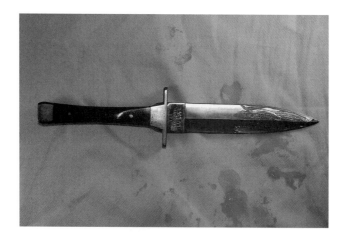

Figure 2.20
Wide aperture.

Figure 2.21
Narrow aperture.

The term f/stop is actually a fraction relating the size of the aperture opening to the lens currently being used and is best explained by the following equations:

$$FFL/f/stop = DoD, \text{ or}$$

$$FFL/DoD = F/stop$$

$$\text{Where: } FFL = \text{Focal length of the lens}$$

$$F/stop = \text{The f/stop selection}$$

$$DoD = \text{Diameter of the diaphragm}$$

In the case of a 50-mm lens, if the camera is set at f/8, the size of the lens opening, also known as the diameter of the diaphragm, is 50/8 = 6.25 mm. Or, a 50-mm lens with a 6.25-mm diaphragm opening is currently set to f/8. 50/6.25 = 8.

Another example: with a 50-mm lens on the camera and an f/4 set on the camera, the diameter of the diaphragm would be 50/4 = 12.5 mm. Or, a 50-mm lens with a 12.5-mm diaphragm opening currently set to f/4. 50/12.5 = 4. Another example: f/22 with a 50-mm lens. The lens opening would be 50/22 = 2.27 mm.

The following are the normal, full f/stop numbers: 1.4, 2, 2.8, 4, 5.6, 8, 11, 16, 22, and 32.

These particular numbers are derived from the physics of light and will be explained later in the discussion of the **inverse square law** as it relates to electronic flash.

For now, however, it is just important to remember that although f/stops are expressed as whole numbers, just like shutter speeds are usually expressed as whole numbers, they are also actually fractions. Therefore, f/8 actually represents 1/8th, and f/4 actually represents 1/4th. Although shutter speeds are fractions of a second, f/stops are fractions of the focal length of the lens.

Notice the relationship between the f/stop numbers and their respective lens openings. F/22, a relatively large number, results in a small lens opening when

Inverse square law: The intensity of light diminishes by the inverse square of the distance change. If the distance light travels is doubled, that distance, 2, is inverted to become 1/2; 1/2 squared = 1/4. As the distance light travels is doubled, its intensity is quartered.

Figure 2.22

Various apertures and their f/stops.

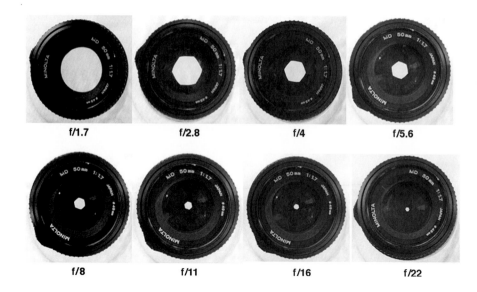

f/1.7	f/2.8	f/4	f/5.6

f/8	f/11	f/16	f/22

used with a 50-mm lens (2.27 mm); whereas f/4, a smaller number than 22, results in a larger lens opening when used with a 50-mm lens (12.5 mm). It should be remembered that f/22 really stands for 1/22nd, and f/4 really stands for 1/4th. Now, we realize the smaller 1/22nd results in the smaller lens opening (2.27 mm), and the larger 1/4th results in the larger lens opening (12.5 mm). The f/stops begin to make sense.

One convention of this text will be to express the term "f/stop" as a fraction, rather than the "f-stop" used by many other texts, which serves as a constant reminder that the f/stops are a ratio of the focal length of the lens currently being used on the camera.

Budding photographers frequently point out that the f/stop numbers are not as simple to comprehend as the shutter speeds. One complaint is that the apparently larger number, the f/22, actually represents a smaller sized aperture, whereas the apparently smaller number, the f/2, actually represents a larger sized aperture. It is then reiterated that the f/22 really is not a large number; it signifies 1/22nd of the focal length of the lens. The f/2 really is not a small number; it signifies $\frac{1}{2}$ of focal length of the lens.

However, because this same issue constantly occurs with every new class of crime scene photographers, it became paramount that a definitive method of making the theory easy to understand was found. Analyzing the types of people likely to be reading this text, it became apparent that readers would likely fall into one of two general categories. Readers would likely belong to some form of law enforcement agency, whether local, state, or a federal agency. Or, they would be students in school preparing themselves for an entry-level career in law enforcement and crime scene processing. The solution to the problem became clear: use examples immediately understood by each respective audience.

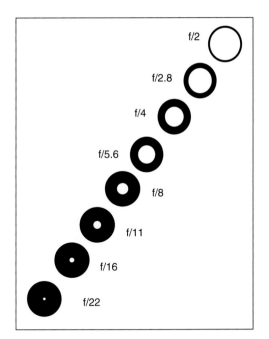

Figure 2.23

Graphical representation of the f/stop continuum relating lens sizes to their respective f/stops.

For those aspiring to their first careers, a pizza would provide the perfect analogy, because most students survive almost totally on this food during their formal educations. Imagine being with a gathering of friends when a pizza arrives at the door.

Now, imagine having to split that pizza with your friends. If your share was an f/2 portion of the entire pizza, you would receive $\frac{1}{2}$ of the pizza, and that is a lot of pizza.

If your share was only an f/22 portion of the pizza, you would receive only 1/22nd of the pizza.

Those already affiliated with a law enforcement agency require a different analogy for the full meaning of f/stops to truly hit home. In this case, it is natural to think in terms of donuts.

Imagine having to split a dozen donuts with your co-workers. If your share were an f/2 portion of the box of donuts, you would receive $\frac{1}{2}$ of the donuts.

If your share were only an f/22 portion of the box of donuts, you would receive only 1/22nd of the total.

When we speak to people in terms they can understand, complex ideas become simple.

The same goes for the amount of light let in by various f/stop selections: f/2 lets in a lot of light; f/22 lets in very little light.

Another important fact to remember about f/stops is that, just like shutter speeds, changing from any one f/stop to an adjacent f/stop number either halves or doubles the light allowed to enter the camera and strike the film. The difficulty is that the f/stop numbers themselves, as numbers that can be halved and

Figure 2.24

The whole pizza (Courtesy of Brigid Reilly, GWU MFS student).

Figure 2.25

Half a pizza: an f/2 portion (Courtesy of Brigid Reilly, GWU MFS student).

Figure 2.26

1/22 of a pizza: an f/22 portion (Courtesy of Brigid Reilly, GWU MFS student).

Figure 2.27

A dozen donuts (Courtesy of Brigid Reilly, GWU MFS student).

Figure 2.28

Half of the donuts; an f/2 portion (Courtesy of Brigid Reilly, GWU MFS student).

Figure 2.29

1/22 of the donuts; an f/22 portion (Courtesy of Brigid Reilly, GWU MFS student).

doubled, do not seem to suggest this. For instance, changing from f/8 to f/5.6 increases the light reaching the film. By how much? By double the light. Look at the graphic again, and notice that f/5.6 is a larger opening than f/8. A 1-stop change of exposure either halves or doubles the light reaching the film. Changing from f/2.8 to f/4 would be half the light, resulting in a −1 stop change.

Figure 2.30

f/8 (Courtesy of Martin Eaves, GWU MFS student).

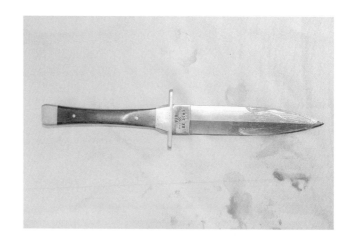

Figure 2.31

f/11 (Courtesy of Martin Eaves, GWU MFS student).

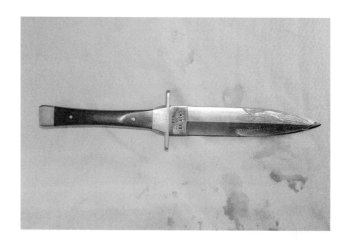

Figure 2.32

f/16 (Courtesy of Martin Eaves, GWU MFS student).

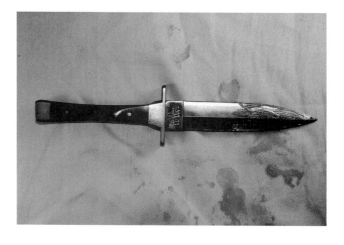

For now, just accept the fact that a change of f/stops that results in the next larger lens opening doubles the light entering the camera; a change of f/stops that results in the next smaller lens opening halves the light entering the camera. Be assured that ultimately the sequence of the f/stop numbers does make sense, but that explanation will come later, when the inverse square law is explained.

Returning to the images of the knives with different exposures, examples of f/stop numbers can now be assigned to them.

An f/8 allows precisely 1 stop more light to strike the film than an f/11 would. An f/16 allows precisely 1 stop less light to strike the film as the f/11 would. These would result in the same +1 and −1 exposures as the changes in the previously mentioned shutter speeds. The photographer has both tools to choose from to arrive at the same goal of having a proper exposure.

Many modern cameras, both film and digital, give the photographer the option of using f/stops that are in smaller increments of the full f/stops given previously. For instance, if the camera has f/stop increments in 1/2 stops, between an f/8 and an f/1↑ would be an f/9.5. If the camera has f/stop increments in 1/3 stops, between an f/8 and an f/11 would be both an f/9 and an f/10. Knowing your camera's capabilities is important, because on different cameras, a change from one f/stop to the next adjacent f/stop may be either a full 1-stop change, a 1/2-stop change, or a 1/3-stop change.

Changing apertures with a digital camera works precisely the same as with a film camera. The same concept of halves and doubles apply. The diaphragm blades are in the lens, and many lenses can be used on both film cameras and digital cameras.

How does one "set" the f/stops on the camera? Many low-end cameras have f/stop selections that can be selected by rotating a ring on the lens. The lens in Figure 2.33 has an f/stop continuum ranging from f/1.8 through f/22. The camera is set to f/8: the "8" is opposite the line doing double duty as the distance mark and f/stop mark.

On more modern film and digital cameras, the photographer uses a variety of dials and buttons to select an f/stop.

Figure 2.33
Camera lens set to f/8.

Film Speeds

Like shutter speed changes and f/stop changes halve and double the light allowed to reach the film, resulting in 1-stop exposure changes, film speed changes can also result in 1-stop exposure differences.

Commonly found film speeds are usually expressed as the following ISO (International Standards Organization) numbers: 100, 200, 400, 800, 1600, and 3200. Digital cameras do not use film but have continued the convention of relating the sensitivity of their digital sensors to light by using the same ISO equivalents.

These numbers are an indication of the film's, or digital sensor's, relative sensitivity to light. Films that require more light for a proper exposure have the lower numbers, like 100; films very sensitive to light, therefore requiring less light for a proper exposure, are designated by higher numbers, like 3200. Many years ago, film was rated by the older ASA (American Standards Association) numbers, but the United States has joined the rest of the world community and uses the ISO numbers now.

Like the shutter speed numbers, the film speed numbers immediately seem easy to understand: 100 ISO film requires more light for a proper exposure; 3200 ISO film can be used outside at night with only minimal light. These ISO film speeds are conveniently doubles or halves of adjacent film speeds. Changing from 100 ISO film to 200 ISO film doubles the film speed number and is an indication that 200 ISO film is twice as sensitive to light as is 100 ISO speed film. This doubling of the sensitivity of the film to light indicates that 200 ISO film would result in a 1-stop exposure increase (+1) compared with using 100 ISO film for the same scene, if all other exposure variables remained the same. Changing from 800 ISO film to 400 ISO film is a halving of the film speed number, resulting in a 1-stop exposure decrease (−1). All things being equal, if the use of 400 ISO film resulted in a proper exposure for a particular scene, changing to 200 ISO film would result in a −1 exposure, and changing to 800 ISO film would result in a +1 exposure.

The different film sensitivities to light are normally a result of different sizes and shapes of the silver halide crystals within the film's emulsion. Silver halide crystals are the light-sensitive component of film. When light strikes a silver halide crystal, it is chemically altered and will become dark when processed into a negative. When the film is processed into a negative, the darkened crystals are fixed on the film, and silver halide crystals that did not react to light are washed away. Very bright light will cause many of the silver halide crystals to become chemically altered, resulting in a dense negative with closely spaced dark crystals. Less bright areas of the scene will darken fewer of the silver halide crystals, resulting in a sparse grouping of dark crystals. The negative, then, represents a reversal of the lighting at the original scene: light areas of the scene produce a dark area on the negative; dark areas of the scene result in a light area of the negative.

Slower speed films, like 100 ISO film, usually have smaller silver halide crystals. With adequate exposures, the result is a dense grouping of dark crystals. This enables the negative to be enlarged greatly before the individual crystals become noticeable, sometimes referred to as apparent **graininess**.

Faster speed films, like 1600 and 3200 ISO films, typically use larger silver halide crystal sizes. Their larger surface area will react to light quicker than the small fine crystals of slow-speed film. This enables fast films to be used in dimmer lighting conditions. The tradeoff is that as the negatives are enlarged, the larger grains become noticeable quicker than slow-speed films. The result is more obvious graininess and a lack of resolution, because fine detail will seem to "come apart" more readily.

This text will avoid a more thorough discussion of the makeup of traditional films for two main reasons. First, to accomplish the goal of good crime scene photography, the evidence technician/crime scene investigator does not need to know the detailed chemistry that is the foundation of film emulsions. Second, the use of traditional film at crime scenes is declining and will eventually be completely replaced with the use of digital cameras. As this manuscript is being written, manufacturers of film, photography paper, and film-based cameras are dwindling. This is not because film, photographic paper, and film-based cameras have ceased to be adequate tools for photography in general. It is because manufacturers are aware that there is a worldwide shift toward digital imaging, and the business of photography will cater to the marketplace. It is not the quality of film, photography paper, and film-based cameras that will lead to their eventual demise; it is their profitability.

If you explore a camera shop, you will see other ISO film speeds different from the ones listed. One can also find ISO 25, 64, 160, and 1000 film speeds. These are best seen as film speed increments of less than a full stop, similar to partial shutter speed and partial f/stop changes.

In the images of the three knives, all things being equal, it would be possible to vary the exposure of a scene or an item of evidence merely by altering the film speeds used to capture the images.

Of course, we do not normally change our selections of film to intentionally produce a different exposure. Exposure changes are much more easily accomplished by changing either shutter speeds or f/stops. However, it is important to know that film speeds adhere to the same increments of halves and doubles as do the shutter speeds and f/stops.

Obviously, with digital cameras, traditional film is not used at all. The digital sensor is now the light receptor, and the image captured by the digital sensor is eventually transferred to some type of digital memory card. The digital memory card, however, is not the equivalent of film, because it only stores the data captured by the digital sensor. Various digital camera manufacturers use many types of digital media. The competition for this niche of the marketplace is intense, and a favorite or preferred media has not emerged victorious at this time.

However, it is important to note that the digital camera sensor can be rated at various ISO equivalents. Whereas the ISO of a roll of film locks the film camera operator to that film as long as it is being used, a digital camera operator has the

Graininess: As images are enlarged, the film's silver halide crystals, which make up the image, begin to separate and may become individually visible. Rather than a composite image being seen, the individual silver halide crystals become evident. These appear as grains of sand.

Figure 2.34
ISO 400 (Courtesy of Martin
Eaves, GWU MFS student).

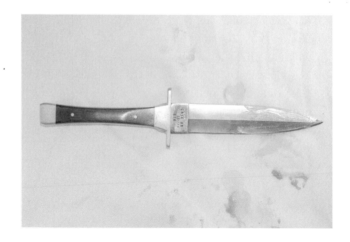

Figure 2.35
ISO 200 (Courtesy of Martin
Eaves, GWU MFS student).

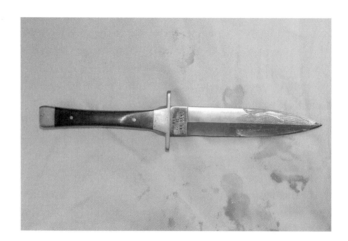

Figure 2.36
ISO 100 (Courtesy of Martin
Eaves, GWU MFS student).

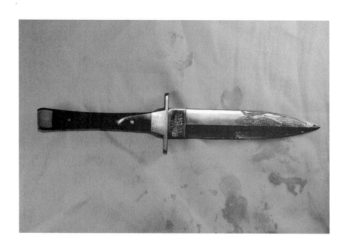

Figure 2.37
Sample film types.

Figure 2.38
Sample digital storage types.

vastly expanded options of changing the ISO equivalent for each shot if desired. With a digital camera, the photographer can frequently set the camera to ISO 100 for one shot, ISO 400 for the next shot, and sometimes ISO 3200, if desired. Digital cameras vary quite a bit with these ISO selections, so you must be familiar with the capabilities of your specific equipment. The digital camera operator has a distinct advantage when determining an exposure combination.

Our ISO film selection, whether real film or a digital equivalent, is usually the first important decision we have to make about the ultimate exposure. That decision-making process usually proceeds as follows.

1. At a particular scene, we access the ambient lighting conditions. Is it bright or dim? If it is bright, a slow ISO film or digital equivalent can be used. If the lighting is dim, a faster ISO film or digital equivalent will be necessary. The ambient lighting dictates the use of an ISO film or digital equivalent that will eventually permit exposures with small apertures, so that the depth of field will be adequate to have most of the scene in focus. If not, an electronic flash needs to be added. The additional light of the flash should be enough light to enable the use of small apertures for good depth of field.
2. The shutter speed is selected next. This will usually be 1/60 of a second, because the lens most frequently used at the crime scene will be a 50-mm lens.
3. The scene is composed through the viewfinder, and the camera meter is activated, usually by depressing the shutter button half way.
4. The meter then determines the last element of the proper exposure triangle. With the camera aimed at a particular scene, the meter knows the ISO setting on the camera and the shutter speed that have been selected. The meter will suggest an f/stop that will properly expose that particular scene with that particular ISO film and that particular f/stop.

If the preceding four steps begin with an ISO film selection decision, how is that decision made? When in doubt, rely on a Cardinal Rule: Maximize depth of field.

How does this relate to film selection? At this early stage of the text, several rules of thumb will be offered.

 Rule of Thumb 2.1: Outside, during the middle of a sunny day (loosely equivalent to 10 AM to 3 PM), select 100 ISO film.

 Rule of Thumb 2.2: Whenever doing **critical comparison** photography, use ISO 100 film.

 Rule of Thumb 2.3: All other times: early in the morning, late in the afternoon, at night, or when photographing indoors, use 400 ISO film.

Critical comparison photographs: These images are sometimes also called "examination quality photographs"—photographs that will be used by laboratory experts or analysts to extract information from the photograph or to facilitate a comparison of the photograph with another item of evidence.

DX coding: The DX codes contain three different types of information: (1) the ISO film speed of the film within the canister; (2) the number of exposures on that roll of film; and (3) the exposure latitude of the film.

Why? Because those film selections will eventually result in the ability to have properly exposed photographs taken with small apertures that maximize depth of field.

How does one "set" the ISO film speed on the camera? On many lower-end cameras, rotating a dial on the camera body sets ISO film speed.

With low-end cameras, pulling up on the silver outer ring of the shutter speed dial allows the various ISO film speeds to be selected by rotating the outer ring. More modern cameras have an assortment of buttons or command dials that can be pressed or dialed to select specific film speeds. The older camera also shows the older ASA rating, whereas now ISO film speeds are used.

Modern film is now usually rated with a **DX coding** system.

DX stands for "data exchange." The film canister is marked with 12 areas that are either black or silver, with the silver areas being electrically conductive. Six data areas are contained in two columns. Depending on whether the areas are black or silver, different information can be conveyed to the camera.

The DX codes contain three different types of information:

1. The ISO film speed of the film within the canister.
2. The number of exposures on that roll of film.
3. The exposure latitude of the film.

Within the camera body of modern cameras are six DX code sensors. Merely loading the film into the camera enables the camera's computer to "read" and set the ISO film speed. Many modern cameras will also be able to automatically rewind after the last exposure has been taken, because the DX code also provides the number of exposures on the roll of film being used.

Light
The Amount of Ambient/Existing Light
Ambient light is the light present at the scene. We either work with it; or, as will be discussed later, we can supplement it with an electronic flash, alternate light sources, or flashlights. Light is a critical component in the proper exposure

Figure 2.39
ISO film speed set to 400.

Figure 2.40
100 ISO.

Figure 2.41
3200 ISO.

equation. For instance, on a bright sunny day, we may be ready to take a photo-graph with the camera set at 1/60 shutter speed, f/22, and 100 ISO film. We have determined that this exposure combination is proper for the lighting of the scene. Then, just before we can trip the shutter, a large dark cloud goes by and covers our intended subject matter in shade. If we were to take the photograph at that instant, without altering our exposure settings for the change of ambient light, our resulting photograph will come back underexposed. Our exposure set-ting was for a fully sunlit scene, but we used that exposure combination for a scene in shade, resulting in underexposure.

So ambient light is a variable in our exposure calculations. We have minimal control over the ambient lighting, and usually vary the other camera exposure variables to ensure we will have a proper exposure. We must, however, be con-stantly mindful of the ambient light's ability to change while we are busy trying to calculate the proper exposure settings.

Figure 2.42

Film canisters with black and silver DX codes.

Figure 2.43

Film canister DX code and its six respective camera body sensors.

We can normally alter the ambient lighting in two ways. As stated previously, we can add electronic flash to the lighting of the scene. Chapter 4 discusses electronic flash techniques. A second method by which we can alter the ambient lighting of the scene is by intentionally creating a shadow over a part of the scene. For instance, if we are photographing an item of evidence in the early morning or late afternoon, the low angle of the sun may create distracting shadows on the evidence. As mentioned in the first chapter, sometimes distracting shadows can be eliminated. Merely block the sun and expose for the evidence within a shadow. The evidence now will not look underexposed; it will look properly exposed, but without shadows.

This ability to block the sun to create a shady area over an item of evidence or a small area of the crime scene obviously will not work on larger areas. Having some lighting control over smaller areas of the crime scene is an important capability.

Figure 2.44

Sun shadows.

Figure 2.45

Sun blocked; scissors in shadow.

The Color of Ambient/Existing Light

Although we do not normally see the different effects that various lighting sources will have on their surroundings, daylight color film will record these different effects. When the task is to faithfully capture colors within a crime scene, we should know only two choices of lighting can be used.

Rule of Thumb 2.4: For accurate color capture, we can only use either midday sunlight, which is approximately the bright yellow sunlight from 10 AM to about 3 PM, or we can use an electronic flash.

Both of these light sources will ensure that various colors within the crime scene were accurately captured on daylight color film. Other lighting conditions will result in various tints to the scene and the evidence within the scene.

Shade, Cloudy, Twilight If the crime scene and the evidence are not lit with the bright midday sun, the color of the light shifts toward blue. Lacking the bright yellow of the sun, we may still have sufficient light intensity for a proper exposure with small apertures, ensuring an adequate depth of field. Although the amount of light might be adequate, we also have to be concerned with the color of the lighting. A bluish tint may ruin an otherwise fine photograph. Worse yet, it may make a photograph unusable in court, because the image is no longer a fair and accurate representation of the scene as we saw it. Minor lighting tints, however, can usually be "fixed" in the wet-chemistry darkroom, with film; and can be corrected with digital software, if captured by a digital camera.

Figure 2.46 is not so much underexposed as it is tinted blue, because daylight color film was exposed in a well-shaded area. The use of an electronic flash for Figure 2.47 captures the true color of the white handkerchief. Please visit the book's companion website (http://books.elsevier.com/companions) 9780123693839 for color versions of these figures.

Figure 2.46
Blue tint from shade.

Figure 2.47
Proper colors with flash.

Sunrise, Sunset Early in the morning or late in the afternoon, it is frequently easy to see that the sun is tinted a distinct orangish red rather than its normal yellow color. This change to the color of the predominant light source will certainly be recorded on film. Unfortunately, the eye-brain combination is such an extraordinary compensator for minor deviations in lighting that we normally do not notice this color shift. As we gaze around, things we know to be white are still seen as white, even though an obvious tint is present. Film, however, will record this change. If color accuracy is critical to a particular image, it is essential to use electronic flash to "correct" for the tint produced by the sun at such low angles. It is always best to capture the image correctly, even if it can later be "corrected" if an improper tint should arise.

Figures 2.48 and 2.49 of a police cruiser taken in the late afternoon demonstrate this effect. The first image by itself may not look too objectionable until the color-corrected image is shown adjacent to it. If not photographing with midday sunlight, electronic flash is our only other option if we wish to capture colors accurately. Despite the fact that such tints may be corrected in the wet-chemistry darkroom using filters during the printing process, this text's goal is to make you a better photographer, not a better darkroom technician.

Tungsten When taking pictures indoors, the predominant light source is frequently tungsten light, or ordinary household light bulbs. If electronic flash is not used indoors, tungsten lighting can produce a yellowish amber tint. Whenever indoors, electronic flash must be used to ensure proper colors are captured. Again, the ambient lighting may be sufficiently intense for a proper exposure, but is it the right color of light? No.

Fluorescent Fluorescent lighting is another indoor lighting variation to tungsten lighting. Many businesses and offices are lit predominantly with this type of lighting. Although it looks like white light to the eye, many types of fluorescent tubes lack much of the red component of white light. Without one part of the

Figure 2.48

Tint from late afternoon sun (Courtesy of the Arlington County Police Department, VA).

Figure 2.49

Color correction (Courtesy of the Arlington County Police Department, VA).

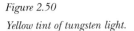

Figure 2.50
Yellow tint of tungsten light.

Figure 2.51
Color correction.

normal white light color spectrum, the opposite of the color that is absent will tend to become the predominant color. Opposite red on the standard color wheel is green. If fluorescent lighting is used to expose daylight color film, a green tint is the usual result.

Some fluorescent lights will produce an amber tint. Fluorescent lighting does not produce truly accurate colors on daylight color film.

Figure 2.52 was taken under fluorescent lighting. Figure 2.53 was taken with an electronic flash.

Color corrections can be made in a wet-chemistry darkroom with filters used during the printing process. Depending on the objectionable tint present on the negative, a contrasting filter can be used to balance the lighting and produce the proper colors on the final print. Nevertheless, it is better to accurately capture the correct color on film in the first place. Having to make a wet-chemistry darkroom color correction correctly implies that the proper color was not originally captured. Although it is comforting to know slight errors can be corrected later, they are still corrections.

Figure 2.52
Tint from fluorescent light.

Figure 2.53
Electronic flash produces true colors.

Digital White Balance   Digital cameras will also render the proper colors of the scene if exposed with midday sunlight or electronic flash. However, a photographer using a digital camera has more control over the accuracy of the final colors appearing in the image. Instead of relying on color corrections in a wet-chemistry darkroom, when objectionable tints may result from the wrong type of lighting, the photographer using a digital camera can set the **white balance** on the digital camera so that it is appropriate for the light in which the image is being taken.

Many of us are familiar with the concept of white balance, because when camcorders first became popular, white balance was one of their features.

Digital cameras allow the photographer to set the appropriate white balance for the ambient light. Figure 2.54 shows a typical set of choices a digital camera menu will offer to the user.

Digital cameras also have the choice of automatic white balance. This setting will provide for proper colors if, and only if, some object is within the scene that is white, or very nearly white. The auto white balance setting will attempt to ensure that those white objects in the scene are, in fact, white. If no white objects are within the scene, the best option is to set the white balance setting for the precise type of light at the scene. Doing so will ensure the original image is properly colored and eliminate the necessity to make color corrections in a postprocessing software program like Photoshop.

EXPOSURE LATITUDE

When discussing film's ability to render accurate exposures, the term exposure latitude needs to be understood. This phrase has a negative connotation and a positive connotation. Considered one way, it is restricting. However, an aspect of film should soothe many concerns we may have about making mistakes

White balance: Instead of relying on color corrections in a wet-chemistry darkroom, when objectionable tints may result from the wrong type of lighting, the photographer who is using a digital camera can set the white balance on the digital camera so that it is appropriate for the light in which the image is being taken.

Exposure latitude: One of the meanings of exposure latitude is that daylight color film has an inherent inability to properly expose items within wide extremes of light. The positive aspect of exposure latitude is to realize that film can have exposure errors corrected in the wet-chemistry darkroom.

Figure 2.54

White balance choices.

Display	Mode
AWB	Auto
☀	Daylight
⌂	Shade
☁	Cloudy, twilight, sunset
💡	Tungsten
▭	White fluorescent light
⚡	Flash

when determining the proper exposure setting for any particular crime scene photograph.

Film Cannot Capture All the Details in Extreme Lighting Conditions

A film's ability to capture details in a scene with extreme lighting conditions is limited. The most frequently encountered extreme lighting condition is found in a scene lit with strong midday sunlight. Usually, deep dark shadows are also present with this type of lighting. The camera exposure settings necessary to properly expose objects in the sunlight will be vastly different than the camera exposure settings necessary to properly expose objects in the deepest shadows. If it were necessary for the photographer to photograph two items of evidence adjacent to each other, but one item of evidence was directly lit by the midday sun and the second item of evidence was in a deep shadow, the photographer may not be able to properly expose both items of evidence at the same time.

Of course, if the photographer were able to take one image of the evidence lit just by the midday sun, that would be easy. If the photographer could recompose on just the second item of evidence in the deep shade, that would also be an easy exposure. In both situations, only one lighting extreme would be present. But if the photographer thought it necessary to include both items of evidence in only one photograph, because he or she wanted to show the distance relationship between the two items of evidence, then an exposure problem must be solved.

One of the meanings of exposure latitude is that daylight color film has an inherent inability to properly expose items within wide extremes of light.

For the sake of discussion, let us presume for a moment that the item in the sunlit area required an exposure combination of 100 ISO film, f/22, and 1/60th of a second shutter speed. If those were the exposure settings used, the item in the sunlight would be perfectly exposed.

Let us also presume the item in deep shade required an exposure combination of 100 ISO film, f/4, and 1/60th of a second shutter speed. If those were the exposure settings used, the item in the deep shade would be perfectly exposed.

If the first exposure combination were used for both items at the same time, the item in deep shade would be 5 stops underexposed and would be so dark that it would not be seen at all. If the second exposure combination were used for both items at the same time, the item in bright sunlight would be 5 stops overexposed and would be so burned-out that it would not be seen at all. Figure 2.55 is an example, although not as extreme as some scenes can be. Had a critical item of evidence, a weapon for instance, been located by the vehicle's tire it would be visible in the first image at the expense of a badly overexposed sunlit area. In Figure 2.56, where the exposure was determined by the sunlit area, the weapon would be completely hidden in shadow.

The first solution that frequently comes to mind is to consider an exposure setting that will be a compromise and, hopefully, properly expose both items at the

Figure 2.55

Exposed for the shade.

Figure 2.56

Exposed for sunlit grass.

same time. The f/stop midway between f/22 and f/4 is f/9.5. That would be $2\frac{1}{2}$ f/stops wider than f/22, and $2\frac{1}{2}$ f/stops smaller than f/4. Is this really an adequate compromise? No! If the item in direct sunlight really needed an f/22 for a proper exposure, by using an f/9.5 it would be $2\frac{1}{2}$ stops overexposed. If the item in deep shade really needed an f/4 for a proper exposure, by using an f/9.5 it would be $2\frac{1}{2}$ stops underexposed. Both would be improperly exposed in the same photograph. The use of a middle-ground f/stop is not the solution.

Solutions exist, however. One is to use a technique called fill-in flash. This technique will be discussed at length in Chapter 4. It should be mentioned here that crime scene photographers frequently use fill-in flash during the daytime. It is the primary method to "even" the lighting extremes when bright areas of a scene and dark shadows are present in the same area. A second solution to this exposure problem is to realize that the exposure problem will exist and then to make exposure corrections either in the wet-chemistry darkroom when film was used or to make corrections when postprocessing a digital image with software programs like Photoshop. Again, it is best to capture the image with a correct exposure originally, rather than to rely on corrections later.

Best to Overexpose Film

It is actually easier to correct overexposures on film. Sometimes up to 3 stops of overexposure can be corrected in the wet-chemistry darkroom, whereas correcting 2 stops of underexposure is frequently the limit for film. This ability of film to have more overexposure than underexposure able to be corrected has led many to remember this Rule of Thumb.

Rule of Thumb 2.5: When using film, if an exposure error is unavoidable, it is better to overexpose film than to underexpose it. Overexposed film is more easily correctable.

Some photographers are in the habit of setting their film cameras so a +1/2 exposure compensation is their "normal" exposure.

Best to Underexpose Digital Exposures

Does the same apply when using a digital camera? Actually, just the opposite is true. Detail in an overexposed digital image cannot be salvaged. However, photographers using digital cameras have a tremendous ability to pull detail out of underexposures. This can be a new Rule of Thumb.

Rule of Thumb 2.6: If an exposure error is unavoidable, it is better to underexpose digital images a bit. More detail can be obtained from a digital underexposure than a digital overexposure.

Exposure Corrections in the Darkroom

The positive aspect of exposure latitude is to realize that daylight color film can have exposure errors corrected in the wet-chemistry darkroom. However, it must be understood that these corrections are limited. Not all overexposures and underexposures can be salvaged in the darkroom. Different ISO speed films have different exposure latitudes. Check the film manufacturer's suggestions for each type of film. There is, however, a Rule of Thumb 2.7 for the exposure latitude for various film types.

Rule of Thumb 2.7: ISO 100 speed film can normally be expected to be correctable for a 2 f/stop overexposure and 1 f/stop underexposure. ISO 400 speed film can normally be expected to be correctable for a 3 f/stop overexposure and 2 f/stop underexposure.

These limits apply to high-volume, next-day film processors. A professional laboratory may be able to correct further extremes of overexposures and underexposures.

The exposure latitude is more restricted when a digital camera is used. Digital overexposures and underexposures are more difficult to correct with digital software. When using a digital camera, the operator needs to be more accurate with exposure calculations. Of course, the ability to instantly see the results of each image captured frequently assures the digital camera operator a "proper" exposure can be captured with the next shot.

Globally Increasing or Decreasing Exposures

Exposure corrections can be applied to the entire image at the same time. Wet-chemistry darkroom personnel have the ability to project different amounts of light through negatives onto light-sensitive paper. Both shutter speeds and apertures control the amounts of light, just as the camera operator varied these controls when capturing the original image. If a photographic print (photograph) would have been properly exposed by projecting light through the negative for 1/60th of a second with an f/8, then changing either the shutter speed or f/stop on the enlarger holding the negative would change exposures just like when the film was originally exposed. If the negative were 1 stop overexposed, the dark room operator could let less light through the negative onto the photographic paper to create a properly exposed print. If the negative were 1 stop underexposed, the dark room operator could let more light through the negative onto the photographic paper to create a properly exposed print. The entire print would show the results of these exposure changes uniformly across the whole print.

The same can be done with digital imaging software when adjusting the exposure levels of digital images. More or less light can be added to the entire image as desired.

Burning and Dodging Selected Areas of an Image

At times, the entire image may not suffer from a uniform exposure problem. Only selected areas of the image may be improperly exposed. In these cases, it is possible to improve the exposure of just these small areas. The area of the print requiring a more "normal" exposure can receive that amount of light.

Areas of the negative that were originally underexposed require more light for a proper exposure. After the "normal" exposure of the entire print, the areas previously underexposed can receive additional light, while the rest of the image is masked so it does not receive too much light. This is called **"Burning."** It adds light to, or burns, areas previously underexposed.

Areas of the negative that were originally overexposed require less light for a proper exposure. During the "normal" exposure of the entire print, the areas previously overexposed can be masked for a part of the normal exposure. This is called **"Dodging."** Light is held back, or dodged, from areas needing less light than the print as a whole.

Digital imaging equivalents of burn and dodge can be made with digital software. Selected areas of an image can be lightened or darkened as desired.

This ability to correct improperly exposed negatives should not make us lax when the original exposure is determined in the field. The most carefully exposed negative, or digital image, will yield the best print. It is better to be able to determine the proper exposure at the crime scene than to be able to correct mistakes in the wet-chemistry darkroom or with digital software. Most would agree that the ideal situation is to become a better photographer and determine the correct exposure when the image is originally captured, rather than become a better technician, able to later correct the exposure mistakes made at the crime scene. Knowing minor corrections can easily be made is, however, enormously comforting to most.

Burning and Dodging: After the "normal" exposure of the entire print, the areas previously underexposed can receive additional light, while the rest of the image is masked so it does not receive too much light. This is called "burning." During the "normal" exposure of the entire print, the areas previously overexposed can be masked for a part of the normal exposure. This is called "dodging."

SHUTTER SPEED AS MOTION CONTROL

MOTION CONTROL TO ELIMINATE BLUR

Generally, three types of movement can be controlled by an appropriate shutter speed:

1. Photographer body movement
2. Subject motion
3. Camera movement

Photographer Motion

If the camera is moved during an exposure, the photograph can be blurred. One type of camera movement can be a result of simply holding the camera improperly. The camera needs to be braced properly to eliminate as much camera movement as possible. Of course, you will be holding the camera with two hands. The right hand holds the right side of the camera body, with the index finger ready to depress the shutter button. The left hand normally cradles the bottom of the camera, with the left thumb and index finger available to either rotate the focus ring, change focal lengths on a zoom lens, or to rotate the f/stop ring.

Frequently, four other means can be used to steady the camera. Rather than having your elbows extended from your body, they need to be tucked in close to your chest. They act like a bipod to help steady the camera. In addition, the camera is usually steadied somewhat by being brought in close to your face and making contact with your brow or forehead. Although some are not keen on this, frequently the camera is also making contact with your nose. Many are not consciously aware this is done until they notice nose oils on the back of the camera or LED screen. The more means used to steady the camera the better.

A proper choice of shutter speeds can eliminate the blur that can still result from hand-holding the camera. As steady as you might think you can hand-hold a camera with the preceding techniques, you cannot eliminate the movement of your heartbeat. You may be able to hold your breath, but you cannot stop your heart from beating during the exposure. Believe it or not, the beat of your heart is enough to blur a photograph. Fortunately, this movement can be compensated for by the proper selection of a shutter speed, which forms another Rule of Thumb.

Rule of Thumb 2.8: To eliminate possible blur from hand-holding the camera, use the shutter speed that is the closest to the focal length of the lens on the camera, inverted into a fraction.

For instance, if the camera had the standard 50-mm lens mounted on it, 50 inverted into a fraction would be 1/50th. Therefore, using the shutter speed that is nearest 1/50th of a second, or faster (if not exactly, then the next fastest shutter speed), would effectively eliminate the possibility of blur from hand-holding a camera with a 50-mm lens mounted on it. The closest shutter speed to 1/50th is 1/60th. Therefore, a shutter speed of 1/60th effectively eliminates the blur that can result from hand-holding a camera with a 50-mm lens on it.

Wide-angle lenses, however, do not follow this rule; 1/60th is normally the slowest shutter speed recommended when hand-holding a 35-mm camera, even with a wide-angle lens on it. Some professional photographers brag about being able to hand-hold a 35-mm camera while using a 1/30th shutter speed. Do not be swayed by their bravado. Adhere to the Rule of Thumb. Crime scene photography is too important to risk blur because of an ego problem. Therefore, resist the urge to use a 1/30 shutter speed if you are using a 28-mm lens on your camera.

Telephoto lenses, however, do follow the rule.

Consider the situation in which you will be hand-holding a camera with a zoom lens set to 80-mm lens; 80 inverted into a fraction would be 1/80th. The closest shutter speed to that is 1/90th, which is $\frac{1}{2}$ a stop between 1/60th and 1/125. If your camera does not have shutter speeds in $\frac{1}{2}$-stop increments, you would have to use 1/125 as the shutter speed to hand-hold a camera with an 80-mm lens on it.

Hand-holding a 100-mm lens requires a shutter speed of 1/100, if your camera has it; 1/125 if your camera does not have the option of a 1/100.

If you are using a 300-mm lens on the camera, you must select a shutter speed of 1/300th, or faster, to avoid blur from hand-holding the camera with that lens on it. Some cameras have a 1/350 halfway between a 1/250 and 1/500. If not, then you would have to use the 1/500th shutter speed.

Hand-holding a 500-mm lens also requires a shutter speed of 1/500th.

If, for some reason, you cannot use the focal length of the lens converted into a fraction as your shutter speed, the other alternative to hand-holding the camera is to mount the camera on a tripod. With the camera mounted on a tripod, movement of the camera is virtually eliminated, and thus any shutter speed can be used without slow shutter speeds blurring a photograph. With the camera mounted on a tripod, some camera movement may still occur when the shutter button is depressed. To avoid this, one of two solutions is usually recommended when the camera is on a tripod.

First, use a remote shutter release cable, as mentioned previously. A remote shutter release cable enables the shutter to be tripped by depressing the plunger at the end of the cable. This mitigates the shock of depressing the shutter, and any associated vibration is absorbed in the length of the flexible cable. No camera movement means no blurred images from camera shake.

Many modern cameras also have a delayed shutter release, usually with a choice of delay times. Frequently, a 10-second shutter delay is set on the camera. In this case, depressing the shutter button directly does not adversely affect the image. If any camera movement was associated with the depression of the shutter button, that movement is dissipated during the 10-second lag time between the depression of the shutter and the shutter actually tripping 10 seconds later. Many cameras also have a 2-second lag time option.

Whenever a camera is mounted on a tripod, it is recommended that one of these features be used. Directly depressing the shutter button can result in a blurred image, even if the camera is mounted on a tripod.

Subject Motion

At times unavoidable motion will be present within the camera's field of view, and it is essential to be able to recognize the people or vehicles in the resulting photographs. Such is the case with surveillance photographs, when the subject(s) of the surveillance will be moving. It also becomes important if the assignment is to photograph a battered child, who will not sit still at your request. Usually, at crime scenes and accident scenes, the scene and evidence within the scene are static, and subject movement is not an issue. But, when it is essential to freeze the motion of anything moving within the scene, knowing how to solve this problem is critical. At an accident scene, for example, a victim may be in the process of being removed from the scene by medical personnel. As they are being wheeled away from the scene, you may want to capture an image or two of their apparent injuries. If their treatment is critical, it would be improper to delay their removal from the scene, so you must "grab the shot" while you walk alongside the moving gurney. With the proper shutter speeds selected, even these kinds of motion can be stopped or frozen.

Consider these recommendations for motion control:

Normally . . .

- 1/125th will "freeze" a walker, enabling the face to be recognized.
- 1/250th will "freeze" a jogger, or someone running, or jumping.
- 1/500th will "freeze" a bicyclist or a slow-moving vehicle (up to about 30 mph).
- 1/1000th will "freeze" a vehicle going approximately 60 mph.
- 1/2000th will "freeze" an airplane powered by propellers.

Figure 2.57
Shutter speeds "freezing" motion.

Of course, fast shutter speeds can freeze motion. Although it will become clearer in Chapter 4, it should also be pointed out here that the use of electronic flash could also freeze subject movement. The upset battered child, bouncing around on a mother's lap, can effectively be frozen in place with electronic flash. Fast flash durations are the reason, and this facet of electronic flash will be covered in Chapter 4.

Another photographic technique can aid in the "freezing" of motion. That is "panning." Panning is the term applied to the photographer's tracking the subject's movement with the camera: in other words, moving the camera to keep the subject in the field of view while the subject is moving. Sports photographers use this technique, in their attempt to capture dynamic motion, while at the same time being able to recognize the athlete diving for a touchdown, arcing over a high bar, or stretching toward the finish line.

This technique is less successful when used for law enforcement purposes. Although this technique will generally ensure that the moving subject is recognizable, because the camera is moving during the exposure, the background will also be blurred. If it is necessary to recognize the subject and to recognize the building they are entering or exiting from, panning is less successful.

When doing moving surveillance, camera motion is unavoidable. Consider the scenario when a suspect will be meeting with known felons in an area also known for its high crime rate. It would be possible to photograph both subjects and the scene with the photographer in the back seat of a vehicle as it drove by. In this case, it would be necessary to have selected a shutter speed sufficient to freeze the motion of the car and the camera within the car.

USING SLOW SHUTTER SPEEDS TO ELIMINATE RAIN AND SNOW

Slow shutter speeds emphasize the blur caused by movement within the field of view. At crime scenes and accident scenes, the last thing we normally want is blurred images. Blur can ruin a photograph, and it can make the photograph ultimately inadmissible in court as evidence. However, there is one circumstance when a slow shutter speed is actually a benefit to law enforcement photography.

When photographing a scene in bad weather, it is possible to "freeze" the raindrops or snowflakes with appropriately fast shutter speeds or with the use of electronic flash. This would certainly document the fact that it was, indeed, raining or snowing at the time photographs were taken. However, these same raindrops or snowflakes would make it difficult or impossible to see the scene very clearly, and the evidence within the scene may also be obscured. If a slower shutter speed is used, the "frozen" raindrops and snowflakes will be captured as blurs. These blurs also effectively mask the scene, and the evidence within the scene.

With a shutter speed that is slow enough, the raindrops or snowflakes can be intentionally blurred so much that they cease to be captured on the film. During a long exposure, what is consistently in the field of view, in the same location, during the entire exposure will be captured on the film. But individual raindrops or snowflakes will not be in the field of view long enough to be recorded by the film, so the result is a scene without the rain or snow.

In these situations, it is recommended that at least one image be taken that does document the fact that it was raining or snowing at the time. The snow, for instance, may be a contributing cause of an accident. Or, it may affect a witness' ability to see what they are reporting to have seen. Crime scene photographers should never allow their photography to be accused of misrepresenting the scene. However, once the fact that it was raining or snowing has been documented, it is no longer necessary to capture the rain and the snow in every remaining photograph. From then on, it is just necessary to document the scene and the evidence within the scene, and the raindrops and snowflakes can then be considered some sort of visual contaminant between the camera and the scene.

For instance, if a tree branch were obscuring the scene from the point of view of one witness, it would certainly be necessary to take a photograph from the witness's point of view, with the tree branch blocking the view. Thereafter, to document the scene and the evidence, viewpoints avoiding the tree branch would be chosen.

Some jurisdictions have large tarps available to cover outdoor crime scenes during inclement weather. Just as it is understandable to protect the crime scene from the deterioration that can be caused by bad weather, so, too, being able to photograph the crime scene without the bad weather interfering with an unobstructed view of the scene and the evidence within it is a distinct advantage.

Practical experiments have shown that shutter speeds in the 2- to 3-second range normally effectively eliminate rain and snow from the image. Such slow shutter speeds obviously require the use of a tripod; otherwise, the scenes themselves will be blurred as well. These slow shutter speeds also let in an enormous amount of light. It can be quite bright even while it is raining or snowing. Sometimes, even when using a slow speed film like 100 ISO, and using the smallest aperture the lens has, like an f/22, the camera's exposure meter may indicate overexposure if the image is captured without any further adjustments.

How can we avoid an overexposure in these situations? The answer is to use "sunglasses" over the lens, just as we would personally use sunglasses at any bright outdoor scene. For the camera lens, these "sunglasses" are filters. Two types of filters may help in this situation.

One type of filter that can block some of the light coming in through the lens is a **neutral density (ND) filter**.

These are usually available in three increments. Neutral density filters typically come with the ability to block precisely 1 stop of light, 2 stops of light, or 3 stops of light. Depending on the level of overexposure present at the scene, the appropriate neutral density filter can be selected as required. Neutral density filters do not affect any colors within the scene; they merely reduce the overall lighting in precise increments.

Neutral density filter:
One type of filter that can block some of the light coming in through the lens. "Sunglasses" for the lens.

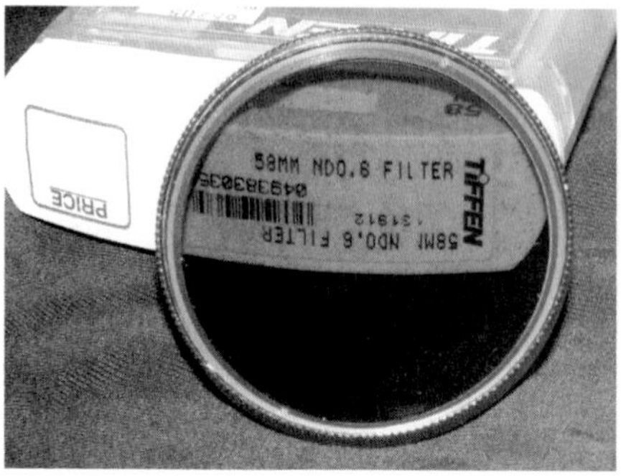

Figure 2.58

A 2/stop, 0.6, neutral density filter reducing the light behind the filter.

Figure 2.59

A polarizer filter reducing the light behind the filter.

Polarizer filter: A filter used to eliminate reflections from glass and water. It also uniformly reduces the light coming through it.

Notice how the 2-stop ND filter in Figure 2.58 reduces the overexposure behind the lens. An ND filter blocking 1 stop of light is usually designated by either a 0.3 or a 2×; 2 stops of light reduction are designated by 0.6 or 4×; 3 stops of light reduction are designated by 0.9 or 8×, depending on manufacturer.

Another filter that may be used in similar situations is the **polarizer filter**. The primary use of a polarizer filter is to eliminate reflections from windows and water and to eliminate glare that may hide the actual colors present at the scene. These aspects will be covered in a subsequent section on filters. Polarizer filters, however, also uniformly reduce the light coming through the lens. Depending on the degree of rotation of the polarizer filter, the degree of light eliminated by the filter can vary from 1.3 stops to 2 stops of light. Therefore, if a rainy or snowy scene is currently overexposed by 1.3 to 2 stops, the polarizer will work nicely to provide a proper exposure.

Figure 2.60

Original snowy scene.

Figure 2.61

Snow eliminated with 2-second shutter speed.

Therefore, if using a 100 ISO film with the camera's smallest f/stop and a 2-second shutter speed still results in an overexposure, adding an ND filter or polarizer filter to the lens may reduce the light striking the film sufficiently to provide a proper exposure. This exposure combination has the additional benefit of eliminating the snow as it is falling.

RECIPROCAL EXPOSURES

The purpose of Figure 2.62 is to explain why a variety of f/stop and shutter speed combinations result in the exact same exposure. Changing one of the exposure variables is permissible if the other exposure variable is also changed to balance the first adjustment. Increasing one exposure variable by 1 stop is balanced by decreasing the other exposure variable by 1 stop. Increasing one exposure variable by 2 stops is balanced by decreasing the other exposure variable by 2 stops.

(a)

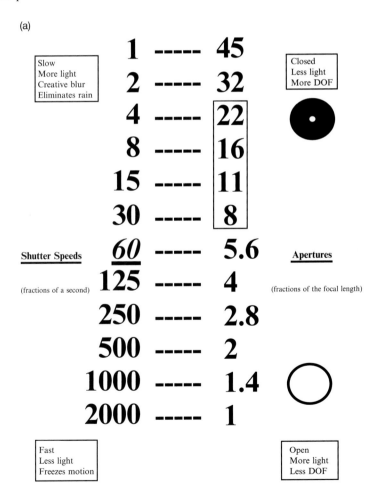

Shutter Speeds		Apertures
Slow / More light / Creative blur / Eliminates rain		Closed / Less light / More DOF
1 ----- 45		
2 ----- 32		
4 ----- 22		
8 ----- 16		
15 ----- 11		
30 ----- 8		
60 ----- 5.6		
125 ----- 4		
250 ----- 2.8		
500 ----- 2		
1000 ----- 1.4		
2000 ----- 1		

Shutter Speeds (fractions of a second)

Apertures (fractions of the focal length)

Fast / Less light / Freezes motion

Open / More light / Less DOF

Figure 2.62a

(a and b) Various combinations of shutter speeds and f/stops.

continued

Reciprocal Exposures

The **same** exposure may be obtained from a variety of different exposure settings. A smaller aperture may be used as long as a longer corresponding shutter speed is used. A wider aperture may be selected as long as a corresponding shorter shutter speed is used. Each change in apertures either halves (with smaller apertures) or doubles (with wider apertures) the light allowed to reach the film. Each change in shutter speeds either halves (with faster SSs) or doubles (with slower SSs) the light allowed to reach the film. Selecting an exposure that halves one variable while doubling the other variable results in the **same** exposure. <u>Why alter exposure settings?</u> To maximize DOF with smaller apertures, or to freeze motion with faster shutter speeds.

In the below example, F-2.8, 1/2000 would be an excellent choice to freeze motion; while F-16, 1/60 would be the best selection to maximize DOF while still being able to handhold a camera with a 50mm lens.

However, all seven combinations would result in good exposures under the same lighting conditions.

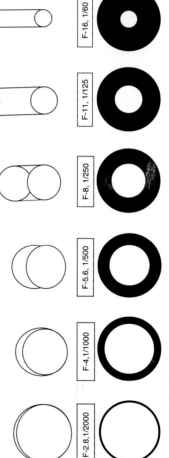

| F-2.8,1/2000 | F-4,1/1000 | F-5.6, 1/500 | F-8, 1/250 | F-11, 1/125 | F-16, 1/60 | F-22, 1/30 |

Figure 2.62b (continued)

(a and b) Various combinations of shutter speeds and f/stops.

Figure 2.62a is designed to be removed from the text and cut vertically down the middle through the middle hash mark.* This will allow the reader to align different shutter speeds with different f/stops. Once any one pairing of shutter speeds and apertures has been aligned, every other pairing of shutter speeds and f/stops are also reciprocal exposures for a particular lighting condition.

Notice in the shutter speed continuum, the *60* is shown in italics. This is an indication that a shutter speed of 1/60th second is the primary shutter speed to be used when hand-holding a 35-mm camera with a 50-mm lens on it; 1/60th of a second should be considered the primary shutter speed used at crime scenes.

Notice in the f/stop continuum, the f/stops 8, 11, 16, and 22 are enclosed in a rectangle. This is an indication that these f/stops produce the best depth of field of all the f/stop choices and should be regarded as the most frequently used f/stop selections. F/stop numbers and their relationship to depth of field will be revisited at length in the next chapter. For now, however, consider this. Because most crime scenes are relatively static, with the crime scene and the evidence within the crime scene fixed in place, the need to control motion is usually secondary to the need to maximize depth of field. Because most of the photos at the crime scene will be taken with a 50-mm lens, 1/60th of a second is the optimal shutter speed to use with that lens. Because it is important to have the best depth of field range when photographing the crime scene and the evidence within it, we will want to be using the smallest apertures as much as possible.

If both are true, and they are, then a new Rule of Thumb 2.9 for crime scene photography exposures can be developed.

Rule of Thumb 2.9: To the extent possible, most crime scene images should be captured with a shutter speed of 1/60th of a second in combination with one of these four f/stops: f/22, f/16, f/11, and f/8.

If we are about to depress the shutter button, and this condition does not exist, we should consider altering our exposure combination until one of the Rule of Thumb exposure recommendations is achieved.

As an example, consider a situation in which a 1/125th of a second shutter speed and an f/11 is about to be used. Aligning the 125 and 11 on Figure 2.62a shows all the other reciprocal exposures possible for the same crime scene lighting conditions. These include: 30 and 22, 60 and 16, 125 and 11, 250 and 8, and 500 and 5.6, among others. The Rule of Thumb recommendation would be 60 and 16. What is the effective difference between the 125 and 11 and the Rule of Thumb recommended 60 and 16? Both will result in the same exposure. The f/16, however, will produce a better depth of field than the f/11 would. Why use

*Figure 2.62a is also found on the companion website (http://books.elsevier.com/companions/ 9780123693839). Print the figure, and cut it down the middle.

an exposure combination that will produce less depth of field? Use the Rule of Thumb and change the exposure to 1/60 and f/16.

Sometimes we may become confused by a camera's design, which indicates only one exposure combination at a time. What is critical to remember is that even if the camera's metering system indicates one exposure combination is proper for a particular scene, many other exposure combinations will produce the exact same exposure value. We must choose wisely from these various possibilities to achieve our desired results. This is the **theory of reciprocity:** for every photo opportunity there are multiple exposure combinations to choose from.

Theory of reciprocity: The same exposure may be obtained from a variety of different exposure settings. A smaller aperture may be used as long as a longer corresponding shutter speed is used. A wider aperture may be selected as long as a corresponding shorter shutter speed is used.

This choice should be deliberate, not arbitrary. We should have a reason to choose one combination over the other. How do we choose? We must decide what we want the photograph to look like: do we want a long depth of field or a short depth of field; do we want to freeze motion, or have the image show creative blur?

RECIPROCITY FAILURE

Rarely, the theory of reciprocity does not work.

Merely doubling the shutter speed time and halving the aperture opening or halving the shutter speed time while doubling the aperture opening does not always result in a proper exposure when the extremes of shutter speeds are used. Because the chemical makeup of film emulsion cannot handle either very long dark exposures or short bursts of very bright light, problems usually occur at shutter speeds longer than 1 second and faster than 1/1000th second. A tendency toward underexposure occurs at both extremes. To compensate for this underexposure, follow these guidelines if you are using either very fast or very slow shutter speeds.

The following guidelines are just that: guidelines. Different film will react differently. The best results come from testing a particular film in the situation you anticipate needing to use it. Then, and only then, will you be confident in your results.

When would such situations occur? The need for long shutter speeds occurs when the ambient lighting is dim. In most of these situations, however, all we practically need to do is add electronic flash. When electronic flash is used, the necessary shutter speeds are very short. This will be elaborated on more in Chapter 4.

However, the use of electronic flash will not be the answer at all times. This text will cover several situations when long shutter speeds are required. For instance,

Table 2.1

Reciprocity Failure Corrections.

If the indicated shutter speed is: 1 sec, adjust the exposure by +1 stop.
If the indicated shutter speed is 10 sec, adjust the exposure by +2 stops.
If the indicated shutter speed is 100 sec, adjust the exposure by +3 stops.
If the indicated shutter speed is 1/1000th sec, adjust the exposure by +1 to +2 stops.

when photographing bloody residues enhanced with the blood reagent, luminol, shutter speeds in the 45-second to 90-second range are typical. When fluorescing trace evidence with an alternate light source (ALS), long shutter speeds may be required to capture the dim fluorescence adequately.

Extremely fast shutter speeds are necessary to "freeze" very fast motion. Fortunately, it is not frequently necessary to "freeze" a suspect flying by in a plane so that his face can be recognized. Fast shutter speeds are also necessary to properly expose brightly lit scenes. In this situation, the proper combination of slow speed film and small apertures is usually all the exposure adjustment that is required.

Do digital cameras also suffer from **reciprocity failure**? No. Reciprocity failure is a result of film emulsion's inability to provide for proper exposures with long shutter speeds, usually longer than 1 second.

With long shutter speeds, many digital cameras do have a well-known problem, although it is not called reciprocity failure. The problem is called **dark noise**.

When a digital sensor is exposed for long time intervals in dim lighting situations, the result is often a "grainy" appearance to the image, where individual pixels are not properly recording colors and light levels accurately. Rather than smooth evenly toned surfaces, the "noisy" digital image appears speckled with odd random colors.

Figures 2.63 and 2.64 were taken with a digital camera in an almost completely dark room. A shutter speed of 7 minutes was required to obtain both images. Figure 2.63 was taken "as is." It clearly shows the speckled "noise" that affects many digital cameras used with long shutter speeds. Figure 2.64 was taken with a camera with a program called automatic noise reduction. The camera's computer can recognize "noise" and fix much of it. The effects of both are more clearly seen with blowups from each image.

The automatic noise reduction program has a great effect on the final product. Although the "noise" has not been completely eliminated, the images with automatic noise reduction are clearly superior to their "noisy" counterparts.

Reciprocity failure: Because the chemical makeup of film cannot handle either very long dark exposures or short bursts of very bright light, problems usually occur at shutter speeds longer than 1 second and faster than 1/1000th second. A tendency toward underexposure occurs at both extremes.

Dark noise: When a digital sensor is exposed for long time intervals in dim situations, the result is often a "grainy" appearance to the image, where individual pixels are not properly recording colors and light levels accurately. Rather than smooth evenly toned surfaces, the "noisy" digital image appears speckled with odd random colors.

Figure 2.63
Noisy: 7-minute exposure.

Figure 2.64
Seven-minute exposure with automatic noise reduction.

Figure 2.65
Crop of Figure 2.63

Figure 2.66
Crop from Figure 2.64 with automatic noise reduction.

THE REFLECTIVE LIGHT METER

THE LIGHT STANDARD: THE 18% GRAY CARD

Some photos can be properly exposed, some can be underexposed, and some can be overexposed. This is easy to see when looking at photographs. But how does the reflective light meter in the camera "know" that the f/stop and shutter speed selected allow the right amount of light to enter the camera? Most cameras have light meters that measure the amount of light being reflected into the camera, but what standard is used to determine whether this amount of light is just right, not enough, or too much?

Figure 2.67, which shows an 18% gray card, is the standard. Many shades of gray exist. Only a gray card that reflects precisely 18% of the light that strikes it is the standard. This particular shade of gray is used because the real world, considering its light shades and dark shades and the various colors found in a "normal" scene, typically reflects 18% of the light that strikes it.

Any author venturing into an explanation of reflected light meters and the 18% gray card would do well to at least briefly acknowledge Ansel Adams' Zone System, formulated in 1939–1940. Often thought of as being an extremely technical and confusing method to determine proper exposures, a simplified selection of relevant ideas will be presented here. Consider a continuum from absolute black to pure white, with various shades of gray in between, where each gradation on the continuum is precisely 1 stop more or less light from an adjacent tone. An 11-step scale will be formed, with absolute black represented by the value 0 (zero) and pure white represented by X (Roman numeral ten). The shade or tone in the center would then be Zone V, which is a shade of gray which matches the 18% reflectance gray card. This is the tone reflective meters

Figure 2.67

A standard 18% gray card (The #429 is my old badge number).

are designed to meter for. Ansel Adams further describes what he calls "a 'normal' range subject. When a reading from the important shadow area is placed on Zone III, the middle-gray value falls on Zone V, and the light surface on Zone VII. . . . Thus the exposure indicated . . . will yield a full-range negative." [†] A full-range negative is produced when the scene is balanced. Mid-gray subject matter is present, with both bright and dark areas present in the same scene. The exposure for this scene would be the same as the exposure for the 18% gray card totally filling the frame.

Camera manufacturers know they will not survive long if they sell cameras that do not produce acceptable results most of the time. After extensive testing of myriad scenes reflecting differing amounts of light, it was determined that the "normal" scene camera owners would likely aim their cameras at was a scene reflecting approximately 18% of the light that struck it. The **18% gray card,** then, is a substitute for a "normal" scene.

If one were to become confused by the variation of objects within any particular scene and were unsure whether the scene were "normal" or not, to ensure a proper exposure was obtained, an 18% gray card can be used as a scene substitute. In other words, the camera can be aimed at the 18% gray card, and a meter reading can be obtained from it.

An imaginary scene can provide an example. Take, for instance, a sunny outdoor scene around noon. A photographer would typically load 100 ISO film into the camera or set ISO 100 as the "film" choice on a digital camera. If a 50-mm lens were being used, set the shutter speed to 1/60th of a second. Then, compose

18% Gray card: An 18% gray card is the standard. Many shades of gray exist. Only a gray card that reflects precisely 18% of the light that strikes it is the standard. This particular shade of gray is used because the real world, considering its light shades and dark shades and the various colors found in a "normal" scene, typically reflects 18% of the light that strikes it.

[†]Adams, A. (1989). "The Negative: The New Ansel Adams Photography Series, Book 2." p. 52. Little, Brown and Company, Boston.

the camera so the 18% gray card totally fills the frame. Ensure the 18% gray card is lit like the scene in question. In other words, if the crime scene is sunny, ensure the 18% gray card is sun-lit; if the crime scene is shady, have the 18% gray card be covered with the same degree of shade. Finally, depress the shutter button half way and take a meter reading to determine the necessary f/stop. Set the camera for that f/stop, compose, and focus on the real crime scene, and a proper exposure should be the result.

LIGHT METERS

Reflective light meters: 35-mm SLR cameras typically have reflective light meters built into them, designed to meter/ measure the amount of light reflected from the scene through the lens into the camera.

Incident light meter: This is a hand-held meter that the photographer will hold near the person or object being photographed. It is designed to measure the amount of light falling on the subject, rather than the light reflected from the subject toward the camera.

Thirty-five–millimeter SLR cameras typically have **reflective light meters** built into them, designed to meter/measure the amount of light reflected from the scene through the lens into the camera.

The reflective light meter is inside the camera body, near the film plane, so it can properly determine the amount of light required to properly expose the film.

Many professional photographers, although not any of the crime scene photographers this author has known, will use an **incident light meter** to determine the lighting required for a proper exposure.

This is a hand-held meter that the photographer will hold near the person or object being photographed. It is designed to measure the amount of light falling on the subject rather than the light reflected from the subject toward the camera. An incident light meter is very precise in controlled situations, where the photographer can stage a variety of lights at different distances and at different intensities to achieve the desired effect. At crime scenes or accident scenes, it is normally too impractical to consider multiple lighting situations.

Several kinds of reflective light meters are available in SLR cameras. Lower-priced, student cameras will frequently only have an averaging or center-weighted meter. As the cost of the camera body goes up, usually the camera will feature other metering options.

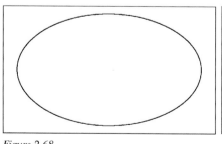

Figure 2.68

Center-weighted meter.

Figure 2.69

Spot meter.

Averaging/Center-Weighted Meters

Averaging/center-weighted meters measure most of the light coming in through the lens, emphasizing the light from the center of the area of view, because that is where most people place the important aspects of their photographs. When taking a meter reading, only the light coming in from the area shown by the ellipse will be considered in determining the meter reading. Most cameras, even if they have other meter options, will have this metering option.

Spot Meters

Spot meters only measure the light reflected from a very small area of the center of the viewfinder. By doing so, the photographer can be very specific about what area of the photograph he or she wants properly exposed, disregarding the rest of the scene. For example, imagine trying to properly photograph an actor on a stage. Some areas of the stage may be brightly lit, and other areas of the stage may be very dim. Should these areas be considered when the concern really is only with the light reflecting from just one particular actor? In this situation, a spot meter would be very useful to determine the light being reflected from just one part of the over-all scene. The crime scene equivalent to this is surveillance photography. Other aspects in the field of view may be irrelevant, so by placing the spot meter only on the subject of interest, just the light reflected from him or her will be calculated into the exposure reading. Light extremes around the periphery will be ignored.

Different cameras have spot meters of various sizes, but the size of the spot meter is usually indicated in the viewfinder, so the photographer can place it carefully where needed.

Matrix Meters

Various matrix-metering systems exist, sometimes with elaborate geometric configurations, which break up the light coming in through the viewfinder into a series of grids and assign different values to the light coming in from the various grids.

These systems are a result of very extensive research and development and are usually found on only the higher-priced cameras. A mini-computer in the camera

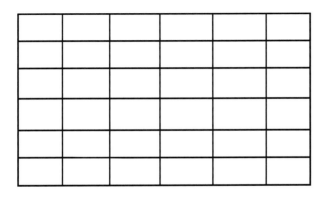

Figure 2.70
Matrix meter.

uses very complex algorithms to "weigh" the differing amounts of light coming in from the various grids. For instance, if the grids in the center of the viewfinder received very limited light reflection and several very bright grids were in one of the upper corners, the computer would recognize that situation as a person in the center of the viewfinder being back-lit by the sun or moon in the distance and recommend an exposure designed to properly expose the dimly lit center of the viewfinder. Matrix metering systems produce very accurately exposed photographs. If a camera has a matrix-metering system, it is normally preferred as the standard metering system.

"NORMAL" AND "NON-NORMAL" SCENES

As stated previously, the reflective light meter built into the SLR camera body is designed to provide proper exposures for scenes that reflect 18% of the light that strikes them back toward the camera. These can be considered "normal" scenes. Reflect on those occasions when you have usually taken photographs outside of your work environment. These include taking photos of family gatherings, vacations, loved ones, pets, and so on. What about them makes them "normal" scenes? They contained a variety of colors and objects that were both black and white and gray. White objects reflect a lot of light, black objects reflect little light, and gray objects reflect some degree of light in between the extremes. Colors also reflect a variety of light intensities. When thrown into one scene, the average of the light reflected back from a "normal" scene toward the camera is 18%.

"Non-normal" scenes reflect more or less light than the typical 18% that SLR camera meters are designed to deal with. The average photographer encounters these types of scenes less frequently. In these situations, the exposure results can sometimes be less satisfactory. We may notice in these situations that our images came back a bit overexposed or underexposed. Why?

Let us image how a reflective light meter "thinks." Designed to provide proper exposures for a scene that reflects 18% of the light that struck it, when the situation in fact does reflect that amount of light, the reflective light meter balances the four exposure variables: shutter speed, f/stops, film speed (or digital equivalent), and the light reflected from the scene. A proper exposure is the result. When the camera knows the ISO rating of the film within it, the shutter speed selected by the photographer, and the camera is composed on a "normal" scene reflecting 18% of the light striking it into the lens, the meter determines the necessary f/stop for a proper exposure. The f/stop recommended completes the four essential elements of the proper exposure triangle. This is what occurs at a "normal" scene.

Remember, however, that each of the four exposure variables comes in 1-stop increments. For example, a 1/60th shutter speed can also be changed to 1/30th (+1) or to 1/125 (−1). An f/11 can be changed to f/8 (+1) or f/16 (−1); 400 ISO film also comes in 200 ISO (−1) or 800 ISO (+1) variants. Although "normal" scenes reflect 18% of the light striking them, other scenes reflect 36% of the light (+1), and some scenes can reflect just 9% of the light that strikes it (−1).

Now consider a "non-normal" scene. Consider the situation where it has been snowing and your children have just successfully built their first snowman of the winter season. As a proud parent, you must document the achievement with a photograph. As the children stand around the magnificent snowman, you fill the camera frame with a scene that is predominantly white. Naïvely, you trust the camera metering system and use the recommended f/stop to take the photograph with your SLR. The result comes back improperly exposed. Your first inclination is to accuse the film developer of incorrectly exposing your perfectly exposed negative. If you return to the film processor and make this complaint, after several minutes of reprocessing your negative, you will probably be provided a print that looks "properly" exposed. You smile to yourself, feel vindicated in your original assumption they botched the original exposure, and leave happily knowing you are, in fact, the camera expert you have always thought you were.

Are you correct on either count? Had your original exposure been the right one? Had they incorrectly processed your correctly exposed negative? No, no. And, no.

Consider your original exposure first. Let us presume, for argument sake, that white snow reflects approximately 80% of the light that strikes it. Here is what the camera meter encountered.

X-ISO film + Y-shutter speed + Z-scene reflectivity (80%)

If meters could talk, this it what it would have said to itself: "Something is wrong. This is not what I am designed to deal with. I am receiving too much light. I am not sure whether this is because the 'normal' scene is lit with too much light or whether the scene itself is reflecting more light than normal. How should I resolve this dilemma? I will presume the photographer is aiming the camera at a 'normal' scene that happens to be lit by an unusual amount of light. I will, therefore, provide a proper exposure for this particular scene as if it was a 'normal' scene lit with a normal amount of light; 80% is too much. I know exposure stop increments are expressed as halves and doubles. Half of 80% is 40%. That is still too much light. Half of 40% is 20%. That is close enough to 18%. I will recommend an f/stop to the photographer that in effect underexposes the scene by 2 stops. That will be the proper exposure for a 'normal' scene lit 'normally.' That will prove I can be trusted to perform as designed."

The "−2 f/stop" is an f/stop 2-stops smaller than would have been appropriate for a "normal" scene lit "normally." The result is an underexposed negative. How much underexposed? Two stops underexposed.

If a film processor "correctly" prints this underexposed negative and the result is an underexposed print, has the processor made a mistake? Technically, no. In this scenario, however, a complaint was made, and the processor took the same negative back into the darkroom and later returned with a print more acceptable to its owner. Is this new "properly" exposed print proof that the originally underexposed print was incorrectly printed? Technically, no. Processing an underexposed negative into an underexposed print is not a mistake. It is certainly not good for the processor's business, but it is not a mistake.

Remember, underexposed and overexposed negatives can frequently be "corrected" in the darkroom, so that more acceptable prints are the result.

Let us examine a scenario where the photographer is presented with an abnormally dark scene. This time, let us use an example related to crime scene photography. Several cars in a neighborhood have had their tires slashed, and a suspect has been caught with a single-edged knife. All the puncture marks in the tires show the blade used had a single sharp edge. Anticipating that a few good photographs might demonstrate this fact to the jury better than just verbally reporting it, photographs of the puncture marks are made. As the photographer composes the first close-up image of a puncture mark and takes a meter reading of the area, let us presume that black tire rubber reflects approximately 5% of the light that strikes it.

If the meter could talk, this it what it would have said to itself: "Something is wrong. This is not what I am designed to deal with. I am not receiving enough light. I am not sure whether this is because the 'normal' scene is lit with insufficient light or whether the scene itself is reflecting less light than normal. How should I resolve this dilemma? I will presume the photographer is aiming the camera at a 'normal' scene that happens to be lit by an unusual dim amount of light. I will, therefore, provide a proper exposure for this particular scene as if it was a 'normal' scene lit with a normal amount of light; 5% is not enough light for a proper exposure. I know exposure stop increments are expressed as halves and doubles. Double 5% is 10%. That is still not enough light. Double 10% is 20%. That is close enough to 18%. I will recommend an f/stop to the photographer that, in effect, overexposes the scene by 2 stops. That will be the proper exposure for a 'normal' scene lit 'normally.' That will prove I can be trusted to perform as designed."

The "+2 f/stop" is an f/stop 2 stops wider than would have been appropriate for a "normal" scene lit "normally." The result is an overexposed negative. How much overexposed? Two stops overexposed.

If a film processor "correctly" prints this overexposed negative and the result is an overexposed print, has the processor made a mistake? Technically, no. But, now, the print is overexposed. Solutions to both overexposures and underexposures in the darkroom have already been explained.

THE "NORMAL" SCENE

The "normal" scene reflects 18% of the light that strikes it. Fortunately, most scenes are "normal." This means that not a lot of exposures need to be corrected in the darkroom: we can trust the camera's built-in light meter to provide for proper exposures.

If pushed to estimate what percentage of crime scene photography involves "normal" scenes, a number has not been indicated in the literature. What is "normal" for the layman taking photographs is certainly not what is encountered by crime scene photographers. A crime scene photographer has to be very aware of when they are about to photograph a "non-normal" scene and how to correct the resultant exposure differences.

"NON-NORMAL" SCENES

Four well-recognized "Non-Normal" scenes will require exposure compensations to provide for proper exposures:

- Predominantly light-colored scenes
- Predominantly dark-colored scenes
- Scenes with a lot of sky in the field of view
- Back-lit scenes or scenes with both bright highlights and dark shadows in the same composition

Exposure compensations are required, because trusting the reflective light meter to provide the "correct" exposure in these situations will lead to unsatisfactory results. Knowing that the meter cannot be trusted when dealing with a "non-normal" scene is the first step. Knowing what exposure compensation is required is the second step. The goal is to capture a properly exposed negative or digital image. It is always preferable to expose the image properly at the time of capture, rather than have to resort to wet-chemistry or digital darkroom corrections.

Predominantly Light Scenes

The nature of crime scene photography presents us with many light-colored scenes and objects. Just as the snowman scenario mentioned earlier resulted in an underexposed image, if the camera meter is trusted, so will all our outdoor crime scenes that include large areas of snow. This will be the result not only when overall images of the crime scene are taken but also when images of individual evidence lying in the snow are taken.

To compensate for this tendency toward underexposure, it will be necessary to manually alter the exposure recommendation offered by the meter. Knowing that snowy scenes are typically underexposed by the meter by approximately 2 stops, take the meter reading as usual, and then manually open the aperture 2 stops wider from what the meter recommended.

For many, this seems counterintuitive. If the crime scene reflects too much light, should we not be reducing the light to ensure a proper exposure? This is the reasoning used by many who do not understand that the reflective light meter is automatically making this compensation for us already. If we use our "gut" feeling and reduce the light even more than what the meter recommends, the result will be greatly underexposed images.

Dirty snow: If the reflective light meter is relied on when taking images of snowy scenes, the result will be "dirty snow." What is "dirty snow?" It is dark or underexposed snow.

A mnemonic is useful here. Remember the phrase, **"dirty snow."** If the reflective light meter is relied on when taking images of snowy scenes, the result will be "dirty snow." What is "dirty snow?" It is dark or underexposed snow. What is the cure for "dirty snow?" Open up 2 stops.

Another predominantly light-colored type of evidence should be mentioned here: Caucasian skin. Crime scene photographers encounter the need to photograph this type of evidence frequently. Whether it is a nude Caucasian cadaver at an autopsy or just a wound or bruise to one part of the body, this type of evidence is frequently encountered. The need to fill the frame of the camera with Caucasian skin recurs frequently enough to recognize the need to properly expose this type of evidence. Although myriad shades of skin are present in all races, "normal" Caucasian skin has been measured to reflect approximately 36% of the light that strikes it. That is precisely 1 stop lighter than a "normal" scene, which the reflective light meter is expecting.

The result: the meter will tend to cause a 1-stop underexposure of Caucasian skin, if it is trusted. The compensation required, then, is to take a meter reading of the Caucasian skin and then open up +1 stop more from what the meter suggests.

Interior overall photographs: Photographing the four walls of a crime scene room; a full 360° view of the room.

Crime scene photographers frequently encounter other light-colored scenes. Most rooms have light-colored walls. When taking **interior overall photographs** of crime scenes and the composition is predominantly of a light-colored wall, a compensation is almost always required.

How much? Interior walls are seldom as bright as snow and are closer to the reflectivity of Caucasian skin, so open up 1 stop from what the flash calculator dial suggests. Flash theory will be covered in another chapter, but the need to photograph light-colored walls occurs so frequently, it is necessary to mention it in this chapter.

Just as snow presented an exposure compensation necessity, so too would scenes with a large amount of sand in them. Although sand may seem a darker shade than snow, its crystalline nature reflects enormous amounts of light toward the camera. Sand requires the same exposure compensation as snow. With a sandy crime scene, take a meter reading, and open up +2 stops.

The need to make an exposure compensation only occurs when the viewfinder is filled with a light-colored scene. It is not required just because a light-colored object is being photographed. If, for instance, a light-colored soccer ball is noticed to have some blood drops on it and it is currently lying on a grassy lawn, the meter may be trusted if quite a bit of grass is also in the viewfinder. It is only when the viewfinder is almost totally filled with the light part of the scene that an exposure compensation is required. If the light-colored object is just one item in the field of view and the meter will be receiving reflected light from other normally toned objects, the meter can be trusted.

What if the situation is "borderline?" In these situations, **bracket**. Bracketing will be discussed shortly.

Predominantly Dark Scenes

Any time the viewfinder is filled with a relatively dark object, such as the black tire mentioned previously, the reflective light meter is not to be trusted. In this case, the mnemonic, "dirty snow," can again help. If light-colored scenes tend towards underexposure, predominantly dark scenes will tend toward the opposite: over-exposure. Take a situation in which a black sweater with a relatively small logo has been discarded by a suspect while fleeing. If the sweater is viewed as just one item in a scene showing the general area where the sweater was found, then an exposure compensation will not be needed. In this case, the sweater is just one dark object among other "normally" toned objects.

Once the sweater has been photographed within the scene, the next series of photographs would be close-up photographs of just the sweater filling the frame of the camera. In this situation, the camera meter will be fooled into recommending an exposure to the photographer that overexposes the sweater.

The black sweater photographed in Figures 2.71–2.73 as metered will come out looking 2 stops overexposed: grayish. Decreasing the exposure 1 stop improves the image, but the sweater still does not look as black as it really was. Only a 2-stop decrease in exposure accurately captures the original look of the black sweater. Unsure whether a −1 or −2 stop exposure is correct for a particular item of dark evidence? Photograph it both ways and compare the images to the sweater.

Large Amounts of Sky

When a large portion of the scene/composition turns out to be the blue sky, realize that the sky reflects more than normal amounts of light toward the camera. Although our eyes adjust well to this bright light source, making it easy for us to forget the sky is much brighter than any objects reflecting normal amounts of light, film and digital sensors will be adversely affected by this exposure situation. Crime scene photographers are presented with this situation every time they take **exterior overall photographs** of crime scenes.

We will take photographs of the building the crime occurred in and frequently this includes quite a bit of blue sky in the background. If a substantial amount of this particular field of view is much brighter than "normal," the meter can only average that large amount of light, add it to the light otherwise reflected from the scene, and come up with an exposure recommendation "suitable for both." The meter will unsuccessfully attempt to provide for a proper exposure of the sky and a proper exposure of the building at the same time.

The result will frequently be an underexposed building, but the sky will look very close to the way we remember it. Of course, having the sky properly exposed at the expense of the exposure to the building is counterproductive. The intent

Bracketing: The act of intentionally taking multiple exposures of the same subject from the same point-of-view.

Exterior overall photographs: These include photographing all four sides of a building in which a crime occurred.

Figure 2.71
Black sweater, as metered.

Figure 2.72
Black sweater, −1.

Figure 2.73
Black sweater, −2.

of the image was to have properly exposed the crime scene building. One of the main purposes of exterior overalls is to show possible points of ingress or egress that suspects may have used. If the building is grossly underexposed, these aspects will be lost.

What is the exposure compensation required to provide proper exposures of buildings when a large amount of sky is also in view? To determine the correct exposure, eliminate the problem. In this case, the sky is ultimately causing an incorrect exposure of the building. So, when taking a meter reading of the building, eliminate the sky. This can easily be done in one of two ways.

First, simply tilt the camera down: lower the camera until the top of the building is at the very top of the field of view, and then take a meter reading without the sky in view. Set the camera for this meter reading, and then recompose back to the original scene with the sky in the field of view. Take the photograph.

Second, if you are using a zoom lens as your primary lens, zoom the lens to its extreme telephoto setting, composing this narrower field of view so it includes only the building façade, and take a meter reading that excludes the sky. Set the camera to that exposure recommendation, reset the lens back to the original focal length, and take the photograph.

In both instances, the sky will then be a bit overexposed, but the building will look properly exposed. As crime scene photographers, this is about as good as we can do. There are, of course, half-frame neutral density filters, which provide shading for only half of the view area. Landscape photographers will position the shaded area so that it covers the sky, and then meter the landscape. The result is a landscape that is properly exposed, and a sky that looks like Figure 2.74. However, not many law enforcement agencies are providing half-frame neutral density filters to their crime scene photographers.

Back-Lit Scenes and Other High-Contrast Scenes

When the primary subject is photographed while being backlit, the primary subject usually ends up looking underexposed in photographs. This is another situation in which a bright light source is in the field of view, similar to the sky

Figure 2.74

As metered, with the sky in view.

Figure 2.75

Metered without the sky in view.

situation. The reflective light meter is designed to properly expose a scene that has a normal range of tones in the scene with the lighting for the scene uniform throughout the scene. When any part of the scene diverges from these uniformities, the meter will have difficulties providing for proper exposures.

In Figures 2.76 and 2.77, a meter reading was taken of the bright sunlit area in the background. This ensures the background will be properly exposed. Without doing this, the background may look grossly overexposed in the resulting image. However, because the subject was not directly sunlit, this exposure combination results in the primary subject being underexposed. This underexposure of the primary subject is compensated for with **fill-flash**. Fill-flash will be completely covered in Chapter 4. For the purposes of this chapter, however, it will be mentioned that any time a scene has both bright and dark areas in the same scene, it will probably be necessary to use fill-flash to ensure both areas receive proper lighting for an overall proper exposure.

A brief explanation here will suffice to explain images Figures 2.76 and 2.77. As with the left image, the exposure for the sunlit background was determined by directly metering that area. The viewfinder was filled with the sunlit area of the scene, and a meter reading was taken. It is important to note that the overall scene was not metered. This would include both the sunlit background and the more dimly lit primary subject. Doing so would result in an improper exposure determination for the sunlit background. Once the exposure for the sunlit background was determined, the camera was set for that exposure combination. A sunny scene would have called for a 100 ISO film. To hand-hold a camera with a 50-mm lens on it would have meant that a shutter speed of 1/60th of a second

Fill-flash: A flash technique to ensure a scene including bright sunlit areas and dark shadows can have both areas properly exposed in one photograph. The sunny areas are metered to properly expose the sunlit areas of the scene. Flash is used to properly expose the shady areas. The flash exposure is balanced for the exposure necessary to properly expose the sunlit area.

Figure 2.76
Backlit scene: primary subject underexposed (Courtesy of Denise Sediq, GWU MFS student).

Figure 2.77
Backlit scene: corrected by fill-flash (Courtesy of Denise Sediq, GWU MFS student).

would be normal. A midday sunny scene frequently requires an f/22, as will become more clear as this chapter continues. Then, to properly expose the primary subject, an electronic flash was set to properly expose this subject. The result is that the sunlit background is properly exposed, while at the same time the primary subject in the foreground is also properly exposed.

There may also be high-contrast scenes, which include both bright sunny areas and dimly lit shady areas in the same area of the same scene. Of course, if you can recompose to eliminate one of the lighting extremes, that is one way to solve the improper exposure tendency. But many times this is impossible. On bright sunny days, the sunlit area is often directly adjacent to an area in deep shadows. When wide extremes in the light levels of a scene exist, we must realize that the film's exposure latitude results in one of the extremes being improperly exposed. Fill-flash is the cure to lighting extremes.

Again, avoid the temptation to merely take a meter reading of the entire scene, believing that doing so will simultaneously cure the underexposure of the shady areas of the scene and the overexposure of the sunny areas of the scene. This "average" exposure fails to properly expose both areas at the same time. For example, in the preceding overexposure of the sunlit area, the exposure was ISO 100, 1/60th of a second shutter speed, and f/5.6. That means that the shady area of the scene required an f/5.6 to properly light that part of the scene. The exposure for the sunlit area of the scene was ISO 100, 1/60th of a second shutter speed, and f/22. That means the sunny area of the scene required an f/22 to properly light that part of the scene. If f/5.6 and f/22 are averaged the result is f/11, which is half way in between. Remember the f/stop continuum: 22-16-11-8-5.6. . . . Taking a meter reading of the entire scene, with both extremes of lighting within it, results in an f/11. Is an f/11 a good compromise? No. At the same time, it underexposes the part of the scene requiring an f/5.6 for a proper exposure and it overexposes the part of the scene requiring an f/22 for a proper exposure. It improperly exposes both parts of the scene by 2 stops.

As mentioned previously, the proper solution is to use fill-flash in this situation.

TOOLS FOR DETERMINING "PROPER" EXPOSURES WITH TRICKY SCENES

How can the photographer determine whether the light reflected from a particular scene is "average" or is a scene reflecting 18% of the light that struck it? Is there a way to be better assured that one is properly metering a particular scene? Does the camera and/or meter somehow alert the user to a problematic scene? These questions, and their concerns, trouble most photographers at one time or another. Even though it is probable that 80% to 90% of crime scene photos are of "normal" scenes and the camera's reflective light meter can be trusted in those

Figure 2.78

Exposed for the shadow area; sunlit area overexposed.

Figure 2.79

Exposed for the sunny area; shadow areas underexposed.

Figure 2.80

Exposed with fill-flash; both sunlit and shadow areas properly exposed.

instances, what can be done when the "non-normal" scene is encountered or when a scene with a "tricky" exposure is encountered?

First, try to remember the exposure compensations recommended previously. If, however, you should "go blank" at any particular time when trying to determine the correct exposure for a complex scene, consider using one of the following "tools" for determining proper exposures.

METER THE 18% GRAY CARD

If any doubt exists about whether a particular scene is reflecting the 18% "normal" amount of light, one can fill the viewfinder with the 18% Gray Card and take a meter reading from it.

If it is lit the same way as the scene is lit, then you can set the camera for an exposure on the basis of the 18% gray card and use that exposure for the scene in question. If the scene is sun lit, the 18% gray card must also be sun lit; if the scene is shady, the 18% gray card must also be covered in the same density of shade. Do not use the 18% gray card when the scene is sun lit but the 18% gray card is in the shade or when the scene is shady but the 18% gray card is sun lit; 18% gray cards come in a variety of sizes, so many photographers carry one in their camera bag, along with other essential equipment.

Official 18% gray cards can be purchased in any camera shop. Some crime scene product distributors produce **photo identifiers** that can also serve as 18% gray cards. See Figure 2.81.

Photo identifiers: A form containing information to document a series of photos:
(1) case number,
(2) address or location,
(3) date and time, and
(4) name of photographer.

This card will be further explained in Chapter 5, but for now, notice it has a substantial area that is gray. Surrounding that area is a black-to-white scale on the right side and pastel and solid colors on the bottom. The left side has inch increments. The total effect: it is a card intended to reflect about 18% of the light that strikes it.

If any particular scene was judged to be confusing or a possible problem for the camera meter to expose correctly, one could fill the frame with this card, meter it, set the camera for that exposure combination, recompose on the scene in question, and take the picture.

METER AN AREA OF GREEN GRASS

All the colors of the visible light spectrum can be converted into various shades of gray. It just so happens that green grass reflects about 18% of the light that strikes it, so it can be used as a surrogate 18% Gray Card if you do not have the 18% gray card handy. Fill the frame with green grass lit the same as the subject matter of concern, take a meter reading, set the camera for that exposure setting, recompose on the chosen subject matter, and take the picture.

Of course, not every crime scene is located near areas of green grass. Also, in some seasons green grass is unavailable. Fortunately, other surrogate 18% gray cards are available.

Figure 2.81

A photo identifier; also an 18% gray card.

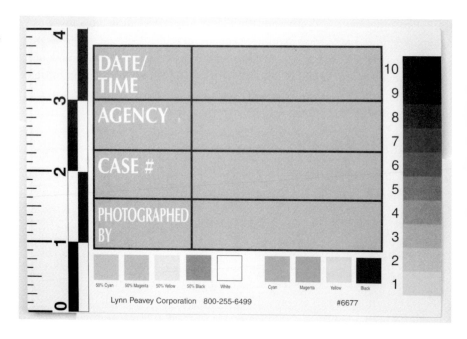

METER WELL-TRAVELED ASPHALT

No green grass in view? How about asphalt? Well-traveled asphalt also reflects about 18% of the light that strikes it. Newly laid asphalt, however, is very black and should not be used as an 18% gray substitute. After asphalt has been exposed to the elements for a while, it turns into a perfect 18% gray card substitute. Fill the frame with it, lit the same way as the scene of interest, take a meter reading, set the exposure, recompose, and take the picture.

Of course, not every crime scene is located near areas of well-traveled asphalt. Some of those crime scenes are also devoid of grassy areas. Is all lost? No.

METER YOUR PALM

When all else fails, you can always meter your own palm. Your palm, regardless of race, reflects approximately 36% of the light that strikes it. Half of 36% is 18%. Remember the theory of halves and doubles? Reflecting one stop more light than "normal," the palm will cause an exposure reading that is 1 stop underexposed, if that exposure reading is trusted. The camera, in effect, says to itself: I expect 18% light; this is too bright. To give the photographer an exposure for 18%, I am going to reduce the exposure by 1 stop. Knowing this is how the meter "thinks," meter your palm and then open up 1 stop to arrive at a "proper" exposure.

Here, too, it is necessary to have your palm lit the same as the scene of interest: sunny scenes require a sunny palm; a shady scene will require the palm be shady also.

It should be mentioned that it is not necessary to properly focus on the 18% gray card, green grass, well-traveled asphalt, or your palm when the meter reading is made. The meter does not care whether the reflected light is in focus or not. All it requires is the right amount of light.

EXPOSURE MODES

There are four common exposure modes: manual, program, aperture priority or shutter priority exposure modes. These modes allow the photographer to choose different options to determine how a proper exposure is obtained. Camera exposure control dials will usually abbreviate them with either "M," "P," "Av," or "Tv."

The other symbols on this exposure dial are directed at the consumer or novice photographer not interested in learning all that an SLR camera system has available in it. These will not be explained in this text. They do not represent true additional camera capabilities; they represent shortcuts to image making.

MANUAL EXPOSURE (M)

With the **manual exposure mode**, the photographer is responsible to manually set the f/stop and the shutter speed. A camera with a manual exposure mode option usually has this choice represented by an "M" somewhere on the camera exposure controls.

Manual exposure mode: With the manual exposure mode, the photographer is responsible to manually set the f/stop and the shutter speed.

Figure 2.82
Exposure modes.

With lower-end cameras, it is also necessary to manually set the film ISO as well. Newer cameras with DX code sensors in the camera body will be able to do this particular aspect automatically. However, the photographer has complete control of the camera's shutter speed and f/stop settings, because these exposure variables have to be manually set by the photographer for each photograph.

Students learning how to make photographs with an SLR camera for the first time are frequently told to use only the manual exposure mode, because this exposure mode makes them responsible for all the critical choices that have to be made when determining the proper exposure for each shot. It is also a good exposure mode to use when photographers wish to make decisions for themselves and not relinquish any creative control to a camera's computer. This is frequently the case with crime scene photographers.

PROGRAM EXPOSURE MODE (P)

Program exposure mode: The camera sets both the f/stop and the shutter speed automatically on the basis of the amount of reflected light coming into the camera.

With the program exposure mode, the camera sets both the f/stop and the shutter speed automatically on the basis of the amount of reflected light coming into the camera. A camera with a **program exposure mode** option usually has this choice represented by a "P" somewhere on the camera exposure controls.

The photographer has no choice over the camera's creative controls. The camera works on preprogrammed algorithms, which select the exposure settings on the basis of the light being read by the meter.

The program exposure mode is sometimes referred to as the "dummy exposure mode." A photographer who only cares about a proper exposure can set the camera to program, aim, focus, and shoot. A proper exposure is almost always a guarantee. There is no creative photographer input.

The program exposure mode is not recommended for crime scene photography because, besides ensuring that there will be a proper exposure for all crime scene images, it is just as important to manage the area that will be in focus (depth of field). Sometimes, it is also important to manage the motion-stopping capability of fast shutter speeds. With most crime scene photography, it is not only important that there be a proper exposure, it is also important that there is a proper exposure with the depth of field maximized. Or, it is important that there is a proper exposure and the motion of a moving subject be eliminated with an appropriately fast shutter speed. The use of the program exposure mode reduces everything to just one narrow concern: exposure. The need for a proper exposure is not to be denied. However, additional concerns are constantly tugging on the mind of the competent crime scene photographer.

The major complaint with the program exposure mode is that its computer algorithms seem to default to f/stops centering around f/4s and f/5.6s and shutter speeds centering around 1/250th of a second and 1/500th of a second. If the crime scene subject matter is static like most crime scene evidence, using faster-than-necessary shutter speeds and wider-than-necessary apertures wastes the full capabilities of the camera controls, just to ensure a proper exposure is obtained.

When the issue is training competent crime scene photographers, the program exposure mode is to be avoided. Proposing it as a viable exposure mode is tantamount to admitting either that (1) there is no one capable to teach crime scene photography properly to your crime scene photographers, or (2) you do not believe your crime scene photographers can understand more advanced crime scene photography methods. Hopefully, neither condition exists within your agency.

APERTURE PRIORITY MODE (AV)

In the **aperture priority exposure mode** the photographer manually sets the f/stop desired, and the camera will automatically set the corresponding shutter speed required to properly expose the photograph for the reflected light coming into the camera.

A camera with an aperture priority exposure mode option usually has this choice represented by an "Av" somewhere on the camera exposure controls. The "Av" stands for aperture value. Care needs to be exercised with this mode, because if small apertures are selected to maximize the depth of field and the lighting is inadequate, the result may be a shutter speed that is too slow for the camera to be hand-held, resulting in blurred photos if the photographer is not aware of the resulting slow shutter speed setting.

Wouldn't a cure for slow shutter speeds be to just put the camera on a tripod and take most images that way? Actually, using a tripod would be enormously inconvenient for most crime scene photography. Certainly, many times a tripod cannot be avoided when taking crime scene images. But, because most crime scene images have to be taken rather quickly, there is just not enough time to take most crimes scene images with the camera mounted on a tripod.

However, when you are in a situation when using a tripod is recommended, using the aperture priority exposure mode is the ideal exposure mode. For instance, when taking a series of images of a bloody shoe print, once the ideal f/stop has been determined, setting the camera on a tripod will allow the camera to use any shutter speed for the series of photographs necessary to completely document the shoe print. This will enable the photographer to vary the exposures without deviating from an optimal f/stop. This topic will be revisited in Chapter 3.

SHUTTER PRIORITY MODE (TV)

In **shutter priority exposure mode**, the photographer sets the shutter speed desired, and the camera will automatically set the corresponding aperture required to properly expose the photograph on the basis of the light reflected into the camera.

Aperture priority exposure mode: The photographer manually sets the f/stop desired, and the camera will automatically set the corresponding shutter speed required to properly expose the photograph for the reflected light coming into the camera.

Shutter priority exposure mode: The photographer sets the shutter speed desired, and the camera will automatically set the corresponding aperture required to properly expose the photograph on the basis of the light reflected into the camera.

A camera with a shutter priority exposure mode option usually has this choice represented by a "Tv" somewhere on the camera exposure controls. "Tv" stands for time value. The shutter priority exposure mode is a very useful mode, because it basically does exactly what the manual exposure mode would do, only faster. The photographer concerned about depth of field as well as a proper exposure will usually select the slowest hand-holdable shutter speed, usually 1/60th of a second when a 50-mm lens is being used. The corresponding aperture is then usually the best to maximize depth of field. In the shutter priority exposure mode, with the camera set to 1/60th, the camera automatically sets this same aperture. The camera just does it faster and without manual adjustments.

Although camera settings that make the photographer responsible for every camera variable are normally recommended, the shutter priority exposure mode efficiently arrives at the exact same position as the manual exposure mode. It does not reduce the photographer's responsibilities; it makes the process more efficient.

BRACKETING

MANUAL EXPOSURE MODE BRACKETING

In the manual exposure mode, the photographer sets both the f/stop and the shutter speed for every photograph. If the camera and film are not DX coded, the ISO film speed was also previously manually set on the camera. When bracketing, the question is whether to alter the f/stop or the shutter speed?

The option to change film speed to achieve a different exposure level, although a technical possibility, is never really an option in the field when using a film camera. It is too cumbersome to change rolls of film during a crime scene and would waste the remaining negatives on the roll of film taken out prematurely. Of course, with a digital camera, ISO film equivalents can be changed as easily as f/stops or shutter speeds. This option represents a great advantage when determining exposures is concerned.

With a film camera, the question remains: which should be changed to increase or decrease the exposure, the f/stop or the shutter speed?

When bracketing for an underexposure, the answer is to remember the Cardinal Rule to maximize depth of field: if the minimum f/stop has not already been used, bracket by decreasing the aperture size. Making the aperture smaller not only results in a dimmer photograph than the first photograph, but it also increases the depth of field. For example, if the original exposure combination were made up of 100 ISO film, a shutter speed of 1/60th of a second, and f/11, the proper way to achieve a −1 exposure would be to change the f/stop to f/16. If the smallest aperture of the camera has already been chosen, then simply use a faster shutter speed. If an f/22 was being used for the original image, change the shutter speed from 1/60th of a second to 1/125th of a second. The faster shutter speed is 1 stop dimmer and can also be used when hand-holding a 50-mm lens.

When bracketing for an overexposure, presuming you are already using the slowest hand-holdable shutter speed for a 50-mm lens, 1/60th of a second, the choice is limited to opening the aperture. In this case, using a slower shutter speed like 1/30th of a second may result in camera movement during the exposure and blur. For example, if the original exposure were 100 ISO film with a 1/60th second shutter speed with an f/11, the only option to achieve a +1 exposure would be to reset the f/stop to f/8.

If the original f/stop used was f/8, one might think that using f/5.6 would be a natural choice. However, think back to the reciprocal shutter speeds and f/stops previously recommended. You will recall that 1/60th of a second was the primary shutter speed recommended for most crime scene photographs. You will also recall that the small apertures f/22, f/16, f/11, and f/8 were the best f/stops to ensure an adequate depth of field was achieved. If you find you are about to use an f/5.6 during crime scene photography, it is strongly urged that you consider two other options before actually using that aperture. This can be considered a new Rule of Thumb.

Rule of Thumb 2.10: Before using an f/5.6, or wider aperture, for a crime scene photograph, consider these options.

1. Use electronic flash. The additional light from the flash will certainly require the use of smaller apertures, which in turn will provide a better depth of field.
2. Use a tripod. Putting the camera on a tripod will allow the use of longer shutter speeds without the risk of blur, and the use of longer shutter speeds will also allow smaller apertures to be used, which will provide a better depth of field.

AUTOMATIC EXPOSURE MODE BRACKETING

In the aperture priority exposure mode, the photographer only has control of the f/stop selections; in the shutter priority exposure mode, the photographer only has control of the shutter speed selections. What if you are using one of these exposure modes and would like to bracket several shots? If all you can control is the f/stop, does changing the f/stop selection in the aperture priority exposure mode result in a change of exposures? If all you can control is shutter speed, does changing the shutter speed in the shutter priority exposure mode result in an exposure change? In both cases, the answer is no.

In the automatic exposure modes, the camera automatically compensates for an adjustment to either the f/stops or shutter speeds so that another proper exposure is still achieved: a reciprocal exposure is the result.

In the shutter priority exposure mode, if the initial shutter speed is 1/60th of a second, will changing the shutter speed to 1/125th of a second result in a −1 exposure? No.

If the shutter speed is changed, the camera's computer will seek an f/stop that will provide a proper exposure for the newly selected shutter speed. For example, if the original shutter speed was 1/60th of a second and the scene lighting allowed the camera's computer to automatically select an f/11 to properly expose that scene, what would happen if the photographer selected a 1/125th of a second shutter speed instead? The camera's computer would select an f/stop that is a reciprocal exposure for the 1/60/f/11 exposure combination. In this case, the new reciprocal exposure would be 1/125 and f/8. Switching from 1/60th of a second to 1/125th of a second would not result in a −1 bracket. It would result in another reciprocal exposure that was the same as the original exposure.

In the aperture priority exposure mode, if the initial f/stop is f/11, will changing the f/stop to f/16 result in a −1 exposure? No.

If the aperture is changed, the camera's computer will seek a shutter speed that will provide a proper exposure for the newly selected f/stop. For example, if the original shutter speed was f/11 and the scene lighting allowed the camera's computer to automatically select a 1/125th of a second shutter speed to properly expose the scene, what would happen if the photographer selected an f/16 instead? The camera's computer would select a shutter speed that is a reciprocal exposure for the 1/125 and f/16 exposure combination. In this case, the new reciprocal exposure would be f/16 and 1/60th of a second. Switching from f/11 to f/16 would not result in a −1 bracket. It would result in another reciprocal exposure that was the same as the original exposure.

So, how is bracketing achieved in the automatic exposure modes?

Many cameras that have automatic exposure modes also have an **exposure compensation dial**. Those cameras that do not have an exposure compensation dial have buttons or other menu choices to allow the photographer the ability to vary exposure choices.

On the right side of the dial in Figure 2.83 is a continuum that proceeds from +2 to −2, with two white dots between each number. This particular camera allows exposure compensation in 1/3-stop increments. Notice the dial is currently set to "0." This is how the camera should be set when there is no need for automatic exposure mode bracketing.

If you were in either the shutter priority exposure mode or the aperture priority exposure mode and you just completed taking a shot at what you determined to be an "optimal" exposure, how do you bracket? For a +1 exposure, rotate the dial to the +1 mark and take a second exposure. For a −1 bracket, rotate the dial to the −1 mark, and take a third exposure.

After this series of brackets, it is critical to remember to reset the exposure compensation dial back to the "0" position. Should you forget to do this, all the remainder of the roll of film will be either overexposed or underexposed, depending on the last exposure compensation dial setting.

Exposure compensation dial: To bracket in one of the automatic exposure modes, this dial is set to the appropriate exposure desired, usually ranging from +2 to −2 stops.

Figure 2.83
Exposure compensation dial.

If the proper exposure triangle only has four variables, and it does, how can the exposure compensation dial effect a change in exposure?

The exposure compensation dial actually resets the ISO film speed to "fool" the camera so an exposure change will be made by the camera. An example is the best way to explain this.

The photographer is using the shutter priority exposure mode with the camera loaded with 100 ISO film and the shutter speed set to 1/60th of a second. When a meter reading was taken by depressing the shutter button half way, the camera's computer determined the proper f/stop for the reflectivity of the scene in question was f/16. This was the first exposure. Wanting to bracket for a +1, the exposure compensation dial was rotated to +1, and the second shot was taken. How does this force the camera's computer to set an exposure combination that results in a brighter exposure? By setting the exposure compensation dial to +1, the ISO setting is actually reset to 50 ISO; 50 ISO film is "slower" film that is less sensitive to light, so it needs more light for a proper exposure. How much more? 50 ISO film would need precisely 1 stop more light than ISO 100 speed film would. If 100 ISO film needed a 1/60th of a second shutter speed with an f/16, 50 ISO film would need 1/60th of a second shutter speed and f/11 as the aperture. Changing exposure from f/16 to f/11 results in a +1 exposure change. Even though the camera now thinks it is loaded with 50 ISO speed film, it is actually still loaded with 100 ISO speed film. The net result: a +1 exposure.

Another example: The photographer is using the shutter priority exposure mode, with the camera loaded with 100 ISO film, and the shutter speed set to 1/60th of a second. When a meter reading was taken by depressing the shutter button half way, the camera's computer determined the proper f/stop for the reflectivity of the scene in question was f/16. This was the first exposure. Wanting to bracket for a −1, the exposure compensation dial was rotated to −1, and the second shot was taken. How does this force the camera's computer to set an exposure combination that results in a dimmer exposure? By setting the exposure compensation dial to −1, the ISO film setting is actually reset to 200 ISO; 200 ISO film is "faster" film that is more sensitive to light, so it needs less light for a proper exposure. How much less? 200 ISO film would need precisely 1 stop less light than ISO 100 speed film. If 100 ISO film needed a 1/60th of a second shutter speed with an f/16, 200 ISO film would need 1/60th of a second shutter speed and f/22 as the aperture. Changing exposure from f/16 to f/22 results in a −1 exposure change. Although the camera now thinks it is loaded with 200 ISO speed film, it is actually still loaded with 100 ISO speed film. The net result: a −1 exposure.

To achieve a true change in the exposure, it is actually necessary to fool the camera about the ISO film speed that has been loaded into the camera.

In the aperture priority exposure mode, the photographer selects the aperture. Changing the exposure compensation dial to a plus or minus number forces the camera to alter the shutter speeds. An example is helpful here also.

The photographer is using the aperture priority exposure mode, with the camera loaded with 100 ISO film and the aperture set to f/16. When a meter reading was taken by depressing the shutter button half way, the camera's computer determined the proper shutter speed for the reflectivity of scene in question was 1/125th of a second. This was the first exposure. Wanting to bracket for a +1, the exposure compensation dial was rotated to +1, and the second shot was taken. How does this force the camera's computer to set an exposure combination that results in a brighter exposure? By setting the exposure compensation dial to +1, the ISO film setting is actually reset to 50 ISO; 50 ISO film is "slower" film that is less sensitive to light, so it needs more light for a proper exposure. How much more? 50 ISO film would need precisely 1 stop more light than ISO 100 speed film would. If 100 ISO film needed a f/16 with a 1/125th of a second shutter speed, 50 ISO film would need f/16 and 1/60th of a second shutter speed. Changing exposure from 1/125 to 1/60 results in a +1 exposure change. Even though the camera now thinks it is loaded with 50 ISO speed film, it is actually still loaded with 100 ISO speed film. The net result: a +1 exposure.

In the preceding example, had the original shutter speed been 1/60th of a second, selecting a +1 on the exposure compensation dial would have forced the camera's computer to change the shutter speed to 1/30th of a second. This, of course, is a shutter speed that is too slow when using a 50-mm lens. Camera shake would almost guarantee blur in the photograph. This is another reason that the aperture priority exposure mode should only be used with caution.

The exposure compensation dial should only be used for bracketing when one of the automatic exposure modes are in use, which are, again, the aperture priority exposure mode, the shutter priority exposure mode, and the program exposure mode. The exposure compensation dial does not produce a bracket when the manual exposure mode is in use. In the manual exposure mode, bracketing is accomplished by manually altering the shutter speed or the aperture.

FLASH BRACKETING

It is also possible to bracket for a brighter or dimmer image when electronic flash is used. This will be explained more fully in Chapter 4. However, for the purposes of this chapter, it will be mentioned that it is often possible to set the flash so that it emits either more or less light than usual. Therefore, the original shutter speed and f/stop can be maintained, and subsequent exposures can be altered merely by altering the flash intensity.

THE F/16 SUNNY DAY RULE

If you look on the inside surface of a box of film, after you take the film canister out, you will frequently find film manufacturer recommendations for proper exposures when using that particular film under different outdoor light levels.

The **F/16 sunny day rule** recommends both an f/stop setting and a corresponding shutter speed for most outdoor photographic situations. The F/16 sunny day rule, however, is just a way to determine a "ball park" exposure combination for different lighting conditions. It is not the best exposure combination if you wish both a proper exposure and you also want to maximize the depth of field. In other words, the net result is just one reciprocal exposure, and it is up to the photographer to choose among the other possible reciprocal exposures when depth of field is also important.

F/16 sunny day rule: This rule recommends f/stop and shutter speed combinations for normally occuring outdoor lighting conditions.

AN F/16 DAY

At midday, on a bright sunny day, when there are crisp distinct well-defined shadows, use an f/16 and convert the ISO film speed to the shutter speed. If you are using 100 ISO speed, convert to a shutter speed of 1/125th of a second. If you are using 400 ISO speed film, convert to a shutter speed of 1/1500th of a second shutter speed. However, if it were midday on a bright sunny day, at this point it is hoped you would only consider the ISO 100 speed film.

If the F/16 sunny day rule suggests using an f/16 and a 1/125th of a second shutter speed with 100 ISO film, is there a better reciprocal shutter speed? Yes.

Figure 2.84
An f/16 shadow.

Changing the shutter speed to 1/60th of a second would allow the use of an f/22 for the same scene, which provides the same exposure and provides a better depth of field.

AN F/11 DAY

On a bright, but hazy, day when the shadows produced are indistinct or what can be termed a "soft" shadow, use an f/11 and convert the ISO film speed to the shutter speed. In this case, the shadows present are ill defined. The shadows form vague shapes, and the item making them cannot be precisely determined. With 100 ISO film, the exposure recommendation would be f/11 and 1/125th of a second shutter speed. Is there a better reciprocal exposure? Yes. Changing the shutter speed to 1/60th of a second would allow the f/stop to be changed to f/16, and this combination is the same exposure with a better depth of field.

AN F/8 DAY

On a bright cloudy day, there may be plenty of outdoor light, but there will not be any shadows. On such a day use an f/8, and convert the ISO film speed to the shutter speed. This results in an f/8 and a 1/125th of a second shutter speed. Although a proper exposure for the scene, a better reciprocal exposure would be f/11 and 1/60th of a second shutter speed.

Figure 2.85
An f/11 shadow.

Figure 2.86
An f/8 day, without shadows.

AN F/5.6 DAY

An overcast day can still produce bright lighting outdoors but will seldom produce any shadows. Another situation for this exposure category would be on a sunny day, when the sun is blocked from the area of interest by a building or something similar. In this case, the item of evidence would be "in" a shadow, but if you stood in the same location as the evidence and looked directly above you, there would be blue sky. Blue sky itself is quite bright. For either of these situations, use an f/5.6 and convert the ISO film speed to the shutter speed.

It is hoped that this last sentence was uncomfortable for you to read. Why? Previously, it was recommended that only the f/stops of 22, 16, 11, and 8 should be really considered for crime scene photography, because they provide the best depth of field compared with the other f/stops. The F/16 sunny day rule variations, however, are only exposure recommendations. They are not concerned with depth of field. Again, it would then be appropriate to change this exposure combination to the better reciprocal exposure of f/8 and 1/60th of a second shutter speed. This exposure combination, at least, conforms to our previous exposure recommendations.

(In a textbook, such as this, one can normally expect to see only the best photographic examples of the theories and concepts explained in the various chap-

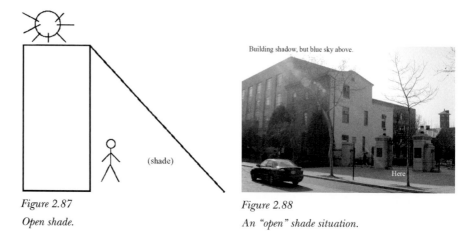

Figure 2.87

Open shade.

Figure 2.88

An "open" shade situation.

ters. Figure 2.88 suffers from lens flare, true, but because it is the most multicolored example I have in my collection, I just had to use it.)

AN F/4 DAY

On a heavily overcast day, use an f/4 and convert the ISO film speed to the shutter speed. There would be no shadows in this situation either. Use the reciprocal exposure combination of f/5.6 and 1/60th of a second shutter speed if the original exposure combination used an ISO film speed of 100. However, recall our previous recommendation about ISO film choices, and you should have been tempted to load the camera with ISO 400 speed film in this situation. ISO 100 speed film was only recommended for bright midday outdoors lighting. Therefore, in this situation, it would be recommended to use the f/4 with 400 ISO film, and, therefore, this would convert to a shutter speed of 1/500th of a second shutter speed if your camera only had full shutter speed choices. This reciprocal exposure is equivalent to 1/60th of a second shutter speed and f/11. Changing from 1/500th to 1/60th is a 3-stop change, requiring a change from f/4 to f/11.

AN F/2.8 DAY

When the crime scene or the evidence about to be photographed is in "deep" shade, with no sky directly overhead, use f/2.8, and convert the 400 ISO film speed being used to the shutter speed, or 1/500th of a second. This exposure combination converts to the reciprocal exposure of f/8 and 1/60th of a second shutter speed.

If a crime occurred under this tree, taking a meter reading would confirm how dim the light is there. Even though it is a bright sunny day elsewhere, there is a remarkable difference in the lighting under the tree.

The f/16 sunny day rule and these variants suggest exposure recommendations that will be close enough to produce an exposure that will be printable.

Because of the exposure latitude of most film, slight underexposures or overexposures will be able to be salvaged in the wet-chemistry darkroom. They may not be the "optimal" exposures, but they will be "close enough." Because they are just guidelines, they should not be used when the reflective light meter is working. The reflective light meter will always be more exact.

The f/16 sunny day rule recommends that the ISO film speed being used be converted to the shutter speed. Therefore, ISO 100 becomes 1/125th; ISO 200 becomes 1/250th; and ISO 400 becomes 1/500th. These are, of course, full shutter speed changes. As already mentioned previously, some cameras allow the photographer to select shutter speeds in 1/2-stop or 1/3-stop increments. If these are possible, the closest incremental shutter speed to the film speed being used should be selected. The shutter speed selected should be as "fast" or "faster" than the shutter speed number. In other words, if you have 400 ISO film in your camera and a shutter speed of 1/350th of a second is available, 1/500th of a second shutter speed should be selected even though 1/350 is actually closer to 1/400th. The 1/500th shutter speed is "faster" than the 1/400th. This is part of the f/16 sunny day rule.

It is important to repeat that the f/16 sunny day rule is just a recommendation for an exposure setting, just as the camera's meter reading is merely one of many reciprocal exposure possibilities. The f/16 sunny day rule does not have anything

Figure 2.89
"Deep" shade: under something.

to do with a crime scene photographer's additional desire to maximize the depth of field; it is just an exposure recommendation. Therefore, once an exposure combination has been determined using the rule, the photographer then has to ask whether that exposure recommendation is the most efficient reciprocal exposure to also maximize the depth of field. An exposure adjustment will almost always be necessary.

CAUSES FOR COMPLETE ROLLS OF FILM WITH EXPOSURE ERRORS

FILM SPEED NOT PROPERLY SET/DX CODE READING ERROR

As mentioned previously, on manual exposure cameras, the film speed has to be set on the camera when any new roll of film is loaded into the camera. If you neglect to set the film speed, the entire new roll of film may be exposed according to the ISO film speed of the previous roll of film placed in the camera. For instance, if the previous roll of film that was used was an ISO 400 film, the camera was probably set for that ISO film speed. Now, if the current roll of film is an ISO 100 film speed, and the camera has not been reset for the new ISO 100 film speed, this entire roll of film will be 2 stops underexposed. ISO 100 film is 2 stops less sensitive to light than ISO 400: half of 400 is 200 (1 stop), and half of 200 is 100 (one more stop). Exposing ISO 100 speed film as if it were ISO 400 speed film will result in underexposures.

If the camera kit is shared by different personnel on different shifts, this situation can become a problem. The last roll of film used by a midnight crime scene photographer using the camera may have been ISO 400 speed film. He then turned over his camera to the day shift. When the camera was first used on the day shift, 100 ISO film was loaded into the camera. However, the day shift crime scene photographer neglected to notice the camera was already set to 400 ISO and proceeded to shoot his entire roll of 100 ISO film with the camera set to ISO 400.

Another possibility is that you may have a camera that reads the DX coding on most film canisters, but somehow you have acquired a non-DX coded film, loaded it into the camera, and presumed the camera has properly set itself for the new film speed. In this situation, the camera usually defaults to ISO 100, but if you used any other ISO film speed, your exposures will all be incorrect.

Finally, the DX film reading contacts inside the camera may have become corroded or dirty, and they are not correctly reading the DX coding on the film canister. Result: improperly exposed film.

Moral: always ensure the camera knows the ISO film speed currently being used. Do not presume this has been correctly done by an automatic camera. Most cameras that have DX coding have a button or dial that can be used to confirm the ISO setting of the film currently in the camera. This is just one more responsibility of the crime scene photographer.

IMPROPER EXPOSURE COMPENSATION DIAL SETTINGS

The use of a camera that has its exposure compensation dial set to anything except "0" will also result in uniform underexposures or overexposure. That is because this is the equivalent of incorrectly setting the ISO film speed.

In other words, if you last bracketed a series of photographs to ensure a photograph required for a critical comparison would be properly exposed and the last photograph of the series was taken at a −2 exposure compensation, the camera may still be set at −2. If some time has gone by between shots, when it is time for your next photograph, you may not remember that a −2 exposure compensation has been set on the camera. If you presume the exposure compensation is set to "0" and use the camera without making a change, all your subsequent photographs will come back underexposed 2 stops.

Before depressing the shutter for that first exposure on a new roll of film:

- Ensure the camera has been properly set for the ISO film speed currently being used.
- Double check the exposure compensation dial to ensure it has been reset to "0."

COMMON FILTERS

UV/HAZE/1A/SKYLIGHT FILTERS

UV/Haze/1A/Skylight filters have two common uses.

Lens Protection

Filters are inexpensive insurance for all the lenses you own. As you use and handle your camera, it swings from a strap around your neck and is otherwise held at the end of your arm. This increases the potential for the camera to frequently bump into a variety of other objects in your environment. Some of these contacts may scratch, break, or dent the front of the lens, requiring an expensive repair or replacement. Having a relatively inexpensive filter mounted on each lens protects the front of those expensive lenses. Replacing the filter that gets scratched, dented, or broken is much less painful than replacing the lens itself. Therefore, an inexpensive filter, usually a **UV filter**, should be placed on each lens you own.

Figure 2.90 shows a dent to the UV filter attached to the lens. Had this been a dent to the lens itself, it would be impossible to use any filter with it, because no filters would be able to screw onto the filter threads of the lens. Now, because only the outer filter is dented, if the photographer wished to use any other filter, all that would have to be done is to unscrew the dented filter. Replacing this dented filter is not necessary. It is still serving its primary purpose of protecting the lens.

Put a filter on each lens you own.

UV filter: A filter used to protect a camera lens from damage. It also filters out UV light from the sky, providing better exposures.

Figure 2.90

Dent to ultraviolet filter instead of the lens itself.

Filtering Excessive Blue from the Sky

Film is very sensitive to both the UV and blue light of the sky. Without a filter, the sky will normally photograph lighter than we remember seeing it. Because film is so sensitive to this light, it frequently is overexposed in photographs. For the sky to record realistically in photographs, it is recommended to use one of several filters. Four filters have basically the same effect on the sky, without significantly altering the way other aspects of the scene will appear. These four filters are the UV filter, the Haze filter, the 1A filter, and the Skylight filter. Their effect is to have the sky record correctly on film. They do this by filtering out some of this light to avoid overexposures. They do not reduce the other light coming through the lens in any significant way, so exposure compensations are not necessary.

These four filters appear to be clear filters, not showing either a colored tint or neutral tinting like polarizer filters and neutral density filters. Figure 2.91 demonstrates this. The blue cloth background appears the same either viewed directly or viewed through the filter.

When composing on scenic views and vast panoramas, these filters will assist in keeping the haze effect from long expanses of air, with the particulate matter suspended in the air, from ruining photographs. Haze can manifest itself as misty vapors. Residents of smog-polluted Los Angeles, California, and Phoenix, Arizona, have an innate understanding of particulates in the air.

In normal crime scene photography, this is not normally a concern. If, however, your task was to document a crime scene with aerial photography, haze may become a problem, depending on the height of the aircraft. By use of a haze filter, you will be able to "shoot through the haze" to visualize the ground structures better.

THE POLARIZER FILTER

A polarizer filter can be a very useful tool for the crime scene photographer.

Although it is not obvious at first glance, the polarizer filter is usually made up of two parts, unlike other filters. It has filter threads, like other filters, which screw

Figure 2.91
Ultraviolet filter.

normally onto the lens. Once secure on the lens, however, the outer part, which holds the polarizer filter itself, can still be rotated.

A word of caution is necessary here. The polarizer filter screws onto the lens like any other filter. If you are looking down on the camera with the lens pointed at the ceiling, the polarizer filter would screw on in a clockwise direction. All filters should be screwed on with the camera in this position. With the lens aimed at the ceiling, you can actually let go of the filter before it has been screwed on, and the filter will remain sitting on the lens. If your habit is to screw on filters with the lens aimed horizontal to the ground, it would be possible to mistakenly believe the filter is properly screwed onto the lens. If you then let go of a filter not properly screwed onto the lens, it can fall to the ground and break.

This becomes a new Rule of Thumb.

Rule of Thumb 2.11: Always screw filters onto the lens when the lens is aimed at the ceiling or straight up. This reduces the chances of dropping the filter.

Once the polarizer filter is properly screwed onto the lens, the outer part can be rotated to reduce or eliminate reflections or glare on a variety of surfaces. It is extremely important that the filter be rotated without unscrewing the filter from the lens. This filter rotation is done while holding the camera normally and looking through the viewfinder. The filter rotation must be done in a counterclockwise direction, with the camera held normally. If the filter rotation is done in a clockwise direction, you may be inadvertently unscrewing the filter at the same time as you are rotating the filter. Too many polarizer filters have been dropped because of this. Always rotate the polarizer filter counterclockwise.

Reflection Elimination

Notice Figure 2.92 shows the polarizer filter with vertical lines running through the filter. Such lines cannot actually be seen. They are placed there to be a visual explanation of the polarizer filter's ability to filter out, or block, polarized light.

Figure 2.92
Polarizer filter.

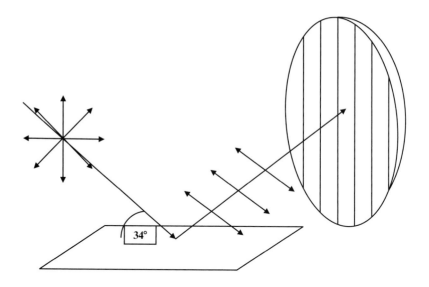

> Unpolarized light vibrates in all directions. When striking water or glass, the light can become polarized in one direction, and show reflections. The reflections can be eliminated by a polarizer filter which absorbs/blocks the polarized light.

On the figure, because the surface reflecting the polarized light is the ground, the polarized light is reflected from it in a horizontal, rather than a vertical, attitude, which is represented by the three diagonal lines coming from the surface toward the filter. Had the light reflected from a building window, the polarized light would have a vertical orientation. In either case, the photographer will be able to see the effect of rotating the filter through the viewfinder. When the filter is rotated so polarized light is blocked by the filter, reflections will be either drastically reduced or totally eliminated. Continuing to rotate the filter would align the filter so the polarized light can come through the filter, and all the reflections will remain visible.

Figures 2.93 and 2.94 show a 34° angle that the light is striking the surface and reflecting from the surface, which is the optimal angle of view of a surface with reflections on it when you want the polarizer filter to eliminate those reflections. As you deviate from that 34° degree viewpoint, the polarizer filter will become more and more ineffective. The crime scene photographer can do nothing about this. The physics of polarized light just has to be accepted. Some solution exists to the problem of distracting reflections.

To maximize the elimination of reflections from glass and water, follow this three-step procedure.

1. Approximate what you feel is a 34° angle to the surface with reflections on it.

Figure 2.93

Glass reflection eliminated with a polarizer filter.

Figure 2.94

Water reflection eliminated with a polarizer filter.

2. Put the polarizer filter on your lens, and rotate the outer element until the maximum amount of reflections are eliminated. If you rotate the outer element beyond the optimal position, the reflections will begin to return. Keep rotating the outer element until the reflections are again diminished almost completely.

3a. At times the reflection has not completely disappeared, which may be merely a function of your not having correctly assumed a 34° angled view of the surface in question. If the surface is a vertical sheet of glass, it is recommended that you "test" your approximation of a 34° angle by leaning to the left and to the right while looking through the viewfinder. By doing so you may achieve a view closer to the magic 34° angle. As you do this, you may see the reflection become more prominent when leaning in one direction and less prominent when leaning in the other direction. Move in the direction that eliminates more of the reflection.

3b. If the reflection is coming from water, that surface is horizontal. To "test" your success of having successfully found the correct 34° angle, move closer to the surface or back away from it while looking through the viewfinder. You may see the reflection become more prominent one way and less prominent as you move the other way. Move in the direction that eliminates more of the reflection.

The polarizer filter will not remove the reflection of the sun itself. If the sun is currently creating an unwanted reflection, the only way to eliminate it is to find a different viewpoint of the surface.

When the polarizer filter has eliminated all the polarizing light reflected from a surface, the result is a severe light reduction the camera is receiving. This is typically a 2-stop light reduction. If you are using a reflective light meter, it is important to remember to re-meter the scene after the polarizer filter has been rotated for its optimal effect. The meter will then ensure the scene is not underexposed.

Enriching Color Saturation

By eliminating polarized light, polarizing filters can also improve the colors in many photographs. Strong sunlight often produces a glare that is sometimes perceptible to film but is not perceptible to the eye. By eliminating this glare, the color saturation of a photograph can be improved. When color accuracy is critical, as when trying to document the exact color of a paint transfer in a fatal hit-and-run accident, this aspect of the polarizer can be an immense benefit.

Figure 2.96 demonstrates how strong sunlight glare can alter colors. The color of the adobe tiles is washed out by the sun's glare. By use of a polarizer filter, the true color of the tiles is revealed.

Skid marks

A polarizer can remove the glare on the pavement at an accident scene, making the skid marks present at the scene more obvious. Up to 10% of the total length

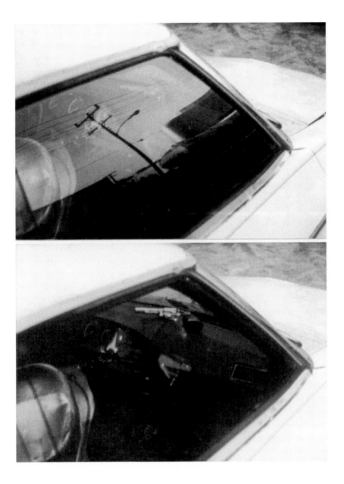

Figure 2.95
Window reflection eliminated with a polarizer filter.

of a skid mark may be a "shadow" mark. This is a less obvious tire mark left by the tire as it was slowing before the tires became totally locked. At times, even this "shadow" mark can be visualized with a polarizer filter. It is important that the driver get "full credit" for the speed he or she was actually driving, and being able to document the full length of the skid mark is one way of doing this.

The effect is visible while looking through the viewfinder while rotating the outer element of the polarizer filter. For this effect, the polarizer should be used at approximately a 34° degree viewpoint to the surface of the pavement. It is recommended you view the skid marks from both ends. The polarizer filter may be more effective from one direction than another.

A polarizer filter should be considered essential equipment for accident scene photographers.

Blue Sky

The polarizer filter can increase the blue of the sky. As mentioned previously, film is overly sensitive to the blue of the sky, resulting in somewhat washed-out skies in many

Figure 2.96

Glare on the left; true colors on the right. Please visit the book's companion website (http://books.elsevier.com/companions/9780123693839) for a color version of this figure.

Figure 2.97

Bottom, polarizer not used; top, polarizer used. Please visit the book's companion website (http://books.elsevier.com/companions/9780123693839) for a color version of this figure.

Figure 2.98

Top, no polarizer filter used; bottom, polarizer filter used. Please visit the book's companion website (http://books.elsevier.com/companions/9780123693839) for a color version of this figure.

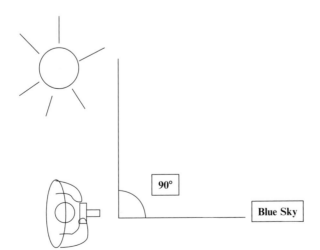

Figure 2.99

Relative angle to darken a blue sky with a polarizer.

Figure 2.100

Polarizer filter: darkens the background cloth.

photographs. The polarizer filter can enrich the blue of the sky and return it to the way we saw it at the original scene. This effect is maximized when viewing the sky with the sun at a 90° angle. Rotate the filter to maximize the effect. See Figure 2.99.

With the sun opposite either shoulder, a polarizer filter will sometimes dramatically deepen the blue look to the sky. The effect is diminished or completely eliminated as the photographer alters this angle and aims the camera more and more either toward the sun or away from the sun.

Like a ND Filter: 1.3–2 Stops Less Light

As mentioned previously, the polarizer filter appears much like a neutral density filter: it is shaded a tone of gray. This suggests that it will also filter "normal" light coming through it, which is true. When set to allow polarized light to come through the filter, the polarizer filter will still filter out approximately 1.3 stops of "normal" light. As the outer filter element is rotated, more or less polarized light is also absorbed by the filter. When set to block all the polarized light, the total light reduction is approximately 2 stops.

Linear and Circular Polarizer Filters

When shopping for a polarizer filter, you will notice there are both linear and circular polarizer filters. Linear polarizer filters are designed to work with manual focus lenses and are usually less expensive. Circular polarizer filters should be obtained for use with lenses with auto-focus capability.

Chapter 6 will be devoted to more exotic filters. The UV filter mentioned previously blocked UV light. Another kind of UV filter transmits UV light and blocks all visible light, which is necessary when photographing evidence only "visible" in the UV light range. When photographing evidence that is fluorescent, it will be necessary to use either a yellow, orange, or red filter on the lens, which will enable the fluorescence of the evidence to be photographed, while the stimulating light of a laser or alternate light source is blocked by one of these filters. An IR filter blocks visible light while transmitting IR light. This filter is used when photographing evidence only "visible" in the IR light range.

Figure 2.101
View of light coming in through the viewfinder.

THE EYE CUP COVER

At the beginning of the chapter, it was explained how the light coming in through the lens serves double duty. A mirror reflected that light up into the pentaprism, where other mirrors allowed the photographer to see the image coming in through the lens. Once this mirror flipped up, when the shutter button was depressed, this same light could travel straight back to strike the film. If light can travel from the lens to the viewfinder, the reverse is also possible.

The lens was removed from the camera in Figure 2.101 to make it easier to see the image of a person walking on the sidewalk across from my office. The light is entering the viewfinder, is bounced around by several mirrors, and we can see the result at the front of the camera. Without a lens to focus the image more clearly, everything is a bit fuzzy. But, certainly, light can get into the camera from two directions.

It is, therefore, appropriate that we end this chapter on exposure by mentioning that light entering the camera through the viewfinder can have an effect on the exposure for any photograph. Normally, our eye is pressed up close to the viewfinder; this possibility is moot. Stray light finding its way around our head to sneak between our eye and the eye cup of the camera and eventually entering the camera is so minimal, it is not a realistic concern. However, many times our eye is not near the viewfinder when an image is captured.

If we are taking a series of bracketed photographs of evidence, the camera is normally on a tripod, with the film plane parallel to the evidence, we have already determined the exposure for the first photograph in the series of brackets, and the camera has been focused. Using a shutter release cable, we take multiple shots, with only the f/stop usually needing to be altered. It is not necessary to look through the viewfinder to do this. Even if the camera has auto-bracketing capability and will take three different exposures with just one push of the shutter release cable button, the photographer is not leaning toward the camera to

Figure 2.102

Eye cup cover on camera strap.

Figure 2.103

Eye cup and frame of viewfinder.

Figure 2.104

Eye piece cover in place.

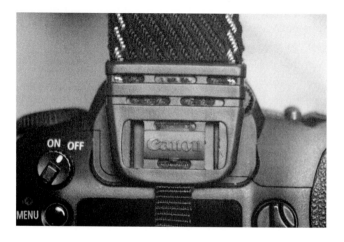

look through the viewfinder. After all, once the shutter is activated, the mirror flips up, and nothing is to be seen through the viewfinder.

If we are about to do a painting with light sequence outdoors at night, trying to properly expose a large dim crime scene with multiple flashes, the camera is again set up as before, without an eye pressed up against the viewfinder. This technique will be fully described in Chapter 4.

In these cases, stray light coming into the camera through the viewfinder can affect the exposure of the camera. Camera manufacturers also know this and usually provide a "tool" to avoid this problem when you first buy a camera. Most cameras sold also come with a camera strap. On most camera straps, there is an eye cup cover. Many people do not even notice it. There is not much to notice: usually just two projections that will eventually slide over the left and right edges of the eye cup frame.

If you find yourself in a situation where you are about to depress the shutter by any means and your eye will not be pressed up to the viewfinder, use the eye cup cover on your camera strap to avoid any exposure irregularities. The result of this extra light influencing the exposure for a particular shot is usually an underexposure. When more light is noticed by the reflective light meter, it adds this to the sum of the light it receives through the lens and will provide a proper exposure for all the light it has noticed. If the "scene" is brighter, the meter will recommend an exposure combination that uses a smaller aperture. A smaller aperture results in an underexposure.

SUMMARY

Determining the proper exposure for an image is an essential element of crime scene photography. This chapter explains the photographic variables related to exposure. Each exposure variable affects the others. One variable can be changed if another is also changed to maintain exposure equilibrium. The same overall exposure can be achieved with different exposure settings. This allows the photographer to capture a properly exposed image with different effects: different motion stopping capability and different ranges of depth of field. First, however, the basic exposure elements must be understood.

Shutter speeds are not only exposure controls. They also are motion controls. The proper choice of shutter speeds not only eliminates the risk of blurred photographs resulting from the movement of the photographer at the instant the shutter is depressed but can also eliminate the blur that can be caused by the movement of subjects and objects within our well-composed photographs. This chapter provides the information required to control motion and eliminate blur in photographs.

For any one exposure setting that the camera may indicate is a good exposure for a particular lighting condition, the photographer needs to be aware that other exposure choices will result in the same exposure setting. Then, from all of the available choices, the photographer must choose the particular combination that will produce the best possible photograph. This choice should be based on the conscious decision-making process of the photographer, not on the basis of an accidental camera setting.

To provide for well-exposed photographs, SLR cameras have to be able to measure the amount of light that is being reflected from the scene toward the camera and into the lens. That standard was explained here. Various light meters are in use in different cameras. These meters determine the adequacy of the light levels. Each was explained here.

The exposure meter is the "tool" used to determine a proper exposure. It indicates whether a "proper" exposure has been established or whether the scene is currently overexposed or underexposed. Different scenes reflect different amounts of light. "Normal" scenes will result in proper exposures. How can "non-normal" scenes be properly exposed? The answers were provided in this chapter.

When the scene of interest does not reflect "normal" amounts of light toward the camera, the reflective light meter in the camera body will not provide a "correct" exposure. Some well-known "tools" can be used to help the photographer determine a "proper" exposure in these cases. This chapter discussed four such "tools."

Four exposure modes are available on some cameras. Two exposure modes are particularly well suited for use at crime scenes, and two exposure modes should be used with caution. Chapter 2 covered these ideas.

Bracketing is the act of intentionally taking multiple exposures of the same subject, from the same point of view. Not every image at a crime scene requires bracketing. Those that are frequently bracketed are images presenting a "tricky" exposure, those images destined for comparisons or analysis, and any image the photographer thinks may benefit from additional exposures. There are different ways to achieve a bracket. Methods to bracket in the manual exposure mode, the automatic exposure mode, and bracketing with flash were all explained.

Eventually, a photographer may be able to develop a feel for different levels of light and will be able to accurately predict the correct exposure for a variety of lighting conditions. The f/16 sunny day rule is the basis for this ability. Frequently encountered light levels require the same exposure settings. The f/16 sunny day rule points out this phenomenon. It is also a way for the beginner to begin educating his or her photographic "eye" to develop an appreciation of how light influences exposures.

It is extremely frustrating to have an entire roll of film uniformly underexposed or overexposed. There are two main explanations for this. This chapter explained these two causes.

Filters are designed to filter the light that passes through them to produce certain effects. Because light gets filtered, filters can affect the ultimate exposures by reducing the light allowed to reach the film. This chapter explained the use of the most commonly used filters. They can serve as lens protection, as light modifiers, and they can produce special effects.

DISCUSSION QUESTIONS

1. What are the four variables that affect exposure? Why does changing one of them require an adjustment to one of the other exposure variables? What are the two main reasons for changing exposure variables while maintaining the same exposure?
2. How do different lights affect daylight color film?
3. What are two meanings of exposure latitude?
4. Shutter speeds affect motion. Which kinds of motion are affected?
5. To determine a "proper" exposure, some standard must be used. What is this standard? Explain "normal" and "non-normal" scenes. Presented with a "non-normal" scene, how can the "proper" exposure be determined?
6. There are various methods for determining "proper" exposures. What are the different exposure modes?
7. Explain the different methods to bracket a shot.
8. Explain the f/16 sunny day rule and its corollaries.
9. Explain the various filters commonly used by photographers and their effects on images.

EXERCISES

The following are composition and exposure exercises only. As such, issues of focus, depth of field, and flash techniques are not considered. You may use an SLR set to auto-focus. Composition and exposure are the only issues in these exercises.

Compare your images with those in this chapter and with other images on the supplemental website of images, referring particularly to the following image folders: Blacks, Exposure, Polarizer, Shutter Speed, Sky, and Whites. If using film, rather than digital, make sure to tell the film processor not to make any exposure corrections when processing the film, because you have intentional overexposures and underexposures, and you do not want these "corrected" during processing.

1. Determine the proper exposure for one item that is totally sun lit and photograph it.
2. Determine the proper exposure for one item that is totally in the shade and photograph it.
3. Freeze a subject walking parallel to the film plane 30 feet from the camera.
4. Fill the frame with a sheet of white paper on which you have written your name. Photograph it after determining the "proper" exposure for it. Bracket +1 and −1.
5. Fill the frame with a black object. Photograph it after determining the "proper" exposure for it. Bracket +1 and −1.

6. Photograph a building facade that is in the shade.

7. Place an interesting object on your vehicle dashboard and photograph it through the windshield with a polarizer filter used to eliminate the reflections on the windshield.

8. From this same position, photograph the same object without the polarizer filter.

9. Place something in a shallow puddle and photograph it through the water with a polarizer filter used to eliminate the reflections.

10. From this same position, photograph the same object without the polarizer filter.

FURTHER READING

Baines, H. (1976). "The Science of Photography." p. 198. Halsted Press, New York.

Davis, P. (1995). "Photography." 7th Ed. pp. 154, 234, 266, 267. McGraw-Hill, Boston.

Duckworth, J. E. (1983). "Forensic Photography." pp. 43–46. Charles C Thomas, Springfield, IL.

Hedgecoe, J. (1996). "New Introductory Photography Course." p. 188. Focal Press, Boston.

Miller, L. S. (1998). "Police Photography." 4th Ed. pp. 63–64. Anderson Publishing Co., Cincinnati, OH.

Mitchell, E. N. (1984). "Photographic Science." pp. 141–150, 351–357. John Wiley & Sons, Inc, New York.

Ray, S. F. (2002). "Applied Photographic Optics." pp. 215–224, 228–231. Focal Press, Woburn, MA.

Siljander, R. P. and Fredrickson, D. D. (1997). "Applied Police and Fire Photography." 2nd Ed. pp. 14–18, 51–54. Charles C. Thomas, Springfield, IL.

Upton, B. L., and Upton, J. (1989). "Photography." 4th Ed. pp. 178, 236. Scott, Foresman and Company, Glenview, IL.

Walls, H. J., and Attridge, G. G. (1977). "Basic Photo Science: How Photography Works." p. 44. Focal Press Limited, London.

Woodlief, T. (1973). "SPSE Handbook of Photographic Science and Engineering." p. 296. Wiley-Interscience Publications, New York.

ADDITIONAL READING

Mannheim, L. A. (1970). "Photography Theory and Practice." Focal Press, London and New York.

Redsicker, D. R. (1994). "The Practical Methodology of Forensic Photography." CRC Press, Boca Raton, FL.

CHAPTER 3

FOCUS, DEPTH OF FIELD, AND LENSES

LEARNING OBJECTIVES

On completion of this chapter, you will be able to ...

1. Explain the terms "resolution," "acutance," and "sharpness."
2. Explain the concept of the "circles of confusion."
3. Explain how to hyperfocal focus on a scene by use of a depth-of-field scale.
4. Explain how to hyperfocal focus on a scene without the use of a depth-of-field scale.
5. Explain how to zone focus on a scene by use of a depth-of-field scale.
6. Explain how to zone focus on a scene without the use of a depth-of-field scale.
7. Explain the focusing adjustment required when using infrared film.
8. Explain the three factors that affect depth of field.
9. Explain the techniques to maximize depth of field.
10. Explain the various designations of lenses: focal length, "fast" or "slow," and the widest aperture of the particular lens.
11. Explain what a reference to a "normal" lens means.
12. Explain the effects on a photograph produced by telephoto lenses.
13. Explain the effects on a photograph produced by wide-angle lenses.
14. Explain the magnification ratios related to macro lenses.
15. Explain the meaning of the term "diffraction" and how to minimize its effects on images.
16. Explain the difference between pincushion and barrel distortion.

KEY TERMS

Aberration

Acutance

Airy disk

Barrel distortion

Circles of confusion

Close-up filter set

Depth of field

Depth-of-field scale
   (DOF scale)

Diffraction

Extension tube

Focal length

Hyperfocal focusing

Hyperfocal focusing
   distance

Macro lenses

1:1 Magnification

1:1 Magnification ratio

Pincushion distortion

Poisson's spot

Resolution

Scientific Working
   Group on Imaging
   Technology
   (SWGIT)

Telephoto lenses

Wide-angle lenses

Zone focusing

FOCUS

A chapter devoted to focus! How can that be? All you have to do is (1) look through the viewfinder, rotate the focus ring until your subject becomes clear, and take the shot, or (2) set the lens to auto-focus, look through the viewfinder, depress the shutter button halfway to have the lens set the focus for the subject, and take the shot.

If it were that simple we could move on to lenses. But it is not.

Although situations certainly exist when focusing with one of the preceding two methods is appropriate, on many more occasions different focusing techniques are better choices.

Snapshot shooters will find it unbelievable, but many times it is best to (1) focus the camera without looking through the viewfinder or (2) look through the viewfinder and focus at a location where no evidence exists.

Experienced crime scene photographers understand the concept of focusing on areas rather than individual objects, but others may be confused.

The best way to introduce these new ideas is to point out the obvious. When arriving at a major crime scene or major accident scene, law enforcement photographers not only photograph individual items of evidence within the scene, but they also photograph the scene itself. The scene itself is an area within which the individual items of evidence are located. The crime scene photographer photographs both crime scenes and accident scenes (areas) and individual objects.

The photographs taken by most people not in law enforcement are very different. Most photographs captured by the general public are of individual items. We photograph our family members, our boyfriends or girlfriends, our pets, a birthday cake, and the turkey on Thanksgiving. Even when the general public photographs more than one individual or object, it frequently is really just "one" thing. Two or three buddies shoulder-to-shoulder are a pair or a trio, one pair or one trio; one grouping. Four bridesmaids lined up for a photograph are more often than not in a tight group or a single line, one group or one line. When we photograph a line of trophies on the mantle, it really is just one grouping of several items.

But the crime scene photographer must spend considerable effort capturing images of medium and large crime scenes, as well as individual items within the crime scenes. These crime scenes differ from the subject matter of most photographs. Foreground detail and background detail are present, and all the items of evidence within this area are present.

A new Rule of Thumb is appropriate here.

Rule of Thumb 3.1: The entire scene, and all the evidence within the crime
scene, should be in focus. If you know a part of the scene will be out of focus if the photograph is captured as composed, you should attempt to recompose the scene so the out-of-focus area is no longer in the field of view.

Ensuring an entire area, from foreground to background, is in focus requires very different focusing skills. It is relatively easy to focus on just one item. How does one, however, focus on an area that is 10′, 15′, or 40′ deep? That is partially what this chapter is about.

Why is ensuring that the entire crime scene is in focus important? Ultimately, we take photographs at crime and accident scenes so that, if needed, they will be admissible in court as evidence. A chapter on the admissibility of photographs in court as evidence is provided later in this text. For now, let us consider just three possible objections a defense attorney might be able to come up with when we try to offer the court a partially out-of-focus image as evidence. The following three "objections" have been created by the author to emphasize the importance of ensuring as much of your image as possible is in focus. The list of court citations in Chapter 10, related to the admissibility of images, both film and digital, do not contain one case in which an image was held to be inadmissible only because it was out of focus. Nevertheless, this author is using these fictitious objections as a means to highlight the importance of focusing to maximize your depth of field, which is one of this author's Cardinal Rules.

"Your Honor, I object to this image. It is clearly out of focus" (in the foreground, or the background, or somewhere). When this photographer was at the crime scene, as he or she looked around the scene, it was all seen in focus. Why are we not afforded the same opportunity to see the scene completely in focus? A partially out-of-focus photograph does not meet the standard of being a fair and accurate representation of the scene. "How can it be fair or accurate if parts of the image are out of focus? Exclude this image!"

"Your Honor, I object to this image. It is clearly out-of-focus (in the foreground, or the background, or somewhere). I submit that the crime scene photographer intentionally blurred part of this scene to hide certain items of evidence from us. He made certain that the evidence he thought was important was in focus. How can we ascertain whether other evidence, possible exculpatory to my client, was intentionally blurred because it did not coincide with his or her premature conclusion that only my client was guilty of this crime? It is common knowledge that photographers intentionally blur parts of photographs to force the viewer to look at just one part of the image area, while hiding other areas of the image area in blur. This photographic trick highlights part of the image and hides other parts. Do not be tricked by this photographer. Exclude this image!"

"Your Honor, I object to this image. It is clearly out of focus (in the foreground, or the background, or somewhere). Either the photographer is trying to trick us by intentionally hiding aspects of the crime scene (as in the objection immediately preceding), or the photographer is incompetent, in which case this image should not be admissible as evidence, because you should not allow inferior representations of the crime scene into this trial. If the (insert your own agency name here) thought an aspect of this crime scene was important enough to photograph and to offer it as evidence in this trial, they would have sent an experienced and competent photographer to do the job. Judge, do not accept mediocre work. A blurry photograph is neither a fair nor an accurate depiction of the crime scene. Exclude this photograph."

Again, these are all fictitious objections. However, they do set the groundwork for the remainder of this chapter, which emphasizes focusing and depth of field.

Focusing on individual items of evidence and entire crime scene areas are two different skill sets. Both are critical.

RESOLUTION, SHARPNESS, ACUTANCE

Resolution: Camera resolution is the ability of the camera system, which includes the lens optics, camera sensor (film or digital sensor), and camera processing software, to distinguish, or "resolve" groups of alternating line pairs as the lines become increasingly thinner and they become increasingly closer together. The classic line pair consists of a black line and a white line. As these become thinner and closer together, at a point the distinction between the black lines and the white lines will become less distinct, and the result becomes a blending into gray.

Before focusing techniques are explained, we must first understand what it is to be "in focus." To do this, it is necessary to differentiate three terms: resolution, sharpness, and acutance. Images are often said to be "in focus" or "sharp" or "clear" as if these words all meant the same. Let us be a bit more precise.

Resolution

Camera **resolution** is the ability of the camera system, which includes the lens optics, camera sensor (film or digital sensor), and camera processing software, to distinguish, or "resolve" groups of alternating line pairs as the lines become increasingly thinner and they become increasingly closer together.

The classic line pair consists of a black line and a white line. As these become thinner and closer together, at a point the distinction between the black lines and the white lines will become less distinct; the result becomes a blending into gray.

A variety of resolution charts are available to measure relative resolution. Figure 3.1 is an example.

One can compare the relative resolution capacity of different lenses when used on the same camera body or the same lenses when used on different camera bodies. One can also compare one manufacturer's camera body with its highest

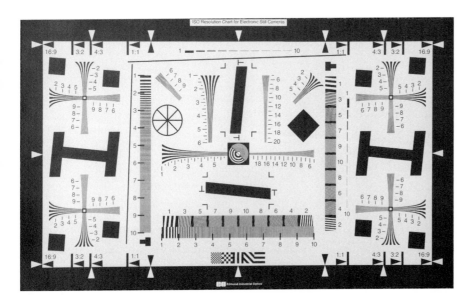

Figure 3.1
Resolution chart.

grade of lens at one particular focal length with another manufacturer's camera body and lens.

Of particular interest in most photography magazines found on grocery store and drugstore bookshelves is the comparison of resolution between film cameras and digital cameras. Digital camera websites are also frequently returning to this question. Law enforcement agencies are also constantly asking, "Do digital cameras have the same resolution as film cameras yet?" The answer to that question can be the deciding factor an agency considers when weighing whether to jump into digital imaging or not.

Although a chapter on digital imaging is included later in this text, this issue is too important to put off and will be partially discussed here. Many may be looking for this answer, and it is certainly difficult to find any authoritative coverage of the issues involved.

The question, "Do digital cameras have the same resolution as film cameras yet?" is a trap. It is asked as if the answer provides the proper foundation for the decision to switch to digital or not. Many do not believe the answer to that question should be the deciding factor in whether any particular agency evolves into digital imaging or not. Two other questions can be asked that may seem to be very similar. One of the variants you might hear is, "Does the resolution of digital cameras meet the published standards recognized within the professional community?" The third variant to the question is, "Can the digital camera my agency provides its crime scene photographers do the job necessary?" Each of these three questions will be examined.

It must be said that film certainly had the "resolution" necessary to capture details of evidence in a photograph so the photos could successfully be compared with known items of evidence, resulting in identifications between the two. How does digital "resolution" match up?

First, one distinction has to be made clearly. When speaking of digital resolutions sufficient so that identifications can be made by latent print examiners or shoe wear examiners or tire track examiners, the question does not simply depend on camera quality. The quality of the printer is an essential part of the mix. A capable camera can have an insufficient image produced by a poor printer. In addition, a camera with insufficient native resolution cannot be rescued by a top-of-the-line printer. Somehow, it must ultimately be discovered what camera "resolution" is necessary when paired with a top-of-the-line printer.

Chapter 3 will limit itself to whether digital cameras have finally achieved the "resolution" necessary for a laboratory examiner to use for identification purposes. After all, if the current crop of digital cameras cannot capture images with enough detail to enable a positive identification between the digital image and a known shoe print or tire track, for instance, many agencies are not going to be too concerned whether digital cameras can meet lesser standards. Most would agree that the current crop of digital SLR cameras can adequately document the overall conditions of crime scenes and can sufficiently record the relative positions of the evidence within the crime scenes. The remaining question is whether digital cameras can substitute for film cameras when capturing images destined to be used for comparison purposes, sometimes called examination quality photographs.

Film Resolution Versus Digital Resolution

Thirty-five–millimeter film is a fantastic tool, and we have become spoiled with it. It was clearly able to document the current conditions of a crime scene or accident scene. It easily captured images of all the evidence within the scene. In addition, 35-mm film enabled the capture of fine details ultimately required by examiners or image analysts to compare the resultant photographs of evidence from the crime scene with actual items of evidence later seized during the investigation of the crime.

The film usually used for most "critical comparisons" or "examination quality photographs" was slow-speed, ISO 100 black-and-white film. Slow-speed film was recommended because it would produce enlargements without graininess and a loss of contrast. Black and white film was usually used because it produced the sharpest images. When taking "critical comparison" photographs, the ability to capture fine detail and show distinct shapes was paramount. Black-and-white 100 ISO film did this the best of all the normal films available.

What is the resolution of 35-mm 100 ISO black-and-white film? What is the digital equivalent?

When looking for an authoritative source for this answer, it is unnecessary to look further than the **Scientific Working Group on Imaging Technology (SWGIT)**.

In the 1990s, the FBI Laboratory began supporting the creation of SWGs (Scientific Working Groups) in the various laboratory disciplines. Some of the other SWGs are SWGDAM (DNA), SWGTREAD (shoes and tires), SWGMAT (materials/trace), SWGDOC (questioned documents), SWGFAST (latent fingerprints), SWGGUN (firearms and tool marks), and SWGDE (digital evidence). SWGIT explains its reason for being:

Scientific Working Group on Imaging Technology (SWGIT): This group was created to provide leadership to the law enforcement community by developing guidelines for best practices for the use of imaging technologies within the criminal justice system.

"The Scientific Working Group on Imaging Technology (SWGIT) was created to provide leadership to the law enforcement community by developing guidelines for good practices for the use of imaging technologies within the criminal justice system."[*]

The multiple guidelines produced by SWGIT are fundamental to the context of this text. It must be stressed, however, that these guidelines are subject to being updated at any time. For current guidelines, check SWGIT's website.

SWGIT has indicated the resolution of 100 ISO black-and-white film is 100 lp/mm.[†] Because the size of a 35-mm film negative is 24 mm × 36 mm, then 24(100) × 36(100) = 2400 × 3600, or 8,640,000 line pairs. If one digital pixel is required for the black line and one digital pixel is required for the white line (even this is debated), the digital equivalent is 24(200) × 36(200) = 4800 × 7200 = 34,560,000 pixels. At the time this text is being produced, a 34-megapixel SLR digital camera does not exist.

- Announced on October 7, 2005, was the Leaf Aptus 75 digital back for medium-format cameras, with 33,899,040 pixels (6726 × 5040 pixels).
- Announced on October 20, 2005, was the KODAK KAF-39000 Image Sensor, featuring 39 million pixels (7216 × 5412 pixels), and the KODAK KAF-31600 Image Sensor, with 31.6 million pixels (6496 × 4872 pixels). Both are only for medium-format cameras.
- Announced on June 19, 2006, was the DALSA Semiconductor 111 million pixel CCD sensor (10,560 × 10,560 pixels) designed for the US Naval Observatory.

If this were the absolute criterion by which digital cameras could be used at crime scenes, no digital SLR cameras could be used yet. At the rate of digital camera improvements, it will be many years before 35 megapixel digital SLRs are routinely available for crime scene work.

Look around you, and you will notice several federal, military, state, and local law enforcement agencies are using digital SLR cameras for crime scene photography, so this must not be the criterion they are using.

Resolution Standards/Guidelines

Are official standards offered for the minimal digital resolution necessary to do a particular job? Yes. It is not a standard used for crime scene photography, however. It only relates to the electronic transmission of 10-print facsimiles. It is, however, the only "hard" standard this author could find related to digital imaging quality.

Several years ago, when Automated Fingerprint Identification System (AFIS) computer networks was developed, it was necessary to develop a standard by which copies/facsimiles (FAXes) of 10-print cards could be sent from one

[*]http://www.theiai.org/guidelines/swgit/index.php
[†]SWGIT, Guidelines for the Field Applications of Imaging Technologies in the Criminal Justice System (Version 2.3 2001.12.06).

jurisdiction to another. One jurisdiction may have a freshly developed latent fingerprint from a crime scene that was tentatively "matched" to a known print in the files in another jurisdiction. As a point of clarification, AFIS systems do not make "matches." They come up with a list of "candidates" that have similarities to a latent print of interest. Only latent print examiners can make identifications. Unfortunately, in the popular media, such a "candidate" is frequently misnamed an "identification." So that a latent print expert could confirm this "match," the 10-print card had to be FAXed to the jurisdiction holding the crime scene print. A minimum standard for the quality and resolution for the FAX machine had to be developed. The National Institute of Standards and Technology (NIST) ultimately decided on 1000 pixels per inch (ppi) at 1:1 as the minimum resolution.

1:1 Magnification ratio: When an item of evidence is life size on a negative.

A **1:1 magnification ratio** occurs when an item of evidence is life size on a negative. As stated previously, the 35-mm negative is 24 mm × 36 mm in size. A single digit fingerprint fits nicely within the bounds of 24 mm × 36 mm.

As long as the FAX could produce 1000 ppi resolution, it could adequately transfer copies of 10-print cards through the wires.

Although probably improper to consider this FAX resolution standard to apply to crime scene photography, just for the sake of discussion, this is the result.

24 mm × 36 mm is roughly equivalent to 1 inch × 1.5 inches. Therefore, $1''$ (1000) × $1.5''$ (1000) would be the necessary resolution for a digital camera to adequately capture a single digit fingerprint; 1000 × 1.500 = 1,500,000, or 1.5 megapixels. Today, when SLR digital cameras are readily available in 6, 8, and up to 16 megapixels, this seems like an easy standard to be bound by. It is a deceiving standard, however. What if the evidence necessary to be captured by a digital camera is larger than $1'' × 1.5''$? If this same standard were used, what would this mean if an item of evidence fits within a $2'' × 3''$ area? Because this area is twice the area of $1'' × 1.5''$, does this not also mean the digital resolution is simply doubled to require a 3 megapixel camera? NO.

Figure 3.2
$1'' × 1.5''$ Fingerprint.

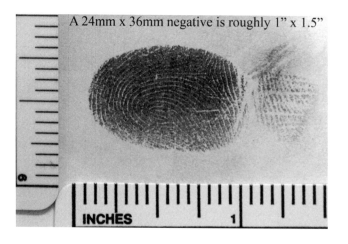

2" (1000) × 3"(1000) = 2000 × 3000 = 6,000,000, or 6 megapixels. Doubling the size of the evidence requires quadrupling the resolution of the digital sensor. That means two adjacent fingerprints would require a 6 megapixel digital camera. It would also mean that if the evidence filled a 3" × 4.5" area, 3" (1000) × 4.5" (1000) = 3000 × 4.500 = 13,500,000, or a 13.5 megapixel camera. If this standard were the required standard for all evidence, it can easily be seen that digital cameras quickly become inadequate to the task.

The NIST standard of 1000 ppi for latent fingerprints, however, is not applied to any other type of evidence. In fact, at the time of this writing, no other objective standards for the digital capture of any other kind of evidence exist. SWGIT does however currently offer a couple of negative statements concerning the use of digital cameras at crime scenes.

"It is recommended that digital cameras not be used for capturing tire impression evidence until research identifying the minimum resolution required to record these impressions is complete."[‡]

"It is recommended that digital cameras not be used for capturing footwear evidence until research identifying the minimum resolution required to record these impressions is complete."[§]

This is certainly discouraging. SWGIT and others, including this author, are currently engaged in the research to determine this requisite resolution. For the time being, this reading of how to determine an adequate resolution of digital cameras used at crime scenes to capture images used for identifications is not satisfying.

When Digital Cameras Can Do the Job

Is there hope? Yes.

What objective criterion can be used to justify the use of digital cameras to capture images that will be used for "critical comparisons" or "examination quality photographs?" Certainly not, "I used the camera I was issued by my agency." With many law enforcement agencies being required to have their Standard Operating Procedures (SOPs) written down to satisfy requirements of the Commission on Accreditation for Law Enforcement Agencies (CALEA), standards should minimally pertain to the adequacy of the equipment used to do the required job.

How can it be objectively substantiated that a particular agency's issued digital camera can do the required job?

A few words of caution must be emphasized here:

1. Do not presume your digital camera has sufficient resolution.
2. Do not use a digital camera on evidence unless you already know successful identifications are possible.
3. Do not "hope" the resolution is adequate enough.

[‡] SWGIT, (Section 9) General Guidelines for Photographing Tire Impressions.
[§] SWGIT, (Section 10) General Guidelines for Photographing Footwear Impressions.

How can an agency meet the challenge? SWGIT provides the answer within its own guidelines. "When documenting major crime scenes . . . conventional silver-based film in 35 mm format or larger is recommended for use as the primary media in this case. . . . It should be recognized that some agencies may wish to use equipment other than that recommended above. In such circumstances, the agencies should demonstrate and document that the selected equipment is adequate to meet the agency's anticipated needs."[*]

How can an agency determine that any digital camera and digital printer combination is adequate for providing images with the resolution necessary to make identifications? Do tests.

Need a Standard Operating Procedure (SOP) for your agency? Consider something like this.

If the issue is whether your agency's digital camera and digital printer can produce images that contain the quality sufficient for an identification to be made between a digital image of a shoe print and a suspect's shoe, have your shoe print examiner pull the evidence from a previous case in which a film photograph of a shoe print on a surface collected as evidence was ultimately successfully identified to a known shoe. This will be your standard for shoe prints, although it would also be necessary to do this several times with several previous shoe print identifications.

Because many possible digital camera–digital printer combinations are possible when such tests are done, one possible combination will be posited as an example. Let us consider the situation in which an agency is currently using a 6 megapixel SLR camera, and a dye-sublimation printer capable of printing at 300 pixels per inch (ppi) is being used. The 6-mp camera has a sensor with 3000 horizontal pixels; 3000/horizontal × 2000/vertical = 6 mp. Because the printer can print 300 ppi, by dividing 300 into the horizontal pixel count of the camera, the size of the image printed at full resolution, considering both the camera and printer, can be determined: 3000/300 = 10 inches. If the camera captures the image at its full resolution, and does not discard any of the digital data by compressing it, and the printer prints all this digital data at the printer's full capacity, then a 10″ print is the result.

It has not yet been determined that a 6-mp camera can, in fact, capture sufficient detail for an identification, which is what the tests are to establish. If the 6-mp camera cannot capture sufficient detail for an identification, other cameras with different resolutions will have to be tested. Not necessarily.

Consider this: A 6-mp camera can be defined as one with 3000 horizontal pixels. If the digital printer can print at 300 ppi, then the 6 mp digital camera can be thought of as a camera that can capture 300 ppi over a 10″ span.

A 10-mp camera (3872 × 2592 pixels) can, therefore, be thought of as a camera that can capture 300 ppi over a 12.9″ span (3872/300 = 12.9).

A 16.6-mp camera (4992 × 3328 pixels) can be thought of as a camera that can capture 300 ppi over a 16.64″ span (4992/300 = 16.64).

[*] SWGIT, Guidelines for the Field Applications of Imaging Technologies in the Criminal Justice System (Version 2.3 2001.12.06).

When the limiting factor is a printer that can print at a maximum of 300 ppi, then different digital cameras, with more and more resolution, cannot ever produce a printed resolution of greater than 300 ppi. The difference will be the size of their prints. The 6-mp camera will produce a 300 ppi print on 10″ of paper; the 10-mp camera will produce a 300 ppi print on 12.9″ of paper; the 16.6-mp camera will produce a 300 ppi print on 16.64″ of paper.

How can this help us?

If we confine each of these three digital cameras to the same sized item of evidence, this is the result. If the object being photographed is 10″ long, and each camera fills the frame with the entire 10 inches, this is the result. The 6-mp camera will spread its 3000 pixels over the 10″ with the result being 300 pixels per inch. The 10-mp camera will spread its 3872 pixels over the 10″ with the result being 387 pixels per inch. And, the 16.6-mp camera will spread its 4992 pixels over the 10″ with the result being 499 pixels per inch.

What may be needed to eventually produce a digital image with sufficient resolution for an identification is a digital camera that can capture more pixels per inch. But, are multiple cameras needed? No.

Consider this: With a 6-mp camera, the largest print that can be made at 300 ppi is a 10″ print because 3000/300 = 10. But, if those same 3000 pixels are placed over an area of only 9″ the result is 3000/9 = 333 ppi, which is the same capture ability as a 7-mp camera. If the object or area to be photographed is only 9″ long, a 6-mp camera acts like a 7-mp camera! A 7-mp camera cannot capture any more detail.

This series may be helpful:

Fill the frame with:

10″ 3000/10 = 300 ppi the capability of a 6-mp camera
 9″ 3000/9 = 333 ppi the capability of a 7-mp camera
 8″ 3000/8 = 375 ppi the capability of a 9.37-mp camera
 7″ 3000/7 = 428 ppi the capability of a 12-mp camera
 6″ 3000/6 = 500 ppi the capability of a 16.6-mp camera
 5″ 3000/5 = 600 ppi the capability of a 24-mp camera
 4″ 3000/4 = 750 ppi the capability of a 37.5-mp camera

What this means is that you can do your tests of the shoe print with only a 6-mp camera. It is not necessary to buy or borrow various digital cameras with higher and higher resolutions to perform your tests. If the 6-mp camera linked with a dye-sub printer does not yield a print containing sufficient resolution for your shoe print examiner to make an identification, use the 6-mp camera like a 7-mp camera and capture a new image to be compared with the known shoe. If no identification is possible at this resolution, use the 6-mp camera like a 9.37-mp camera, and capture a new image. Continue these tests until the shoe print examiner notices the minute details, defects, cuts, gouges, or imbedded debris necessary to make the identification.

Once an identification can be made, repeat these tests with other similar items of evidence. The first set of tests may have been conducted with evidence that is easier or more difficult to make an identification. The examiner can decide the number of tests necessary for him or her to feel confident the camera resolution and printer combination are sufficient for any likely evidence coming into the laboratory as casework.

The preceding explanation of how to conduct your tests is not a suggestion that a 6-mp camera is, in fact, sufficient to capture all the evidence found at crime scenes. For example, if the tests suggest a 12-mp camera produces sufficient resolution for an identification with one type of evidence, then if the evidence is in fact 7″ or less, the 6-mp camera will certainly work. If the evidence encountered is larger than 7″, then you are faced with a choice. You can capture overlapping 7″ segments of the evidence with a 6-mp camera, or you can consider obtaining a higher-resolution camera.

Tests like these will have to be done with different types of evidence normally encountered in different sizes:

- The size of three simultaneous latent fingerprints
- The size of a full palm print
- A shoe print 10″ or 12″ long
- A tire track, typically photographed in 18″ to 20″ segments

Your tests may reveal another conclusion. Your tests may convince you that your digital camera is insufficient for a particular job. If you reach a "failure" point, when an identification cannot be made, the evidence is too large for your digital camera to resolve all the detail within it.

Solutions for "failure," when your current digital camera cannot produce an image that can be identified to a known item of evidence:

1. Consider segmenting the evidence. With a print of a previous shoe print that had been identified, for example, crop the image to make two halves of the original: toe to the shoe midpoint and heel to the shoe midpoint. Then, see whether an identification can be made. If not, segment the shoe into thirds (sole, arch, heel), and see whether that size of evidence allows sufficient resolution for an identification.

    Your tests may convince you that whereas film could capture the detail necessary for an identification, your current digital camera can only capture the same resolution in 5″, 6″, or 7″ segments. If film could capture the detail necessary for an identification when photographing a 20″ segment of a tire track, three 7″ digital images may be necessary to accomplish the same thing, which may be unrealistic when four tire tracks are at the crime scene, each one exceeding 8′ long.

2. Consider a digital camera with more resolution.
3. Use film for that type of evidence.

If this procedure is followed, and you are eventually challenged in court about the wisdom of using your particular digital camera to capture a particular type of evidence, you can then reply that your choice of digital camera was based on your agency's tests indicating your digital camera could, in fact, resolve the detail necessary to make positive identifications in similar types of evidence in compliance with SWGIT guidelines. Defense attorneys will not be asking too many follow-up questions after this answer, because doing so would only make you and your agency look more competent, and that is certainly not their intent.

Acutance

Acutance refers to the camera's ability to render a sharp edge of the subject as a sharp edge in the photograph.

This is obviously part of what we think of when we speak of sharp focus and good resolution. Acutance means a clear and precise distinction exists between the edge of one object and the beginning of another adjacent object. Referring back to resolution and the line pairs it deals with, if a white line were adjacent to a black line, the edge between them would be crisp and well defined, with no overlap between the two.

Unfortunately, we take this for granted. However, with imaging systems, achieving this goal is often difficult. Before camera lenses existed that had to capture such fine detail, telescopes and microscopes had to deal with the same lens quality issues. In the 1880s instrument maker Carl Zeiss began collaborating with physicist Ernst Abbe to perfect the optics of microscopes. The Carl Zeiss name is still found on some of the best camera lenses made. Ernst Abbe's name is so closely related to quality optics, that the term **aberrations**, denoting the defects in optics quality that he worked to eliminate, is actually derived from his name.

The concept of edge sharpness will be revisited when diffraction is covered. Images of sharp-edged items will be shown that do not replicate the sharp edges of the object. The causes and possible solutions to this problem will be discussed later in this chapter.

In Focus or Being Sharp

When we think of an image being in focus, we usually think of light coming into the camera through the lens and converging at the film plane at a precise point. As light makes its way through the lens, it becomes a smaller and smaller circle, which looks like the converging ellipses in Figure 3.3.

Figure 3.4 shows the light focusing in front of the film plane, then continuing on until it strikes the film as a small circle. In this case, the image will be out of focus.

Figure 3.5 shows the light unfocused until some distance after the film plane. Of course, the light cannot travel through the film plane to focus behind it. But, when it strikes the film plane, it is also a small circle. In this case, again, the image will be out of focus.

These simplistic graphics are presented to introduce the concept of **circles of confusion**.

Acutance: The rendering of a sharp edge in the subject as a sharp edge in the photograph.

Aberration: A term encompassing a variety of causes that result in poor image quality. They come from imperfect lens design and construction. Because perfect lens construction is impossible, the best lenses can only minimize aberrations.

Circles of confusion: As long as the light, reflected from a single point in space, enters the lens and remains a circle, rather than coming together at the film plane at a specific point, the image is "confused" and out of focus.

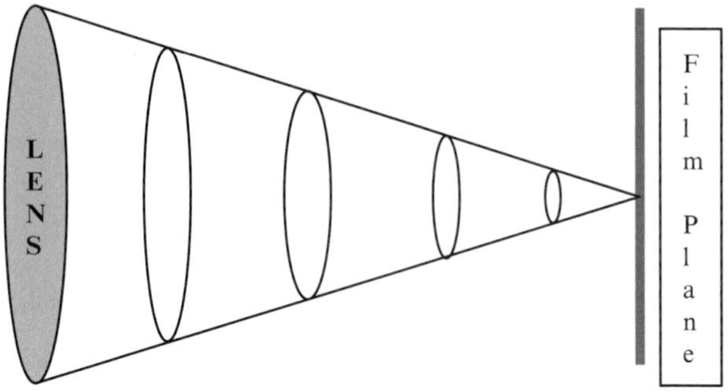

Figure 3.3

Light rays converging at film plane.

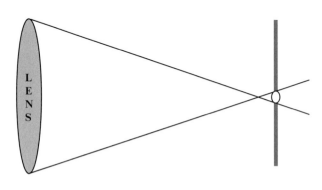

Figure 3.4

Light rays converging before the film plane.

Figure 3.5

Light rays converging after film plane.

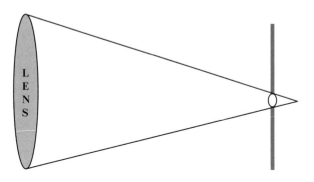

Simply stated, as long as the light, reflected from a single point in space, enters the lens and remains a circle, rather than coming together at the film plane at a specific point, the image can be considered "confused" and out of focus. It does not matter whether the light would have become a point in front of or after the film plane. If a circle of light is formed on the film plane, the result is an out-of-focus image.

Of course, the concept of **depth of field (DOF)** is temporarily being ignored, which is the extended area in front of and behind the plane of precise focus that still appears to the eye to be in focus.

Consider this explanation of the circles of confusion to be the purely mechanical definition of critical focus. We will get to depth of field soon.

Depth of field: The extended area in front of and behind the plane of precise focus that still appears to the eye to be in focus.

MANUAL FOCUSING

When the word "focusing" is mentioned, most people think of the manual focusing technique. To manually focus a camera, one simply rotates the focus ring while looking through the viewfinder until the object/person of interest comes into focus, which is the way most people focus a camera. However, whenever more than one object/person, at different distances from the camera, is of interest, then we need a different strategy to ensure all the items/people are in focus at the same time.

Focusing on just one item at a time is the perfect situation for using the manual focusing technique, which presumes, of course, that the item is composed with the film plane parallel to the item of evidence. That Cardinal Rule certainly applies here. If, and only if, the entire item of evidence is composed with the film plane parallel to it, will manual focusing be successful. If the item of evidence is (mistakenly) composed from a diagonal point of view, the manual focusing technique may not be successful.

Figure 3.6 shows a photo of a semiautomatic pistol taken with the film plane parallel to the top surface of the weapon. Because the entire surface of the pistol is the same distance from the camera, it is completely in focus. We expect an entire item to be in focus when a photograph is taken. But, sometimes our expectations are not met. Figure 3.7 shows the same pistol. In this shot, however, the photographer assumed a diagonal point of view to the pistol. The focus point was the near end of the muzzle. The muzzle is closer to the photographer than the pistol's grip. It is evident that as you look further along the pistol, from front to rear, the weapon begins to become "softer." The entire pistol is not in focus. The carpeting just beyond the grip is noticeably "softer" also.

Manual focusing on single items of evidence is best done with the film plane parallel to the top surface of the evidence.

Figure 3.6
Gun, film plane parallel.

Figure 3.7

Gun, oblique camera viewpoint.

AUTOMATIC FOCUSING

Many newer cameras have auto-focusing capabilities. Basically, when the shutter button is depressed half way, these cameras usually focus on the object in the center of the viewfinder. Different camera manufacturers have different methods of indicating just what this auto-focus point is. Frequently, it is designated by a small rectangle or circle in the center of the viewfinder. Consult your camera manual for these variations.

Caution, however, should be used when using auto-focus. The improper use of auto-focus is often the explanation for out-of-focus photographs. How can this be?

An example may help. At a homicide by throat slashing, a crime scene photographer tried to photograph the gaping wound to the victim's neck, and this crime scene photographer was in the habit of using the auto-focus capability of the camera. All of the photographs came back with the chin of the victim in focus and the wound out of focus. What happened? The auto-focus locked on the chin, without the crime scene photographer noticing it. The proper use of the auto-focus capability of the camera would have resulted in a better image. But, in this case, the crime scene photographer relied on the automation of the camera to do his job for him. The camera did, in fact, attain focus on part of the scene, just not on what the photographer was really interested in.

Many teachers of basic photography techniques require their students to use only manual camera settings. It is best for the student to learn to be totally responsible for the net product of their efforts. Relinquishing responsibility for any of the camera's functions to a camera's automatic functions is not the proper way to learn photography. It is better to have well-trained crime scene photographers than to have "intelligent" cameras. Teachers often tell students they can either teach them photographic theories and principles, or the students can just leave their "smart" cameras in class and then go home. It is much better for the photographer to learn how to use every facet of the camera kit than for the photographer to rely on the camera to make certain decisions for him.

Moral: The photographer is responsible for the final quality of the image. Blaming an automatic feature of the camera equipment for a marginal or unacceptable photograph is a poor excuse. If you choose to use any feature of a camera, you must use it correctly and be responsible for the results.

The auto-focusing capability of a camera can be used effectively; it is just important to be aware of what the camera has locked on to before the shutter is fully depressed capturing the image.

When to Use Auto-Focus

When many purchasers are willing to spend additional money on an auto-focus camera, how can anyone justify a recommendation not to use it? It is certainly a valued feature of many cameras. However, many general-purpose camera purchasers also demand a camera with auto-exposure, or program exposure mode. It is said that many purchasers of cameras want the camera to automatically make all the "complicated" decisions for them, so they can concentrate on more "creative" aspects of photography. Crime scene photographers usually fall into a different class. The images we capture can help put someone in prison. The images we capture can help convict a defendant ultimately sentenced to death. These images usually are more purposeful than those captured by the general public.

When would the auto-focus mode be appropriate for crime scene photography? Certainly, when only one main subject is in the field of view, auto-focus may speed things up a bit without having the photographer relinquish any of his or her responsibility.

In a surveillance situation, when it can be anticipated that the subject will be moving toward the camera or away from the camera, predictive auto-focus may be the only way to capture the image. Predictive auto-focus enables the camera to track a moving subject at different times at different distances from the camera. Presuming that the subject is moving at a constant rate, the camera's computer can determine the anticipated distance change in a given amount of time and automatically focus the camera for the position the subject will be at in the next instant. It would be impossible to hold the camera and continually make fine refinements to the focusing in situations like this, which is one time the technology must be used to capture the desired image.

An experienced photographer knowledgeable about when auto-focus is likely to be unsuccessful in attaining accurate focus can also use auto-focus. With more awareness on the part of the photographer imaging the slit neck in the homicide situation mentioned previously, auto-focus could have been successful. It just takes a photographer aware of what the camera is doing.

Difficult-to-Focus Subjects for Auto-Focus

When is auto-focus likely to fail? Some auto-focus cameras require subject matter that has prominent vertical elements, and some require prominent horizontal elements. Check your camera's manual. In either case, when the subject matter lacks distinct detail for the camera's focusing system to lock onto, the auto-focus capability of a camera may fail, which can usually be noticed by the focusing system continually searching for something to grab onto, moving throughout its focusing continuum and never locking at any one distance. Frequently encountered examples of this type of subject matter are a uniformly blue sky or an interior wall with a single color on it.

If this occurs, usually two options exist: Attempt to find a substitute object to focus on at approximately the same distance from the camera. Focus on it, by depressing the shutter button down half way, keep the shutter button partially depressed, swing the camera around to the primary subject, and depress the shutter button the rest of the way. Or switch to manual focus.

Auto-focus may also fail in dim, low-light conditions. Under these conditions, some cameras may have the capability of emitting an infrared pattern that projects onto the subject, and the camera then focuses on this infrared pattern. If your camera does not have this capability, switching to manual focus may be your only alternative.

Conversely, bright lights or strongly backlit scenes may "confuse" a camera's auto-focus system. If a surface is reflecting specular elements, the auto-focus system may fail.

When nearby and distant elements are near the center of the viewfinder, the auto-focus system may not be able to decide which is the intended target and give up trying. One example is a caged animal in a zoo. If the bars or wiring are too close to the animal in the field of view, the camera may not be able to decide if the relatively close cage or the distant animal is the prime subject matter. In these situations, you may have to switch to manual focus.

Finally, if two distinct types of subject matter are near each other, the auto-focus system may successfully focus on one of them when it was really the other object you were concerned with. Remember the chin and slit neck situation mentioned earlier. Double check that the auto-focus system is focusing on what is important to you.

Or rely on manual focus and feel assured you will infrequently focus on an aspect of the scene unimportant to you.

FOCUSING WITH A ZOOM LENS

Many crime scene photographers have a zoom lens as their primary lens. Frequently encountered are zoom lenses like 35 mm to 80 mm or 28 mm to 105 mm. Many years ago, a distinct advantage existed to the use of a prime lens with only one focal length. Such lenses were clearly sharper and had fewer lens aberrations.

Lens design and construction has significantly improved to the point where such a negative stigma is attached to using a zoom lens. Indeed, some excellent zoom lenses are on the market.

If a zoom lens is regularly your lens of choice, or if it is what your agency has provided you for the job, how can you use it to maximize your focusing task? As mentioned previously, many zoom lenses include a wide-angle range and a tele-photo range. One of the known benefits of telephoto lenses is their magnification capability. Would you rather focus on an object when it is small and difficult to see, or when it is larger and more clearly defined? Obviously, the latter. If a zoom lens is the tool you are currently using, use it efficiently. To focus on any subject, zoom the lens to its telephoto setting to use its magnification capability. Focus. Reset the lens to the focal length you want to use, and be comforted in the knowl-edge that once you have focused the lens at any focal length setting, selecting a different focal length does not de-focus the lens. Know the tool, and use it appropriately.

PRE-FOCUS

This author once worked with a crime scene technician who would enter a crime scene, raise the camera to his eye, and then move forward and backward until the subject became sharp. However, under one circumstance, this is exactly the best way to focus a camera. Have the camera pre-focused, and then move the camera forward and backward until the subject matter becomes sharp.

What is this circumstance? It is when you intend to take a series of bracketed images of a fingerprint. In this circumstance, you will set the camera so it captures a 1:1 magnification of the fingerprint, which is when the single fingerprint fills the frame in the field of view, also ensuring the fingerprint is life-sized on the neg-ative. Recall Figure 3.2. A single digit fingerprint nicely fills the frame when magnified to 1:1.

Figure 3.8 shows the size of a single digit fingerprint on a negative. Almost per-fect sizing. Another way to think of 1:1 magnification is to think of filling the frame with a single nickel. Figure 3.9 does the same thing. The nickel is life-sized on the negative. Of course, there is not much need to fill the frame with nickels at crime scenes, but we certainly photograph many fingerprints.

Why pre-focus? For two reasons.

One alternative is to put your camera on a tripod, set the tripod down an arbi-trary distance from the fingerprint, and then begin focusing. You may focus and focus and focus, and then eventually realize the camera is too close to the finger-print to ever achieve proper focus. Every lens has a close focusing distance. Closer than this distance, the lens cannot focus. Placing the tripod down at this distance is a waste of time and makes you look like you do not understand your own equip-ment. Not a good way to earn the respect of your peers.

Figure 3.8
Fingerprint on negative.

Figure 3.9
1:1 Nickel.

If you place the tripod too far from the fingerprint, the camera will eventually be able to focus. However, if you are further than need be, when you do achieve focus, the fingerprint will be smaller than optimal. Remember the Cardinal Rule, fill the frame? Here is when it counts. If the camera is further from the fingerprint than necessary, the fingerprint is smaller on the negative. If the fingerprint is not as large as possible, you are using too many film silver halide crystals or too many digital pixels capturing non-fingerprint detail. Why waste your sensor's ability to capture detail on irrelevant space?

Pre-focus the camera to its closest focus distance. One end of the focusing continuum is infinity; the other end is the lens' closest focusing distance. This can be done without looking through the viewfinder. Just look down on the lens and adjust the focusing ring until it stops at the short end of its distance scale. Having pre-focused the camera, move the camera mounted on a tripod toward the fingerprint until the fingerprint comes into focus while you look through the viewfinder. In this situation, all other focusing techniques are second best.

HYPERFOCAL FOCUSING

Many times the crime scene photographer must capture entire crime scenes and areas in focus rather than individual items of evidence. As mentioned earlier, focusing on areas requires a different skill set. In the normal sequence of crime scene photography, we frequently must capture images of large outdoor crime scenes. If the crime occurred indoors, we frequently begin by photographing the exterior of the building in which the crime occurred and the surrounding areas by which any suspects could have gained ingress and egress. These are large areas.

What is the proper technique to maximize the depth of field with large exterior crime scene photographs?

It might be good to revisit our definition of depth of field. Depth of field is the variable area, from foreground to background, of what appears to be in focus to the eye. It is a variable area, because the photographer can select camera variables that restrict the depth of field to just one plane or just one distance from the photographer. Or the photographer can select camera variables that maximize the depth of field, ensuring large areas are in focus when the photographic print is eventually made.

This is made more complicated because the camera usually only shows just one plane, or one distance from the camera, to be in focus at a time. When looking through the viewfinder, the photographer usually only sees one area of the scene in focus at a time. The well-trained crime scene photographer, however, will be confident that when the image is printed, more than what they saw through the viewfinder will be in focus.

Figures 3.10 to 3.12 show this ability to manipulate the area in focus to suit the photographer's needs. Figure 3.10 shows a restricted depth of field, with only the front of the numbered evidence markers in focus. Figure 3.11 transfers this narrow area that is in focus to the rear numbers. Finally, by choosing the correct camera variables, the photographer can maximize the depth of field range and have all the numbers in focus at the same time, as in Figure 3.12.

Here, a new Rule of Thumb will be offered to the reader.

Rule of Thumb 3.2: If you are composing on an area in your image, the entire area, from front to back, should be in focus.

Why include it in your image if you know it will be out of focus? Another way of saying the same thing is this: If you know an area will be out of focus, exclude it from your composition. Do not give anyone the opportunity to criticize your photographs because areas are out of focus.

This may be the first time this author's pedagogical style has been so blatant. The crime scene photographer is capturing very important images. The consequences to the defendant are enormous. The responsibility of the photographer is, therefore, just as weighty. The first chapter mentioned the photographer's full field-of-view responsibility. Do not take this lightly. Purposefully include in your photographs only the details you wish in them; exclude from your images as much as is irrelevant as possible. Once you have decided what will appear in your field of view, you now have the obligation to ensure it is well exposed. Achieving a proper exposure is not sufficient. It is just as important to ensure it will be in focus. Certainly you will not be considered much of a photographer if your well-exposed detail is out of focus. So, too, having in-focus detail improperly exposed is just as problematic. Photographers have the responsibility to compose, expose, and focus properly on their subject matter.

Figure 3.10
Wide aperture, focused on #1.

Figure 3.11
Wide aperture, focused on #15.

Figure 3.12
Small aperture, all in focus.

Hyperfocal focusing is the technique to maximize the depth of field when infinity is composed in the background. Infinity (∞) is normally considered an unimaginable large number or quantity. On a camera's focusing ring, however, infinity is more mundane.

In Figure 3.13, the top two rows of numbers are distance indicators, both in feet (top row) and meters (2nd row). Notice that infinity (∞) is the next figure after the 10-meter mark. As the distance indications move from left to right, they seem to be getting progressively closer together. The space between the 2-meter mark and the 3-meter mark is larger than between 3 meters and 5 meters. The space between 5 meters and 10 meters is smaller yet, which suggests that if an actual number had been printed on the distance scale in place of the infinity symbol, it would be perhaps 20 meters or so. What this means, for all practical purposes, is that as far as the camera is concerned, infinity is the equivalent of 60 to 80 feet and beyond.

Thinking like a camera, then, we can place photographs into three rough categories:

1. Large scenes, with infinity in the background.
2. Medium distance scenes, where infinity is not in the background.
3. Close-up images of individual items of evidence.

Hyperfocal focusing is the focusing technique for the first category, when maximizing depth of field is a primary concern. If you accept Rule of Thumb 3.2 and wish to ensure that infinity must be in focus if it is in the field of view, you may be tempted to actually focus on infinity.

Figure 3.16 shows a camera focused on infinity. The focusing line is aligned with the infinity symbol, which is the most efficient way to ensure infinity will be in focus, right? Wrong.

Hyperfocal focusing: The technique to maximize the depth of field when infinity is composed in the background.

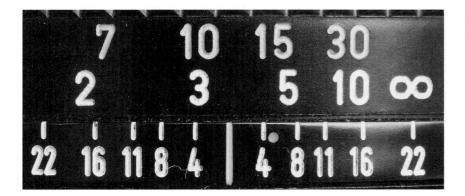

Figure 3.13
Focusing ring.

Depth of field scale (DOF scale): A tool to enable the photographer to determine the DOF, the area that will appear to be in focus, when the image is processed.

Figure 3.16 also shows a third row of numbers beneath the distance rows. These are pairs of numbers on either side of the focusing point called the **depth of field scale (DOF scale)**.

Many lens manufacturers include this very handy scale on their lenses. Unfortunately, a trend to omit this scale on many newer lenses seems to exist. Do not fear. This text will provide suggestions for hyperfocal focusing whether or not your particular lens actually has a DOF scale.

What is the DOF scale? The DOF scale is a tool to enable the photographer to determine the DOF, the area that will appear to be in focus when the image is processed. Remember, when looking through the viewfinder, only one distance from the camera appears to be in focus at any time. The photographer must somehow know that more than just one distance will be in focus when the image is printed, depending on the camera controls selected.

It was previously indicated that only one point exists where the light coming in through the lens will come together to form an exact point on the film plane. It was previously indicated that if the light coming in through the viewfinder formed a circle when it reached the film plane, the result would be an out-of-focus image. Now is the proper time to refine that simplistic explanation of focusing. To do this, we must revisit the concept of the circles of confusion.

The notion that being in focus only occurs when light meets at a single point on the film plane is a very mechanical explanation. In fact, the eye is a very remarkable sensor. It can actually detect some of the light striking the film plane as small circles as being in focus, which means that instead of a single distance from the camera appearing to the viewer as being in focus, an area before the point of exact focus appears to the viewer to be in focus and an area behind the point of exact focus appears to the viewer to be in focus. Some of the light striking the film plane as small circles is perceived to the viewer as being in focus, just as the light striking the film plane at a single point.

Research has shown that circles striking the film plane as large as 0.025″ to 0.033″ in diameter will appear to the eye to be in focus. You may not be able to imagine circles with that diameter range, so it is difficult to grasp this concept with just this information. Graphics help immensely, and perhaps they will make the concept easier for you to understand.

Although Figure 3.14 cannot exist in reality, because light cannot travel past the film plane, it does make it somewhat easier to understand DOF. Consider the small shaded circles to have a diameter of 0.25″ to 0.033,″ which is the resolution limit of the eye. Light coming together at the film plane that is not in perfect mechanical focus, forming circles of this size, or smaller, will still appear to the eye as being in focus. Naturally, the point at which light actually intersects will be seen as being in focus. For the sake of discussion, let us consider this distance as being 20′ from the camera. Light coming from a distance of about 22′ will not intersect at the film plane. It will actually intersect before the film plane, cross over, and continue until it strikes the film plane as a small circle. Imagine

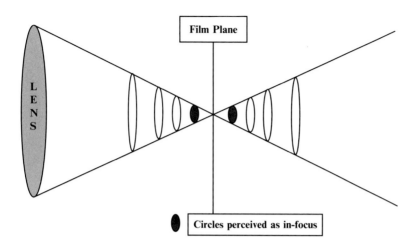

Film Plane

Circles perceived as in-focus

L E N S

Figure 3.14

In-focus circles of confusion; wide aperture.

this small circle as the left shaded circle in the graphic. Light coming from a distance of 19′ will not intersect at the film plane. If it could, it would intersect after the film plane. At the film plane, it is still a circle. Imagine this small shaded circle as the right circle in the graphic.

DOF is the range, in front of and behind the plane of exact focus, of what also appears to the eye to be in focus. Therefore, not only will the item at 20′ appear in focus, but also anything between 19′ and 22′. Anything closer than 19′ and anything beyond 22′ will appear to be out of focus, because at the film plane they would create a circle of confusion larger than what the resolution limit of the eye will detect as being in-focus. These distances are depicted as ellipses in the graphic. If light forms a circle larger than 0.025″ to 0.033″ on the film plane, the image appears to the eye to be "confused" or blurred.

In Figure 3.14, the light is coming from the extreme edges of the lens, intending to suggest the widest aperture of the lens. Let us consider this aperture to be f/2 for the time being. What would occur if the light entered the lens when a smaller f/stop is being used? Figure 3.15 shows this situation. The circles of confusion are the same size in both Figures 3.14 and 3.15.

Notice that as the aperture becomes smaller, the circles of confusion, perceived by the eye to be in-focus, move further away from the film plane, which is the explanation for the larger DOF produced by using smaller apertures. Figure 3.14 showed the equivalent of an f/2 aperture. Figure 3.15 shows the equivalent of perhaps an f/11 aperture. The use of an f/11 would have more appear to the eye as being in focus. In this case, the DOF range would be something like 12′ to 36′ when focused on 20′.

One of the camera's variables that affect DOF is the f/stop selection. With this information, let us return to the DOF scale. The DOF scale is designed to be used in conjunction with the distance scale. The DOF scale is composed of pairs of f/stop numbers radiating outward from the point of exact focus. Some lenses do

Figure 3.15

In-focus circles of confusion; smaller aperture.

not show as many f/stop numbers as are depicted in Figure 3.17. Some lenses do not have a DOF scale at all. Figure 3.17 is a good training tool to explain DOF.

Once the f/stop for a proper exposure has been determined, the photographer can then refer to the DOF scale to determine exactly how much of the crime scene will be in focus when the image is printed. With Figure 3.17, it is evident that the photographer has focused at about the 12′ distance. Aligning the DOF scale with the distance scale, the photographer can then see what the DOF range will be. The distance that falls between the pair of f/stop numbers on the DOF scale is the DOF range. With each f/stop, then, we see the following DOF ranges, when the camera is focused at 12 feet:

1. f/22: 6′ to ∞
2. f/16: 7′ to 30′
3. f/11: 8′ to 25′
4. f/8: 9′ to 20′
5. f/5.6: (not shown)
6. f/4: 10′ to 14′

Wider apertures result in narrower DOF ranges; smaller apertures result in progressively longer DOF ranges.

It is a great tool, if you have it. But, what if your lens does not have a DOF scale? Here is how to hyperfocal focus without a DOF scale. To begin, consider Figures 3.16 and 3.17, side by side.

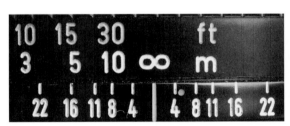

Figure 3.16

Focused on infinity.

Figure 3.17

DOF scale: pairs of numbers opposite the orange line.

Figure 3.16 shows one way to ensure infinity will be in focus. Simply focus on the infinity symbol. Now, by use of the DOF scale, you can determine what will be in focus when the print is made. If the lighting enables an f/22 to be used for a proper exposure, the distances between the two 22s will be in focus when the print is made. Figure 3.16 shows this to be 12′ to infinity. Not bad. Notice all the distance between the infinity symbol and 22 on the right side of the DOF scale, which is, in effect, wasted DOF.

Hyperfocal focusing is the technique to maximize the DOF, when infinity is in the field of view. To hyperfocal focus, first determine the f/stop required for the lighting at the scene. Once the f/stop has been determined, align that f/stop, on the DOF scale with the infinity symbol on the distance scale. Presume it is a bright sunny day, and the meter reading indicates an f/22 can be used to properly expose the scene. Figure 3.17 shows the result. The infinity symbol is now aligned with the right 22 on the DOF scale. The DOF range will be the distance between the two 22s as they align with the distance scale. Figure 3.17 indicates the new DOF range is 6′ to infinity. That is a net gain of 6′ that will be in focus when the image is printed.

With both focusing techniques, infinity will be in focus. With hyperfocal focusing, you have also maximized the DOF. Hyperfocal focusing increases the DOF range in the foreground.

No other focusing technique will provide better DOF when infinity is in the background.

Let us examine the hyperfocal focusing range with f/16, f/11, and f/8.

This may perhaps be best understood in Table 3.1.

Notice that the point of exact focus, the **hyperfocal focusing distance**, is double the short end of the DOF range. We will use that fact momentarily.

Having a lens with a DOF scale is certainly very convenient. However, as mentioned earlier, many recently manufactured lenses do not have this handy tool. In many situations having a DOF scale would make the job easier on a crime scene photographer. It comes down to this. Is there a way to hyperfocal focus without a DOF scale? Yes.

Hyperfocal focusing distance: When hyperfocal focusing, this is the point of exact focus. It is usually double the short end of the DOF range.

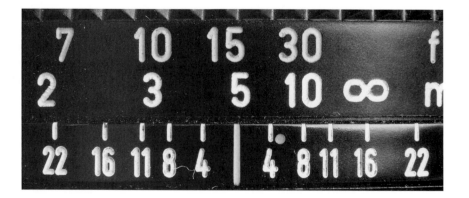

Figure 3.18
Hyperfocal focusing, f/16.

Figure 3.19
Hyperfocal focusing, f/11.

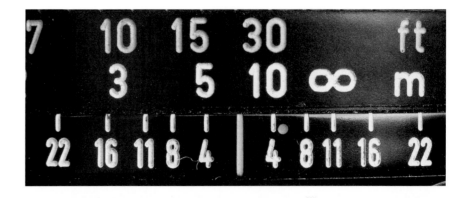

Figure 3.20
Hyperfocal focusing, f/8.

Table 3.1
Hyperfocal Focusing Ranges

F/stop	DOF range
f/22	6' to ∞, when focused at 12'
f/16	8' to ∞, when focused at 16'
f/11	12' to ∞, when focused at 24'
f/8	15' to ∞, when focused at 30'

Lens construction standards are so uniform that even if your particular lens does not have a DOF scale, you can still use the theory of hyperfocal focusing to maximize your DOF when photographing large exterior crime scenes when infinity is in the background. How? First, determine the f/stop required for the lighting of your scene. Then focus at one of the following distances appropriate for your f/stop:

- If using an f/22, focus the lens at 12'.
- If using an f/16, focus the lens at 16'.
- If using an f/11, focus the lens at 24'.
- If using an f/8, focus the lens at 30'.

This is important enough to memorize which distances to focus at for the f/stop required for the lighting at the scene.

You can focus the lens by using the distance scale, by aligning the focusing point of the lens with the correct distance on the distance scale. Or you can find a surrogate item somewhere in the crime scene the prescribed distance you need and focus at. Remember, hyperfocal focusing is the optimal focusing technique for this particular focusing situation. Any other focusing technique will result in either the background (∞) being out of focus or less of the foreground being in focus. Neither of these options is desirable.

Why are hyperfocal focusing ranges for the other f/stops not provided? Remember, these are images of large outdoor crime scenes, when it is necessary to have infinity in the field of view. Using wider apertures results in a smaller DOF range, making it more difficult to have any evidence relatively close to the camera be in focus while maintaining infinity in-focus at the same time.

Some writers feel the DOF scale on lenses does not really ensure that the DOF range indicated will truly be in focus. In answer to such critics, here is a recommendation. With your particular camera and lens, set up a series of test distances. Test the f/22 hyperfocal focusing range by placing objects 6′ from the camera and 8′ from the camera. Focus at 12 feet. Take the photograph. When the photograph is printed, check to see whether the object at 6′ looks to be a bit "soft" or slightly out of focus. If so, hopefully the object at 8′ looks sharp. If the object at 8′ looks sharper than the object at 6′, just bump up the hyperfocal focusing recommendations one notch.

- Consider the f/22 DOF range to be 8′ to ∞, when focused on 12′.
- Consider the f/16 DOF range to be 12′ to ∞, when focused on 16′.
- Consider the f/11 DOF range to be 15′ to ∞ when focused on 24′.
- Consider the f/8 DOF range to be 20′ to ∞ when focused on 30′.

The preceding effectively "calibrates" the hyperfocal focusing concept for your specific camera equipment.

ZONE FOCUSING

Infinity is not always in the background of outdoor crime scenes. An obstruction, like the wall of a building, may be present 30′, 40′, or 50′ from the camera. If, when you look through the viewfinder, infinity is not in the background, hyperfocal focusing is not applicable. How do you maximize the DOF when infinity is not an issue?

For instance, on a bright sunny day, you may want to photograph a scene between you and a wall about 30′ away. Nothing in the field of view will be farther away than the 30′ distance of the wall. Now, it is not important to ensure that ∞ is in focus. We just need to ensure the wall, as the background of the photograph, is in focus.

Zone focusing: The technique to maximize the DOF for an area when infinity is not in the background.

Zone focusing is the technique to maximize the DOF for an area when infinity is not in the background.

The zone focusing technique is a bit like the hyperfocal focusing technique. This technique also relies on your lens having a DOF scale. Like the hyperfocal focusing technique, the first step is to determine the f/stop required for the lighting at the scene, which is done by simply taking a meter reading of the crime scene. The second step is to determine the distance of the background of this particular crime scene, which may be a real obstruction, like a wall, or it may be the distance from your camera you decide will be the top of your field of view when you compose this image. Look at Figures 3.21 and 3.22.

Both figures show the same area. Figure 3.21 shows the scene, with the top of the image 30′ from the camera. Figure 3.22 shows the same area, but the wall is 30′ from the camera. In both cases, it is not necessary to have anything beyond 30′ be in focus.

With zone focusing, it is necessary to determine the distance of the background. Manually focus on the wall to determine its exact distance. Focus on the wall and then look at your distance scale and note the distance shown. Had your crime scene been Figure 3.22, this distance would be 30′.

If no wall or other fixed object is in the background, determine the distance of the top edge of your composition. Consider Figure 3.21. Make a mental note of what is at the top of your composition, then swing your camera up so that area is now in the center of your viewfinder, and manually focus the camera. To make this easier, consider the previous suggestion to use the telephoto end of your zoom lens. Set the focal length to your len's most extreme telephoto focal length. For instance, if your zoom lens has a 28-mm to 105-mm zoom range, set the lens to 105 mm. That will magnify all areas in your viewfinder. Focus at the distance that will eventually be the

Figure 3.21
Zone focusing: top 30′ away.

Figure 3.22
Zone focusing: wall 30′ away.

top of your composition when viewed with a 50-mm lens setting. Check the distance scale of your camera to see the exact distance of this area. Again, if Figure 3.21 was your crime scene, 30′ should show on your distance scale. That distance will then be aligned with the f/stop required for the lighting of the scene on the DOF scale.

If the lighting will permit the use of an f/22, align the right 22 over the 30′ mark. See Figures 3.23 through 3.26 for the alignment of the depth of field scale and the distance scale to zone focus at 30′ with f/22, f/16, f/11, and f/8.

- If the lighting requires an f/22, the DOF range is 5′ to 30′.
- If the lighting requires an f/16, the DOF range is 6.5′ to 30′.
- If the lighting requires an f/11, the DOF range is 9′ to 30′.
- If the lighting requires an f/8, the DOF range is 11′ to 30′.

In each of these lighting conditions, it would be important to make sure the bottom edge of your photograph does not include anything in the field of view before your DOF range.

- If using an f/22, do not include detail before 5′.
- If using an f/16, do not include detail before 6.5″.
- If using an f/11, do not include detail before 9′.
- If using an f/8, do not include detail before 11′.

Figure 3.23
Zone focused at 30′ with f/22.

Figure 3.24
Zone focused at 30′ with f/16.

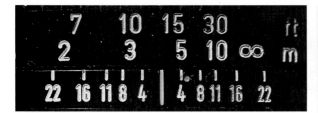

Figure 3.25
Zone focused at 30′ with f/11.

Figure 3.26
Zone focused at 30′ with f/8.

Simply compose your image so the bottom of your field of view matches the close end of your DOF range, which is done by raising and lowering your camera while you look through the viewfinder.

No other focusing technique maximizes the DOF better when it is not critical that infinity be in focus.

Some insightful readers may now be imagining a situation where raising the camera to one of these prescribed near end distances forces the top edge of the field of view to include more than the 30′ we began this discussion with. Aligning the top of the field of view at 30′ may force more into the near end of the field of view than the preceding distances. What to do then? These situations may occur at real crime scenes, depending on the lighting at the scene. Although it is a Cardinal Rule to attempt to maximize the depth of field at all times, nothing guarantees that everything in your field of view will be in focus for every image you capture. That is our goal, and sometimes this goal is not achievable, particularly when the lighting becomes dimmer. The question remains: what should we do? If the choice is between obviously out-of-focus foreground or obviously out-of-focus background, this author usually opts for the out-of-focus background, because it is slightly less noticeable.

FOCUSING BY THE RULE OF THIRDS

What if the camera equipment does not have the DOF scale on the lens? Many lenses do not have this feature. How should we zone focus then?

First, it needs to be pointed out that the DOF range is not usually the same in front of and behind the plane of exact focus. For instance, if you are focused on a point 10′ from the camera, while using an f/16 for the lighting, the DOF range will be almost 14′ in total. This 14′ is not divided by 2 resulting in 7′ in front of the 10′ focus point being in focus, and 7′ behind the 10′ focus point being in focus. It certainly would be nice if this was the case, but it is not the case. Usually, more of the DOF range exists behind the point of exact focus than in front of it.

At this point, many other authors insert a dictum that the depth of field range extends one third of the total range in front of the point of exact focus, and two thirds of the total range behind the point of exact focus.

In fact, this is often presented as a Rule of Thumb.

 Rule of Thumb 3.3: the DOF extends one third in front of the point of exact focus, and two thirds behind the point of exact focus.

Although a handy rule of thumb in some circumstances, it is certainly incorrect in many instances.

In all hyperfocal focusing situations, it can be immediately seen that the DOF range behind the point of exact focus is much more than double the DOF range in front of the point of exact focus. Recall that an f/22 will result in a DOF range of 6′ to infinity when focused at 12′. That is 6′ in focus in front of the exact point of focus and 12′ to infinity in focus behind the point of exact focus. Certainly not a one-third and two-thirds ratio.

When zone focusing with background distances at 30′ or beyond, it is again clear that a one-third and two-thirds ratio does not apply. With an f/22, the DOF range is 5′ to 30′ when focused at 9′. That is 4′ in front of the point of exact focus and 21′ behind the point of exact focus: almost a one-sixth and five-sixths ratio.

However, when the maximum background distance in the field of view is reduced to between 5′ and 20′, the one-third and two-thirds ratio is pretty closely approximated.

When closer than 3′, the ratio approaches ½ and ½, which is certainly the DOF ratio for all close-up photographs. As much will be in focus in front of the plane of exact focus as will be behind it. This is extremely important to remember if you photograph fingerprints on a curved surface. Do not be misled by the Rule of Thumb one-third and two-thirds DOF range in this situation.

Finally, if you take many photographs through a microscope, you should know that the DOF range changes again, and this time the DOF range approaches two thirds in front of the plane of exact focus and just one third behind the plane of exact focus.

Back to the rule of thirds as a focusing technique: This author has crunched the numbers at each distance between 3′ and 30′ and with the f/stops f/8 through f/22. The results follow. Do not be too concerned with these numbers and percentages. It is not necessary to memorize them. A very easy method to zone focus will soon be indicated. This information is presented to further indicate that the one-third and two-thirds ratio of the area in focus around the point of exact focus is not very precise.

- If 30′ is the background distance, the average ratio of foreground-to-background area in focus is 20% in front and 80% behind.
- If 20′ is the background distance, the average ratio of foreground-to-background area in focus is 28% in front and 72% behind.
- If 15′ is the background distance, the average ratio of foreground-to-background area in focus is 28% in front and 72% behind.
- If 10′ is the background distance, the average ratio of foreground-to-background area in focus is 36% in front and 64% behind.
- If 5′ is the background distance, the average ratio of foreground-to-background area in focus is 40% in front and 60% behind.
- If 3′ is the background distance, the average ratio of foreground-to-background area in focus is 40% in front and 60% behind.

How does this help us zone focus if our lens does not have a DOF scale? Representative images of each of the preceding scenes will make it more obvious.

Figure 3.27
Zone focused at 30'.

Figure 3.28
Zone focused at 20'.

Figure 3.29
Zone focused at 15'.

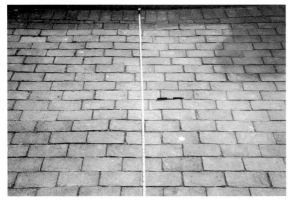

Figure 3.30
Zone focused at 10'.

Figure 3.31
Zone focused at 5'.

In each of these images, a pen was placed at the prescribed distance indicated in the preceding bulleted list. Even as the precise ratio of what is in front of the plane of exact focus and what is behind the plane of exact focus changes with each distance, one very obvious similarity exists in each of the photographs. The pen appears to be positioned at approximately the midpoint of each image. The pen appears to be positioned halfway between the top of each image and the bottom of each image. If fact, it is not, it just appears that way, because of the perspective view we have of each scene.

How can we put this information to use at crime scenes?

If you have a DOF scale, use it. That is the easiest way to maximize the DOF whether you are hyperfocal focusing or zone focusing. But if you are in a zone focusing situation, and your lens does not have a DOF scale, the quickest, most accurate method of focusing to maximize the DOF is to focus at the distance that is midway between the top of the composed image and the bottom of the composed image.

Crime scene photography is already a very stressful situation. Rather than making crime scene photography procedures more difficult, a good text should make them easier to understand and apply. The best way to do zone focus is to focus at a distance that appears to be midway between the top and bottom of the composed image.

Let us make this a revised Rule of Thumb.

Rule of Thumb 3.4: When attempting to maximize the DOF with crime scenes ranging from 5' to 30', the most effective way to do this is to focus at a distance that appears to be midway between the top and the bottom of the composed image.

INFRARED AND ULTRAVIOLET FOCUS ADJUSTMENTS

We have been dealing with variations of focusing visible light until now. After all, when we view the real crime scene and meter it through the camera viewfinder, we are dealing with visible light. However, crime scene photographers sometimes have the need to photograph evidence currently lit by other forms of light: sometimes ultraviolet (UV) light, sometimes infrared (IR) light, and sometimes fluorescent light. Chapter 6 will deal with more on these topics, but for the purposes of this chapter dealing with focusing issues, a brief discussion on focusing for unusual lighting is warranted.

Outside the visible light spectrum, UV and IR are forms of light that are not visible to the eye. However, by use of the proper photographic equipment, we can still capture images of evidence stimulated by light in these ranges. However, if we cannot see evidence lit by UV or IR light, how are we to ensure such images will be in focus?

Fortunately, lens manufacturers usually provide a handy "tool" for us to work with. Most lenses have an IR correction point indicated on the distance scale.

In the following images, the number "2" is a distance indication in feet; the 0.6 is a distance indication in meters. The 3rd line is flanked by the DOF scale numbers. The fourth line shows an indication of which f/stop number has been set on the camera, in this case an f/11. Figure 3.32 shows that the camera has been focused at 2′. In both images, notice the small dot just to the right of the right "4" on the DOF scale, which is the IR focus correction indication. To adjust focus for IR light, the photographer must first determine the correct focus while viewing the scene or object with visible light, which is what Figure 3.32 shows. Once focus has been determined for visible light, whatever distance has been focused on is then rotated to the IR red focus dot. In this case, it is the 2′ point on the distance scale. The 2′ point on the distance scale is then rotated so that the 2′ point is now aligned with the IR dot. Figure 3.33 shows this adjustment has been made.

If you normally use a zoom lens, the IR adjustment indicators will look somewhat different, allowing you to adjust for IR light and the particular focal length you have chosen to set your camera at.

Figure 3.34 shows a zoom lens focused for a 5′ distance in visible light. In this case, the 5′ indicator is aligned with the exact focus mark. It would not matter which focal length was chosen; when focused for visible light, this is the result. When it is necessary to focus for IR light with a zoom lens, it is essential that whatever focal length of lens is being used, the IR adjustment duplicates this focal length choice. With Figure 3.35, we can see five (5) IR correction choices, and in this case, the 5′ distance mark has been adjusted so it is opposite the 50-mm indicator.

These adjustments have to be made whenever you have your camera loaded with IR film. A few digital camera sensors have the ability to detect IR light, and these cameras can manually or automatically focus for IR light just like visible light. These cameras are the exception, however, so you have to know your equipment. It is usually clear if any particular digital camera has this capability.

If you were to photograph an item of evidence with UV light, a similar focus adjustment is necessary. For instance, a severe bruise or bite mark may have healed and can no longer be seen with visible light. It may be possible to photograph it with UV light, because UV lighting can sometimes visualize a deep muscle bruise

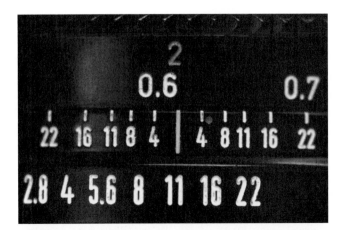

Figure 3.32
Focused on 2′–visible light.

Figure 3.33
Focused on 2′–IR light.

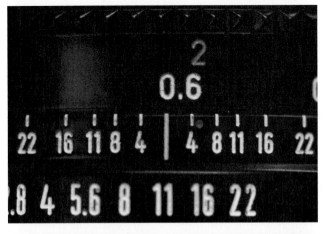

Figure 3.34
Focused on 5′–visible.

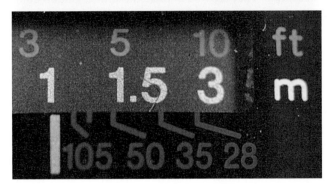

Figure 3.35
Focused on 5′–IR light.

that is no longer readily apparent in visible light. In this case, however, the focusing adjustment is just the opposite as is necessary with IR light. Unfortunately, no UV correction mark is shown on any lens like an IR correction mark. How then can we make this necessary adjustment? First, focus on the evidence with visible light, which might be the surface of skin bitten or bruised several weeks or months previously. If you were going to do an IR focus adjustment, notice the distance to the right that you would have to rotate the distance scale until it aligned with the IR correction point. Now, for a UV adjustment, rotate the distance scale this same distance, but do it to the left instead of to the right.

We previously mentioned bracketing exposures when it was uncertain what the correct exposure would be. When photographing evidence in the IR or UV range of the electromagnetic spectrum, it would be wise to bracket with your focusing technique also. Determine what is the probable focusing adjustment for either IR or UV and capture that image. Take another photograph with the focus ring rotated just a bit more to the right, and take another photograph with the focus ring rotated just a bit more to the left of the original focus position. This should be done with film cameras. If you are using a digital camera, it will immediately be evident if the focusing has been correct and whether fine-tuning with the focus is required.

OUT-OF-FOCUS IMAGES?

It is unusual to teach a course on crime scene photography without receiving several photographs from a student that are out of focus, which is pointed out to the student, and they are asked to retake the photographs. When this is done, the photographs sometimes come back out of focus again. It becomes clear that the student is unaware their vision needs correction. If they have never seen anything with clear sharp focus, they do not know the difference and turn in what looks to be the best to them. However, several students have tried to defend their out-of-focus photographs by suggesting it must have been an "equipment malfunction."

This is how the conflict can be resolved. The student is set up with a digital SLR camera on a tripod, with the film plane parallel to the object they have had trouble with. They are asked to manually focus the camera as best they can and take the photograph. With the camera in the same position, they are told to switch to the auto-focus mode. They are told to press the shutter button down half way and listen closely to determine whether they can hear the camera readjust focus. They then capture that image. Comparing the two images usually can convince them focusing by eye results in an out-of-focus image whereas the camera's auto-focus results in a better image. Solution: the student is told to go get glasses or have their current prescriptions updated.

Sometimes the student swears their vision is good, but whenever they look through the camera viewfinder, everything always looks blurry, which may actually happen for a very good reason. This is an age when most people make a purchase and leave the directions, or in this case the camera manual, in the box, go out, and start taking pictures. Most cameras have a feature to allow individuals to adjust the viewfinder for their particular eye quality. It is called the diopter adjustment dial. It is usually located just to the right of the viewfinder. See Figure 3.36. It may be a dial or thumb slide that moves up and down. In either case, it allows the photographer to adjust the viewfinder for his or her own eyes. Usually, it is recommended to select an object, focus on it as best you can, and then adjust the diopter both ways to see whether doing so improves focus for your eye.

It is, more or less, the equivalent of trying on various strengths of reading glasses and is often the solution to some focusing issues.

DEPTH OF FIELD

FACTORS AFFECTING DOF

Three camera variables directly affect the DOF range. They are:

- F/stop choice
- Lens choice
- Camera to subject distance

We will examine each in turn.

F/Stop Selection as a DOF Variable

As previously explained, f/stops are a primary exposure control. Along with shutter speeds, they are one of two primary camera controls that affect exposures.

In addition to being an exposure control, f/stops also are one of the camera variables that directly control the DOF range. This cannot be overemphasized.

DOF, by definition, is the variable range, from foreground to background, of what appears to be in sharp focus.

Figure 3.36
Diopter dial.

When considering DOF, it is important to remember that it is possible to maximize the DOF range, and it is possible to minimize the DOF range. Obviously, because one of the Cardinal Rules is to maximize the DOF as much as possible, that is our main concern with crime scene photography.

However, understanding the effect of minimizing the DOF range is valuable. This text will offer two situations when minimizing the DOF will be a benefit to the goal of the crime scene photographer. Both relate to intermediate objects/obstructions between the photographer and the subject matter considered the primary subject. In such a situation, if the photographer can create a small DOF range and place it on a distant subject, intermediate details will fall outside of the DOF range and become blurred and out of focus. This intentional blurring of intermediate detail can be so pronounced as to make the intermediate detail so blurred as to become almost invisible. Making an intermediate obstruction so blurred as to become almost invisible allows us to see the primary subject better. The two situations when this is a manifest advantage are:

- In daytime surveillance situations, positioning the camera behind a bush or other obstruction is necessary to prevent the photographer from being seen by the subject of the surveillance. In this case, having the bush in front of the camera is both good and bad. It is a good visual obstruction in that it makes it difficult or impossible for the subject of the surveillance to see the photographer. It can be bad if the same bush makes it difficult for the camera to acquire a good image of the subject of the surveillance.
- In a heavy rain or snowstorm, the rain or snowflakes can be so thick as to make it difficult to see and photograph the crime scene and the evidence within the scene. Because the crime scene is usually fixed and is not moving, ways exist to blur the falling rain and snow flakes, sometimes to such an extent that they cease to appear in our photographs.

Both of these situations will be discussed later.

What is the effect of using a wide aperture? With a wide aperture, the DOF range can be very short. One object/person can appear to be in focus, with everything else, both in front of and behind that object/person, appearing to be out of focus. This short DOF range can be placed in the foreground, in the mid-ground, or in the background. Wide apertures like f/2, f/2.8, and f/4 create very short ranges of DOF.

Figure 3.37 shows the effect of using an f/1.8, a very wide aperture, while focusing on the #1 in the photograph. The #1 is obviously in focus, whereas the focus quickly deteriorates behind the #1. Figure 3.38 moves the point of focus to the #2. Now, both #1 and #3 appear "soft" or out of focus. Figure 3.39 moves the point of focus to the #3. Both #1 and #2 appear "soft" or out of focus.

Figure 3.40 shows the effect of using an f/22, a very small aperture, on the same scene. By zone focusing, the entire area appears to be in focus. All three numbers are sharp.

Figure 3.37
Focused on #1.

Figure 3.38
Focused on #2.

Figure 3.39
Focused on #3.

Figure 3.40
Zone focused.

Recall Figures 3.15 and 3.16. When the aperture is wide, the circles of confusion perceived by the eye to still be in focus are relatively close to the point of exact focus. The DOF range, then, will be very small. When the aperture is very small, the circles of confusion perceived by the eye to still be in focus are relatively far apart. The DOF range, then, will be very long.

With either wide or narrow f/stops, we would be able to focus on an object/person at any distance, and they would be in perfect focus in the resulting photograph. As the two graphics show, with either a wide or narrow f/stop selection, the light rays can be brought into focus where the lines meet at the film plane. As the light rays travel from the lens opening, on the left, to the point of perfect focus, they are converging down their respective cones, until the light rays meet at the film. At the film plane the light rays are in perfect focus.

The DOF range will increase as f/stops are changed, moving on a continuum from f/2 to f/22. F/stops representing large lens openings produce relatively short DOF ranges, and f/stops representing small lens openings produce relatively long DOF ranges, which is purely a function of the way they bend light to converge at the film plane and the relative distance of the circles of confusion from the point of exact focus at the film plane.

In summary, considering an f/stop range from f/2 to f/22, which is common with most lenses, the f/2 results in the shortest DOF range, and the DOF range progressively increases as one moves toward the f/22 lens opening.

Lens Choice as a DOF Variable

How can the lens choice affect the DOF? Examine Figure 3.41.

The f/stop number has a relationship to the focal length of a lens. This equation shows the relationship: FFL/f-stop = DoD, where . . .

- FFL equals the focal length of the lens
- F/stop is the particular f/stop number
- DoD is the diameter of the diaphragm or the size of the lens opening.

This equation can also be written as either FFL/DoD = f/stop, or DoD(f/stop) = FLL.

1. The diameter of the diaphragm can be determined by dividing the FFL by the f/stop.
2. The f/stop number comes from dividing the FLL by the size of the lens opening.
3. The FLL is the product of multiplying the f/stop number and size of the lens opening.

This graphic is a visual demonstration that, at any f/stop, wide-angle lenses will have a better DOF range than "normal" lenses, and "normal" lenses will have a better DOF than **telephoto lenses**.

Another way of saying this is to state that the same f/stop number will result in different lens openings, depending on which focal length of lens is being used.

To make this clear, let us begin by examining an f/stop of f/8. Notice the relative DoDs that result with different lenses. In the example in Figure 3.41, lenses of 24 mm, 50 mm, and 100 mm are compared, but the theory would hold for any lenses.

Telephoto lenses: Lenses with focal lengths greater than 50 mm are regarded as telephoto lenses.

- The DoD for a 24-mm lens, set at f/8, would be 3 mm: FFL/f/stop = DoD, or 24/8 = 3 mm.
- The DoD for a 50-mm lens, set at f/8, would be 6.25 mm: FFL/f/stop = DoD, or 50/8 = 6.25 mm.
- The DoD for a 100-mm lens, set at f/8, would be 12.5 mm: FLL/f/stop = DoD, or 100/8 = 12.5 mm.

Because the DoDs of each lens is a different size, as light travels from the front of the lens and diaphragm opening to the film plane, figures similar to the graphic are produced. Even though the same f/stop has been selected for all three lenses, the effective size of the opening into the light will be entering the camera will change. As we have already determined with our examination of f/stops, the DOF range is improved with smaller lens openings, which is shown by the pairs of small ellipses shown in all three examples. They represent the circle of confusion perceived by the eye to still be in focus. In each example, the pairs of small ellipses

Figure 3.41

DOF as a function of lens choice I.

Depth of FIeld as a Function of Lens Choice

Since: F-Stop# = FLL/DoD (Focal Length of the Lens/Diameter of the Diaphram)

Then: DoD = FLL/F-stop#

Therefore, if 24 mm, 50 mm and 100 mm lenses are set to F-8, this is the result:

a.   The DoD for = 24 mm lens <u>at F-8</u> = 3 mm

b.   The DoD for = 50 mm lens <u>at F-8</u> = 6.25 mm

c.   The DoD for = 100 mm lens <u>at F-8</u> = 12.5 mm

(a)

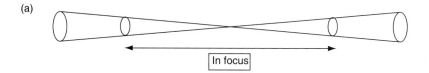

(b)

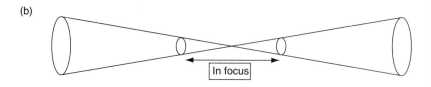

(c)

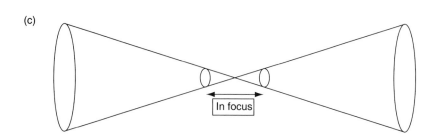

appear to be different distances from the point of exact focus or the point where the lines intersect in the middle of each figure. As the ellipses get farther apart, the DOF range increases.

Therefore, because wide-angle lenses produce smaller lens openings than "normal" and telephoto lenses, the resultant DOF range will be greater. Because "normal" lenses produce smaller lens openings than telephoto lenses, their resultant DOF range will be greater.

Camera to Subject Distance

How can camera to subject distance affect DOF? As you will recall, by hyperfocal focusing with an f/22, the resulting DOF range is 6′ to infinity. No other f/stop will provide more DOF when hyperfocal focusing. When zone focusing with a background of 30′, that same f/22 will result in a DOF range of 5′ to 30′, quite a substantial reduction in the DOF range. When zone focusing with a background of 10′, that same f/22 will result in a DOF range of 3.8′ to 10′. Again, the DOF range has been reduced. Lets jump to the situation where you need to capture a close-up photograph of a .45-caliber cartridge casing at a crime scene. Using close-up equipment, which will be explained later in this chapter, one can fill the frame with the cartridge casing. An f/22 can also be used in this situation. You would expect a pretty good DOF by using the f/22, would you not? If you find yourself nodding in agreement to that last sentence, you would be wrong.

A .45-caliber cartridge casing is approximately ¼″ wide. That means that the surface the casing is sitting on is ¼″ farther from the film plane than the top of the casing that is focused on. Surely the DOF range with an f/22 would cover this distance. Wrong again.

Figure 3.42 shows a .45-caliber cartridge flanked by two scales. The scale on the left has been raised up to the level of the top of the casing; the scale on the right has just been laid next to the casing. Figure 3.43 shows the same positioning of scales, but this viewpoint makes the difference in height between the two scales clear. As you examine Figure 3.42 more closely, do you notice any difference in the two scales? If you have very good eyes, you might notice that the word "inches" appears a bit "soft" or slightly out of focus. Why? Because the DOF range in a close-up situation like this is less than the ¼″ difference between the left scale and the right scale.

Lets make this more obvious by cropping in on the photograph.

Figure 3.44 is an extreme crop of Figure 3.42. Figure 3.44 shows the mouth of the casing and the background fabric. Various stria are plainly visible on the mouth of the casing, because focus was adjusted for the top of the casing. Notice in this image that the fabric is more obviously "soft" from being a bit out of focus.

Figure 3.45 is the same composition as Figure 3.44, but in this image, the focus was adjusted to be on the right scale, laying on the surface of the cloth. Now, the

Figure 3.42

.45 Casing with two scales.

Figure 3.43

.45 Casing with two scales.

Figure 3.44

Focused on the top of the casing.

fabric is much sharper, but the stria on the mouth of the casing has become blurred. Why? The difference is simply because the DOF range with an f/22 does not encompass both distances when close-up photographs are taken.

An f/22 can provide a DOF range of 6′ to infinity when we hyperfocal focus, and this same f/22 has less than a ¼″ DOF range when extreme close-ups are taken.

This should make us extremely cautious whenever we take close-up photographs, and this caution is well deserved. As we get closer and closer to our subject matter, our DOF range begins to collapse.

It would now be good to relate two seemingly unrelated topics: focusing on infinity and close focusing on small objects. They are related. Focusing on infinity wasted DOF in the foreground; hence, our recommendation is to hyperfocal

Figure 3.45
Focused on the fabric.

focus instead. Both focusing on infinity and hyperfocal focusing ensured infinity would be in focus. What is the practical difference? Hyperfocal focusing provided for more in focus in the foreground.

With Figure 3.44 still fresh in our minds, we may see the link and realize that focusing on top of the cartridge casing may actually be wasting some of the available DOF range. If, in fact, the DOF range for extreme close-ups is approximately ½″ in front of the plane of exact focus and ½″ behind the plane of exact focus, then focusing on top of the cartridge casing wastes ½″ of the DOF range! It would actually be more effective to focus a little down the side of the cartridge casing. That way, the near DOF range could include the top of the casing, and the far DOF range would proceed further down the casing, which would be extremely relevant when a fingerprint is on the cartridge casing and we want to make sure as much of the fingerprint is in focus as possible, which would also be applicable whenever any critical evidence is on a curved surface and maximizing the DOF range was important. In these situations, bracketing by changing the focus point instead of the exposure value would be warranted. But, certainly, just automatically focusing on top of the cartridge casing, as we are all instinctively prone to do, would actually be counterproductive to the crime scene photographer wishing to maximize the DOF for a close-up image.

TIPS TO MAXIMIZE DEPTH OF FIELD

The DOF is the variable range, from foreground to background, of what appears to be in acceptable focus. That is its definition. When photographing crime scenes, accident scenes, and the evidence within those scenes, can you think of any time when we would intentionally want part of those photographs to be out of focus? For the most part, we should follow the Cardinal Rule: maximize DOF whenever possible. An out-of-focus photograph may not be acceptable in court as evidence, which will be emphasized again when legal matters concerning photographs are discussed. However, this cannot be overstated. Maximizing the DOF should be one of our constant concerns when engaged in crime/accident scene/evidence photography.

Here are several techniques to ensure the DOF is maximized. We have already dealt with three of them.

1. Hyperfocal focus when taking photographs of large outdoor crime scenes when infinity is in the background.
2. Zone focus when taking photographs of large crime scenes but infinity is not in the background, which will require having a DOF scale.
3. Focus by the rule of thirds when taking photographs of large crime scenes when infinity is not in the background, and you do not have a DOF scale.

A few more ideas will help ensure the DOF is maxed out most of the time.

Select the ISO film speed to maximize the DOF.

Because the f/stop selection is one of the critical factors affecting the DOF, and the small f/stops (f/22, f/16, f/11, and f/8) always provide for better DOF ranges than the wide f/stops (f/2, f/2.8, f/4), how do we choose a film speed that will ensure the use of these small apertures?

Commonly available film speeds are ISO 100, 200, and 400. Faster films, like ISO 800 and ISO 1000, are generally more expensive. So, what is the slowest film speed that will allow photography with the small apertures?

 Rule of Thumb 3.5: On bright sunny days, ISO 100 film will allow photography with the smaller f/stops. Otherwise, on cloudy or dark days, at night time, or indoors when using electronic flash, use a faster ISO film like ISO 400.

Some agencies have taken this recommendation, and modified it, and supply their crime scene photographers with only ISO 200 film, because "it is the perfect compromise between ISO 100 and ISO 400 films, and has the benefits of both." They say, "Why complicate things, when a satisfactory compromise exists to using two different kinds of film?" Budget bureaucrats who have never worked as crime scene photographers usually make such statements.

Although it is true that the difference between ISO 100 speed film and ISO 200 speed film is only 1 stop, and the difference between 400 ISO speed film and 200 ISO speed film is only 1 stop, a 2-stop difference still exists between the ISO 100 speed film and the 400 ISO speed film. Having films with a 2-stop difference available under certain circumstances is a huge benefit; however, it is all moot now. With the steady trend toward digital imaging at crime scenes, all crime scene photographers will soon have digital 35-mm SLR cameras as their standard tool and will be able to select any ISO film equivalent for each and every image. So, taking a defensive position on this issue is not needed. Soon, the crime scene photographer will be able to select the ISO equivalent for each shot.

However, why was ISO 100 speed film recommended for bright sunny days? If you recall the f/16 sunny day rule, it recommended using an f/16 on a bright sunny day and converting the chosen film speed into a shutter speed. If ISO 100 film was used, the shutter speed became 1/125th of a second. By the theory of reciprocity, f/16 and 1/125 could be changed to f/22 and 1/60th of a second. Better DOF and the same exposure.

For other lighting conditions, ISO 400 film was recommended, because in those circumstances, one would be able to close down the aperture 2 stops from what a 100 ISO film could provide. For example, on an overcast day, an f/8 was recommended with a 1/125th of a second shutter speed if using 100 ISO film. By reciprocity, this could be changed to 1/60th of a second shutter speed and f/11. When an ISO 200 speed film is used, the f/8 would be paired with a 1/250th second shutter speed, and reciprocity would allow this to be changed to 1/60th of a second shutter speed and f/16, a 1-stop improvement. However, if ISO 400 film was being used, the beginning point would be f/8 with a 1/500th of a second shutter speed, which by reciprocity could be changed to 1/60th of a second shutter speed and f/22; the same exposure with a better DOF.

In 1988, this author was attending the Virginia Forensic Science Academy, a 10-week school put on by Virginia Forensic Laboratory personnel and trainers to teach crime scene investigators how to properly identify, process, and package evidence from crime scenes so that when it does eventually arrive at the laboratory, their examiners and analysts could do the most with it. The school included a crime scene photography component. During this training, the photography instructor, Norm Tiller, was attempting to explain this concept of selecting film ISOs to ensure one would eventually be able to select small apertures for each photograph, thereby maximizing the DOF range for each shot. This synthesis of Norm's ideas became "Tiller's Rule of Film Selection to Maximize DOF."

When considering loading any particular ISO film into the camera, before actually loading it into the camera, do this. Dial in the proposed ISO film speed into the camera. Set the shutter speed to 1/60th of a second. Set the f-stop to f/8. Take a meter reading. If the meter reading indicates a proper exposure or an overexposure, it is OK to actually load that film into the camera. That is because an f/8 is one of the small apertures we had been told to use mostly. And, if overexposed, the solution was to close down the aperture even more, ensuring an even better DOF range. If the exposure reading resulted in an underexposure, repeat the above sequence with a faster film speed until the exposure reading eventually results with a proper exposure or overexposure with F/8 and 1/60th of a second shutter speed.

This is still good advice when using a film camera.

However, as previously stated, with digital imaging this is all moot, because we can select a different ISO film speed equivalent for each and every shot we take. With film, it was more critical. We normally learned to live with the film we had loaded into the camera. When we changed from photographing evidence in a brightly lit outdoor area and moved inside to photograph more evidence, we frequently had half of a roll of film left. Typically, we would not switch out our film from the 100 ISO "outside" film to a 400 ISO "inside" film. Now, with digital imaging, we can.

Finally, it should be mentioned that many cameras have a DOF preview button. This needs some initial explanation. Most modern lenses are designed so that when mounted onto the camera body they are automatically set to the widest aperture the lens has. You may select a different f/stop to take a particular image, but until the shutter is depressed, the lens remains at its widest aperture. Once the shutter button is depressed, the lens shuts down to the selected f/stop. The shutter will immediately open up to its widest aperture after the image is captured. Two main reasons exist for this lens design feature. It is easier to see and compose the image when the aperture is wide and the scene is well lit. It is also easier to focus the camera when the aperture is wide and the scene is well lit.

If the lens has an f/stop range from f/2 through f/22, the lens is designed to be at the f/2 aperture all the time the camera is composed and focused. Then, just as the shutter is depressed, the lens momentarily shuts down to the pre-selected aperture as the image is captured. Immediately, the lens will return to the f/2 aperture so the next image can be composed and focused on.

The down side to this lens design is that the DOF with an f/2 is minimal, and we will only see one distance in focus, while the foreground and background will appear to be out of focus in the viewfinder.

To enable the photographer to be able to see the true DOF range, the DOF preview button allows the lens to shut down to the preselected f/stop before the picture is taken. The main complaint is that by having the lens shut down to an f/22 aperture, for example, so much light is cut out from the viewfinder that it is frequently too dark to see the now apparent DOF range. Many find this situation to be too frustrating to effectively use. So, many do not use the DOF preview button at all.

The DOF preview button is usually located low on the camera body to the left of the lens. Some cameras have this feature on the low right side of the camera body. Figure 3.46 shows one DOF preview button as is; Figure 3.47 shows the DOF preview button activated.

LENSES

LENS DESIGNATIONS

Walk into a camera store, and you will be amazed by the variety of cameras and lenses available. How do you intelligently compare one lens with another on the shelf right next to it? Each lens has aspects you need to be aware of. These are some of the most frequent variations.

Figure 3.46
DOF preview button, normal position.

Figure 3.47
DOF preview button, set to check it.

Focal Length

The **focal length** of the lens is, by definition, the distance in millimeters (mm) between the optical center of a lens and the film plane when the camera is focused on infinity (∞).

This aspect of the lens being focused on infinity is important, because many lenses extend in length as their focus changes from infinity to their closest focusing distance.

Figure 3.48 shows a 50-mm lens when focused on infinity; Figure 3.49 shows the same lens when focused at its closest focusing position. As the lens is focused, various optical elements within the lens may change positions. When focused on infinity, the lens is at its most compact configuration. The optical center of the lens elements would actually be farther than 50 mm when the lens is focused at any other point. It should be mentioned here that many modern lenses focus internally, without any extension of the end of the lens. The definition of focal length was formed before this modern variation of lenses. It would still be correct, because the lens is always at its most compact configuration.

As lenses of different focal lengths are considered, it should now be obvious that their relative lengths will differ with the focal length of the lens, which is easily demonstrated by seeing how a zoom lens changes configurations as it is set to different focal lengths. The lens in the following images, by the way, focuses internally, so it does not matter that the focus has not been dialed to infinity.

The focal lengths of these five images begins with 28 mm on the left, then proceeds through 35 mm, 50 mm, 70 mm, and finally 105 mm. As we would expect, the length of the lens progressively gets extended more and more.

Figure 3.55 is a Sigma 800-mm lens. It is a monster. But, if you want to get up close and personal with a bad guy 400 feet away from the camera, there is no alternative.

A 50-mm lens has 50 mm in distance between the optical center of the lens and the film when the lens is focused on infinity. With a 100-mm lens, that distance is

Focal length: The distance in millimeters (mm) between the optical center of a lens and the film plane when the camera is focused on infinity (∞).

Figure 3.48
Focused on infinity.

Figure 3.49
Focused for close detail.

100 mm between the film and the optical center of the lens. That is why telephoto lenses are longer than most other lenses.

"Fast" and "Slow" Lenses

When shopping for a lens, the dealer will sometimes refer to a lens as being "fast" or "slow." Lenses themselves do not have "speeds," but they do affect the ultimate shutter speed that is usable for a particular photograph.

What the lenses for 35-mm cameras do have within them is an adjustable diaphragm, which can be opened to a wide aperture or closed down to a smaller

Figure 3.50
Zoom lens set to 28 mm.

Figure 3.51
Zoom lens set to 35 mm.

Figure 3.52

Zoom lens set to 50 mm.

Figure 3.53

Zoom lens set to 70 mm.

Figure 3.54

Zoom lens set to 105 mm.

Figure 3.55

An 800-mm lens (Courtesy of Tom Beecher, Photografix, Richmond, VA).

aperture to vary the amount of light that can pass through the lens toward the film. Adjusting this diaphragm is usually referred to as selecting the f/stop. F/stops are one indicator of the amount of light that is allowed to reach the film.

Together, the shutter speeds and the f/stops control the total quantity of light that is allowed to ultimately reach the film for an exposure. The light intensity can be controlled by volume, f/stops; and time, shutter speeds.

If a larger volume of light can be transmitted by any particular lens, a proper exposure may be obtained by using a shorter or "faster" shutter speed. If a smaller volume of light can only be transmitted by a particular lens, in the same lighting conditions, that volume will have to be transmitted for a longer time to arrive at the same exposure. It is basically a balancing act. With more light volume or larger diaphragm openings, less time or shorter shutter speeds will be needed for a proper exposure. With less volume or smaller diaphragm openings, more time or longer shutter speeds will be required for the same proper exposure.

An f/stop of f/2 is regarded as the pivotal f/stop for designating lenses as being either "fast" or "slow." If a lens has a wider maximum lens opening than f/2, it is called a "fast" lens, because by letting in a larger volume of light, less time, or a "faster" shutter speed is needed for a proper exposure. An f/1.8 would be an example of a "fast" lens. "Fast" lenses are more expensive, because photographers value them for enabling nonflash photographs in dim light, thereby maintaining a natural look to the scene. This aspect of "fast" lenses, valued by most professional photographers, does not have the same value to crime scene photographers, because wide apertures produce the worst DOF ranges, and we are usually concerned with maximizing our DOF range.

If a lens has a smaller lens opening than f/2 as its maximum lens opening, it is called a "slow" lens, because it will require the shutter to remain open longer for the same exposure. An f/4 is regarded as a "slow" lens.

This becomes important when shopping for lenses. The same focal length lens may be available with different maximum apertures. Expect to pay more for the "faster" lenses. If purchasing lenses for your agency, be comforted that "slow" lenses are perfectly suited for crime scene work, and you will save your agency some money.

Lenses Are Designated by Their Widest Apertures

Because of the premium placed on the widest aperture of any particular lens, all lenses are designated by their widest aperture. It is marked on the lens barrel, and it will be marked on the packaging material. One might go shopping for a 50-mm f/1.4 lens that is regarded as very "fast" but have to settle for a more reasonably priced 50-mm f/2 lens.

Zoom lenses will be designated by their widest aperture at each extreme in their focal length. A typical zoom lens that provides for focal lengths between 35 mm and 105 mm may be designated as a 3.5/4.5 lens. That means the widest aperture at the 35-mm setting is f/3.5 and the widest aperture at the 105-mm setting is f/4.5. Zoom lenses, therefore, are not normally regarded as "fast" lenses.

THE "NORMAL" LENS

Different lenses may make the scene viewed in the viewfinder appear to be different than the eye sees the scene. A 50-mm lens is considered to be the "normal" lens for a 35-mm SLR camera, because it portrays the scene in exactly the same way as the eye saw it. (It must be mentioned that different camera formats will have different "normal" lenses. For instance, the "normal" lens for a medium-format camera will be an 80-mm lens. However, most crime scene photography is done with the 35-mm camera format.) A distant object viewed through a 35-mm camera with a 50-mm lens looks the same distance away as when we directly look at it. A near object, viewed through a 35-mm camera with a 50-mm lens looks to be the same distance from us as when we look directly at it. The relative distance between the near and far objects looks the same to the eye as when viewed through a 35-mm camera with a 50-mm lens on it.

Determining the "normal" lens for a camera system is "officially" supposed to be determined by using the diagonal of the film negative; 35-mm film is 24 mm × 36 mm, and Pythagorus would remind us that the hypotenuse of any of these right triangles would be about 43 mm. He would be correct. For some reason, 35-mm camera manufacturers have always called the 50-mm lens the "normal" lens for this camera format. With the 50-mm lens on the camera, view the scene through the viewfinder. Immediately look up above the camera at the scene. Nothing should have changed. Close objects will look the same distance through the viewfinder as they look to the eye. Distant objects will look the same through the viewfinder as they look to the eye. The 50-mm lens has the world look the same when looking through the viewfinder and when looking at the world without a camera in front of your face.

However, some new cameras have viewfinders with magnification ratios. The Canon EOS 5D digital camera has 0.71 magnification when looking through the viewfinder. Essentially, that camera will give the viewer a "normal" view (50 mm) through the viewfinder only when set at the 70-mm lens focal length; 0.71 mag. × 70-mm lens = 49.7-mm equivalent.

The 50-mm lens, then, is the focal length of choice when it is a crime scene photographer's intention to offer the photograph of the crime scene in court and maintain that the photograph is "a fair and accurate representation of the scene." This deserves to be Rule of Thumb 3.6.

Rule of Thumb 3.6: The use of a different focal length lens will introduce lens distortion into the photograph. Wide-angle lenses will elongate the distance between the foreground and background. Telephoto lenses will compress the distance between the foreground and background. Only the 50-mm lens will capture the image without this distance distortion.

One other aspect of the 50-mm lens is that it only provides a view of the world that is about 46° of the field of view. Items farther to the left and to the right will not be included in the field of view unless you back up. Then they will appear smaller, because they are now farther away. If, however, it is our intent to eliminate extraneous objects from our field of view, all we need do is move closer to the primary subject until the extraneous items are outside of the 46° viewpoint.

Figure 3.56 shows various angles of view of selected lenses. Lenses range from an ultra-wide fisheye lens that has a 180° view of the world to a super-telephoto 1200-mm lens with just about a 2° view of the world. Lenses with focal lengths less than 50 mm are referred to as wide-angle lenses; lenses with focal lengths more than 50 mm are referred to as telephoto lenses.

TELEPHOTO LENSES

Lenses with a focal length longer than 50 mm are called telephoto lenses. They can range from 60 mm to 1200 mm. The optical centers of these lenses are farther from the film plane than a 50-mm lens' distance. This difference produces distinct

Figure 3.56

Various angles of view of some lenses (Courtesy of Jeff Robinson-scamper.com).

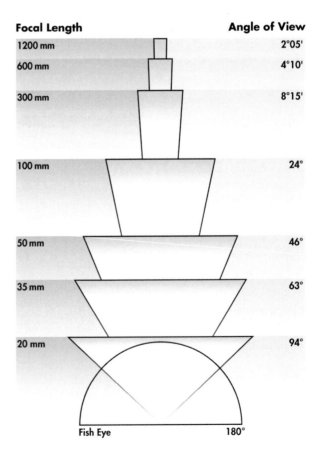

Focal Length	Angle of View
1200 mm	2°05'
600 mm	4°10'
300 mm	8°15'
100 mm	24°
50 mm	46°
35 mm	63°
20 mm	94°
Fish Eye	180°

changes to a photograph, compared with the 50-mm lens look. These differences are as follows:

- Magnification
- Narrower field of view
- Compression of foreground to background distances
- Narrower DOF range

Magnification

Telephoto lenses are perhaps best known for their ability to magnify distant objects. This telescopic effect can be determined by dividing the focal length of the telephoto lens by 50 mm. Therefore, a 100-mm lens would be the equivalent of a 2× telescope; a 300-mm lens would be the equivalent of a 6× telescope. The distant subject would be magnified by two times and six times, respectively. This magnification ability is perhaps best suited to surveillance photography, when it is necessary to identify distant subjects by photographing them while the photographer maintains a safe distance so they will not be noticed.

Compare two scenes in Figures 3.57 and 3.58, one taken with a 50-mm lens, the other an enlargement of the first.

Figure 3.57
50-mm view, subject 400′ away.

Figure 3.58
50-mm view, subject 400′ away, cropped and enlarged.

In the center of Figure 3.57, you may be able to notice a small vague dark-jacketed person. It is extremely difficult to do so, because the person is standing 400′ from the camera, and the image was taken with a "normal" 50-mm lens. A 50-mm lens is certainly not sufficient to resolve the detail of the subject. Figure 3.58 confirms this. It is a crop and enlargement of Figure 3.57, showing the subject "better." Resolution is not adequate to make out the subject, other than to suggest the subject may possibly be a blonde female. Individual characteristics are lacking altogether.

Figure 3.59 is an image of the same subject standing the same distance from the camera at night. The difference here is that an 800-mm lens was used. Although the subject appears to be underexposed, she is certainly much larger. Figure 3.60 shows the same image digitally cropped and enlarged, and the under-exposure has been corrected. The resolution is sufficient to make out the face and to determine the letters and numbers on the license plate held by the subject. It appears the tag has the frame with a name on it: Jerry's. Certainly, this focal length lens has adequate magnification for an identification of the individual. For this reason, Rule of Thumb 3.7 has been established.

 Rule of Thumb 3.7: In surveillance situations, to be confident magnification will be adequate to recognize either a subject's face or to read a license plate, you

Figure 3.59

800-mm view, nighttime, subject 400′ away, cropped and enhanced.

Figure 3.60

Cropped further, enlarged, and enhanced.

should use a lens that has a focal length of 2 mm of lens per foot of distance between the camera and the subject.

With Figure 3.59, an 800-mm lens was used for a distance of 400′ between camera and subject.

To understand this Rule of Thumb better, examine the following images. On the campus of The George Washington University, one might expect to run into George several times. One of his statues is life-sized. Figure 3.61 shows George viewed by a lens that had 2 mm of lens for each foot between camera and statue. From 50′ away, a 100-mm lens was used. George is the size on the image that is recommended for surveillance situations. When his face is cropped and enlarged, we would still expect to be able to recognize it. It should be pointed out that these images were taken with midday lighting. As the lighting gets more and more problematic, the difficulties would increase. But, even in a nighttime situation, 2 mm of lens should be able to resolve distinguishable features. Figure 3.62 was taken from the same position, but with a zoom lens adjusted to about 50 mm: 1 mm of lens per foot of distance. The face is a bit smaller in the image. We would expect enlargements that had the face recognizable to be just a bit more difficult, especially in darker conditions. Figure 3.63 was taken from the same position, but with the zoom lens adjusted to about 24 mm: 0.5 mm of lens per foot of distance. The face is smaller yet, compounding any attempt to resolve the face so it is recognizable. Because capturing the facial features in surveillance situations is so critical, we cannot take chances of losing the detail we seek to capture in the first place. The use of 2 mm of lens per foot of distance guarantees this detail will be obtained.

From any viewpoint of an item of evidence, if it is desired to fill the frame with critical aspects of the object, moving closer to the object matter is usually advised. However, at times this is not practical. Other critical evidence may be in the way, prohibiting moving closer to the subject matter of the photograph. Or it may not be safe to step closer to the item of interest. At an arson scene, the floor may be weak just ahead of where the crime scene photographer is currently standing. In these situations, one should consider using a zoom lens in its telephoto setting, which will magnify the area of interest.

Care should be maintained that this use of a telephoto lens merely magnifies or enlarges critical details of the evidence. As explained previously, use of a telephoto lens can distort the relative perspective between objects. Therefore, a zoom lens cannot be used to show "a fair and accurate representations of the scene," specifically the distance between two or more objects. The use of a zoom lens to enlarge an individual element of the scene is permissible.

An example is helpful. At a remote fire scene with fatalities, an electric clock on a wall suggested the time of the fire when the fire damaged the electric cord and stopped the clock. The floor around the clock had completely collapsed. The clock could only be viewed from another room that had a wall also burned away. To photograph the clock, a telephoto lens was used. The clock was magnified and

Figure 3.61

2-mm of lens per foot of distance.

Figure 3.62

1-mm of lens per foot of distance.

Figure 3.63

0.-5 mm of lens per foot of distance.

filled the frame of the viewfinder. The telephoto lens was used for its strength: magnification. No other scene details were included in the composition to avoid the tendency of telephoto lenses to compress foreground-to-background distances.

Narrower Field of View

As can be seen in the preceding photographs, telephoto lenses also take in a narrower field of view than the 50-mm lens. Although the 50-mm lens takes in 46°, a 100-mm lens takes in only 24°, a 300-mm lens takes in only about 8°, a 600-mm lens takes in only about 4°, and a 1200-mm lens takes in only about 2° view of the scene. Recall Figure 3.56. This narrower field of view can be useful; by eliminating extraneous items to the left and to the right, the photographer can force the viewer to concentrate on the intended primary subject, rather than being distracted by a myriad of objects in the field of view.

Compression of the Foreground and Background

Because the background will be magnified with a telephoto lens, making it appear enlarged, the relative perceived distance between the camera and the background is necessarily compressed. Objects appear closer together than they really were. This perceived distance between the foreground (the photographer's position) and a magnified object in the background appears to be shorter.

Consider how this might affect skid marks at a fatal accident scene. If the photographer frames the scene from the beginning of 100′ of skid marks leading to where the vehicles are found resting together and uses a 50-mm lens to capture the image, the resulting photograph will duplicate the scene as the eye had seen it. If a 100-mm lens is used instead, the vehicles in the background will appear to be closer to the photographer than they had been in reality, and the perceived length of the skid marks will appear to be shorter than they originally were, which distorts reality. The photograph taken with a telephoto lens could not be used in court as "a fair and accurate representation of the scene as it appeared." An unscrupulous defense attorney, representing the driver who skidded, may be tempted to hire a photographer to take photos of the scene using a telephoto lens. The resulting shorter skid marks can be interpreted by the jury as suggestive of less speed just before the accident.

Figures 3.64 and 3.65 demonstrate this idea, although they are images of double yellow lines instead of skid marks. Figure 3.64 shows the length of the double yellow lines when photographed with a 50-mm lens, which is also the length of the lines as seen by the eye. Figure 3.65 shows the same double yellow lines photographed with a 100-mm lens. This lens compacts the apparent true length of the lines, making them appear shorter. Crime scene photographers need to understand that using telephoto lenses with vertical evidence will alter their "fair and accurate" appearance, so that they are not tempted to use these lenses in these situations. They also need to be aware of the effect of compacting foreground-to-background distances in case photographs presented by the defense look radically different from their own, so they can explain the differences to the judge and jury in court.

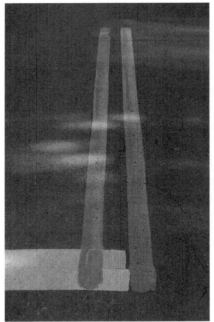

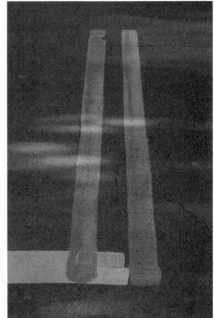

Figure 3.64
50-mm lens with double yellow lines.

Figure 3.65
Same double yellow lines with 100-mm lens.

Because of this compression effect, telephoto lenses are used in law enforcement mostly for their magnification qualities: to make it easier to recognize distant objects. The use of telephoto lenses to document overall crime scenes and overall accident scenes is strongly discouraged because of the relative distance distortion they create.

Narrower Depth of Field

As mentioned earlier, depth of field is the variable area, from foreground to background, of what appears to be in sharp focus. Another effect of using a telephoto lens is a reduction in the depth of field area. When using a telephoto lens, less appears to be in focus than when using a 50-mm lens. Why is this?

The same f/stop will produce different diameters of the diaphragm with different focal length lenses. Recall Figure 3.41.

Should you use an f/8 with a 24-mm lens, a 50-mm lens, and a 100-mm lens, the resulting aperture size will be different with each lens. With a telephoto lens, the diameter of the diaphragm will be the largest compared with either a normal lens or a wide-angle lens. The telephoto lens would, therefore, have its circles of confusion, which are still perceived to be in focus to the eye, closer to the plane of exact focus, which is the film plane. The result is a shorter DOF range.

When used in surveillance situations, this reduced DOF range is usually not a problem, because we will be focusing on relatively small areas or on individuals.

Even the widest apertures used with telephoto lenses will produce some DOF when distant scenes or subjects are focused on.

WIDE-ANGLE LENSES

Lenses with a shorter focal length than 50 mm are called **wide-angle lenses**.

Wide-angle lenses can range from a 35-mm lens to a true fish-eye lens, which can capture 180° of detail. The optical centers of these lenses are significantly closer to the film plane than a 50-mm lens' distance. This difference produces distinct changes to a photograph compared with the 50-mm lens look. These differences are:

- Wider field of view
- Elongation of foreground-to-background distances
- Increased depth of field

Wider Field of View

Wide-angle lenses naturally capture wider views of the scene. The relative increase in the field of view can be very close to what a normal 50-mm lens would capture when using a zoom lens set to about 40 mm or 45 mm. Remember that the true diagonal of a negative that is 24 mm high and 36 mm long is 43 mm. Therefore, when the scene is viewed from a zoom lens set at about 43 mm, the scene should look exactly the same as when you remove your eye from the viewfinder. If not, some magnification ratio may be set into the viewfinder optics. Check your camera's manual.

A 35-mm lens, and wider lenses, should begin to show more and more of the scene in the viewfinder, which becomes extremely handy when we have the duty to photograph wide objects, like the exterior façade of a large building. With just a 50-mm lens, we would have to take many more photographs. With a wide-angle lens, we would be able to capture the same overall detail with fewer images. The same would apply when faced with the need to capture interior overall photographs of the room that contains a crime scene. If we want to capture the full 360° view of all the walls, using a wide-angle lens would reduce the number of photos required. This idea will be revisited in Chapter 5.

Figure 3.66 shows the view of a kitchen while standing at the doorway and while using a 50-mm lens, which is the typical 46° view that a 50-mm lens offers. Figure 3.67 was taken from the same position, but with a 28-mm lens. Much more of the scene has been captured from left to right. With moderate wide-angle lenses, such as 35-mm and 28-mm lenses, the only perceived difference in photographs is this wider field of view. As much wider focal lengths are used, usually at around 20 mm and wider, another effect can also be noticed. Vertical lines at the left and right edges of the image begin to bow outward more and more.

Wide-angle lenses: Lenses with a shorter focal length than 50 mm.

Figure 3.66

Room photographed with a 50-mm lens.

Notice how the left and right edges of Figure 3.68 appear to be bowed outward. This image was captured with a 24-mm lens. The bowing outward of vertical lines at the edges of an image is referred to as barrel distortion and is used for its creative effects by many photographers. Crime scene photographers, however, usually try to avoid this effect.

Figure 3.67

Same room photographed with a 28-mm lens.

Figure 3.68

24-mm Lens producing barrel distortion.

Avoiding the widest of the wide-angle lens choices can accomplish this. Again, the 35-mm and 28-mm lenses are used most by crime scene photographers, because a distinct gain exists in the left-to-right field of view, without incurring the barrel distortion wider-angle focal lengths can sometimes produce.

At times, the barrel distortion is replaced by having the tops of vertical elements at the edges of the image appear to be leaning inward. See Figure 3.69.

Elongation of Foreground and Background

A wide-angle lens seems to make the background appear farther away than it really was. Going back to our example of an accident with skid marks, if the wrecked cars are now appearing to be farther away, the skid marks running from the camera's position to the vehicles appear to be longer than they really were. A juror may interpret longer skid marks as an indication of greater speed. Again, when "a fair and accurate representation of the scene" is the desired goal of the photographer, wide-angle lenses should not be used.

Figures 3.70 and 3.71 are the same double lines previously considered in the section on telephoto lens effects. Figure 3.70 is a duplicate of Figure 3.64: a 50-mm lens used to capture the double yellow lines. Figure 3.71 shows the same double yellow lines captured with a 28-mm lens. The lines appear to be stretched longer than they appeared to the eye when the photograph was taken, which is certainly not an accurate representation of the scene as it appeared to personnel at the scene. Figure 3.70 could be used in court; 71 should not be used in court.

As with telephoto lenses, one can use a wide-angle lens for its strength, without incurring its weakness. With telephoto lenses, their strength is magnification. Their weakness is perspective distortion in the form of foreground-to-background compression. The strength of a wide-angle lens is its ability to capture more detail to the left and to the right. The weakness of the wide-angle lens is perspective distortion in the form of foreground-to-background elongation. How can we use

Figure 3.69
Edges tilting in with wide-angle lens.

Figure 3.70
50-mm Lens with double yellow lines.

Figure 3.71
Same double yellow lines with 28-mm lens.

its strength without also having our image suffer from its weakness? If we are careful with our composition, by carefully eliminating any foreground-to-background detail in the field of view, a wide-angle lens can be used for its ability to capture more details to the left and right.

Figures 3.72 through 3.74 show foreground-to-background distance variations with different focal length lenses. Figure 3.72 was taken with a 28-mm lens; Figure 3.73 was taken with a 35-mm lens; Figure 3.74 was taken with a 50-mm lens. All were photographed from the same location. Because the wide-angle lenses show the buildings apparently farther away from the photographer, a viewer of only one of the images may incorrectly assume the distance between photographer and the buildings was greater that it is in reality, which is the distance distortion to be avoided with wide-angle lenses. The 50-mm lens may not show as wide a scene as desired. Can we have wide views without foreground-to-background distortion? Yes. See Figure 3.75.

Figure 3.75 shows the necessary composition adjustment required to have the benefits of the wide view with a wide-angle lens without having any foreground-to-background distortion. Eliminate the foreground. Just raise the camera up until the bottom edge of the viewfinder includes the bottom of the structure you want to photograph. Better to have a bit more sky in the composition than distorted foreground.

Figure 3.72
Buildings and foreground,
28-mm lens.

Figure 3.73
Buildings and foreground,
35-mm lens.

Figure 3.74
Buildings and foreground,
50-mm lens.

Figure 3.75
Building without foreground,
35-mm lens.

For those with a keen eye, you may also notice that Figure 3.75 looks a bit different than Figure 3.73. The edges of the building are no longer tilted in quite as much. Although it is not within the scope of this text to serve as a how-to book on Photoshop processing techniques, letting the reader know what is available is relevant to this text. Photoshop CS2 enables some lens distortions to be corrected. Figure 3.75 has had some of its wide-angle lens distortion corrected so the building sides and trees at the left and right of the image are not leaning in toward the center quite as much.

Increased Depth of Field

Another effect of using a wide-angle lens is an increase in the depth of field. There will appear to be more in focus between the foreground and the background than when using a 50-mm lens. This is again a result of the equation, FFL/f/stop = DOD. The same f/stop used with different lenses will produce different diameters of the diaphragms: 100 mm/f/8 results in a DOD of 12.5 mm; 50 mm/f/8 = 6.25 mm; and 24 mm/f/8 = 3 mm. Light coming through a smaller aperture will result in the circles of confusion, perceived by the eye to be in focus, to be farther from the plane of exact focus at the film plane. Recall Figure 3.41.

Unfortunately, this increased DOF is rarely usable by crime scene photographers, because at the same time a wide-angle lens is causing the perspective distortion previously mentioned. Despite our Cardinal Rule to maximize DOF, we cannot use wide-angle lenses for most of our photography.

MACRO LENSES

Telephoto lenses provide magnification of distant objects, but if the need is to magnify small objects to fill the frame of a photograph, other types of magnification are needed. To put this into perspective, consider that the normal 50-mm lens, when it is focused to its minimum focusing distance, will fill the frame of the viewfinder with an object that is approximately 6″ × 9″. See Figure 3.76.

Many 50-mm lenses have about 18″ as their closest focusing distance. If you position the camera closer than 18″ to the object being photographed, the lens will not focus. Therefore, anything smaller than 6″ × 9″ will require additional enlarging to truly fill the frame.

Magnification with a 1:1 Ratio

Another way this is usually explained is by saying that a typical 50-mm lens only has a 1:7 magnification ratio. That means that an object the size of a negative,

FOCUS, DEPTH OF FIELD, AND LENSES 187

photographed with a 50-mm lens at its minimum focusing distance, will be 1/7th of its real size on the negative or a print made from that negative without enlargement. Or seven objects the size of a negative will fill the frame when photographed with a 50-mm lens focused to its closest focusing distance. Figures 3.77 to 3.79 will help clarify this.

A nickel is about the size of a negative. Figure 3.77 shows this. Figure 3.78 shows that the nickel photographed with a 50-mm lens at its closest focusing distance looks small. Figure 3.79 shows seven nickels aligned vertically, and they fill the frame nicely when photographed with a 50-mm lens focused to its minimum focusing distance, which does not comply with our Cardinal Rule to fill the frame whenever possible.

Some sort of magnification is required to make a small item of evidence appear larger when photographed, rather than just attempting to enlarge the small nickel-sized object in the darkroom. If magnification is required, how much magnification should we have available? Most law enforcement agencies seem to have decided that providing their crime scene photographers with equipment that produces a **1:1 magnification** is optimal. A 1:1 magnification is obtained when an item of evidence about the size of a nickel is life-sized on the negative.

1:1 Magnification: When an item of evidence about the size of a nickel is life-sized on the negative.

An item the size of a nickel will fill the frame when viewed through the viewfinder. Figure 3.80 shows what a nickel will look like through the viewfinder, and how a print of that image will look, if photographic equipment providing a 1:1 magnification ratio is used.

If equipment with a 1:2 magnification ratio is used, the nickel-sized object will be one half of its real size on the negative. If equipment with a 1:3 mag-

Figure 3.76

50-mm lens' minimum focusing distance: 6″ × 9″.

Figure 3.77
Nickel on negative.

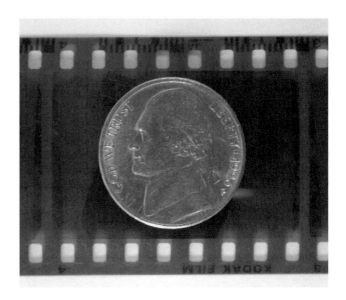

Figure 3.78
50-mm View of a nickel at minimum focusing distance.

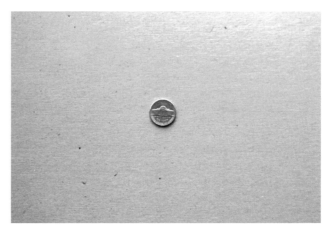

Figure 3.79
1:7 Magnification Ratio of a 50 mm Lens

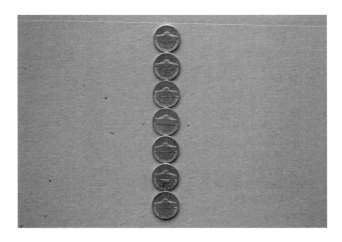

nification ratio is used, the nickel-sized object will be one third of its real size on the negative. If equipment with a 1:7 magnification ratio, a typical 50-mm lens, the nickel-sized object will be 1/7th of its real size on the negative, like Figure 3.78.

The only thing special about a nickel is that it almost perfectly fits the size of a negative. Perhaps a better object to be concentrating on would be a single-digit fingerprint. See Figures 3.81 and 3.82.

Like the nickels, Figure 3.81 shows that about seven fingerprints, stacked vertically, fill the frame with a 50-mm lens. When equipment providing a 1:1 magnification ratio is used, a single fingerprint fills the frame nicely, as in Figure 3.82.

These concepts about magnification ratios are made more confusing because lens manufacturers will advertise that a lens has "macro" capabilities, charge a premium price for it, but when the literature is closely examined, the lens turns out to have only a 1:5 or 1:4 magnification ratio, which is only a slight improvement from what a 50-mm lens provides without any help.

For crime scene work, the macro capability desired is a 1:1 magnification ratio, but a 1:2 ratio is acceptable. Anything less will have small objects appearing small on full-sized prints or will require the darkroom operator to spend a lot of time enlarging small items of evidence. The smaller the evidence is on the negative, the more it will have to be enlarged to have the item appear to fill the frame of a print made from that negative.

Forensic laboratory experts will want a life-sized image to work with to compare to evidence recovered from a suspect. The more the image on the negative has to be enlarged to accomplish this, the more loss of detail possible from the enlargement process. Therefore, it should be our constant goal to ensure the evi-

Figure 3.80
1:1 Magnification ratio of a nickel.

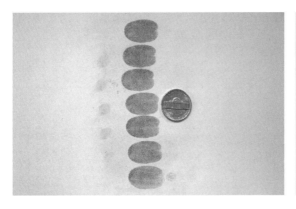

Figure 3.81
1:7 Magnification ratio of a fingerprint.

Figure 3.82
1:1 Magnification ratio of a fingerprint.

dence appears as large as it can be on the negative when the image is originally captured.

How can these magnification ratios be obtained?

Macro Options

Three common alternatives exist to achieve a 1:1 magnification ratio.

- Use a true macro lens
- Use close-up filters
- Use an extension tube

Macro Lenses

Macro lenses: A lens designed to provide either a 1:1 or 1:2 magnification ratio.

Macro lenses providing a 1:1 or 1:2 magnification ratio can be very expensive. Macro lenses made by Nikon and Canon can cost approximately $300.

Macro lenses, however, produce the clearest, crispest close-ups of the three options discussed in this text. This should be your first choice if you have any influence in the purchasing decisions of your agency.

Extension Tubes

Extension tube: A supplemental lens, inserted between the primary lens and the camera body, to move the optical center of the lens farther from the film plane. This results in magnification.

Second on your shopping list of equipment providing 1:1 magnification ratios should be an **extension tube**.

An extension tube is a supplemental lens, inserted between the primary lens and the camera body, to move the optical center of the primary lens farther from the film plane. This results in magnification. Some manufacturers provide differing lengths of extension tubes to choose from, providing differing magnifications. Extension tubes can cost approximately $75 to $100. The various extension tubes can also be stacked, providing more magnification when used together. Some manufacturers make extension tubes that are completely without any lens elements. You can stick you finger through the center of the tube. Some manufacturers incorporate lens elements into the tube to assist with the magnification process. Another way to vary the magnification possibilities is to use the extension tube with a zoom lens rather than a prime lens with just one focal length. By altering the focal length you can control the amount of magnification desired. Figure 3.83 shows the alignment of camera body, extension tube, and a zoom lens. Figure 3.84 shows the magnification of the extension tube used with the zoom lens set at 105 mm.

Close-Up Filters/Diopters

A more cost-effective choice, if the budget is limited, which provides magnification ranges from 1:6 to 1:2, is the supplemental **close-up filter set**, which is a set of three filters with differing magnifications, which can be stacked for different magnification ratios.

The individual filters are usually designated a +1, +2, and a +4. The +1 and +2 can be stacked for the equivalent of a +3. The +4 and +2 can be stacked for the equivalent of a +6. When more than one filter is used, it is recommended to screw the highest number onto the primary lens and apply the weaker filter last. Some manufacturers suggest that all three filters should not be used together. Their reasoning is that each addition of glass between the evidence and the film or digital sensor can potentially degrade the quality of the resulting image.

Another consideration is that many agencies recognize the wisdom of purchasing the best lenses that are currently affordable. After all, the quality of the final image depends greatly on the quality of the lens that captured the image in the first place. But, sometimes, when the decision to purchase close-up filters is made, that same reasoning gets lost. To be consistent, and logical, if it makes sense to purchase good quality lenses, it is just as important to purchase good

Close-up filter set: A set of three filters with differing magnifications, which can be stacked for different magnification ratios, usually from 1:6 to about 1:2.

Figure 3.83
Camera, extension tube, and zoom lens.

Figure 3.84
Magnification with the extension tube and a 105-mm lens.

quality close-up filters. Putting inferior glass on the end of an expensive high-quality lens only serves to degrade the primary lens and its resultant image.

Close-up filter sets cost approximately $30 to $50, depending on manufacturer. Being able to stack the lenses in different combinations, changing the magnification ratio to match the size of the object, makes them quite versatile. The sequence below shows a 50-mm lens used without any close-up filters (Figure 3.85), and then a +1 (Figure 3.86), a +3 (Figure 3.87), a +5 (Figure 3.88), and a +7 (Figure 3.89) used with a stack of nickels.

LENS OPTICAL PROBLEMS

In the late 1970s, when this author was taught crime scene photography techniques, the instructor stated that whenever critical comparisons, now called examination quality photographs, were to be taken, the following Rule of Thumb was to be used.

Rule of Thumb 3.8: Never use the two smallest or the two widest apertures on the lens' f/stop continuum when critical comparison photographs are taken.

The preceding Rule of Thumb was certainly an interesting admonition, so it was fair to ask, "Why not?" The answer received was unsatisfying. He said he was not sure exactly why, but that was what he was taught when he first learned crime scene photography, so he was passing on the best information he could.

Researching optics and camera lens design suggests the answer. The optics used in camera lenses have the same optical problems suffered by microscopes and telescopes. Despite camera lens manufacturers' advertising claims, these lens problems can only be minimized through modern lens design and materials; they cannot be totally eliminated. Unfortunately, this infor-mation is not readily available from the current crop of resources (books, journals, magazine articles, etc.) dedicated to issues related to crime scene photography.

Where are microscopes and telescopes most frequently used? The fields of medicine, the "hard" sciences (biology, chemistry, and physics), and astronomy use microscopes and telescopes. The theories of lens optics most properly belongs in physics courses and physics textbooks. That is where the answers to the Rule of Thumb above can be found.

Three problems with camera lenses have been identified.

- Aberrations
- Diffraction
- Distortion

Aberrations

Lens aberrations, or defects in an image produced by a lens, account for a reduction in the resolution of an image. Most, but not all, involve the inability of a lens

Figure 3.85

50-mm Lens used without close-up filters.

Figure 3.86

50 mm +1.

Figure 3.87

50 mm +3.

Figure 3.88

50 mm +5.

Figure 3.89

50 mm +7.

to focus light at a precise point on the film plane. The result is frequently an impression of "softness" to the image and a loss of clarity of the details in the image. The word "aberration" is derived from the name of Ernst Abbe, a German physicist, who collaborated with lens maker Carl Zeiss in 1866 to try to solve these problems as they occurred in microscopes in the mid-nineteenth century.

Most lens aberrations occur at the widest apertures of a lens. These lens aberrations are usually "cured" by stopping the lens down 2 stops to 3 stops. For instance, if the widest aperture for a particular lens was f/2, then stopping down to f/2.8, and then from f/2.8 to f/4 would almost totally eliminate image problems associated with the aberrations noticed when using an f/2. Actually, it is impossible to totally eliminate lens aberrations. Manufacturing the perfect lens is beyond the skill of lens designers; the best that can realistically be hoped for is that the aberrations are minimized to the point that they do not have an obvious effect on the image.

This is the foundation of Rule of Thumb 3.8 mentioned previously to avoid the widest apertures of any lens when critical comparisons are being captured. However, as far as crime scene photography is concerned, we are usually not tempted to use wide apertures because of our all-consuming concern for maximizing the DOF.

What are the aberrations related to the widest apertures of a lens?

Chromatic Aberrations

As light travels through the various lens elements, the different colors in white light will refract differently. The primary colors of red, green, and blue will separate and converge near the film plane at different locations, rather than at one precise point. If these colors do not focus at the same point, the image is noticeably "softer" and the colors may appear to be a bit out of line. Rather than all the colors coming together to form the proper color of an object, a fringe color may be noticed as an unsharp outline of the subject, which can occur two different ways: as a longitudinal chromatic aberration, or as a lateral chromatic aberration. The longitudinal chromatic aberration involves light coming into the lens from directly in front of the camera, which is called on-axis light. The lateral chromatic aberration involves light coming into the lens from the sides of a scene, which is called off-axis light. In both situations, the primary colors will refract/bend to different degrees and come together near the film plane in different areas.

Figures 3.90 and 3.91 show both the longitudinal and lateral form of chromatic aberrations.

Designing color-corrected lens elements is a partial solution to these aberrations. Stopping down the aperture is also required to totally "eliminate" this aberration from appearing in an image.

Spherical Aberrations

Similar to chromatic aberrations, the spherical aberration is caused by the different points of light that will converge near the film plane, but this is not related to

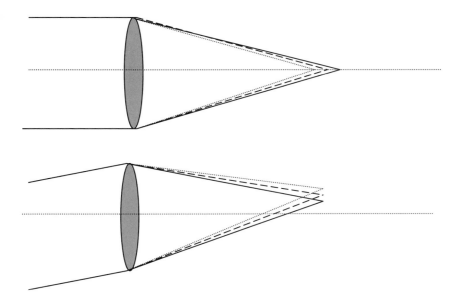

Figure 3.90
Longitudinal chromatic aberration.

Figure 3.91
Lateral chromatic aberration.

the colors of white light being refracted by the lens elements. With spherical aberrations, the difference is attributed to the various parts of the lens the light comes through: the outer perimeter of the lens, the midpoint of the lens, or various intermediate distances between the two. With light from different areas of the lens focusing at different points near the film plane, the result is a loss of crispness to the focused image. Again, the image can appear "soft" or a bit out of focus. See Figure 3.92.

Again, the "cure" for spherical aberrations is to stop-down the aperture. Do not use the widest apertures when critical comparison photographs are being taken.

Coma

As spherical aberrations appear somewhat similar to longitudinal chromatic aberrations, coma may seem similar to lateral chromatic aberrations. With coma, light coming into the camera from an off-axis location is focused at different areas of the film plane, depending on what area of the lens it came from: the perimeter, the midpoint, or intermediate distances between those two. See Figure 3.93.

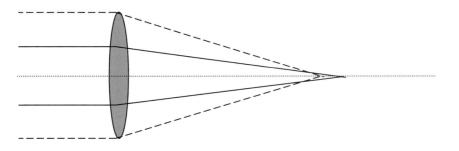

Figure 3.92
Spherical aberrations.

Figure 3.93
Coma.

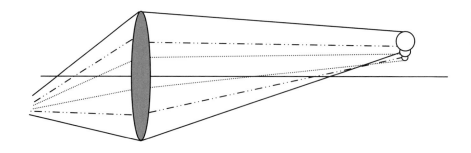

With coma, the light from a single point in space forms circles of various sizes at the film plane. These overlapping circles suggest the vague shape of a comet with a small trailing tail, hence the term "coma." The result of coma is a lack of sharpness, contrast, and resolution on the film plane. The "cure" to coma is to use a smaller diaphragm opening or a smaller f/stop.

Other aberrations include astigmatism and curvature of field. Because these are minimized by better lens design and not aperture selection, they will not be described here.

When a lens is adversely affected by aberrations, it is called aberration limited. A lens can also be diffraction limited.

Diffraction

Diffraction: The bending of light when it strikes an edge.

Diffraction is the bending of light when it strikes an edge.

The effect of diffraction is a loss of resolution, a loss of edge sharpness, and a loss of clarity in an image. Diffraction is most severe when using the smallest apertures of a lens' f/stop continuum. For instance, if a lens has f/stops ranging from f/2 through f/22, the f/16 and f/22 apertures would be most severely affected by diffraction, which is the other basis for Rule of Thumb 3.8. Diffraction is the reason small apertures are not recommended when images to be used for critical comparisons are being captured. When critical comparisons are the issue, the aperture of choice should be f/11.

What about the Cardinal Rule to maximize the DOF? When trying to capture critical comparisons it will be necessary to modify this Cardinal Rule a bit. At times, an image will suffer more from diffraction than it will suffer from a lack of the largest DOF range as provided by the smallest apertures. This cannot be overemphasized. As desperate as we are for the sharpest, clearest, most in-focus image we can get, when trying to capture critical comparison images, use of the smallest apertures of a lens is counterproductive.

It is perhaps this emphasis on the adverse effects of diffraction that partially separates this text from many others. If the reader acquires an understanding of diffraction, it may change the way many capture critical comparisons, also called examination quality photographs. Because of the importance of this subject matter, it will be best to do a little historical review.

Some time ago the concept of diffraction, the bending of light when it strikes a barrier, was not accepted by the world at large. In the late 17th century, the influence of Sir Isaac Newton (1643–1727) was very pervasive. Newton thought that light was made up of small particles, which was called the Corpuscular Theory of Light.

Three graphics will most clearly explain this prominent theory of light in the late 17th century.

Figure 3.94 shows the Corpuscular Theory of Light applied to light striking a linear opaque barrier. The opaque barrier will, of course, prohibit discrete light particles from traveling any farther, but those light particles that do not strike the opaque barrier continue in a straight line until they do strike a surface. On the surface struck by the light that was not stopped by the opaque barrier, the light is uniformly bright, from the very edge of the shadow outward. The area within the shadow has a very clear and distinct beginning point, and it is uniformly dark throughout the shadow area. How could anyone believe otherwise? When Sir Isaac Newton subscribed to it, who in their right mind would challenge such an idea? But remember, the world was thought by all to be flat at one point also.

Figure 3.95 shows a variation on this theme. Rather than light striking the edge of a straight opaque barrier, what would be the result of light striking a round opaque barrier? This graphic shows the result, according to the Corpuscular Theory of Light. The circular barrier, of course, will block some of the discrete light particles, but those avoiding the round barrier will continue until they do strike a surface. On that surface, the light will form a very clearly defined bright area. A very clearly defined circular shadow will be formed by the lack of light in that area. The area lit by the light will be uniformly bright throughout its area; the area in shadow will be uniformly dark. Again, this is all very straightforward. Who could doubt it? Certainly not reasonable men.

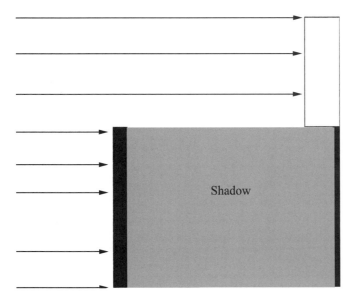

Shadow

Figure 3.94

Light striking a linear opaque barrier.

Figure 3.95

Light striking a round opaque barrier.

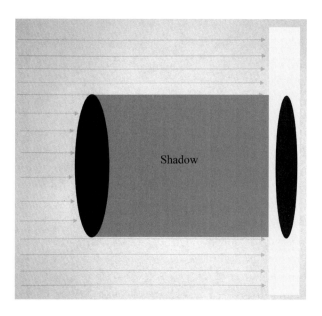

Figure 3.96 shows another variation on this theme. This graphic shows light coming through a circular aperture and striking a surface. Of course, the light blocked by the sides of the circular aperture and the frame holding it never get to the surface. The circle of light striking the surface is uniformly bright throughout the circular area. A clear and distinct edge is present to the circle of light on the surface, with uniform shadow outside the circular area.

Why does this author bother to waste your time with such obvious graphics? Because Newton was wrong. On all three counts.

Figure 3.96

Light coming through a circular aperture.

The champion of the Wave Theory of Light in the latter half of the 17th century was Dutch physicist Christian Huygens (1629–1695). Disagreeing with the universally respected theories of Sir Isaac Newton, however, was not a politically correct position, and not many agreed with Huygens. The Wave Theory of Light is depicted by the simple Figure 3.97.

The Wave Theory of Light postulates that when light strikes the edge of a surface it bends. Just as a wave of water would create an expanding wave after striking an object in a stream, light follows the same model. On striking an edge, an expanding wave of light is formed, with portions of the light wave striking the area otherwise lit by direct light rays and portions of the light wave bending into the area that would otherwise be the totally dark shadow area. Perhaps you are thinking to yourself that this appears to be patently false. Many in the latter half of the 17th century also thought the same.

The Wave Theory of Light did not enjoy popularity or respect among most physicists in the 1700s or the early 1800s. This continued until the French Academy of Science announced that the grand prize for physics at its 1819 meeting would be given to the best paper on diffraction. The chairman of the committee to judge the grand prize was Dominique Francois Jean Arago, no friend of the wave Theory of Light. Another committee member was Simeon Denis Poisson, another well-known enemy of that theory. Augustin Fresnel presented a paper on diffraction he was sure would finally win over his fellow physicists. Poisson was determined that Fresnel would not win the competition.

Poisson was well respected during this time frame, having won a previous grand prize for his work on electricity. It must be said that he was indeed brilliant. To his

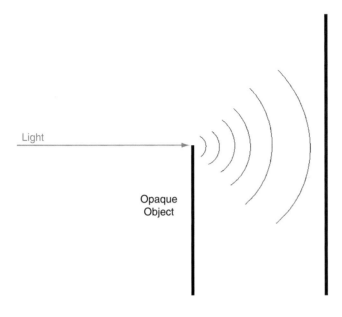

Figure 3.97
The wave theory of light.

Light

Opaque
Object

great credit, Poisson had seriously considered the Wave Theory of Light and its necessary consequences. He formulated a theory about an aspect of the Wave Theory of Light that even the proponents of the theory had not anticipated. In fact, being so fundamental to the Wave Theory of Light, Poisson thought he had finally been able to prove absolutely that the Wave Theory of Light could not correctly describe how light acted. Poisson announced that if the Wave Theory of Light were correct, if light struck an opaque disk, it would bend around the disk and eventually come together as a point of light in the middle of the dark shadow formed by the disk. In addition, if the Wave Theory of Light were correct, then this point of light, formed inside the shadow area, would be just as bright as if the disk were not there. A perfect refutation of the despised theory. Indeed, it did seem to be an accurate consequence of the theory. Because it could not possibly be correct, Poisson had used the theory itself to prove it could not be correct.

Committee Chairman, Dominique Arago, a friend of Poisson, and a proponent of Newton's Corpuscular Theory of Light, was fortunately also a good scientist. He proposed a test of Poisson's theory. When the test was conducted, the disputed circle was, in fact, seen by all. Poisson's genius had ultimately proven the theory he had been so opposed to. To Poisson's chagrin, this small bright circle was thereafter called **"Poisson's spot."**

Poisson's spot: The bright spot in the center of a circular shadow from light striking a circular opaque barrier and bending around the edges and meeting in the center of the shadow.

Many examples of "Poisson's spot" can be found in physics and optics textbooks. This author knew that one such image had to be included in this textbook, because it is fundamental to the concept of diffraction. The solution became clear. This author sought out The George Washington University Physics Department's professor for Optics and Astronomy courses, Dr. J. Roger Peverley. A joint project to create classic diffraction images we could both use in our courses was proposed. Dr. Peverley agreed to the project.

Before these images are revealed, it makes sense to first clarify the logical consequences that diffraction would have on all the three graphics previously presented to demonstrate the position of the Corpuscular Theory of Light.

If diffraction and the Wave Theory of Light did explain how light would react when it struck a linear opaque barrier, light would be bent in an expanding arc from each point on the opaque barrier that was struck by the light. Not just one wave, but multiple waves would be generated. Think of the wave peak as being bright, and think of the depression of the wave as being dark. Multiple waves would be sent out. On the surface eventually struck by the light, in some areas multiple peaks will superimpose over each other. These areas will be very bright. In some areas, multiple wave depressions will coincide. These areas will be very dark. Physics texts call the multiple peak areas *areas of constructive interference* and the multiple depression areas *areas of destructive interference*. The net effect is progressive light and dark waves. Where these waves strike an area already lit by direct light, unaffected by diffraction, the result is areas actually brighter than areas lit only by direct light.

Where the waves bend into the shadow area, a masking of the hard, sharp edge between where the lit areas and the shadow areas would have been exists. An image of the diffraction pattern created when light strikes a linear opaque edge will be shown momentarily.

Poisson's spot is the result of light bending around an opaque disk. An example of Poisson's spot will also be shown.

What about the effect of diffraction and the Wave Theory of Light on light coming through a circular aperture? This is directly related to trying to capture an image with a small aperture like f/22.

This work is mainly attributed to Sir George Airy (1801–1892), the 7th Astronomer Royal. Rather than an image similar to Figure 3.96, instead, when light from a single point comes through a lens with a small aperture, the light strikes the edges of the diaphragm and radiates outward as multiple waves. These will be circular waves rather than the linear waves produced by straight opaque barriers. A bright spot will certainly be in the center of the surface on which the image is formed. Radiating outward from the central bright area are progressively dimmer and dimmer areas of alternating light and dark rings. See Figures 3.98 and 3.99.

Figure 3.98 is one of the images Dr. J. Roger Peverley and this author collaborated on together. We used a Metrologic Helium–Neon Laser Model ML 869 projecting its ruby laser beam through 1-mm and 0.3-mm apertures, which is the diffraction pattern first described by Sir George Airy. It has been called the **airy disk** ever since.

Airy disk: When light from a single point comes through a small circular aperture, the diffraction of light striking the edges of the aperture forms a figure with a bright central feature, with progressively dimmer radiating rings that are light and dark.

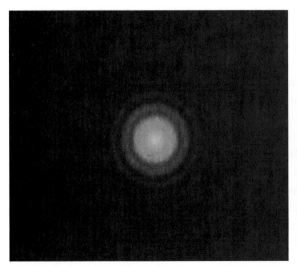

Figure 3.98
The airy disk: diffraction pattern through a circular disk.

Figure 3.99
Enhanced in Photoshop.

Suspecting that there may be more to the dimming wave rings than is shown in Figure 3.98, the bright interior of the image was brought into Photoshop to isolate it from my proposed enhancements and then the remainder of the image was lightened. Figure 3.99 is the result of this enhancement. The alternating light and dark rings extend farther than is immediately obvious. Figure 3.100 shows the airy disk as a 3-D profile.

The bright central spot of Figure 3.98 is represented by the high central peak of Figure 3.100. The alternating light and dark rings are shown in both 2-D and 3-D in the image and graphic.

This concept of the airy disk needs to be elaborated on for its full significance to be appreciated. This single airy disk is a product of light reflecting from a single point in the scene. When we have previously thought about light reflected from subject matter in a crime scene, most of us have thought that the light would be focused by the lens so that it formed a single point on the film plane. However, instead of a single pinpoint on the film plane, a shape like the 3-D airy disk is formed on the film plane. Should there be another point of interest in the scene that is very close to the original point, these two airy disks will be very close on the film plane. They may be close enough that their outer rings overlap. When these rings overlap, their light intensity increases. See Figures 3.101 and 3.102.

Figure 3.101 shows the peripheral rings overlapping and increasing in intensity. If the two points in the crime scene reflecting light to the camera are even closer, Figure 3.102 results. If the two points are close enough, the airy disks for each begin to merge, and the result is that the eye does not resolve two distinct points any longer. Critical comparison photography involves just this: an attempt to capture and distinguish minute fine detail, often with the detail very close together. The fingerprint ridge is adjacent to the valley next to it. Shoe print and tire track photography is attempting to capture edge detail in the nicks, cuts, gouges, and imbedded debris within the treads, which is all edge detail.

The effect of diffraction is a loss of resolution and the diminishing of edge definition. To see this directly, recall the resolution chart in Figure 3.1. Figure 3.103 is a crop of the central portion of Figure 3.1.

Figure 3.103 contains the most minute detail of the entire resolution chart, with black lines approaching 0.1 mm in width at the point indicated as "20" (20 is the abbreviation for 2000 LW/PH, line widths per picture height). This chart is usually used to test differences in various camera resolutions or the differences in various lenses used with the same camera. In this instance, we will use it to demonstrate various resolutions obtained when using the same camera and the same lens, but with different apertures. The camera used was a Canon EOS 5D digital camera, with a Canon EF 24-105-mm 1:4 L IS USM lens. The camera was placed on a tripod. The focal length was set to 50 mm, and the complete resolution chart filled the frame. Images were captured with an f/22, f/16, f/11, and f/8.

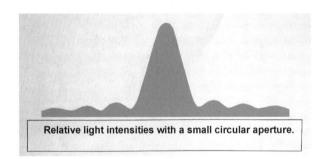

Figure 3.100
Single discrete point.

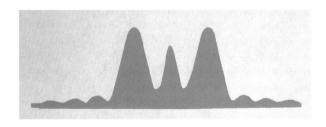

Figure 3.101
Two discrete points close together.

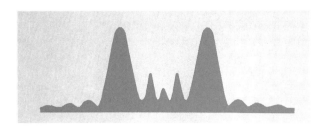

Figure 3.102
Two discrete points closer together.

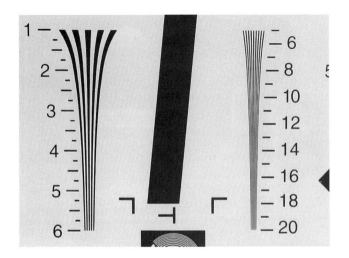

Figure 3.103
Crop of resolution chart, image #1.

Figure 3.104 was the f/22 shot; Figure 3.105 was the f/16 shot; Figure 3.106 was the f/11 shot. The f/8 shot was virtually the same as the f/11 shot. Detail in Figure 3.104 blurs just past the #18. Detail in Figure 3.105 is discernible almost to the #20. Detail in Figure 3.106 is noticeable to the #20 hashmark, with the parallel lines themselves more distinctly "resolved." Also obvious should be the quality of the hashmarks and the numbers themselves. The f/11 image is clearly the best of the three images.

These differences are solely attributable to the changes of the aperture.

This is the model used to emphasize the importance of using f/11 when critical comparison photographs will be captured. Imagine taking an image of a partially smeared fingerprint, where a ridge ending is located at the very edge between clear detail and smeared detail. If that particular ridge ending can be resolved, an extra point of identification is part of the information used to decide about a possible identification. If your photography technique is even a little off from optimal, that edge detail will not be resolved. Your technique is the crucial variable that will decide whether that point of identification is captured. Can you afford mediocre technique? This is when the theory of diffraction matters the most. In this case, it is not focus or the depth of field range that is critical. The resolution of edge detail is critical, and that can only be optimized by the use of the appropriate aperture for such a shot. For examination-quality photographs, do not use the smallest aperture of the lens. Open up 2 to 3 stops, to either f/11 or f/8.

Figure 3.104
Extreme close-up, with f/22.

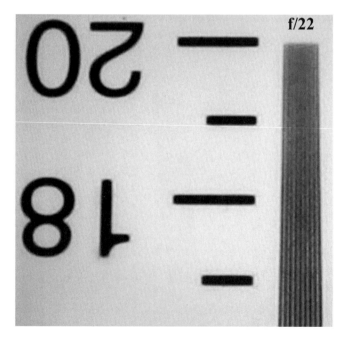

Figure 3.105
Extreme close-up, with f/16.

What about Poisson's spot? Arago did not have a laser to test his theory; Dr. Peverley did. Not only is the bright spot found at the center of the shadow area, constructive and destructive interference waves can be seen throughout the shadow area. Also evident is the lack of a distinct edge to the circular shadow area.

Figure 3.106
Extreme close-up, with f/11.

In addition, constructive and destructive interference waves can be seen in the area that should be uniformly lit by direct light. Diffraction waves affect both the shadow and lit areas of the image. See Figures 3.107 and 3.108.

Figure 3.107 is a magnetized ballbearing at the tip of a pinpoint. Figure 3.108 is the diffraction pattern created by the laser beam, Poisson's spot. In addition to the diffraction patterns inside and outside the shadow area of the ballbearing, take a moment to see something unusual about the light pattern associated with the pinpoint. This strange phenomenon will be returned to soon.

Recall our discussion of the Newtonian Corpuscular Theory of Light as it strikes a linear barrier? Figure 3.94 showed what would be the result if that theory of light were accurate. Diffraction, however, creates a very different pattern. Figure 3.109 shows what should happen, according to the theory of diffraction. The vertical line is the classic divide between shadow on the left and the lit area on the right of the line. Light will enter the shadow area, and constructive and destructive interference waves will be visible in the area lit by direct light rays. Theoretically, no clear and distinct edge should be present between the shadow area and the lit area. Figure 3.110 is the result.

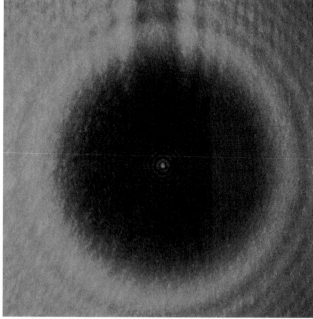

Figure 3.107
Ball-bearing, lit by laser.

Figure 3.108
Poisson's spot.

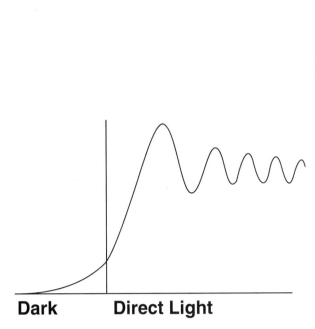

Dark   **Direct Light**

Figure 3.109

Graphic: diffraction around a linear opaque edge.

Figure 3.110

Diffraction pattern around a linear opaque edge.

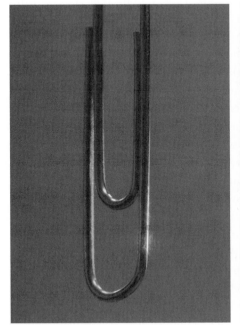

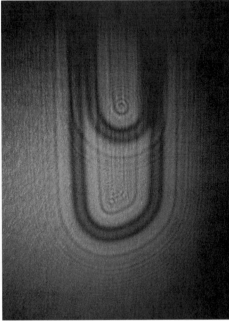

Figure 3.111

Paper clip lit by laser.

Figure 3.112

Diffraction pattern of paper clip.

We will conclude this discussion of diffraction by showing just two more images. Figures 3.111 and 3.112 show a paper clip lit by laser light and the diffraction pattern that results.

Again, the constructive and destructive interference waves can be seen in the area that should be uniformly lit just by laser light. Another very interesting phenomenon can also be noticed. This brings us back to the pinpoint from which the ballbearing was previously suspended by magnetism. Both the pinpoint and the paper clip show light areas inside what should be solid shadows where they blocked the light of the laser; these are variations of Poisson's spot. Not only will an opaque circular object bend light into its interior, forming a bright spot in the center of the shadow area, opaque linear objects will also bend light into the shadow area from both sides of the object. The result will be bright lines in the area that should be solid shadow. A cursory search of the physics literature has not resulted in this phenomenon being pointed out or named. Until another term can be found in the literature, this author will refer to these as Poisson's lines.

Let us summarize the effects of diffraction.

- A loss of resolution
- A reduction in the clarity of sharp edges

The cause of diffraction is light bending from the diaphragm blades when small apertures are used to capture an image. The cure to diffraction is to open the aperture up by 2 stops or 3 stops. When capturing critical comparisons is the task at hand, use an f/11 or an f/8 to avoid the image degradation caused by diffraction.

It must be reiterated that diffraction is not a type of aberration. Whereas quality lens manufacturing can diminish the effects of most of the aberrations, quality lenses will still suffer from diffraction if the smallest apertures are used. Diffraction is based on the physics of light, not lens manufacturing improvements. The only "cure" to the degradation of an image's sharpness because of diffraction is to open the aperture at least 2 stops from the smallest aperture of the lens. Most of the time, maximizing the depth of field will be our primary concern at crime scenes. When capturing examination-quality photographs, minimizing the effects of diffraction is more important than using the smallest aperture for the best depth of field.

A specialty lens was recently being demonstrated to show its remarkable depth of field range. Although not a lens designed to be used at crime scenes, it was a part of a digital imaging work station intended for use by examiners working with latent prints on curved or multilevel surfaces, as well as other types of evidence. Figure 3.113 exhibits an "impossible" depth of field range. A .45-caliber casing has been placed on top of a tennis ball, and a scale was placed on the surface supporting the tennis ball. Both the head stamp of the casing

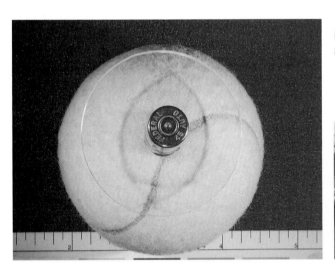

Figure 3.113

"Impossible" depth of field range (Courtesy of Tom Beecher, Photografix, Richmond, VA).

Figure 3.114

Fodis Pro 2D macro lens (Courtesy of Tom Beecher, Photografix, Richmond, VA).

and the scale are in focus. Try that with your current macro lens. Figure 3.114 shows the Fodis Pro 2D Macro Lens responsible for the great depth of field in Figure 3.113.

The issue was dodged each time specifications on the lens were requested. A comment was made, however, that caught the attention of this author. It was mentioned that the aperture of the lens was opened up 2 stops from the lens' smallest aperture "because of diffraction." Diffraction is not just an abstract idea found in dated physics books. Many professional photographers routinely acknowledge its effect on images. Crime scene photographers will also benefit from a knowledge of its effects and should modify their critical close-up photography procedures because of it.

Aberrations can degrade images captured with the widest apertures of a lens; diffraction can degrade images captured with the smallest apertures of a lens, which can also be expressed by saying lenses can be aberration-limited and diffraction-limited. The apertures 2 stops to 3 stops in from the extreme f/stops of a lens are called the "sweet spots" of a lens. Depending on the lens, there may be two or three "sweet spot" apertures. If this is so, the Cardinal Rule to maximize the DOF still applies. If a lens' "sweet spots" include f/4, f/5.6, f/8, and f/11, the Cardinal Rule suggests that f/11 is the aperture that both acknowledges the effects of diffraction, and maximizes the DOF from the available apertures.

Distortion

Two other effects lenses can have on an image are barrel distortion and pincushion distortion. These are usually effects caused by the use of wide-angle and telephoto lenses.

Barrel Distortion

Barrel distortion is usually an effect of using a wide-angle lens.

The result is that straight lines appearing at the edges of an image may appear to be bent outward a bit. Photoshop and other imaging software programs may be able to correct for barrel distortion. Otherwise, not much can be done to avoid this type of distortion other than trying to avoid composing linear objects at the periphery of an image. If it is your current task to compose an entire building that is the site of a crime scene, this may be impossible. See Figure 3.115. The top and right side of this building appears to be bowed outward.

Pincushion Distortion

Pincushion distortion is usually an effect of using a telephoto lens.

The result is that straight lines appearing at the edges of an image may appear to be bent inward a bit. Again, some imaging software programs may be able to

Barrel distortion: Usually a result of using a wide-angle lens, an image may show straight lines at the edge of the image bowed slightly outward.

Pincushion distortion: Usually a result of using a telephoto lens; image may show straight lines at the edge of the image bowed slightly inward.

Figure 3.115

A building showing barrel distortion.

correct for this distortion. It only occurs at the periphery of an image captured when a telephoto lens is used. Figures 3.116 through 3.118 show the same spacing between buildings, but it is composed on the left side of Figure 3.116. It is in the center of Figure 3.117, and, therefore, no distortion is present at all, even when the same lens is used. The spacing has been composed on the right side of Figure 3.118. Figures 3.116 and 3.118 show marked bending of lines toward the center of the image.

Figure 3.116
Pincushion distortion on the left edge.

Figure 3.117
No pincushion distortion.

Figure 3.118
Pincushion distortion on the right edge.

Many times, the eye may not even notice these distortions, because we have become so accustomed to the effects of these lenses. They are, however, inaccurate representations of a scene that we may be asked to explain in court. It is best to be familiar with our "tools" and the effects they will have on our crime scene images.

SUMMARY

When a layman speaks about wanting a camera with good resolution so that the images produced by it are clear and sharp, they are usually oblivious to the precise meanings of such terms as "resolution," "acutance," and "sharpness." This chapter explained each of these terms. Also, when it is asked whether a particular camera has sufficient resolution to be used as the primary camera at a crime scene, this chapter provides the means to answer that question.

Most laymen focus their cameras on individual objects. Because photographing crime scenes also requires that large and small areas fall within the depth of field range, methods to focus on areas, rather than on individual objects, were discussed and explained.

Depth of field was defined, and the factors that affect depth of field were explained.

Lens types were distinguished, and their differences explained. The effects of each lens type were presented.

Lenses can suffer from aberrations and diffraction, despite the high quality of modern lens-manufacturing techniques. Each was explained. Because diffraction can adversely affect our most critical images, suggestions for minimizing its effect were provided.

DISCUSSION QUESTIONS

1. Briefly explain the nuances of each of these terms: resolution, acutance, and sharpness.
2. Automatic focus has difficulties locking in on some types of scenes. Explain two situations where it may be better to use manual focusing.
3. Explain the concept of the circles of confusion. How are they related to different apertures? How are they related to different focal lengths?
4. Explain hyperfocal focusing. Include an explanation of how to use the technique when you do have a depth of scale on your lens, and when you do not.
5. Explain zone focusing. Include an explanation of how to use the technique when you do have a depth of scale on your lens, and when you do not.
6. Explain the camera variables that affect depth of field, which maximize depth of field, and which minimize depth of field.

7. Explain which focal length of lens is most appropriate for most crime scene photography and why this is so.

8. Most of the time depth of field is the most important concern. When does diffraction most affect crime scene photographs? What is its effect? What is the "cure?"

EXERCISES (ALL NON-FLASH SHOTS)

1. Create a single-digit fingerprint on an outside surface that is fully sunlit. Place the camera on a tripod, pre-focus the lens so it will produce a 1: 1 or 1:2 magnification ratio, include a scale, and take a set of three exposures at 0, +1, and −1.

2. Hyperfocal focus on a large outdoor scene where infinity is in the background.

3. Zone focus on an area in front of a building façade that is 30′ from the camera.

4. Zone focus on an area in front of a building façade that is 20′ from the camera.

5. With the widest aperture of the lens, focus on an object 20′ away. Use a 50-mm lens.

6. With the smallest aperture of the lens, focus on the same object 20′ away. Use a 50-mm lens.

7. With the smallest aperture of the lens, focus on the same object 20′ away. Use the widest-angle lens available.

8. With the smallest aperture of the lens, focus on the same object 20′ away. Use the longest telephoto lens available.

9. Place a coin on the sidewalk with the mint designation or date plainly visible. With the camera on a tripod, determine the best exposure using an f/22.

10. Same as No. 9, but use an f/16.

11. Same as No. 9, but use an f/11.

12. Same as No. 9, but use an f/8.

13. Enlarge Nos. 9 through 12 so the mint letter or date is greatly enlarged. Which seems to have the best definition? Why?

FURTHER READING

Baines, H. (1976). "The Science of Photography." Halsted Press, NY.

Born, M., Brown, E. (2003). "Principles of Optics." Cambridge University Press, Cambridge, MA.

Ditchburn, R.W. (1991). "Light." Dover Publications, NY.

Katz, J., and Fogel, S. J. (1971). "Photographic Analysis: A Textbook of Photographic Science." Morgan & Morgan, Inc., Hastings-On-Hudson, NY.

Landt, A. (1998). "Lenses for 35 mm Photography the Kodak Workshop Series." Silver Pixel Press, Rochester, NY.

McDonald, J. A. (1992). "Close-Up & Macro Photography for Evidence Technicians." 2nd Ed. Phototext Books, Arlington Heights, IL.

Mitchell, E. N. (1984). "Photographic Science." John Wiley & Sons, Inc, NY.

Paduano, J. (1996). "The Art of Infrared Photography." Amherst Media, Inc., Amherst, NY.

Paduano, J. (1998). "Wide-Angle Lens Photography." Amherst Media, Inc., Amherst, NY.

Ray, S. F. (2002). "Applied Photographic Optics." Focal Press, Woburn, MA.

Sheppard, R. (1997). "Telephoto Lens Photography." Amherst Media, Inc., Buffalo, NY.

Woodlief, T. (1973). "SPSE Handbook of Photographic Science and Engineering." Wiley-Interscience Publications, NY.

Walls, H. J. and Attridge, G. G. (1977). "Basic Photo Science: How Photography Works." Focal Press Limited, London.

White, L. (1995). "Infrared Photography Handbook." Amherst Media, Inc., Amherst, NY.

CHAPTER 4

ELECTRONIC FLASH

LEARNING OBJECTIVES

On completion of this chapter, you will be able to . . .

1. Explain the meaning of flash guide numbers.
2. Explain the meaning of the flash sync speed.
3. Explain why it is important for the flash head to duplicate the camera and lens viewpoints.
4. Explain the basics of the manual flash exposure mode and explain some of the drawbacks with the manual flash exposure mode.
5. Explain the inverse square law of light. Explain how flash intensities can be determined when the distance the flash travels is known. Explain how the f/stop numbers are derived from these different flash intensities.
6. Explain the various ways that the flash output could be reduced or softened to provide for precise flash control at different distances.
7. Explain the basics of the automatic flash exposure mode.
8. Explain the basics of the dedicated flash exposure mode.
9. Explain the benefits that these modes have over the manual flash exposure mode.
10. Explain how to bracket in all three flash exposure modes.
11. Explain how to use fill-in flash to obtain a proper exposure with various lighting conditions.
12. Explain how to take photographs using the oblique flash technique.
13. Explain what kinds of situations would benefit from the use of bounce flash.
14. Explain how to compensate for the light loss resulting from the use of bounce flash.
15. Explain when painting with light (PWL) would be the preferred photography technique.

16. Explain how to PWL, when it is possible to flash in toward the scene from two sides.

17. Explain how to PWL, when it is only possible to flash in toward the scene from one side.

KEY TERMS

Automatic flash expo-
   sure mode
Dedicated flash expo-
   sure mode
Fill-in flash
Flash calculator dial
Guide number

Hard shadows
Inverse square law
Manual flash mode
"Normal" room
Oblique flash
PC cord and remote
   flash cord

Painting with light
Positive dust print
Soft shadows
Snow print wax
Sync speed

Many times, the ambient lighting is not adequate for a proper exposure with the small apertures that will guarantee the best depth-of-field ranges. Not only is the amount of light an issue, the quality of the light is also important. With the wrong type of lighting, color film will show unwanted tints, even if the amount of light will ensure the use of small apertures. Electronic flash is the solution to both of these issues. Adding electronic flash to a scene will increase the light levels so that small f/stops can be used. Electronic flash is also color balanced for daylight color film. If you recall, daylight color film will provide proper colors if lit with midday sunlight or electronic flash.

Early in the morning, late in the afternoon, at night, or any time photographs will be taken indoors, these are occasions that electronic flash will benefit the crime scene photographer.

GUIDE NUMBERS

Guide number: The guide number is a means to discuss a flash's relative output power. It can also be used to determine the f/stop to properly expose an object at a particular distance by using the formula: GN/distance = f/stop.

When electronic flash units are compared, it is common to inquire about a particular flash's **guide number**.

The guide number is a means to discuss a flash's relative output power. One way to compare several flash units is to compare how much light they are capable of throwing into a scene. Several common presumptions come into play when the topic of guide numbers comes up. Guide numbers can be discussed on the basis of the number of feet or meters that need to be lit. It is a common presumption that feet will be the unit of distance when guide numbers are discussed. Guide numbers also vary by the ISO film speed being used. The common presumption is that ISO 100 film will be used to compare various flashes. We will be thinking in terms of the light that can adequately expose 100 ISO film over certain distances (thought of in feet).

A car can be compared by horsepower or by the miles per gallon it is expected to get. A flash unit's guide number is the means by which we can compare various flash units as to their light output.

Knowing a particular flash unit's guide number was once necessary to determine the exposure settings for the camera. Which camera variables were affected? Because it is often dim when the need to use flash arises, 400 ISO film is commonly used in such circumstances; 400 ISO film is more sensitive to light, so it can more easily be exposed in dimmer lighting situations. Except when critical comparisons are being photographed, 400 ISO film is normally the film speed of choice in dim lighting conditions. If critical comparisons are the issue, ISO 100 speed film is still recommended, because of its ability to be enlarged for comparisons without the graininess that may be noticeable with 400 ISO film. With digital cameras, it is still recommended to use the slower ISO 100 equivalent film speed, because it will enlarge more without pixilation and will show fewer digital artifacts.

Shutter speeds are not normally an important exposure factor in most electronic flash situations. In flash situations, the optimal shutter speed is usually the sync speed. This topic will be fully discussed shortly. At this point, it is sufficient to state that when using flash, shutter speeds are not as variable as they are when non-flash images are being exposed.

Returning to the concept of the proper exposure triangle, now only two exposure variables are left: light in the form of the electronic flash and the f/stop. The guide number of the flash unit allows the f/stop to be determined for any particular flash unit depending on the distance between the flash and the subject matter. Electronic flash units today have a calculator dial that tells the camera operator what f/stop to use for objects at different distances. Before electronic flashes became so user-friendly, it was up to the photographer to do some mental math to determine the proper f/stop to use for any particular distance between the photographer and the subject matter. The formula used was:

$$GN/distance = f/stop$$

The guide number divided by the distance between the flash and the subject matter equals the required f/stop for a proper exposure.

This will only make sense if some real numbers are used in an example. Flash units very commonly issued to crime scene photographers have a guide number of 120 or higher. If the photographer were 10 feet from the object to be photographed, then $120/10 = 12$. Because most cameras do not have an f/12 as a choice, this defaulted to f/11. With a 120 guide number flash unit, an f/11 would be required when photographing an object 10 feet away.

As a quick comparison guide, here are some common flash guide numbers (GN):

- GN 160: Found on very strong and very expensive flash units.
- GN 120: Considered the "norm" with many law enforcement agencies.
- GN 80: Less powerful but still usable as a crime scene flash unit.
- GN 56: A weak flash unit not recommended for crime scene work
- GN 36: The strength of the built-in flash units on some cameras.

If all these flash units were to be used for the same object 10 feet from the photographer, these would be the corresponding f/stops required for a proper exposure:

- GN 160: f/16
- GN 120: f/11
- GN 80: f/8
- GN 56: f/5.6
- 217GN 36: f/3.5 (about half-way between an f/4 and f/2.8)

We now have a basis for comparing these flash units by their relative flash outputs. A flash with a GN of 160 is so powerful it requires a small aperture, f/16, to properly expose an object 10 feet away. The flash with a GN of 56 puts out less light, so it requires a wider aperture at the same distance. It, therefore, goes without saying that crime scene photographers should not use weaker flashes that require wider apertures. Those wider apertures will result in poorer depth of field.

Fortunately, none of this mental math is required now, because the calculator dial on most flash units directly indicates what f/stop is recommended at any given distance, eliminating all the previously required math.

Today, the preceding formula still has one use. If you have lost or misplaced your flash unit's manual and you do not know your flash unit's GN, you can determine it this way.

Make sure the flash is set for 100 ISO film. Notice which f/stop is recommended for a subject 10 feet away, multiply that f/stop number by 10, and that is the guide number of your flash unit.

Of course, other ISO film speeds will frequently be used when the light is dim. When flash units are compared and guide numbers are being discussed, this convention is used to avoid all the confusing variables.

You may be asked to recommend new flash unit replacements for your department. If your agency has had success with its current flash units, perhaps you will not need more powerful units. Perhaps many have complained about the output of your current flash units. Knowing what your current flash units are rated is the first step to getting a more powerful unit.

FLASH SYNC SPEEDS

Sync speed: The shutter speed to use whenever an electronic flash is also used; the shutter speed that has the shutter completely open when the flash fires.

The **sync** (short for synchronization) **speed** is the shutter speed at which cameras should be set whenever an electronic flash is used.

Each camera has one particular shutter speed that has been designated as its sync speed, which actually is a benefit to the photographer, because it reduces the number of photographic variables that have to be considered and manipulated to ensure that a proper exposure has been correctly calculated when using flash.

Check your camera's shutter speed dial if it is an older camera and actually has one. The sync speed is usually indicated by a different color, a lightning bolt may be alongside one of the shutter speed numbers, or an "x" may be alongside a

Figure 4.1

Sync speed indicated next to the 60.

particular shutter speed. In Figure 4.1, the lightning bolt arrow next to the 60 indicates the sync speed for this particular camera is 1/60th of a second.

When all else fails, read the manual, obtain a duplicate manual from a dealer, or download a duplicate manual from the Internet to determine your particular flash unit's guide number. If no resource is available to determine the particular sync speed of your camera, you can presume that 1/60th of a second will work as the sync speed for your camera. Why? Because, even if another shutter speed has been designated as the particular sync speed for a camera, 1/60th of a second will also work. It is not slow enough to incur blur problems from hand-holding the camera and it is fast enough that ambient light present will not normally overexpose the image.

The sync speed is the shutter speed that has the shutter completely open when the flash unit fires. The shutter opens, the flash fires, and the shutter closes shortly thereafter. Because the shutter is open when the flash fires, the entire scene receives the light from the flash, and the scene reflects this light into the camera before the shutter closes.

The use of a shutter speed that is faster than the sync speed is to be avoided. In effect, using a faster-than-sync shutter speed means the shutter will be closing when the flash unit fires. That means a part of the shutter will be covering the film or digital sensor when the flash fires, and that part of the image will receive no light at all. Look at the examples in Figures 4.2 to 4.5. Some cameras allow the photographer to take photographs at faster-than-sync shutter speeds. Some newer cameras will not let this happen. They will default to the sync shutter speed even if you accidentally or intentionally set a faster-than-sync shutter speed.

Figures 4.2 through 4.5 show progressively faster and faster shutter speeds. More and more of the image is covered by the shutter in the sequential images. In this case the shutter travels from the bottom to the top. The shutter on some cameras moves from left to right. If you are careful, you will never find out which way your shutter travels by having your images come back covered by the shutter.

Figure 4.2

Sync + 1 (Courtesy of Denise Sediq, GWU MFS Student).

Figure 4.3

Sync + 1.3 (Courtesy of Denise Sediq, GWU MFS Student).

Figure 4.4

Sync + 1.6 (Courtesy of Denise Sediq, GWU MFS Student).

Figure 4.5

Sync + 2 (Courtesy of Denise Sediq, GWU MFS Student).

Of course, if you are still using a film camera, you can simply open the back of the camera and push the shutter button to find out whether it travels vertically or horizontally.

What would happen if you accidentally used a shutter speed slower than the sync speed of your camera? In that case, the shutter opens, the flash fires, and the shutter remains open for the designated time before closing. If the shutter speed is not too slow, no adverse effect will usually be seen. Typically, because it is dim when we are usually using flash, ambient light is not enough to overexpose the film while the shutter remains open. If, however, the shutter speed is very slow, two negative effects can be noticed. If the shutter speed is slower than the normal shutter speed of 1/60th of a second, the image may appear blurred from camera movement.

One other effect may be seen, which is interesting, but not desirable. If enough ambient light is in the scene and movement is present within the scene, you may capture a "ghost" in your image. The shutter opens and the flash fires, freezing any movement within the scene. However, if someone is moving in front of a backlit scene, when the subject moves, the background they previously blocked with their body can now reflect light into the camera. If flash froze the subject in this area, they will be seen in that location. If they move, however, and now light can reach the film from an area they were previously blocking, then that background is now also exposed on the film. When a person is in an area and we can see the background in the same area, it appears like we are looking "through" them. A semitransparent person is a ghost. See Figure 4.6.

Figure 4.6
A ghost.

In Figure 4.6, the left side of this subject's head is semitransparent from this effect. Eerie, to be sure, but not an effect we would want on an image going to court.

To simplify, whenever using electronic flash, use only the sync shutter speed. Do not be like one crime scene photographer this author knew, examining his latest set of homicide photographs, who noticed one third of each indoor image was covered by a dark area. When all his victim body shots were covered by these same mysterious dark areas, he was overheard to say, "They misprinted my photographs again." The wet-chemistry photo lab cannot correct flash photographs that have been taken with the wrong shutter speed.

With non-flash photography, shutter speeds were a major exposure control. Shutter speeds were also the primary control of motion within the scene. With flash photography, shutter speeds no longer serve either purpose. Exposure is now a function of the intensity of the flash unit, with the shutter speed chosen only so the flash fires when the shutter is open. Motion control is now fully taken over by the flash unit's duration. The use of fast shutter speeds must be avoided when using electronic flash. If a shutter speed is faster than the sync speed, the shutter will be partially obstructing the film/digital sensor when the flash fires. Fast flash durations now freeze motion. Figure 4.7 is an example of a photographer using a fast shutter speed with flash to freeze the motion of a playing child.

Figure 4.7

Flash freezes motion faster than sync; shutter speed has the shutter closing when the flash fires.

The jumping child is frozen in mid air by the fast flash duration. A faster-than-sync shutter speed only ensures the shutter will be closing when the flash fires.

Many years ago, the common sync shutter speed was 1/60th of a second. Newer camera models are incorporating faster and faster sync shutter speeds. First, some camera makers made a change to 1/125th of a second sync shutter speed. Today, it is not unusual to find top-of-the-line cameras featuring 1/200th and 1/250th of a second sync speeds. Are these a marked improvement? Having the shutter close more quickly after the flash has fired does limit the effects strong ambient lighting can have on the exposure. Because most flash is used in dim lighting conditions, this has a minimal effect on most crime scene photography. It is not recommended to let the flash sync speed be a major consideration when purchasing a camera.

Before leaving the subject of sync shutter speeds, one other idea should be discussed, even though it rarely applies to crime scene photography. When shopping for a new camera, the literature may indicate a particular camera has the option of "rear-curtain" sync. Obviously, this differentiates this camera from one having only "front-curtain" sync. What does this mean? For this discussion, let us presume the flash that works with this camera has a normal flash duration of 1/1000th of a second. Let us also presume the sync speed of the camera is 1/125th of a second. If the shutter will be open for 1/125th of a second, the quicker flash can fire early during that time period or later during that 1/125th of a second time period. Most cameras are designed to have "front-curtain" syncs. That means as soon as the camera's shutter has completely opened, the flash will fire. The shutter will then remain open until the full 1/125th of a second has elapsed, and then it will close.

Rear-curtain sync cameras have the shutter open, some time passes, and then just before the shutter is ready to close, the flash fires. What is the difference? This applies only to creative photography, when the photographer wishes to have creative blurs appear in the photograph. Imagine trying to photograph a speeding car so that its red taillights appear to be elongated streaks behind the car. Over the course of a slow shutter speed, the taillight will streak completely across the image, from left to right. Now it is up to the photographer to determine whether the car will have those streaks in front of the speeding car or trailing behind the car. Naturally, the latter makes more sense. If the camera, however, had only "front-curtain" sync, the shutter would open and the flash would fire immediately, freezing the car at the beginning of the frame, with the red streaks appearing in front of the car. A "rear-curtain" sync would have the shutter open, the car would move across the field of view with the taillights streaking nicely, and just as the car reached the opposite side of the frame, the flash would fire, freezing the car, which is a neat feature, but hardly applicable to crime scene photography.

SET THE FLASH FOR THE FILM USED

Just as it was critical to set the camera for the ISO film speed that was just loaded into it, it is just as critical to set the flash for the film used. Many flash units still require this to be done manually, so this step must not be forgotten. It is the job of the flash calculator dial to suggest an f/stop on the basis of the distance from flash to subject. This can only be done correctly if the flash "knows" what film is being used.

Just as many cameras can "sense" the ISO film speed loaded into it by its DX sensors, so too, some flash units communicate with the cameras they are connected to and would be able to electronically detect the ISO film rating of the film loaded into the camera body. Nevertheless, it is ultimately the responsibility of the photographer to make sure this has happened. How can this be done? No dial or gizmo on a flash unit indicates to the photographer that this has successfully occurred. This will ultimately be done by the photographer's experience with the flash and camera equipment. With any ISO film in the camera and the guide number of the flash known, some results will be common. For example, with 400 ISO film and a 120 guide number flash unit, the f/stop for a 10-foot flash-to-subject distance should be f/22. If the calculator dial suggests anything else, either the photographer failed to manually set the ISO film speed on the flash unit or the flash unit has incorrectly "sensed" the ISO film speed in the camera.

Despite the fact that this is a short section of the chapter, it is nevertheless important. Unless the flash "knows" the ISO film speed being used, all images captured with the flash may be incorrectly exposed.

MANUAL FLASH MODE

Manual flash mode: In the manual flash mode, the full power of the flash is used each time the flash is fired. Because the flash is used to photograph objects at different distances, the usual way to alter exposures is to change f/stops.

Just as non-flash photography has various exposure modes (manual, aperture-priority and shutter-priority exposure modes) flash photography has several ways to determine a proper exposure. **Manual flash mode** was the first method of determining the correct exposure when an electronic flash unit was used.

In the manual flash mode, the full power of the flash is used each time the flash is fired. Because the flash is used to photograph objects at different distances, the usual way to alter exposures is to change f/stops.

Film is usually chosen because the light is dim, so frequently 400 ISO film is used when using the manual flash mode. The one obvious exception (at least it is hoped that this exception is now obvious) is when critical comparison photographs or examination-quality photographs will be captured. In those cases, it is recommended to use 100 ISO film or the digital equivalent.

The shutter speed will be the sync speed.

The only remaining variables of the proper exposure triangle are the lighting at the scene, which will be the flash unit and the f/stop selection.

If the full power of the flash is used when 10 feet from a subject, and the full power of the flash is used at 20 feet, and every other distance, it is the f/stop selection that would have to be responsible for the ultimate exposure.

The formula GN/distance = f/stop applies here. The **flash calculator dial** eliminates the necessity to do any mental math now.

On the basis of this equation, the flash calculator dial now relates various distances with different f/stops. As mentioned previously, it is first necessary to "tell" the flash calculator dial which ISO film has been loaded into the camera. Once that is "known," the flash calculator dial suggests a particular f/stop for every distance between flash and subject. See Figures 4.8 and 4.9.

Figure 4.8 shows how one calculator dial is manually set for 400 ISO film. The white triangle is adjusted opposite the 400 mark. Once that has been done, examine the face of the calculator dial, shown in Figure 4.9. Notice three rings of numbers are around the dial: the outer ring is f/stops, the second ring is distances in feet, and the third ring is distances in meters. Notice f/22 is aligned with the 10' distance, f/16 is aligned with 15", f/11 is aligned with 20", and f/8 is aligned with 30'. In the manual flash mode, those f/stops should be used for the indicated distances. By extrapolation, if an object were midway between any two distances indicated on the dial, the f/stop midway between the indicated f/stops should be used. At 12.5" use an f/19, at 17.5" use f/13, at 25" use f/9.5. The obvious question in many minds is now this: if f/22 is appropriate for a 10' distance, how can closer images be captured without overexposing the image, because f/22 is the smallest aperture on most lenses?

Most, but not all, flashes have some method of diminishing the full power of a manual flash. For this particular flash unit, the calculator dial has several vari-power settings ("vari" being short for variable).

Figure 4.10 shows one particular flash's options for variable power. It shows the flash currently set for full power. Other options include 1/2 power, 1/4 power, and 1/16 power. Decreasing the full manual flash power would enable the crime scene photographer to work at closer ranges than 10' with an f/22. In this case, selecting the 1/16 power option would enable the photographer to work at a 2" distance between flash and subject matter, close enough for most crime scene needs. It should, perhaps, be clarified that this means the flash-to-subject distance, not the camera-to-subject distance. The camera can be as close as necessary to fill-the-frame with the subject matter. The flash must be kept so it is no closer than 2" with this particular flash unit.

With the flash pictured here for demonstration purposes, a Vivitar 285 HV unit, adjusting the calculator dial must also be accompanied with an adjustment to the flash unit's sensor eye. (This flash unit is chosen here because, despite the fact it

Flash calculator dial: In the manual flash mode, it suggests different f/stops to properly expose objects at different distances from the flash.

Figure 4.8

Setting the calculator dial for the ISO film being used.

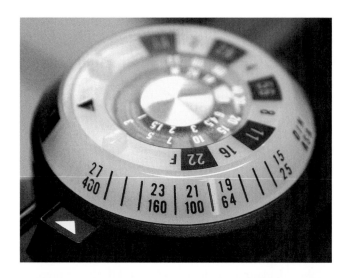

Figure 4.9

A calculator dial face.

Figure 4.10

Vari-power increments.

has been around for a long time and many newer flash units are currently available, it is still in common use with many law enforcement agencies, and it clearly shows manual flash exposure determinations.) See Figures 4.11 through 4.13.

The right side of the Vivitar 285 HV flash is seen in Figure 4.11. Figure 4.12 shows a better look at the flash's sensor eye, and Figure 4.13 shows the side of the sensor eye with the manual flash set to ½ power. It can also be set to M (full manual), ¼, and ¹⁄₁₆, like the calculator dial. Newer flashes have buttons and dials to accomplish the same adjustable settings. The point is that depending on the necessary work distances, a good-quality manual flash unit should be able to adjust to those distances.

Showing the flash unit's sensor eye when discussing manual flash may be confusing to some. In the manual flash mode, the flash unit's sensor eye is actually disabled, or covered, so it cannot "sense" reflected light. However, with this particular flash unit, variable adjustments to the full power of the manual flash are manipulated on the side of the sensor eye. The flash's sensor eye is only relevant to the automatic flash exposure mode, which will be discussed soon.

Vari-power, or the ability to reduce the full manual flash output, needs further explanation. If we can reduce the full power of a manual flash, we must have some idea about its full power capabilities. That will have to be done by reading the manual for your particular flash. One variable, of course, will be the guide

Figure 4.11

The side of a Vivitar 285 HV flash.

Figure 4.12

Two views of the sensor eye.

number for your particular flash unit. Otherwise, some additional commonalities exist. When it is stated that the full power of the manual flash goes off each time the flash unit is fired, this usually means that the flash fires at its normal duration. The intensity of a flash is related to the duration of the flash. For example, many flash units fire at full strength with their duration at 1/1000th of a second. That is the most power they are capable of. As we have seen, however, their intensity can often be reduced. Reducing the intensity of a flash is the same as reducing its duration. So, if full power is 1/1000th of a second, changing the vari-power to ½ power would change the flash duration to 1/2000th of a second. Changing the vari-power to ¼ in effect changes the flash duration to 1/4000th of a second, and changing the vari-power of the flash to 1/16 changes the flash duration to 1/16,000th of a second. Faster flash is dimmer flash. Many modern flash units can be set to increments of 1/64th power or even 1/128th power. Those are some very fast flash durations.

Recall that it was previously stated that the flash now controls motion, whereas with non-flash photography it was the shutter speed that controlled motion. At 1/1000th of a second, or faster, the flash can, indeed, stop most motion. A fast shutter speed is not necessary to freeze motion when the flash is being used.

One other aspect of the manual flash mode also needs to be clarified. If the calculator dial suggests different f/stops for different distances, are the exposure error possibilities of the camera meter when exposing light-colored and dark colored-subjects still relevant for manual flash? In other words, when using manual flash, must the photographer still compensate for the tendency of light-colored objects to be underexposed and for the tendency of dark-colored object to be overexposed? No.

Manual flash provides the right light for the distance. It does not matter whether the subject is white, black, or gray toned. Manual flash provides the right light for the distance. The reflectivity variations of the subject are irrelevant when using manual flash. This knowledge can be put to good use many times. Camera meters and flash sensors can often be fooled by the reflectivity of certain subjects. If this can be anticipated, just switch to the manual flash mode, and the problem has been solved.

Figure 4.13

Sensor eye increment set to ½ power.

The manual flash calculator dial exposure recommendations also presume the flash is aimed directly at the subject or object to be photographed, which is because direct flash will light the subject or object on the side facing the camera, and this front lighting most effectively reflects light toward the film or digital sensor. If the photographer were to remove the flash from the camera with a remote flash cord and begin lighting the subject or object more and more obliquely, obviously there will be incrementally less and less light reflected directly toward the film or digital sensor. The calculator dial cannot account for this, so its exposure recommendations only apply to the direct light situations. In situations where oblique lighting best captures the details of the subject matter, other methods to determine the proper exposure will be needed, or other flash modes, yet to be discussed, will have to be used.

THE "NORMAL" ROOM

The theory of manual flash also depends on another presupposition. In an effort to ensure flash equipment works properly under the conditions most people will use it, flash manufacturers put a lot of research and development money into discovering when and where most of their customers will use their equipment. Because every possibility cannot be covered, flash manufacturers uniformly decided to design their equipment for the most prevalent situation. Research indicated most people use flashes when they take photographs of family and friends in their homes. Therefore, the flash unit has been designed to work properly in the man-

"Normal" room: A "normal" room is the common living room or bedroom. It is about 10′ × 12′ in dimension, it has white ceilings about 9′ or 10′ high, and it has light-colored walls, rather than walls painted with any other color.

ual flash mode when it is used in a **"normal" room**. What is a "normal" room, and why does this make such a difference?

A "normal" room is the common living room or bedroom. It is about 10′ × 12′ in dimension, it has a white ceiling about 8′ or 9′ high, and it has light-colored walls, rather than walls painted with any other color.

When not in a "normal" room, the flash calculator dial exposure recommendation will be inaccurate. Why? What is it about this "normal" room that the flash calculator dial depends on?

When the flash goes off, most of its light travels in a direct line from the flash to the subject or object photographed. But some of the light also moves upward and laterally, and eventually this light will strike a ceiling and/or a wall and bounce inward toward the subject. Manual flash presumes this additional bounce light will be present. The calculator dial's recommendation of one f/stop to be used at any particular distance presumes this additional light will be present. When anyone uses manual flash in any other circumstance, the calculator dial will be incorrect. Just three locations are most frequently a problem.

Do not trust the calculator dial when using manual flash if you are in a room smaller than a "normal" room, if you are in a room that is larger than a "normal" room, or if you are outside. In these situations, compensation will have to be made to the exposure recommendation suggested by the calculator dial. In a room smaller than a "normal" room, there will be more reflected light from the ceiling and side walls, so the exposure will have to be reduced from what is recommended. This would apply to kitchens, bathrooms, or other small rooms. The exposure reduction necessary is normally ½ to 1 full stop. Bracket to be sure. In rooms larger than a "normal" room, there will be less light reflected from the ceiling and side walls, so the exposure will have to be increased from what is recommended. This would apply to very large rooms, auditoriums, parking garages, and the like. The exposure increase necessary is normally ½ to 1 full stop. Bracket to be sure. Outside, there will be no light reflected from ceilings or walls, so the net effect is a much dimmer scene. Begin by adding 1 stop of light as an exposure compensation when using manual flash outside. Bracket. Rule of Thumb 4.1 applies when using manual flash outside.

 Rule of Thumb 4.1: Manual flash used outside is at least 1 stop less bright than what the calculator dial suggests.

A couple of examples may be helpful. In a "normal" room, when the camera is loaded with 400 ISO film and the flash unit has a guide number of 120, if the object being photographed is 10′ away, the calculator dial will indicate the proper f/stop is f/22. If that same object is the same distance from the camera but on a kitchen or bathroom floor, reduce the manual flash power to ½ or ¼ power to compensate for the increased reflections likely to be caused by being in a smaller room. If that same object is the same distance from the camera but outside on the ground at midnight, increase the aperture to f/16, and bracket, to compensate for the lack of any reflections from walls or a ceiling.

Photography students will frequently question how colored walls or walls with various types of wallpaper will affect a "normal" room. If the walls and ceiling are not painted white or off-white, the results of almost every other variation will be diminished light, and compensations will have to be made.

THE FLASH HEAD

Several aspects of the flash head itself have to be understood.

On-Camera Versus Off-Camera Flash

Most camera bodies have a hot-shoe on the top of the camera body, so the flash unit can be directly mounted to the camera. For many photographers, this seems like the optimal placement of the flash, particularly after what was just written about direct manual flash; having the flash unit connected to the hot-shoe will definitely ensure the light from the flash strikes the subject matter directly rather than from an oblique angle. However, examine Figure 4.14.

The flash head is not exactly aimed the same direction as the camera's lens. They are both aimed straight ahead, with the flash head almost parallel to the line of sight of the lens. When the subject matter to be photographed is some distance from the camera, the light of the flash will properly light the same area. Does it

Figure 4.14
Flash mounted to hot shoe.

not make sense, as the subject matter to be photographed gets closer and closer to the camera, the flash will eventually be aimed above the actual target of interest? Look at Figures 4.15 and 4.16.

The target is the Post-It note on the wall. This makes it obvious the flash unit is aimed above what the camera is imaging. Want to see the effect? Figures 4.17 and 4.18 show the effect nicely.

Figure 4.17 was taken 3′ away from a blank wall. First, notice the flash hot spot is above the center of the image. Then notice the underexposed area at the bottom of the wall. With the flash aimed higher than where the camera lens is aimed, one would expect this. Figure 4.18 was taken 6′ away from the same wall. The flash hot spot is still a bit high in the field of view. A small portion of the bottom of the wall is still a bit underexposed, although most people would not notice this fact if it were not pointed out.

Because of these occurrences, a new Rule of Thumb can be suggested.

 Rule of Thumb 4.2: When you are 5′ or closer to your intended subject matter, remove the flash from the camera's hot-shoe, so it can more accurately be aimed at your intended target.

Most flash units do not permit the flash head to be aimed downward. This was the basis for this Rule of Thumb. Now some flash units do permit the flash head to be lowered a bit. Figure 4.19 shows a Canon 550EX Speedlight, which does allow for a slight downward tilt. Despite this, the Rule of Thumb does still apply to most shots closer than 5′.

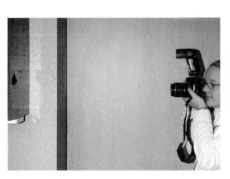

Figure 4.15
3′ away (GWU MFS Student Tahnee Nelson).

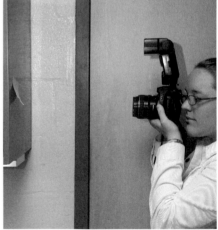

Figure 4.16
1.5′ away (GWU MFS Student Tahnee Nelson).

Figure 4.17
3' away.

Figure 4.18
6' away.

When the flash unit is removed from the camera's hot-shoe, it must be connected to the camera body with either a **PC cord** or **remote flash cord**.

The term PC cord should be explained, although now it is taken for granted when a need to use off-camera flash exists. Two German shutter manufacturing plants are normally credited with developing a camera shutter system that would allow a remote flash to fire while the shutter was open: Prontor and Compur. The PC cord is an electronic cord, frequently coiled, that allows a remote flash to remain electronically connected to a camera so that the flash will fire when the camera

PC Cord and Remote Flash Cord: The PC cord is an electronic cord, frequently coiled, that allows a remote flash to remain electronically connected to a camera so that the flash will fire when the

Figure 4.19
The Canon 550EX Speedlight,
with downward tilt.

camera shutter button is depressed. The PC initials are used to denote the German shutter manufacturing plants that first created this capability: Prontor and Compur. Remote flash cords are used with modern cameras and flashes. Besides enabling the flash to sense when the shutter is tripped, the remote flash cord can also let the flash "know" what ISO film setting has been selected. It also signals the flash unit when the camera's zoom focal length has been changed, so the flash may also accommodate its flash throw for the new focal length selected.

shutter button is depressed. See Figure 4.20. Remote flash cords are used with modern cameras and flashes. Besides enabling the flash to sense when the shutter is tripped, the remote flash cord can also let the flash "know" what ISO film setting has been selected. It also signals the flash unit when the camera's zoom focal length has been changed, so the flash may also accommodate its flash throw for the new focal length selected.

A question naturally arises. If having the flash mounted on the hot-shoe is discouraged, how can the photographer maintain the direct flash lighting that the manual flash system presumes? The flash connected to the camera by a PC or remote flash cord can still be aimed directly toward a subject in front of the photographer. Even if the flash is held at arm's length to the side of the camera, the path the light takes to the front of most subjects is still relatively direct. It is only when the flash angle to a relatively close object is more from the side than from the front that problems relying on the calculator dial's exposure recommendations begin to develop.

It was mentioned earlier that most flash units are not designed to be used closer than about 2′ to the subject matter. It is the PC or remote flash cord that allows the camera to get as close as necessary to fill the frame with small evidence, while at the same time moving the flash farther away to avoid an overexposure. Each photographer usually goes through a learning curve to get the feel of visualizing the subject through the viewfinder while holding the flash at arm's length while properly aimed at the subject. Many photographs have been ruined by allowing the flash to wander away from the necessary direction while the photographer was busy composing and focusing on the subject. It is not a natural position, but fortunately most photographers pick it up without too much difficulty. The trick is to force yourself to hold the flash with the elbow locked straight. If you allow your elbow to bend, the flash frequently creeps closer than the 2′ distance to the subject, and overexposure is usually the result.

Figure 4.20

A camera and flash connected by a remote cord.

Two well-known flash systems enable the flash to be closer to the subject than this normal 2′ distance. Several manufacturers make Ring-Lites. These are round flash units intended to be attached to the end of a camera's lens. Canon's version features circular twin flash tubes that can fire at even power or varied between them over a 6-stop range. Therefore, they can uniformly light a small object without any harsh shadows resulting from flash units held at low oblique angles. A variation on that theme is Canon's Macro Twin Lite, which features two flash heads that are also attached to the end of the camera's lens. Two separate flash heads can be swiveled around the lens, aimed separately, and even removed from their holder and mounted off-camera. See Figures 4.21 and 4.22.

Some might ask about possible benefits of using a flash bracket to remove the flash from the camera hot-shoe. See Figure 4.23. Many flash brackets are on the market. Their primary purpose is to remove the flash unit from the camera hot-shoe to reduce the possibility of the red-eye effect that comes from using a flash unit close to the camera lens. Red-eye occurs when the light of the flash enters the eyes of subjects, reflects off the blood vessels at the back of the inside of the eye, and travels directly back toward the camera. Flashes used on the camera's hot-shoe are notorious for creating this effect. So are cameras with built-in flashes. The solution is to separate the flash from the axis of the lens. The closer the flash is to the camera's lens, the higher the probability red-eye will occur. The flash bracket is normally designed to move the flash far enough off-axis that red-eye is eliminated. A secondary benefit of the flash bracket is to make holding the camera and the off-camera flash less awkward.

The flash bracket does, indeed, satisfy both of these needs. Many crime scene photographers sometimes use them. If you believe it benefits your particular shooting style, then by all means, use one. The only down side is there will be many times that the flash will have to be positioned at other angles, and taking the flash off the flash bracket and returning it to the flash bracket may prove

Figure 4.21

A Canon Macro Ring Lite MR-14EX (Courtesy of Canon, USA).

Figure 4.22

A Canon Macro Twin Lite MT-24EX (Courtesy of Canon, USA).

Figure 4.23

A flash bracket to hold the flash off-camera.

inconvenient for some. Because of this, many flash brackets allow easy detachments and then allow quick reattachments to the bracket.

Flash Head and Camera Orientation

Looking through the camera's viewfinder, we notice that the field of view is rectangular, just as is the negative or digital sensor. The camera is frequently held horizontally. Or it can be rotated to provide a vertical viewpoint of a subject or scene. Likewise, the electronic flash head also throws light into the scene in a rectangular fashion. If the camera is held horizontally, the flash head should be held horizontally; if the camera is held vertically, the flash head should be held vertically. With the flash mounted on the camera's hot-shoe, it automatically corresponds to the camera's orientation. It has just been suggested that the flash will often better light subject matter that is close to the camera if the flash is off-camera on a PC or remote flash cord. When the flash is disconnected from the camera, it is essential to ensure the camera and flash orientations correspond. Otherwise, the light from the flash will not adequately light the scene, and underexposed edges of the image will be noticed.

Figure 4.24 shows a body photographed with the camera held vertically to accommodate the body's orientation. The photographer, however, forgot to also turn the flash head vertically. The result is that the flash did not adequately cover the viewpoint of the camera, so the top and bottom of the image are underexposed. Figure 4.25 attempts to demonstrate this misalignment of camera and flash head. Figure 4.26 shows the correct flash technique: both the camera and flash head are positioned vertically, so lighting is even over the entire area viewed by the camera.

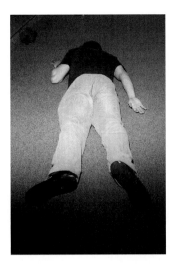

Figure 4.24
Camera vertical; flash
horizontal.

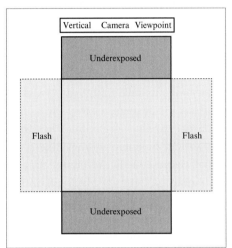

Vertical Camera Viewpoint

Underexposed

Flash   Flash

Underexposed

Figure 4.25
Camera vertical; flash
horizontal.

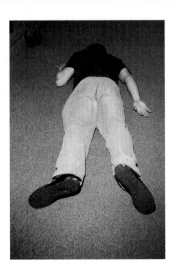

Figure 4.26
Both camera and flash vertical.

Flash Head and Focal Length

Many agencies provide their crime scene photographers with a zoom lens as the standard lens. With a zoom lens, it may be possible to view the scene or an item of evidence with a wide-angle focal length, with the lens set to 50 mm, or with a short telephoto focal length. Having these options makes a zoom lens very versatile. If dim lighting requires a flash be used to obtain a proper exposure, it is then necessary to have a flash that has the same versatility. In this case, if the lens is zoomed in or out to either a wide-angle setting or telephoto setting, it is necessary to have the flash head also zoomed in or out to the equivalent focal length.

Some flash units, however, only have one flash head setting. In such a case, this kind of flash should only be used with the zoom lens set to the 50-mm focal length. With this kind of flash, the camera's lens should never be set to either the wide-angle or telephoto range. The effect can be similar to the horizontal/vertical misalignment of camera body and flash head just discussed.

Many flash heads, fortunately, are adjustable to accommodate various zoom lens focal length options. Some of these flash heads require a manual zooming: the photographer must extend or compress the end of the flash head to match the focal length of the lens selected.

Figures 4.27 and 4.28 show a manual flash head compressed to accommodate a wide-angle lens setting, set to "normal" to properly light a scene with a 50-mm lens, and extended to accommodate a telephoto lens setting. When the flash head is compressed for a wide-angle lens setting, the light of the flash is distributed wider than is normal. Because the light from the flash is being forced to light a wider area, it is a bit dimmer straight ahead. The flash intensity is not as effective at lighting objects directly in front of the camera. When the flash head is elongated for a telephoto lens setting, the light of the flash is narrower than it would be at the "normal" setting, and because of this, the light travels farther than it would otherwise. Of course, if you are using a telephoto lens setting simply to crop in on a specific area of a crime scene that is not further from the camera, it would not be necessary to alter the flash head to the telephoto setting. Anytime a wide-angle lens is used, the flash head must be adjusted to its wide-angle setting, so that the edges of the field of view now seen by the film or digital sensor will be properly exposed.

Some flash heads automatically zoom in and out as the focal length of the lens is adjusted. An electronic signal is sent from the camera body to the flash, and adjustments are automatically made. These types of flashes usually zoom internally, so their exterior shapes do not change. See Figure 4.29 as an example of what these flashes would look like as different focal lengths are selected. Again, it is possible to manually set these changes also. Many flash units also can be used with a wide-angle flash diffuser. This can be a diffuser that fits into a slot

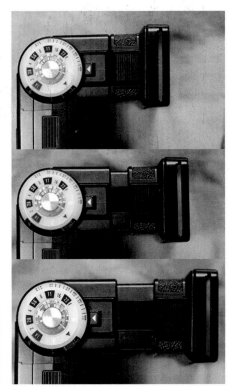

Figure 4.27
Manual flash head variations I.

Figure 4.28
Manual flash head variations II.

near the flash head, clips over the end of the flash head, or can be retracted from the flash head itself to cover the flash face. Flash diffusers can be used with extreme wide-angle lenses. Or they can be used merely to soften the light of the direct flash when simply diminishing the flash intensity does not produce the effect desired. Very close-up images on reflective surfaces might benefit from such diffused lighting. Injuries or wounds to skin would be another situation when diminishing the glare of direct light may be a benefit. Figure 4.30 shows a wide-angle flash diffuser.

PROBLEMS WITH THE MANUAL FLASH EXPOSURE MODE

The first type of flash available was manual flash. There was no choice. Today, two other electronic flash exposure modes exist. These other modes were created to solve the perceived problems that the manual flash exposure mode had. They can definitely be thought of as improved flash modes. However, occasions still occur when the use of the manual flash mode will be recommended as the best mode of

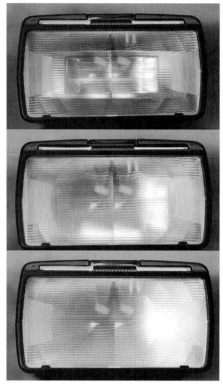

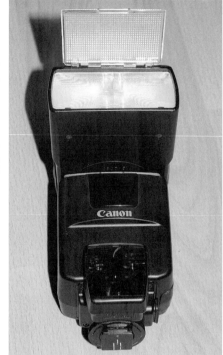

Figure 4.29
Automatic flash head adjustments.

Figure 4.30
A wide-angle flash diffuser.

the available choices. During the course of this chapter, we will point out when one particular mode may be better than the others. To begin, the perceived problems with the manual flash exposure mode should be detailed.

Frequent f/Stop Manipulations

In the manual flash exposure mode, the full power of the flash goes off each time the flash is used. At 3′ from an object, the full power of the flash goes off. At 10′, the full power of the flash goes off. At 20′, the full power of the flash goes off. At every distance, the full power of the flash goes off. How is exposure controlled at all these different distances? By altering the f/stop selection. This means, every time the photographer moves from one position to another position, the photographer has to manually alter the f/stop selection to maintain a proper exposure. This keeps the photographer very busy. Would it not be an improvement if this constant need to alter f/stops could be eliminated or at least diminished? Working a major crime scene is complicated enough. Perhaps a way of simplifying the assurance of a proper exposure can be made?

Battery Consumption

With the full power of the flash going off with every photograph, the batteries in the flash unit do not last long. Replacing batteries frequently can be a very expensive proposition, not to mention an inconvenience. The manual flash exposure mode is notorious for its battery consumption.

One may consider using rechargeable batteries or supplemental battery packs to prolong the life of the batteries, but these represent additional costs too. The basic premise continues: the manual flash exposure mode is hard on batteries.

If a flash mode could be created that does not consume the batteries so quickly, that would be a distinct improvement.

Electronic Flash Recharge Times

If the full power of the flash is used with each photograph, the electronic flash must recharge completely after each use. One must wait for the flash to completely recycle, because if the flash is used again before it has had a chance to completely recycle, only a partial flash fires, and an underexposed image is the result. So, we must wait for the flash to indicate it has completely recycled. Usually, a "ready" light is on the flash somewhere to indicate it has completely recharged. As the batteries near the end of their normal life, this recycle time begins to get longer and longer. Very inconvenient. A new Rule of Thumb relates to this.

Rule of Thumb 4.3: If the flash recycle time exceeds 30 seconds, it is time to replace the flash batteries. Major crime scenes are not the time to be waiting too long for the flash to recycle.

Perhaps a new flash exposure mode can be developed that "solves" all three of these problems at the same time? We will get to that momentarily.

BRACKETING IN THE MANUAL FLASH EXPOSURE MODE

Photography students should be reminded that talking about electronic flash is still talking about light, and, therefore, we are still talking about exposure "stops." The theory of halving and doubling the light intensity from one exposure situation to another exposure situation is still valid.

When bracketing with an electronic flash, you cannot alter the shutter speed as with non-flash photography. We must look for other ways to change the exposure.

Manual Flash Bracketing for a +1

If the beginning exposure was, for example, ISO 100 film/1/60th/f/16, we can bracket by changing f/stops, just like bracketing in the non-flash manual exposure mode. If the original f/stop selection was f/16, to bracket for a +1 exposure, the f/stop can be changed to f/11.

Bracketing can also be achieved by altering the intensity of the flash output. How would you be able to get more power than full manual flash? Full manual flash is the most power the flash unit is designed to produce. How would you get more than "all?" In this situation, the +1 can only be accomplished one way: you can move the flash unit closer to the subject. How much closer? To determine this distance, we need to reexamine the calculator dial.

If you recall, the calculator dial is designed to relate two variables: distances and f/stops. In the manual flash exposure mode, one f/stop can be used for every different distance the flash can be from the subject matter. Notice the following alignments on the face of the calculator dial:

- f/8 is more or less aligned with 30′.
- f/11 is aligned with 20′.
- f/16 is more or less aligned with 15′.
- f/22 is aligned with 10′.
- f/22 is aligned with 7′, when the Vari-Power ring is set to ½.
- f/22 is aligned with 5′, when the Vari-Power ring is set to ¼.
- When the Vari-Power ring is set to 1/16 (Figure 4.32), the f/22 is opposite the 2′ distance.

To achieve a +1 exposure, change the flash-to-subject distance:

- When using an f/8, move to the 20′ distance.
- When using an f/11, move to the 15′ distance.
- When using an f/16, move to the 10′ distance.

Figure 4.31

The flash calculator dial.

Figure 4.32

Calculator dial set to 1/16 power.

- When using an f/22, move to the 7' distance.
- When using an f/22 at ½ power, move to the 5' distance.
- When using an f/22 at ¼ power, move to the 3.5' distance.

These are flash-to-subject distances. These changes can be made by either having the photographer actually move to the prescribed distance or sometimes the distance change can be effected merely by the photographer moving the flash closer to the subject while the photographer remains at the same position. When only short distance changes are indicated, these can be done simply by moving the hand holding the flash closer. For example, to change from the 7' to the 5' distance, the photographer can step forward or remain at the 7' distance while extending their hand holding the flash unit to the 5' distance.

Manual Flash Bracketing for a −1

If the beginning exposure was, for example, ISO 100 film/1/60th/f/16, we can bracket by changing f/stops, just like bracketing in the non-flash manual exposure mode. If the original f/stop selection was f/16, then to bracket for a −1 exposure, the f/stop can be changed to f/22.

Or, we can bracket by altering the intensity of the manual flash output. Many flash units will allow the manual flash exposure to be reduced by changing the calculator dial to ½, ¼, or progressively smaller increments. A −1 exposure is achieved by setting the manual flash unit to ½ power. A −2 exposure is achieved by setting the manual flash unit to ¼ power. Some manual flash units can reduce their manual flash output down to 1/128th power.

Of course, a −1 flash bracket can also be achieved by keeping the power of the flash at full manual while moving the flash unit farther from the subject, in the opposite increments mentioned in the preceding.

Enterprising photographers have used a variety of other techniques to reduce the flash intensity. Collecting variations on a theme makes one a more versatile photographer. Students are always told that the more tools they have on their bat-belts, the more choices they will have to solve problems that may develop. Options are good. Some of these may prove useful to the reader also.

When using the manual flash exposure mode, placing one layer of a hand-kerchief over the flash head effectively reduces the manual flash output by about 1 stop. Caution should be used when applying this concept. Trying two layers and three layers of handkerchiefs over a flash head, thinking the light diminished would be uniform with each layer, may seem logical. Not so. Do not use more than one layer of a handkerchief. Many photographers carry a hand-kerchief and rubber band in their camera kits, just for this possibility. A variation is to use a layer of toilet tissue. Again, multiple layers of toilet tissue cannot be counted on to yield uniform reduction in lighting levels. White paper towels are too thick.

No handkerchief handy? You can always use your finger to diminish the flash output.

Experience has shown that holding one finger over the flash head effectively reduces the manual flash output by about 1 stop. One might think that this would cause a dark area in the middle of the image, which is the effect if you are too close to your subject. However, the light coming from either side of the finger spreads out and eventually meets in the middle. Do not use this technique if you are about 8′ or closer to your subject. If it is not obvious why Figure 4.34 shows the finger a short distance away from the flash, allow other's experiences to save you from some pain. The electronic flash puts out a large quantity of light: enough to be quite hot to the touch, perhaps not enough to inflict real burns with just one flash, but certainly enough to be uncomfortable. A small distance between flash and finger is warranted.

Simply changing the flash zoom head from "normal" to "wide" will reduce the flash intensity, even if the focal length of the lens used is not a wide-angle lens. Using a 50-mm lens but setting the flash head to "wide" will soften the light by about ½ stop.

The wide-angle flash diffuser, designed for use with wide-angle lenses, can also be used to simply diffuse the flash when used with other focal length lenses. Putting the wide-angle flash diffuser over the flash head when using a 50-mm lens still diffuses the light wider than normal while decreasing its intensity straight ahead. The wide-angle flash diffuser reduces the flash output straight ahead by about another ½ stop.

Figure 4.33
One finger over the flash I.

Figure 4.34
One finger over the flash II.

Just as many filters can be used over the lens of the camera, so too a variety of filters are designed for use with some flash units. Figure 4.35 shows several flash filters designed to be used with Vivitar flashes. The upper middle flash filter is a neutral density filter that reduces the flash output by 2 stops. The other flash filters are designed for other creative purposes, not relevant to crime scene photography.

Finally, flash will be softened if it is bounced off of another surface before arriving at the scene of interest. A section on bounce flash is provided later in this chapter. It is mentioned here just for the sake of completeness.

THE INVERSE SQUARE LAW

The f/stop numbers are a strange sequence of numbers to be sure. It was previously promised that the f/stop numbers themselves would be explained so they make sense. It is now time for that. Manual flash is based on the relationship between the intensity of an electronic flash and the different distances between the flash and the subject matter being photographed. Recall the equation GN/distance = f/stop. When the light of the flash was brighter (closer to the subject), a smaller aperture had to be used for a proper exposure. When the light

Figure 4.35
Flash filters.

of the flash was dimmer (further from the subject), a wider aperture had to be used for a proper exposure. With a 120 GN flash:

- $120/11' = f/11$
- $120/60' = f/2$

Inverse square law: The inverse square law states that as light spreads outward from a point of light source, its intensity varies inversely by the square of the distance it travels. Specifically, if the distance light travels is doubled, the intensity of the light at this new distance is quartered.

The **inverse square law (ISL)** explains the f/stop numbers, finally.

Strictly speaking the inverse square law states that as light spreads outward from a point light source, its intensity varies inversely by the square of the distance it travels. Although an electronic flash unit is not strictly a point light source, because the flash head itself is larger than such a theoretical point, the inverse square law also explains light emanating from a flash unit for all practical purposes.

As light travels away from its source, it does not do so and remain a straight line, as would a laser beam. Normal light expands as it travels outward. Its geometric shape would more closely conform to an expanding cone of light. As light spreads a certain distance, it is relatively weaker at any one point at the new distance traveled. The inverse square law allows us to precisely determine its relative intensity at any given distance.

Examine Figure 4.36. It is the basis of electronic flash exposure determinations in the manual flash mode.

Notice at distance "D," the area lit by the flash unit would be the single rectangle, and it is one unit. As the distance the light travels is doubled, to "2D," the area the light covers is now four times as great. Covering four times the area, at any one point, the light is only 1/4th as intense. This makes sense. As light travels farther, it is dimmer.

Not only is it dimmer, the inverse square law indicates exactly how much dimmer the light is. At twice the distance, the light is 1/4th as bright.

Let us consider an example. If at 10′, an f/22 produces a proper manual flash exposure, what would be the proper f/stop to use for an object 20′ away? In other words, because the light will be dimmer after it travels farther, how much do we have to alter the original exposure settings to achieve a proper exposure at the new distance? Because the light will be dimmer, we expect that the aperture will have to be opened up for a proper exposure. Just how much will it have to be opened up?

The inverse square law can be expressed as an equation: $I = 1/D^2$. The intensity of light (I) equals the inverse of the distance change squared (D^2). Because the distance is doubled, from 10′ to 20′, D = 2. I (intensity) = $1/(2)^2$. In this case, $I = {}^1/_4$. Recall the concept of exposure stops being either halves or doubles. Each halving of the light intensity is a −1 stop change. In this case, the intensity has diminished and is currently $^1/_4$ as bright as it was. Because $^1/_2 \times {}^1/_2 = {}^1/_4$, then changing from 10′ to 20′ results in a 2-stop reduction of the light. If the original f/stop was f/22, and the light intensity has diminished 2 stops, it will be necessary to open the aperture 2 stops to maintain a proper exposure. At 20′ it will be necessary to use an f/11 instead of the original f/22.

The Inverse Square Law I

2D

¼ I

½ x ½

1D

1I

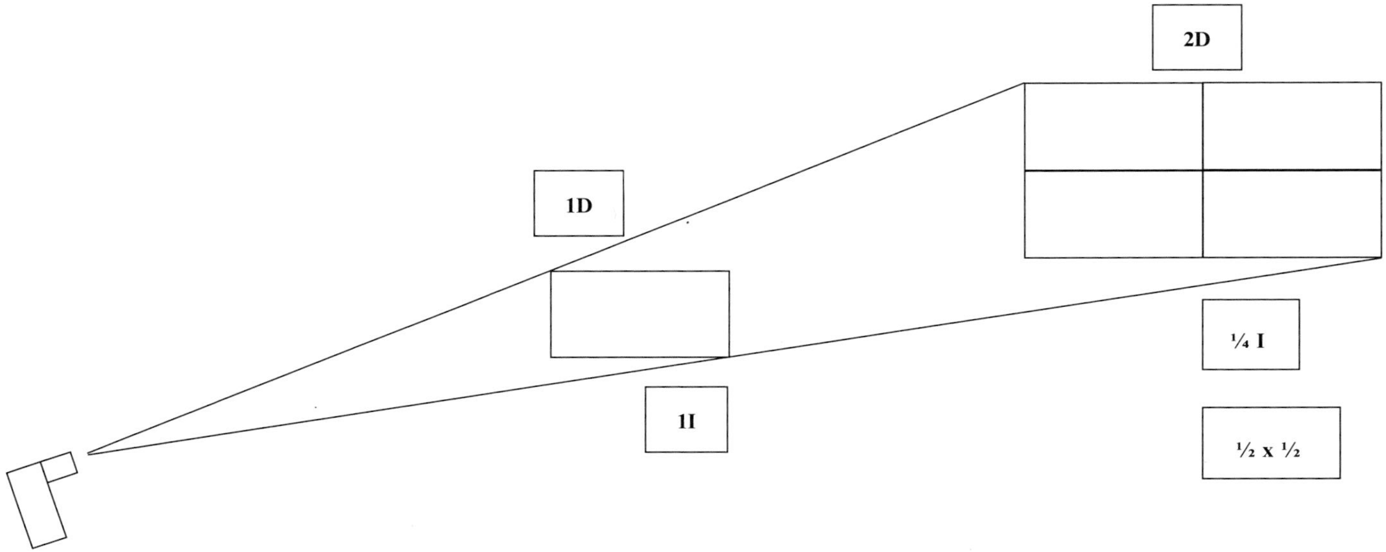

The **Inverse Square Law** shows the relationship between the *Distance* light travels and the *Intensity* of the light at different distances. As the *Distance* light travels is doubled, its *Intensity* is quartered. The light at 2D covers four times the area as at 1D, therefore the light is ¼ as bright at any single point. To determine this, take the distance change, 2, and invert it (make it a fraction). 2 becomes ½. Square that (½ x ½), and the result is the *Intensity* of the light (¼ I) at that new *Distance*(2D). The *Intensity* of light at any *Distance* change can be expressed by this equation: $I = 1/D^2$.

Figure 4.36
The inverse square law I.

Of course, just checking the flash calculator dial would also indicate this, so mental math is not required. Recall Figure 4.31.

The inverse square law indicates these changes. The result: doubling the distance that light must travel results in a quartering of the light at that new distance.

The equation, $I = 1/D^2$, allows us to determine the new intensity for any distance change. Another equation allows us to determine the distance change if a new intensity is known. $D = \sqrt{1/I}$. The inverse square root of an intensity change is the new distance.

Many photography books mention the inverse square law is the basis for the derivation of the f/stop numbers. Although they state this fact, they do not explain it. Figure 4.37 is the beginning of the explanation. If doubling the distance light must travel results in a quartering of the light output ($\frac{1}{2} \times \frac{1}{2}$) at the new distance, it may make sense to ask at what distance would the light only be 1 stop less bright? For many, this may seem like a trivial question, because the answer seems obvious. Half the distance between 10′ and 20′ is 15′. That is certainly true, but the question still remains, at what distance would the light be exactly ½ as bright? The equation, $D = \sqrt{1/I}$, offers the solution. The inverse of ½ is 2. The square root of 2, then, is 1.414D. Because the original D was 10′, the new D is 14.14′. Had we not expected the answer to be 15′? Is this just a rounding issue? No. 14.14 does not round to 15. If anything, it rounds to 14. Along with the original D, we now have a sequence of three distances: 1D, 1.414D, and 2D. Does this sequence ring any bells? For many, this significance may not yet be clear.

If not, let us ask another interesting question. At what distance would the light be 1/8th as bright as the original distance? See Figure 4.38. Realize that 1/8th is just another way of expressing $\frac{1}{2} \times \frac{1}{2} \times \frac{1}{2}$, or three stops less bright than the original distance. We must return to the equation $D = \sqrt{1/I}$. The inverse of 1/8 is 8; the square root of 8 is 2.82 or 2.8D rounded off. And 10′ × 2.8 is 28′.

Now, let us look at the progression: 1D, 1.4D, 2D, 2.8D. Does this series of numbers suggest anything? Add this to the mix: each number in the series represents an exact 1-stop diminishing of the light from the previous number. If the series continued, would you be surprised to find out that the next number in the series was 4D? Or, the next number after that was 5.6D? And, the next number was 8D? And the next rounded off to 11D? See Figure 4.39. Followed by 16D, and 22D?

These are the f/stop numbers. Each "larger" number (further distance) represents $\frac{1}{2}$ the light of the previous number, which is the same relationship the f/stop numbers have with each other. The inverse square law is the foundation for the derivation of f/stop numbers.

If a manual flash was being used, the photographer was standing at exactly 11′ from an item of evidence, and the immediate task was to bracket a series of images of the evidence, other than the obvious changes in composition, bracketing could be done by taking three images from 8′, 11′, and 16′. This presumes, of course, the proper exposure had already been determined for the 11′ distance. If a photogra-

The Inverse Square Law II

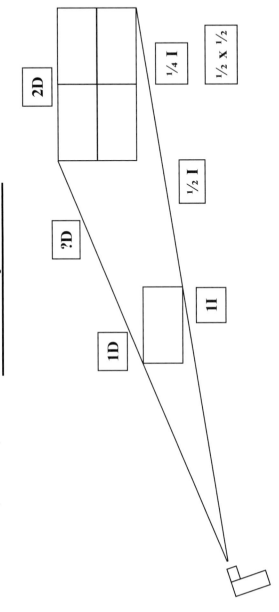

If the light at 2D is ¼th as bright, we now realize this is a 2/stop diminishing of the light. Each 1/stop decrease of the light is half the light of the original. A 2/stop diminishment of the light is ½ x ½, or ¼ of the light. The question now is "At what **Distance** would there only be a 1/stop decrease of the light: ½ the light?" If, for example, 1D is 10 feet from the flash, and 2D is 20 feet from the flash, at what **Distance** would the light be just ½ the **Intensity** as at 10 feet? We want to say 15 feet, don't we? But is this correct?

This is how to determine the **Distance** when the **Intensity** of the light is known. Assume we knew the **Intensity** was ¼th as bright at a certain point, and we needed to determine the **Distance** that corresponded. Pretend we don't already know the answer: 2D. Beginning with the **Intensity** change, ¼, invert it. The result is 4/1 or 4. Take the square root of 4, and that is the distance change, or 2. The **Distance** change would be 2D, but we already knew that. Well, if this works with an answer we know, it will also work when we don't know the answer. So, consider ½ I. Invert the ½ and that is 2/1 or 2. Take the square root of 2, and that will be the new **Distance**. D = √1/ I. In this case, it will be 1.4D, or 14 feet. The progression from 1D to 1.4D to 2D is a curious one, to be sure!

Figure 4.37

The inverse square law II.

The Inverse Square Law III

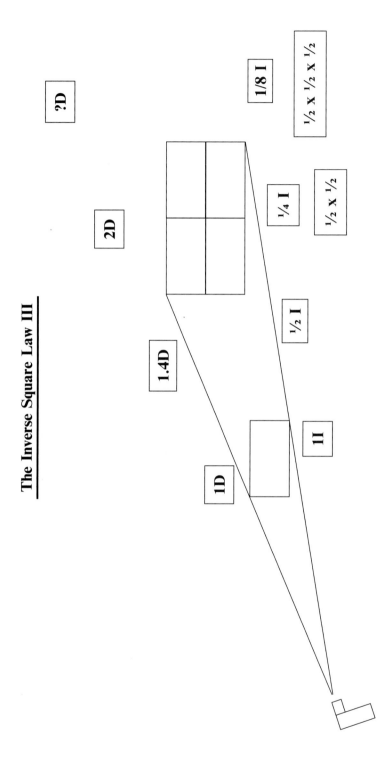

Now, let us ask ourselves at what **Distance** would the **Intensity** of the light be 3/stops less bright or 1/8th as bright? Take the inverse of 1/8 and that is 8/1 or 8. Determine the square root of 8, and that is 2.8. So, at 2.8D, or 28 feet, the **Intensity** of the light would be just 1/8th as bright. Does the sequence 1D, 1.4D, 2D, and 2.8D look familiar? It should. Maybe one more example will make it more apparent.

Figure 4.38
The inverse square law III.

The Inverse Square Law IV

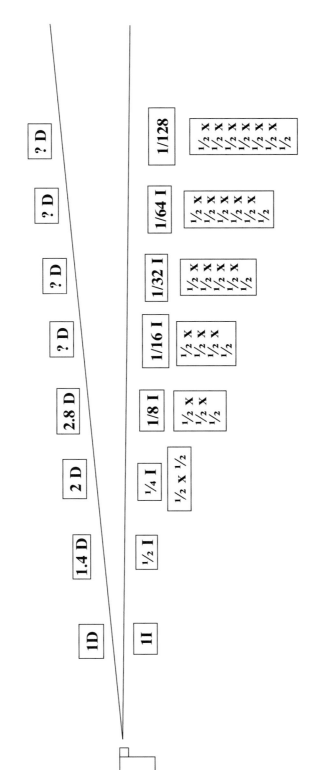

What is the inverse square root of 1/16? 4! What is the inverse square root of 1/32? 5.6!
What is the inverse square root of 1/64? 8! What is the inverse square root of 1/128? 11!
Now look at the progression: 1D, 1.4D, 2D, 2.8D, 4D, 5.6D, 8D, and 11D! That series you have to recognize! It is the f/stop sequence!
The Inverse Square Law is the foundation for the derivation of the f/stop numbers! So, the f/stop numbers are exact increments that describe precisely a reduction of the light intensity by ½ at each step. Or, the f/stop numbers can be considered as distances at which the light is reduced by precisely 1/stop.

Figure 4.39

The inverse square law IV.

The Inverse Square Law V

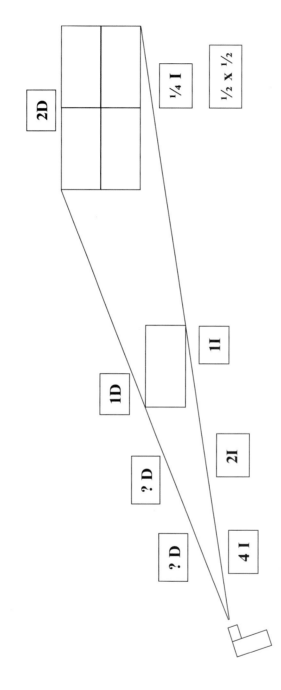

Now, rather than having the *Intensity* of the light diminishing, let us look at two examples of the light becoming brighter than the beginning point. At what *Distance* would the light be twice as bright as at 10 feet? Or, at what distance would the light be exactly 1/stop brighter? Find the inverse of 2, which is ½, or .50. Find the square root of .50, and that is 0.7.; 0.7D is 7 feet.

At what *Distance* would the light be four times as bright as at 10 feet? Or, at what distance would the light be exactly 2/stops brighter? Find the inverse of 4, which is ¼, or .25. Find the square root of .25, and that is 0.5.; 0.5D is 5 feet.

Notice one other thing. Doubling the original *Distance* quarters the light *Intensity*.
Halving the original *Distance* quadruples the light *Intensity*.

Figure 4.40

The inverse square law V.

pher is standing at one of these distances (1.4′, 2′, 2.8′, 4′, 5.6′, 8′, 11′, 16′, 22′), moving to any other adjacent distance will result in precisely 1 stop more or less light on the image. The f/stop numbers can also be considered as distances. They are the precise distances light will travel to be more or less bright by exactly 1 stop.

The inverse square law also works if the need is to determine either distances or intensities that are closer or brighter than any particular starting point.

If 10′ was the starting point, where would the light from a manual flash be precisely twice as bright or 1 stop brighter? $D = \sqrt{1/I}$. The inverse of 2 (twice as bright) is ½ or .5. The square root of that is 0.7D, which is 7′. Where would the light from a manual flash be precisely four times as bright or 2 stops brighter? $D = \sqrt{1/I}$. The inverse of 4 is ¼ or .25. The square root of that is 0.5D, which is 5′.

AUTOMATIC AND DEDICATED FLASH EXPOSURE MODES

As good as the manual flash exposure mode is at helping the photographer determine the precise exposure in many situations, it does have some negatives associated with it. Manufacturers of flash units developed two other flash exposure modes specifically aimed at resolving these negatives.

AUTOMATIC FLASH EXPOSURE MODE

Many newer flash units have an **automatic flash exposure mode** capability.

These flash units have a small sensor eye located somewhere on the front of the unit. Recall Figure 4.12. The sensor eye measures the amount of light reflected from the scene, and when it has determined that enough light has been reflected toward it for a proper exposure, this unit can actually cut short the flash duration to prevent overexposures. Light travels so fast (the **speed of light** is frequently expressed as 300,000,000 meters per second, or as 186,000 miles per second) that the flash may still be actively projecting light, while some of this light has reached the scene, has reflected off of the scene, and has returned to the camera and flash unit: all seemingly in the same instant.

If the right volume of light has been received for a proper exposure, the flash shuts off early. Early? The full manual flash duration is 1/1000th of a second. If the amount of light for a proper exposure has reflected off of the scene and has been sensed by the automatic flash sensor near the camera, the sensor can stop the flash so it would be shorter than the normal 1/1000th of a second. How much reflected light is required? When the sensor receives 18% of the light that reflects from the surface of the scene, the sensor will shut off the flash.

Is this beneficial? Yes. Actually three benefits exist.

- Rather than emitting all the power of the flash, like the manual flash exposure mode does, the automatic flash exposure mode conserves the flash power not

Automatic flash exposure mode: The sensor eye on the flash unit measures the amount of light reflected from the scene, and when it has determined that enough light has been reflected toward it for a proper exposure, this unit can actually cut short the flash duration to prevent overexposures.

Speed of light: The speed of light is frequently expressed as 300,000,000 meters per second, or as 186,000 miles per second.

consumed when the flash goes off. The unused battery power results in quicker flash recycle times. Because all of the flash power has not been expended, the saved power enables the flash to be ready faster for the next photograph opportunity. The flash unit needs to recharge not from zero, but from 20% to 80% of a full charge, depending on the distance between the flash and the scene. Therefore, it takes much less time to fully recharge.

• This unused battery power prolongs the life of the batteries as well. They just last longer because they do not have to fully recharge the flash unit every time the flash is fired.

• The automatic flash exposure mode also uses flash ranges, rather than just 1 f/stop for every different distance. For instance, in the manual flash exposure mode, an f/22 may be just right for a 10′ distance when using an ISO 400 film and a flash rated with a 120 GN. Any closer, and the photograph is overexposed; any farther, and the photograph is underexposed. With an automatic flash unit, the range for an f/22 might be 2′ to 11′. Anywhere in that range, the f/22 would produce the proper exposure. At 11′ the flash works just like the manual flash exposure mode would. But at closer distances, the flash durations get shorter and shorter, recycle times are quicker, and the battery life improves.

Different flashes have different numbers of automatic flash exposure modes. Some will have two auto ranges, some will have three auto ranges, and some can have four auto ranges.

Examine the particular calculator dial in Figure 4.41. This flash unit has four auto ranges, represented by the four shades of gray in this black and white photo, which are really the colors purple, blue, red, and yellow. The purple auto range would require an f/22. Notice the f/22 is aligned with the purple box at the 10′ range. Also notice a line coming from this position, running to the right, which continues past the 3′ point. This indicates the purple auto range covers the distance from 2′ to 11′ when using an f/22. At the 11′ distance, the purple auto mode works exactly like the manual flash exposure mode. The full power of the flash goes off, none of the flash power is retained by the flash, and the flash has to recycle from zero power back to fully charged, which can make the recycle time quite long.

At 7′, however, reflected light from the scene will reach the sensor quicker, and the flash will be shut off early. Because the change from 10′ to 7′ is about a 1-stop difference, the flash duration will change from the normal 1/1000th of a second to about 1/2000th of a second (half the original). The unused power will enable the flash to recycle quicker, and the batteries will last longer. Also, when the photographer moved from the 10′ distance to the 7′ distance, he or she did not have to change the f/stop setting.

At 5′, the time it takes for 18% of the light to reflect from the scene to the flash sensor is shorter, so the flash sensor will shut the flash unit off quicker. This time, the flash duration is approximately 1/4000th of a second. More unused flash power has been retained by the flash unit, and it recycles quicker.

Figure 4.41

Calculator dial and auto flash exposure ranges. Please visit the book's companion website (http://books.elsevier.com/ companions/9780123693839) for a color version of this figure.

At 3.5′, the flash duration is extremely fast, and the flash is ready for another shot almost immediately. Again, f/22 can be used, which saves the photographer from making an exposure adjustment to the camera.

The blue auto mode works the same way, except with it, an f/16 has to be used. The blue auto range is indicated as between 2′ and 15′.

The red auto mode works the same way, except with it, an f/8 has to be used. The red auto range is indicated as between 5′ and 30′.

This flash unit has another auto mode, but it is not recommended for crime scene work, because it has to be used with an f/4, which does not provide a good depth of field.

Because the concept of depth of field has just come up again, this would be a good time to point out the obvious: the best f/stop for depth of field distance, f/22, just happens to be the worst f/stop for flash distance, 2′ to 11′ in the auto flash exposure mode. When using flash, we are continually having to compromise depth of field for proper lighting distances. With a 120 GN flash, an f/8 will only get us 30′ of light from our flash. Because it is our continued goal to have our crime scenes both properly exposed and in focus, when using flash, a new Rule of Thumb is relevant.

Rule of Thumb 4.4: When using flash, the world beyond 30′ does not exist, because beyond 30′, the scene will either be underexposed or a wider aperture will have to be used, which results in a smaller depth of field range.

This presumes two things: you are using a 120 GN flash, and you are using a single flash burst to light your scene. Later in this chapter, you will learn how to use multiple flash bursts to light large crime scenes. If you have a flash unit with a stronger guide number, the 30′ distance may be extended to 40′ or so.

Rule of Thumb 4.4 simply means that if you need to light a scene longer than 30′, wider apertures will be required, which begins to severely impact your depth of field range. Of course, you would not need a long depth of field range if your only goal were to light a single object 60′ away. In that case, the calculator dial previously shown indicates an f/4 will do the job. If, however, it were your goal to show a wide area around an item of evidence that is 60′ away from the camera, the f/4 would not provide the depth of field required.

Another aspect of the purple, blue, and red auto flash modes shown on the calculator dial needs to be pointed out. Parts of the three flash ranges overlap: f/22 is appropriate for a distance between 2′ and 11′; f/16 is the correct f/stop for 2′ to 15′; and, f/8 is right for 5′ to 30′. Between 5′ and 11′ all three colored auto modes could be used and would produce proper exposures. In these situations, how should any particular auto mode be selected? When all else is equal, maximize DOF. Therefore, between 2′ and 11′, use the purple auto mode. Only between 11′ through 15′ should the blue auto mode be used. And, only between 15′ and 30′ should the red auto mode be used.

If the automatic flash exposure mode has the preceding three benefits, is there any downside to using this flash mode? Yes. Because a sensor eye is reading reflected light from the scene, there will be times the sensor eye might be "fooled" about the proper light to properly expose some scenes. The auto flash exposure mode can be "fooled" by the same situations that will "fool" the exposure meter in the camera body when used for non-flash shots. Overly bright scenes will reflect more light than a "normal" scene, causing the flash to be shut off early, and underexposures will be the result. Overly dark scenes will reflect less light than "normal" scenes, causing the flash duration to be prolonged, resulting in overexposures. The flash sensor can be "fooled" exactly the same ways the camera body exposure meter was fooled under certain circumstances. What can be done in these situations? You can either remember the exposure recommendations earlier suggested for non-flash photography, or, in these situations, you can switch to the manual flash exposure mode, which does not make its exposure recommendations on the basis of reflected light.

However, these situations are the exceptions to the rule. The automatic flash exposure mode will benefit the crime scene photographer under most situations and should be used as the primary flash mode most of the time.

We previously learned how to bracket in the manual (non-flash) exposure mode and the automatic (non-flash) exposure modes. We also learned how to bracket in the manual flash exposure mode.

If you recall, in the manual (non-flash) exposure mode, bracketing was accomplished by manually changing the f/stop selection. For example, if the original exposure was 100 ISO film/1/60th/f/16, +1 and −1 exposures would be done with f/11 and f/22, respectively.

In the automatic (non-flash) exposure modes, bracketing was accomplished by changing the exposure compensation dial to +1 or −1. This applied to the program automatic exposure mode, the aperture-priority automatic exposure mode, or the shutter priority automatic exposure mode.

In the manual flash exposure mode, bracketing was accomplished by changing the f/stop settings, by changing the flash intensity, or by changing the flash-to-subject distance.

Now, if we are using, for example, a Vivitar 285 flash in the auto-purple or auto-blue mode, how is bracketing accomplished? (Other automatic flashes will behave similarly.)

Let us look at a couple of practical examples. Because we are using flash, we probably have ISO 400 film loaded in the camera. The purple auto mode allows an f/22 to be used from 2′ to 11′. The blue auto mode allows an f/16 to be used from 2′ to 15′. Automatic flash exposure modes provide ranges within which 1 f/stop can be used. Exposure is controlled within these ranges by the flash sensor eye reading the reflected light coming from the subject and then altering the flash duration so that an 18% exposure is obtained. If you are in the close end of the flash range, the flash duration will be cut short when 18% reflected light has been received. The flash duration will be increased as the flash-to-subject distance increases, until at the far end of the flash-to-subject range, the flash duration is the same as a full manual flash. In other words, in the automatic flash exposure mode, the flash duration can be less than the full manual flash output, or it can be the equivalent of the full manual flash output. Automatic flash output cannot be more that the manual flash output.

To bracket in the auto-purple mode when using an f/22, a +1 is simply accomplished by switching to f/16. In the auto-blue mode, if the prescribed f/stop were f/16, change to f/11 for a +1 exposure.

For a −1 bracket when in the auto-purple mode, because we are already using the smallest aperture of the lens, an f/22, we must move to another distance, outside the purple auto flash range, according to the inverse square law. For example, because 11′ is the far distance of the purple auto flash range, consider that 11′ distance as an f/11. If you were using an f/11 and needed to dim the lighting precisely 1 stop, how is this to be done? This would be accomplished by changing the f/stop to f/16. So, if you were to move from 11′ to 16′, you would reduce the flash effect on the scene by precisely 1 stop. Or, because 11′ is the far end of the f/22 auto flash range, and it is equivalent to the manual mode exposure mode, you could put one layer of a handkerchief over the flash head to reduce the flash output by 1 stop.

If you were not at the extreme end of the auto flash range, and needed to achieve a −1 bracket, it is recommended that you switch to the manual flash mode and determine a −1 exposure for the precise distance you currently were at. For instance, if you were 7′ from your subject, the correct manual flash setting would be ½ power. Changing the flash to ¼ power would be a −1 flash exposure.

In the blue-auto mode, an f/16 is recommended according to the calculator dial. Bracketing can be achieved simply by switching to both f/11 and f/22.

Some flash units also have a flash exposure compensation dial or button, which is very different from the camera body exposure compensation dial. That, you will recall, works by telling the camera that the ISO film speed is different from what is actually loaded in the camera. The flash exposure compensation dial/button works differently. It works by affecting the flash duration/intensity. A −1 flash exposure compensation will shorten the flash duration, much like the manual flash mode vari-power setting has the ½ manual power setting, which will decrease the flash duration by 1 stop. If the full manual flash fires normally in 1/1000 second, ½ manual flash will fire at a duration of 1/2000 seconds. Similarly, in the auto flash exposure mode, if the flash unit has a flash exposure compensation capability, setting it to a −1 reduces the flash intensity to 1/2000 second, if it is used at 10′.

However, if full flash fires at 1/1000 of a second and it is impossible to have more than full flash, how can a +1 flash exposure compensation setting work? The sensor eye cuts short the flash duration if the flash unit is used at distances less than the maximum auto flash range. If auto-purple works within a 2′ to 11′ range, using the flash at any distance from 2′ to 7′ will cut short the flash duration at least 1 stop less than full flash. This unused flash power is conserved by the flash and is the reason auto flash has quicker recycle times than full manual flash. At 7′ from your subject, you can be using the auto-purple mode with an f/22 and you would be able to use a +1 flash exposure compensation. This would, then, be the equivalent of full manual flash: a +1 flash intensity if at 7′.

If in the auto-purple mode, and standing 5′ from your subject, the flash duration will be cut off the equivalent of 2 stops early. This unused power is what allows the flash to recycle faster than if you were standing at 11′. If you were photographing a gun lying in the snow at night, you should anticipate an exposure error from the flash sensor being "fooled" by the reflectivity of the snow. Snowy scenes "fool" the camera meter with non-flash photographs and cause underexposures. The same thing will happen with the flash in the auto mode. The sensor eye will read the light reflecting from the snowy scene and cut the flash off early, resulting in an underexposure. The compensation necessary is the same with auto flash as with the camera meter. Bracket toward +2. If using auto-purple, you could actually set the flash exposure compensation dial/button to +2, because at 5′, the equivalent of 2 stops of light currently is unused by the flash. For example, in the auto-purple mode, being 5′ from a normal subject would cause the flash duration to be reduced to /14000th of a second. If you set the flash compensation dial to +1, it would be changed to 1/2000th of a second; and if set to +2, it would be changed to 1/1000th of a second.

Of course, these are specific recommendations for bracketing with a Vivitar flash. Other flash manufacturers' auto flash units will react similarly.

A brief recap of automatic flash exposure mode might be helpful:

- Auto-purple: use an f/22 within the range of 2′ to 11′.
- Auto-blue: use an f/16 within a range from 11′ to 15′.
- Auto-red: use an f/8 within a range from 15′ through 30′.
- Auto-yellow: not recommended for crime scene work because it uses an f/4, which does not provide a good depth of field.

What, if any, are the negative aspects of automatic flash exposure mode? First, the sensor can be "fooled" by the reflectivity of very light-toned surfaces and very dark-toned surfaces. In these situations, exposure compensations will be required in the same way as exposure compensations were required when the camera body exposure meter was "fooled" in similar situations.

The flash head sensor eye is designed to provide the proper exposure when it is relatively close to the film or digital sensor. If the flash is removed from the camera's hot-shoe and used with a PC or remote shutter cord, it must still be aimed relatively directly toward the subject. The automatic flash sensor eye cannot provide good exposures when the flash is used obliquely to the subject.

Automatic flash exposure mode is a distinct improvement to the manual flash exposure mode. Is there another flash system that is even better?

DEDICATED FLASH EXPOSURE MODE

With a **dedicated flash unit**, the camera and the flash unit are designed to work together closely; frequently they are made by the same manufacturer. Rather than relying on a sensor eye on the front of the flash unit, a dedicated flash's light output is controlled by the light meter in the camera body. Although the light from the flash is being projected, some light has already reached the scene, reflected off the scene, entered the camera through the lens (TTL), and has been metered inside the camera body to determine when enough light has been received for a proper exposure. At times, this is expressed as a flash system relying on light being reflected into the camera and read by the camera meter after the light has bounced off the film (OTF). When this light level has been obtained (18%), the camera's light meter signals the flash unit, and the flash duration is cut off early.

This represents a vast improvement in exposure determination because the light is metered inside the camera body, next to the film or digital sensor where it is critical, rather that at the automatic flash's sensor eye. Many flash photographs are taken with the flash unit removed from the camera body, further distancing the auto flash sensor eye from the film or digital sensor area. In these instances, a dedicated flash unit will always provide for more accurate exposures than automatic flash units.

Dedicated flash exposure mode:
Rather than relying on a sensor eye on the front of the flash unit, a dedicated flash's light output is controlled by the light meter in the camera body.

Dedicated flash units also work within various ranges. When a particular f/stop is selected, the flash range for that f/stop is shown on the flash.

In the examples shown in Figures 4.42 and 4.43, a dedicated flash is used with the camera set to f/8 and f/22, and the flash ranges for both are shown.

Dedicated flash exposure modes have the same attributes as the automatic flash exposure mode:

- If used at the far end of the flash range, it is the equivalent of the manual flash light output.
- If used at closer distances within the indicated flash range, the flash will recycle more and more quickly.
- If used at closer distances within the indicated flash ranges, the life of the batteries will be significantly improved.
- If used within the indicated flash range, the same f/stop can be set once and not have to be reset at different distances.
- The exposure determination will be incorrect when photographing scenes with predominantly light-toned subjects or with predominantly dark-toned subjects.

However, the dedicated flash mode is not adversely affected if it is used obliquely to the subject, because the meter in the camera body will prolong or cut short the flash duration when it receives the correct amount of light. Being right next to the film or digital sensor is the ideal position to be to determine proper exposures for most "normal" situations.

Flash units featuring dedicated flash mode operation are the most expensive flash units. Many dedicated flash units will still offer the option to switch to manual flash exposure mode when precise exposures need to be determined or when reflectivity issues may adversely affect the dedicated flash's ability to properly determine the correct exposure.

Figure 4.42
Dedicated flash set for f/8: range 5'–40'.

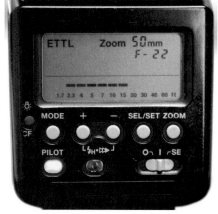

Figure 4.43
Dedicated flash set for f/22: Range 2.3'–15'.

Dedicated flash units will invariably work with a flash exposure compensation dial/button that changes the flash durations. Many offer a flash exposure compensation from +3 through −3, sometimes in $\frac{1}{2}$-stop or $\frac{1}{3}$-stop increments. Of course, the + brackets are not really available if the flash is used at the farther end of the flash-to-subject working distance, which is the equivalent of the full manual flash output, and the dedicated flash output cannot be greater than the full manual flash output. It can always be less, but it can never be greater.

If a flash unit has all three flash exposure modes, using the dedicated flash exposure mode is recommended for most flash photography instances.

BUILT-IN FLASH UNITS

Many SLR cameras have built-in flashes. This text will not spend much space addressing this type of flash, other than to point out some of the reasons they should not be relied on at most crimes scenes.

Perhaps foremost in their list of negative aspects is their low guide numbers. Many feature guide numbers around 35, indicating that at a 10′ distance an aperture of f/3.5 would be appropriate. Nowhere else in this text will you find a recommendation to use an f/3.5 as the optimal aperture for most crime scene work. The depth of field would be horrible. If it is necessary to capture just one item of evidence at a relatively close distance, perhaps the built-in flash can be used for its sheer convenience. However, convenience is often a reason for doing something when no other reasons can be imagined. Using a stronger removable flash would result in smaller apertures, which is normally preferred.

Being so close to the lens, red-eye will almost always result when photographing people looking toward the camera.

Built-in flashes also emit light in a very direct line toward the subject. Direct flash is not an "attractive" type of light. It is not a flattering light for people, and it frequently produces harsh hot spots on many surfaces. Many other aspects of creative and professional photography do not apply to crime scene photography, but this aspect of built-in flashes also makes them an unpopular type of light. Even getting the flash off the camera with a PC or remote flash cord goes a long way toward reducing the negative aspects of direct flash.

FILL-IN FLASH

This text has already introduced the reader to non-flash exposure concepts, and now the basics for determining proper exposures with a flash unit have been presented. Discussing **fill-in flash** incorporates both aspects: parts of the scene will be lit with relatively strong ambient light, and an electronic flash unit will supplement that light to produce softer lighting in areas currently in the shadows produced by the stronger ambient lighting.

Fill-in flash: Parts of the scene will be lit with relatively strong ambient light, and an electronic flash unit will supplement that light to produce softer lighting in areas currently in the shadows produced by the stronger ambient lighting. The fill-in light is intentionally set so that it is not as bright as the ambient light. The fill-in light is set to be exactly 1 stop less bright than the ambient light.

Recall the concept of exposure latitude. Film cannot record fine detail in both shadow areas and well-lit areas. In these situations, we can use the ambient lighting to properly expose those areas that are brightly lit and use the electronic flash to properly expose those areas that are dimly lit. The trick is to determine the proper balance between the two, so the scene does not look as if it were lit by two suns or two flash units lighting the scene from two different directions.

The lighting recommendations for fill-in flash are borrowed from professional photographers who do portraiture photography. In a photographer's studio, the primary light is referred to as the key light, or main light, which is frequently a large light box or flash aimed into an umbrella. It is frequently raised up higher than the subject's face, and it is directed to light the subject from a high diagonal angle, which is the strongest light for a portraiture setup. Lighting one side of the subject's face, this key light naturally produces shadows on the opposite side of the subject's face. These shadow areas are normally too dark for an effective photograph. To throw some light into these areas, a secondary fill-in light is used. This can also be a light box or flash directed into an umbrella, so that harsh light does not strike the subject's face. But, the fill-in light is intentionally set so that it is not as bright as the key light. The image would look strange if the subject's face was illuminated with equal light intensities from two different directions. Photographers have experimented with many different key-fill light ratios to determine a pleasing effect, and one of the most frequently used ratios is 1:2. This means the key light is set as desired to light one side of the subject's face, and then the fill-in light is set to be exactly 1 stop less bright. The fill-in light is half as bright as the key light. This lighting situation produces a shadow on the side opposite from the key light, but details of the face can still be seen and recognized in the shadow area.

How is this ratio established? We must first determine just what intensity of light the bright parts of the scene are receiving, which is done by taking an exposure reading from the bright part of the scene. When this is done, it is important to meter only the bright area of the scene. Do not take a meter reading from an area of the scene that includes both the bright and dim areas of the scene.

For example, according to the f/16 sunny day rule, we would expect the sunlit area of a scene to require either an f/16 or f/22 for a proper exposure. An area in deep shade might need an f/5.6 or f/4 for a proper exposure. If we made the mistake of metering an area that was half sunlit, and half dimly lit, the meter would do its best to average these lighting extremes. This average might result in a meter reading of f/9.5, halfway between f/22 and f/4. Actually, this f/9.5 would light neither area adequately. It would be 2½ stops too bright for the areas of the scene requiring an f/22, and it would be 2½ stops too dim for those areas of the scene requiring an f/4 for a proper exposure. The highlights will be overexposed, and the shadow areas will be unacceptably underexposed. Film, by itself, does not have the exposure range to capture details under conditions that include both bright highlights and dark shadow areas.

Flash is strong enough to be the primary light source when required. In the fill-in flash situation, the flash is delegated to being just a supplement to the bright ambient light at the scene. It is normally recommended to balance the two light sources so that the light coming from the flash unit is 1 stop less bright than the bright ambient light in the scene. Having two different main lights, coming from two different directions, would produce a confusing result. It is more natural to have one of the light sources be considered as the key light and the other light source be delegated to be a fill-in light, adding just enough light to the shadow areas to provide for proper exposure there.

Figure 4.44 has been exposed for the sunny part of the scene, and that area looks properly exposed, but the shady areas are too dark. Figure 4.45 has been exposed for the shady areas, and those areas look properly exposed, but the sunny area is now overexposed. Figure 4.46 was taken using fill-in flash. Both areas of the scene appear properly exposed. The scissors in the shady area can easily be seen, as can the cigarette butts, but the shadows have not been obliterated with too much flash.

This is the fill-in flash sequence:

1. Because the scene is brightly lit, use 100 ISO film or set the digital camera to 100 ISO.
2. Meter the light in the bright part of the scene. Two things must be remembered. First take this meter reading with the camera set at the camera's sync shutter speed. After all, flash will eventually be used. Second, fill-the-frame with the sunny area of the scene when taking this meter reading. You are not interested in the meter reading of the scene half sunlit and half in shade. You want to set the camera exposure settings to properly expose for only the sunlit part of the scene. When selecting this sunlit area, care should be used to avoid certain subjects. Do not take a meter reading on sunlit vehicles within the scene. Glare from painted sheet metal would produce an erroneous meter reading. Also avoid extremely light-toned sunlit areas or extremely dark-toned sunlit areas. The best areas to meter would be sunlit green grassy areas or sunlit well-traveled asphalt. Even if these areas will not be in the composed part of the scene of interest, they will provide the proper starting point.
3. This meter reading will suggest an f/stop that corresponds to the sync shutter speed. Set the camera for this f/stop.
4. Set the flash unit to the manual flash mode. We do not want the automatic flash's sensor eye or the dedicated flash's camera light meter to read glare or other overly bright lights at the scene and cut the flash duration off prematurely.
5. Now it is necessary to position the flash unit and determine the ratio of flash lighting to ambient lighting. To determine the position for the flash unit, recall that the flash calculator dial associates two variables: distances and

Figure 4.44
Exposed for the sunny area.

Figure 4.45
Exposed for the shady area.

Figure 4.46
Exposed using fill-in flash: both areas look good.

f/stops. Because an f/stop recommendation has just been obtained from the meter reading of the sunlit part of the scene, notice what distance the flash calculator dial indicates is appropriate for that f/stop. Using the flash calculator dial, position the flash unit at this distance. That may mean you step in closer to the primary subject, or take a step backwards.

6. Next, realize that actually taking the photograph from this position is the equivalent of taking a full flash photograph. Or is it? Although we definitely want to set the flash so that it is 1 stop less bright than full flash, do we need to adjust the flash any further? Remember the concept of manual flash used in a "normal" room? If not in a "normal" room, the manual flash will be either brighter or dimmer. Outside, manual flash is already 1 stop dimmer, because it is not receiving the extra light that white walls and a white ceiling would reflect back toward the subject. Therefore, the flash is now set to be 1 stop less bright than the sunlit areas of the scene.

7. Ensure the flash is aimed at the area of the scene where it is most needed. Sometimes having the flash unit mounted on the camera's hot-shoe, on top of the camera, is not the most appropriate place for it. Consider turning the camera upside-down, so the flash is underneath the camera, or, take the flash off the camera, connect it to the camera with a PC cord or remote flash cord, and aim the flash where it will do the most good.

8. Take the photograph.

When done properly, not only will the sunlit area of the scene be properly exposed, but the area in the deep shadows will also reveal details not otherwise seen. Although these details are visible, it will be noticed that the sunlit area is not overexposed. Both look well exposed.

Figure 4.47 was exposed for the sunlit area of the scene. The sunny areas look well exposed, but the shady area is underexposed. Figure 4.48 was exposed for the shady area of the scene. The flashlight is visible, but the sunlit areas are obviously overexposed. Figure 4.49 has been exposed with fill-in flash. Both the sunlit areas and the shady areas of the scene look properly exposed. The flashlight is visible, but the shadow it is in has not been blasted by too much flash.

A backlit scene is also one with wide extremes of light in it. If a person is sitting near a window, and the scene outside is included in the field of view, the camera meter will overreact to the bright outside area in the background of the photograph and cause an underexposed subject. This will also occur if a person outside is facing you, and the sun or bright sky is behind them. If your composition will include a lot of the scene around your primary subject, and that area is brightly sunlit, the subject who is shade-lit will be underexposed if the camera's meter is set according to the sunlit area. The proper use of fill-in flash can ensure both areas of the scene are exposed adequately.

Figure 4.47
Exposed for the sunny areas.

Figure 4.48
Exposed for the shady areas.

Figure 4.49
Exposed using fill-in flash: both areas look good.

When the idea of fill-in flash is taught, because two very differently lit areas are within the same composition, it is usually pointed out that either part of the scene can be overexposed, properly exposed, or underexposed. To make sure students fully understand the ability to control the exposure level of either area separately, students are frequently required to arrange three photos so that the interior subject is properly exposed in all three, but the exterior should be overexposed in one photograph, properly exposed in another, and underexposed in the third. Then, the students are told to maintain a proper exposure for the exterior while underexposing the interior in one photograph and overexposing the interior in the final image.

It is usually pointed out that Hollywood frequently uses this ability to expose scenes differently for various effects. Look at some of the old time "film noir" genre films. Those are gangster movies or other films where it seems like the entire film occurs in the night or under dark rainy conditions. The movie cinematographer uses the dark dreary lighting conditions to set the mood of the movie. At times, the filming of the movie continued after the sun began to rise, but to continue the same feel for the movie, the camera operators merely underexposed the sunlit scenes. If you look closely, when you catch a rerun of these films, people are running around night time street scenes, but strangely suspicious shadows are around them. A sunny day can be underexposed to simulate night time, but the shadows remain.

The sequence of Figures 4.50 to 4.52 demonstrates the ability to separately expose the dark part of the scene and bright part of the scene. The interior exposure of the swan was kept constant, while the exposure for the exterior was varied. Obviously, Figure 4.51 is the only image of the series that should be considered acceptable. This does not usually happen without knowing and using the fill-in flash technique.

The alternative is to merely determine the exposure for the interior subject, with no concern for the exterior exposure at all. However, images like Figures 4.50 and 4.52 will frequently be the result. A good crime scene photographer should know his or her equipment and how to use it. Again, when properly done, fill-in flash ensures that both parts of the scene are properly exposed, both the area that had been sunlit and the area that was backlit.

It is fair to ask which flash exposure mode is best suited for fill-in flash. Although some very expensive flash units offer automatic fill-in flash settings, this author believes the manual flash exposure mode is most appropriate for fill-in flash. This technique is used when bright parts of the scene are present. The automatic flash mode and the dedicated flash mode offer the possibility of reducing the flash duration when bright light affects the flash sensor or the light meter in the camera body. If the flash durations should be altered, the 1:2 key/fill flash ratio may be altered, and that is not desirable. Only the manual flash exposure mode gives us complete control of this lighting condition.

Figure 4.50
Exterior intentionally underexposed.

Figure 4.51
Exterior properly exposed.

Figure 4.52
Exterior intentionally overexposed.

Whenever lighting conditions include bright sun-lit areas and dark shadows, the use of fill-in flash should be considered. If the sun is producing noticeable shadows, the use of fill-in flash will benefit most photographs.

OBLIQUE LIGHT, BOTH FLASH AND NON-FLASH (FLASHLIGHT)

Many times, lighting the three-dimensional textures and patterns of evidence is necessary. Tire tracks, shoe prints in dirt, bite marks, fingerprints in window putty, indented writing, and dusty shoe prints are just some examples of the range of three-dimensional impressions the crime scene photographer may encounter. Some may consider dusty shoe prints to be an example of a two-dimensional pattern, although they are actually a physical medium sitting on the surface of another medium.

To successfully photograph these types of evidence, some form of side lighting is often recommended. Electronic flash is usually the light of choice, although a good flashlight can perform well in certain situations. Despite the fact this is a chapter on flash techniques, when a flashlight can do as good or better, that will be pointed out.

DIRECT LIGHT VERSUS OBLIQUE LIGHT

When three-dimensional (3D) evidence has to be photographed, directly lighting the evidence is actually counter-productive to producing the best photograph possible. When 3D type of evidence, such as a shoe print in dirt, is directly lit, the light from the flash gets into all the nooks and crannies of the imprint and actually washes out some of the 3D aspects of pattern and depth.

Figure 4.53 shows a shoe print in dirt lit by direct flash. It is very difficult to see the impression because the flash effectively washed out the detail within the impression. Figure 4.54 is the same impression lit by **oblique flash.**

The side lighting creates shadows, which makes it easier to see the 3D aspects of the pattern. Depth and texture are evident now. Oblique lighting is absolutely necessary to show the detail of the shoe pattern.

Side lighting, or oblique lighting, provides the shadows necessary to better visualize 3D aspects of evidence. However, the shadows produced by an oblique flash are also very dark. Details within this shadow area may not be visible. The contrast between the brightly lit areas and the shadow areas may be too extreme for the film. This was explained in reference to fill-in flash.

These very dark shadows are called **hard shadows.**

Details within them cannot be seen. We are now faced with a paradox. If oblique flash is not used to light 3D shoe print details, all detail may become washed out. If oblique flash is used to light a shoe print, our own lighting technique may be the cause of important details being lost within the "hard" shadows produced. Is there a solution to this problem? Yes.

Oblique flash: Electronic flash can be removed from the camera's hot-shoe but still be attached with a PC or remote flash cord. It can then be held at the side of 3D evidence to produce shadows within the texture or pattern of the evidence. These shadows allow the details of the evidence to be seen rather than being washed out by direct flash.

Hard shadows: Shadows created by oblique flash. They are very dark, and details necessary for the identification of the impression may not be seen within them.

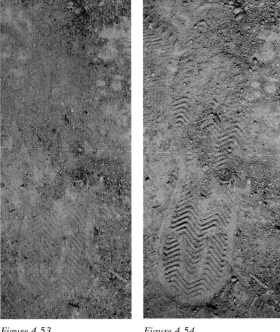

Figure 4.53
Direct flash.

Figure 4.54
Oblique flash.

The shadows produced by oblique flash have both good and bad effects. How can we retain the good effects and eliminate the bad effects? The use of a reflector solves the problem. When a reflector is used with oblique flash, the light from the flash strikes the impression and continues past it until the light strikes the reflector. The light then bounces back toward the impression, lighting areas previously covered by "hard" shadows. The reflected light is not strong enough to wash out the shadows, but it is strong enough to light details within the shadow area. Shadows are called **soft shadows** when details within them can be seen.

Hard shadows hide detail; soft shadows reveal detail.

To teach this in a classroom, an old shoe is photographed on the floor.

Figure 4.55 shows the effect of an oblique flash producing a hard shadow on the opposite side of the shoe. The carpeting within the hard shadow area is impossible to see. Having a shadow obscure areas around the evidence may seem trivial to us, however. But shadows are also produced within the shoe print. If part of the shoe print is hidden from view by our flash technique, it is time to alter our flash technique.

Figure 4.56 shows the result of the use of a reflector to correct the hard shadow effect. Light has bounced off of the reflector and has lit the areas previously within the hard shadows. The carpeting beyond the shoe is now visible because it is now in a soft shadow. The shadowing has not been eliminated. It is now, however, possible to see details within the shadow area. Again, lighting an area

Soft shadows: When a reflector is used with oblique flash, the light from the flash strikes the impression and continues past it until the light strikes the reflector. The light then bounces back toward the impression, lighting areas previously covered by "hard" shadows. The reflected light is not strong enough to wash out the shadows, but it is strong enough to light details within the shadow area. Shadows are called "soft" shadows when details within them can be seen.

Figure 4.55
Shoe with a hard shadow.

Figure 4.56
Shoe with a soft shadow.

around the shoe does not seem that important. If shadows within the shoe print have also been lightened, that would be important. The shoe was altered to demonstrate the importance of bouncing light into the shadows within the shoe print. That is where shoe print examiners need to be able to see nicks, cuts, gouges, and embedded rocks. To make this point, a name was printed three times inside the shoe tread.

Figure 4.57 shows all three "Teds" lit with direct light. The name was put on three different vertical surfaces, so that light coming from various sides of the shoe would tend to hide one or the other in a hard shadow. Figure 4.58 shows the hard shadow in a close-up of the shoe lit by oblique flash with no reflector. One of the "Teds" is now hidden within the shadow. Consider the name to be an identifying nick, cut, gouge, or embedded rock. Hidden evidence cannot be used by anyone for identification purposes.

In these situations, it is critical to ensure a reflector is used to be able to "soften" the shadows, so details can be seen within the shadow area. Figure 4.59 shows the same area with a reflector used to bounce light back into the shadow areas. Within a soft shadow, the third "Ted" is now visible.

What kind of reflector was used to create this miracle? Reflectors can be as simple as a single sheet of report form paper. We always have these handy. However, this requires one or two hands to hold it so it reflects light back into the impression, and frequently we are doing this kind of photography without help.

Figure 4.57
Three Teds.

Figure 4.58
Teds with hard shadow.

Figure 4.59
Teds with soft shadow.

We usually are busy holding the flash off-camera on a PC or remote flash cord and do not have extra hands available. At times, simply folding the paper so it would stand on its own is all that is needed. This has worked numerous times. But the slightest breeze defeats this plan. Pictured here is a home-made reflector that effectively bounced the light where it was needed without anyone having to hold it.

Figures 4.60 and 4.61 show this reflector. It is not patented, and this author hereby gives all readers of this book permission to copy it. The back is tilted down a bit to facilitate the light bouncing into the impression rather than missing the print. A little tape, cardboard, and white paper, and you will be set. You will be the envy of your entire agency.

THE OBLIQUE FLASH SEQUENCE

At what angle should the flash be held? This angle is best determined by viewing the imprint from directly overhead while you move a flashlight through various angles, watching as different shadows are created with the different flashlight positions. This will vary with the depth of the imprint, so this procedure will always have to be performed to determine that the optimal shadowing has been obtained.

Figure 4.60
Homemade reflector 1.

Figure 4.61
Homemade reflector 2.

From how many locations around the impression should the flash be aimed at the impression? Because the camera is always mounted on a tripod so the film plane can be positioned parallel to the impression, the best answer is to fire a series of flashes between each of the three tripod legs. The series is, of course, a set of three brackets: a 0, a +1, and a −1 from three different angles. Not that we do not trust our ability to determine the proper exposure for this situation, we take these bracketed shots so that the shoe print examiner has a wealth of options from which to choose. All nine of these images have a scale placed alongside the impression; not just alongside of the impression: it is placed on the same plane as the depth of the impression. If the shoe print is ½″ deep in the dirt, a depression alongside of the impression that is ½″ deep needs to be dug so the scale is the same distance from the film or digital sensor as is the bottom of the impression. Only in that location can the scale enable a life-sized print to be made.

Before those nine photographs are taken, one image should be taken without the scale in view. It is always wise to take at least one image before we have added anything to the scene. Defense attorneys may ask what has been hidden by the scale, and if the original scene was conveniently found with the scale already in it.

Whenever the camera is on a tripod, the shutter button should not be manually depressed. Doing so may generate some camera movement and result in blur. Use a shutter release cable or a delayed shutter setting. Many cameras now offer the photographer a 2-second delay or a 10-second delay. This brief time delay allows the camera to settle down after the shutter has been pressed to avoid camera shake and blur.

Earlier it was mentioned that three series of brackets should be taken from three different positions. Some may ask what the beginning exposure is. Determining this is not as simple as it sounds. Of course, because this is an example of a critical comparison photograph, or examination quality photograph, the film that is recommended is 100 ISO black-and-white film. Because a flash will be used to light the imprint, it makes sense to use the sync shutter speed.

The f/stop that is best for critical comparisons is f/11. But, would f/11 provide a proper exposure in this particular situation? If a dedicated flash unit were used, then the light meter in the camera body would ensure a flash duration proper for all the other variables. That would really simplify things. If you do not have a dedicated flash, is automatic flash or manual flash the better flash mode to use? The automatic flash exposure mode is not an option. Having the flash sensor eye so far away from the film or digital sensor makes no sense. With the manual flash mode, how is the f/stop to be determined? Remember that the flash calculator dial only applies to direct flash, not oblique flash.

The only way to determine whether the f/11, normally suggested for critical comparisons, provides a proper exposure for oblique flash situations is to do a series of test exposures. Fortunately, these have already been done for you, under the two extremes this technique will have to be used: at midnight and noon.

At midnight, with the only light available being a single flash unit, a shoe print was placed in dark dirt, and it was photographed at every f/stop between f/2 and f/22. The flash used had a GN of 120, so this recommendation will have to be modified if your flash unit is different. These are some of those images.

We are in luck. The f/11 not only is a good aperture for critical comparisons, it also happens to be the f/stop that provides the best exposure of all the f/stops when photographing shoe prints in dirt at midnight.

Brackets are also recommended. How? The +1 bracket is achieved simply by changing the aperture to f/8. That is also appropriate for critical comparisons. It is not, however, recommended to change the aperture to f/16 for the −1 bracket. The f/16 may show negative effects from diffraction. Instead, change the flash to f/11 at ½ power, or place one layer of handkerchief over the flash head. Either of those would be an acceptable −1 bracket.

Taking photographs of a shoe print in dirt at noon is the other end of the lighting spectrum. Initially, one would think that the additional ambient light would make this photographic situation problematic. Not so. At midnight, the initial exposure combination was 100 ISO film, sync shutter speed, and f/11. What is to prevent this combination from being used at noon? Our only fear is that at noon, with its bright ambient light, this same exposure combination may lead to an overexposure. So, we should think of ways to ensure the use of the f/11 does not lead to overexposure. Can the f/16 sunny day rule help us find a solution? A brief restatement:

- Use f/16 with bright light producing well-defined shadows.
- Use f/11 with light producing soft-edged shadows.
- Use f/8 for overcast days without any shadows.
- Use f/5.6 when in an "open" shadow but with blue sky above.

It would seem that all we would have to do is get the shoe print in dirt out of direct sunlight, and then an f/11 would not overexpose the impression. This works not only in theory, but in practice as well. All that is necessary is to ensure that sunlight does not directly strike the impression, which can be done simply by blocking the sun with our body, or have anyone present assist by blocking the sun with their body. When this is done, the flash then becomes the primary light, and all the other photographic variables as were used with the shoe print at midnight apply.

Before leaving the topic of photographing shoe prints in dirt, one other issue should be discussed. How long should the PC or remote shutter cord be? Most PC cords are about 3′ long. A couple of sources, including William J. Bodziak, author of the well-known *Footwear Impression Evidence* text, have recommended that the flash should be held 5′ to 6′ from the shoe print. Why is this? Mr. Bodziak explained that having the flash farther from the shoe print produces a more even lighting situation. The closer the flash is to the impression, the more the lighting will vary from the toe to the heel. This does make sense. According to the inverse

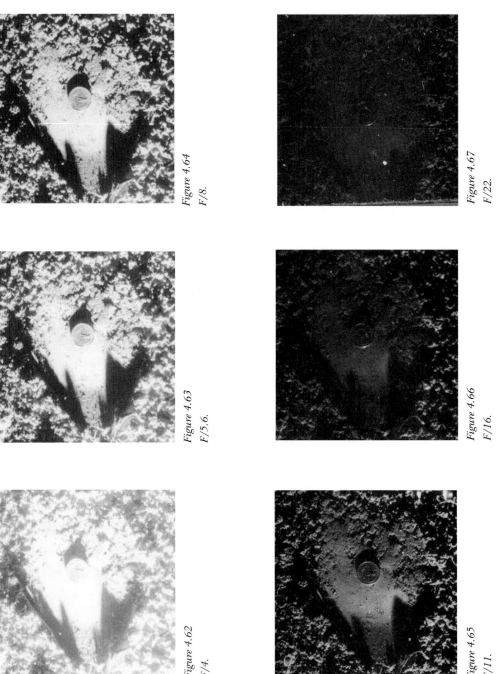

Figure 4.64
F/8.

Figure 4.67
F/22.

Figure 4.63
F/5.6.

Figure 4.66
F/16.

Figure 4.62
F/4.

Figure 4.65
F/11.

square law, a 1-stop difference is between 1.4′ and 2′. Another 1-stop difference exists between 2′ and 2.8′, between 2.8′ and 4′, and between 4′ and 5.6′. That means if the shoe print was 12″ long, and the flash were 3′ from the impression, the side of the impression closest to the flash can be more than 1 stop brighter than the part farthest from the impression. This can make it harder for the examiner to see details within the impression. Removing the flash a bit farther reduces the flash intensity difference between toe and heel. With the flash 5′ to 6′ from the impression, the difference in light intensity from toe to heel is only about two thirds of a stop, which is much less problematic for the examiner.

To get to this 5′ to 6′ distance, some PC or remote flash cords can actually be connected to form one longer cord. If your cords cannot be connected, another way to mitigate the flash intensity between toe and heel is to simply avoid aligning the flash so it must travel from toe to heel.

Figure 4.68 shows the tripod setup when the tripod does not have a reversible center stem. The camera is above the impression, which runs parallel to an imaginary line connecting the two front legs. Figure 4.69 shows the camera and tripod setup when the center stem is reversible. The impression is again lined up parallel to an imaginary line connecting the two front legs. Figure 4.70 shows the three flash positions with the tripod set up either way. One flash will come from the front, aimed to strike the impression from the side, the narrowest distance across the impression, rather than from toe to heel, the farthest distance from one end to the other. The other two flash angles strike the impression from diagonals. These angles are also shorter than the full toe to heel distance. Until your agency supplies its crime scene photographers with 5′ cords, this setup may lessen the adverse impact of different light intensities over different distances.

SHOE PRINTS IN SNOW AND SAND

Photographing shoe prints in snow adds some interesting complications to the normal photography of shoe prints in dirt. To begin with, snow can sometimes be on the verge of melting, making it imperative that the photography be done quickly and accurately the first time, because there may not be anything remaining after a short time span. Figure 4.71 shows the results of a practical exercise for students in a basic crime scene photography course regularly taught at a local police academy. It was during the summer, so we simulated snow by crushing ice in a blender. The room temperature immediately began melting the slush we had created, so we doubted we would be able to document any worthwhile detail. We captured three quick photographs of the slush before treatment with **snow print wax**, then sprayed the impression with the wax, and took two more photographs before everything melted away.

Snow print wax is used with shoe and tire tracks in snow because it provides contrast; the white subject matter makes it difficult to see pattern and details within the imprints. We were pleasantly surprised to notice the snow print wax almost visualized the brand name, "Spalding," before the slush completely melted away.

Snow print wax: Used with shoe and tire tracks in snow, it provides contrast for photography, because the white subject matter makes it difficult to see pattern and details within the imprints.

Figure 4.68

Tripod position: center stem not reversible.

Figure 4.69

Tripod position: reversible center stem.

Figure 4.70

Flash directions indicated with arrows.

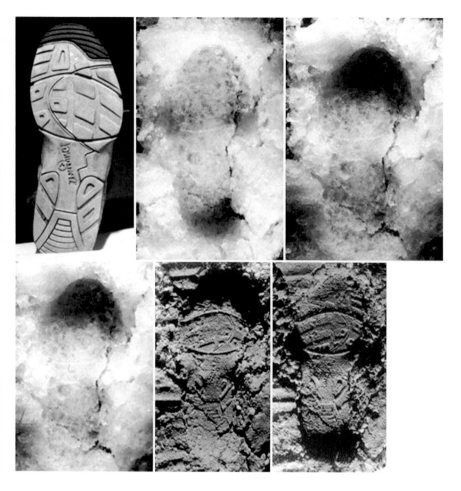

Figure 4.71
Snow print wax on "snowy"
shoe print.

The snow itself, as white evidence, has its own exposure issues. If either the automatic or dedicated flash modes are used, their sensors will read the excessive reflected light and tend to underexpose the images. This, in itself, would not be too bad, because such white-on-white subjects are very hard to see with any details visible. A little underexposure may even benefit the photographs. Because the actual amount of underexposure may vary, many opt for the manual flash exposure mode when photographing snowy impressions.

Oblique flash normally produces shadows that make the 3D texture and pattern easier to see. With snow, light will bounce around each of the vertical surfaces and tend to wash out the shadows needed to visualize fine details.

For these reasons, the normal "as is" photographs are not expected to be very useful, although they will still be done. Because of this white-on-white contrast problem, the next thing usually done is to add color to the snow prints to increase contrast, so that more details can be seen. The color itself is unimportant. It is just necessary to diminish the white-on-white look of the original impression. Some have used a variety of spray paints, and some have even twirled a fiberglass filament fingerprint brush, loaded with black powder, over the impression. Both of these ideas will produce a better result than the original photographs. If, however, snow print wax is available, this is the optimal material to use. It provides a good contrast with its red wax for the next round of photography. It also helps prepare the shoe print for casting, by covering the shoe print with a waxy barrier that helps prevent melting of the snow when the snow print wax begins to warm with its exothermic chemical reaction.

Spraying the snow print wax over the impression should be a two-step process. It is recommended to begin with a spray from just one direction. Spraying the impression from just one direction darkens some of the impression, and leaves other areas without snow print wax. The effect is very close to the shadowing produced by oblique flash. Figures 4.72 and 4.73 show this effect. Although fine details may not be seen with this procedure, it certainly shows gross pattern very well. Another sequence of photographs is captured after spraying the shoe print with snow print wax from one side.

After these photographs are taken, the impressions should be sprayed with snow print wax from all sides, so that all vertical surfaces within the impression are covered with wax. This can be done while walking around the impression and spraying into the impression. Although the wax can now be quite thick and may be masking fine details within the impression, the final round of photographs should now be taken.

Photographing shoe prints in sand is very similar to photographing shoe prints in snow. Sand is highly reflective, frequently resulting in an underexposure if the dedicated flash mode is used. Like snow, vertical surfaces of the sand will bounce light within the impression, minimizing the shadows produced with oblique flash.

Figure 4.72

Shoe in snow combo I.

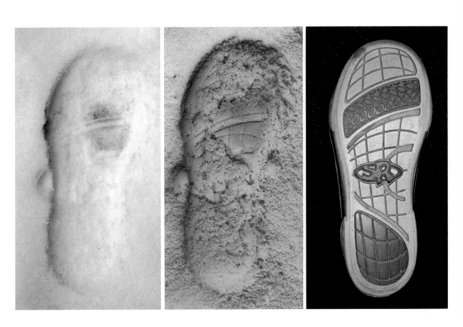

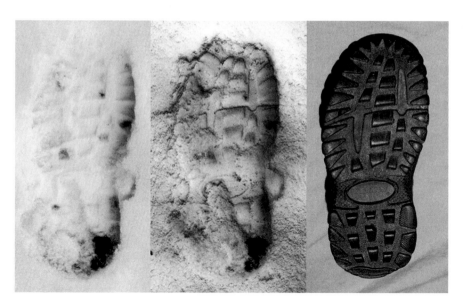

Figure 4.73
Shoe in snow combo II.

For these reasons, the manual flash mode is again recommended. To help with contrast, because the normal shadows produced by oblique flash will tend to be washed out a bit, consider adding a colored spray. Snow print wax will do double duty with sand. Spraying it from one side adds contrast for the next series of images, and the wax will serve to hold the fine sand particles together during the casting process.

TIRE TRACKS

The photography of tire tracks shares many aspects of the techniques needed to properly photograph shoe prints. However, some additional considerations exist. Instead of an impression perhaps 10″ to 12″ long, now we may be faced with several very long impressions. The first task is to ensure we document the full circumference of each tire mark present. Normally, unless we are dealing with very wide tire marks, which would suggest an oversized tire, we can anticipate the full circumference of a tire to be 7½′ or shorter. Of course, many times only segments of a tire track may be present. But, you should know what to do with 34′ of a tire mark. No more than 7½′ usually needs to be documented.

That 7½′ will be broken down into five 1½′ segments. Before dealing with any individual segment, the entire length needs to be documented. Mark five 1½′ segments with lettered or numbered evidence markers and photograph these together in one photograph, as in Figure 4.74.

For example, the right front tire mark can be considered one item of evidence, regardless of its length. If it is longer than 1½′ long, we will begin segmenting the entire length of the tire tract into 1½′ segments. To make this easier, a tape measure will be laid down next to the impression. Different-scale increments will be

Figure 4.74

Tire track combo (Courtesy of GWU MFS Student Sara Dziejma).

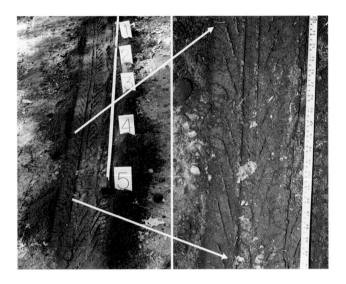

included in the photographs of each segment. No more than five 18″ segments need to be documented. Each of the five segments will be photographed and casted, and each will be given their own designations. If the front right tire mark will be listed as evidence item #17, the five segments will be #17A, #17B, or #17.1, #17.2, etc. After the entire length has been photographed, you will begin photographing segments. The first segment documented will be the segment with the beginning of the tape measure alongside of it. This segment will also have either the evidence marker "1" or "A" beside it.

Although each segment is 18″ long, we do not want to miss any portion of the tire track where the different segments meet, so it is suggested to build in a 1″ overlap between segments. That means each 18″ segment will be bordered by a 1″ safety margin, making each photograph cover 19″ of the tire track. It will be your preference whether the 1″ overlap will be on the left or right side of each 18″ segment. A 1″ overlap is not necessary on both sides of every 18″ segment. The segments will be 0″ to 18″, 18″ to 36″, 36″ to 54″, 54″ to 72″, and 72″ to 90″. The 1″ overlap can be on either side of a segment, so the first segment can be either from 1″ before the 0 mark on the tape measure to 18″, or it can be from 0″ to 19″. Just make sure to continue the 1″ overlap consistently with each 18″ segment.

Like a shoe print, each segment will get photographed 10 times:

- The first without a scale. (The tape measure only helps differentiate segments. It is not on the same plane, is it?)
- A scale will be added at the same plane as the depth of the impression, and three brackets will be taken with the flash coming from between two of the tripod legs.

- The flash will be moved so it comes between two different tripod legs, and another three brackets will be taken.
- The flash is moved so it comes between the last two tripod legs, and three more brackets will be taken.

At this point the tripod can be moved to the next position, to photograph segment "2" or "B." If you wish the 1″ overlap to be on the right side of each segment, in the viewfinder you should be able to see 0″ to 19″, 18″ to 37″, 36″ to 55″, 54″ to 73″, and 72″ to 91″.

BITE MARKS

As the 3D nature of the impression becomes less pronounced, the positioning of the light becomes more and more critical to visualizing the detail of the impression. As the impression becomes shallower, many tend to favor the light of a flashlight over the light from an electronic flash. One can be successful taking bite mark impression photographs and indented writing impression photographs with a flash unit, but you may occasionally lose an image because the flash was not aimed at the precise angle needed to visualize the detail optimally. Between the time you had determined the appropriate angle to light the impression by using a flashlight and the time you had interposed the flash unit, some slight variance in angle was introduced, and the photograph was less than successful. Many other photographs were successful. Because many hate losing any shot to this difference in the flash angle, some have considered ways to increase their success rate. What was the cause of the occasional unsuccessful image?

Obviously, because most flash units do not project a constant "modeling" light, the photographer could not see that the flash angle was improper. A "modeling" light is a light similar to a flashlight, which is constantly on. Because it constantly produces the shadows that will eventually be captured when the image is taken, you have confirmation the angle is still appropriate for the shot. Without a "modeling" light, we are left to hope the appropriate flash angle has been determined and that it has not changed in even the slightest degree.

The solution for many was to begin using a flashlight instead of a flash in those situations where the angle of the light is very critical to the success of the image. With a shoe print or tire track in dirt, minute changes on the flash angle did not make a difference. As the depth of the impression becomes shallower, the angle of the light becomes more and more critical.

It should not be suggested that every use of an electronic flash should be eliminated in these situations. Many photographers produce some very fine images by using an electronic flash with impression evidence. It is suggested you try using a flashlight in these situations to see whether your success rate improves also. This author's did; perhaps yours will also.

Because bite mark photographs are critical comparisons, the camera should be placed on a tripod to help ensure critical focus is maintained while you are busy maintaining that critical angle with the flashlight. In these situations, ISO 100 film is suggested, and f/11 is the recommended f/stop. Before the series of close-up photographs is begun, a midrange photograph, showing the bite in relation to a known body part, is taken. This can be a joint or other body part that does not compromise the modesty of the victim. This midrange photograph is required to establish just where on the body all the close-up photographs are coming from.

Then, take a series of close-up photographs. This is the perfect time to select the aperture priority exposure mode. If the relatively dim flashlight causes the need for longer shutter speeds, the camera on a tripod ensures the long shutter speeds do not result in blur. Actually, slow shutter speeds are a benefit when using a flashlight as your light source. A flashlight often has a characteristic beam of light that is not uniform. During a slow shutter speed, quickly panning the flashlight beam over the evidence causes the distracting flashlight beam to be transformed into an even lighting all over the evidence, rather than producing streaky lighting. Figure 4.75 shows the typical flashlight streak that results from not moving the flashlight during the exposure.

Figure 4.76 shows another frequent problem that occurs with bite marks on skin. Because the skin is often curved, it is hard to properly focus on subject matter, which is different distances from the camera, and it is difficult to uniformly light a curved bite mark. For this reason, the upper and lower arches are usually considered different items of evidence, because it is difficult to compose and properly light both arches in the same image. It is still wise to show them together in one photograph, although this may not be the best image of either arch. Then, each arch is photographed separately with a series of brackets. The camera can meter the bite mark lit by the flashlight to get the beginning shutter speed. Take that photograph without a scale. Add a scale, then bracket with another photograph at the "proper" exposure, and add a +1 and −1 exposure. Consider also a −2 image. Some have

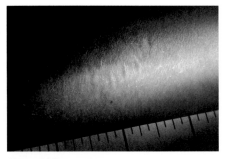

Figure 4.75
Flashlight streak.

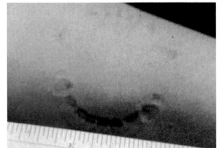

Figure 4.76
Bite on curved skin.

noticed that most bite mark photographs show more detail within the bite mark when the images are a bit underexposed. Give the odontologist working the case more "tools" to work with, and add a −2 exposure to your collection. Figure 4.77 shows a series of images bracketed toward a −2 instead of the normal +1 and −1 brackets.

Figure 4.78 shows one arch photographed with the flashlight moving during the exposure. The normal flashlight streak is gone, and the entire area looks evenly exposed.

Although this is a text on photography, the crime scene photographer often has other duties at crime scenes. This may include "collecting" the bite mark by other means than photography. Of course, before touching the bite mark, the area should be swabbed to collect any saliva present for DNA tests. After that, if the bite marks have not pierced the skin, consider casting the 3D bite mark with a silicone rubber compound like Mikrosil. Although this is a dental-grade silicone

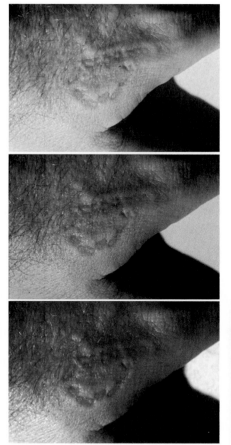

Figure 4.77
Bracketing toward −2.

Figure 4.78
Flashlight panning.

rubber designed to be put into patient's mouths, your agency may have prohibitions to using such techniques around broken skin or open wounds. In addition to casting the bite mark, you may also powder and tape lift the impression. See Figure 4.79. Parts of this image have been flipped and rotated to align them all side-by-side for easier viewing.

Of course, not only skin can be bitten by suspects. Figure 4.80 shows two purses that were stolen during a burglary. As the suspect crawled out the window to flee the scene, he temporarily held both purses in his mouth. He "tooled" each purse with his teeth. A suspect was later positively identified to the purses by his bite marks.

Many times, food products may be partially eaten, with bite marks remaining on the unconsumed portions. These can be photographed, swabbed, and casted. Figure 4.81 shows one such example.

INDENTED WRITING

With very shallow 3D types of evidence, like indented writing, the angle of the flashlight becomes even more critical. An almost parallel angle of the flashlight to the paper on which there is indented writing is required to properly visualize the fine impressions. Again, one can be successful in photographing indented writing with a flash. Some flash shots may come out improperly, because at times the light from the flash may "miss" the indented writing. You may increase your success rate by switching to a flashlight.

One aspect of visualizing indented writing must first be considered before we discuss the proper exposure for this situation. When the words have been printed, and even with cursive, many vertical and horizontal aspects of many letters are present. If the light of the flashlight goes down a depression in paper, the

Figure 4.79

Combo: Castone, Mikrosil, photograph, powder, and tape lift.

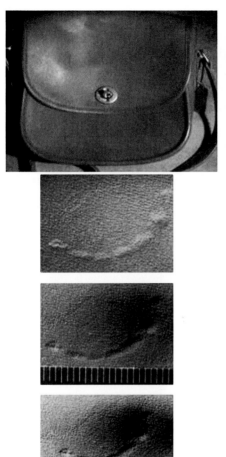

Figure 4.80

Stolen purses (Courtesy of
the Arlington county
Police Department,
Arlington, VA).

Figure 4.81
Bite mark on candy bar.

effect is similar to lighting a shoe print in dirt with a direct flash. If the depression is fully lit, it will be difficult to see from the camera's position above the paper. When the light from the flashlight strikes the depression from a perpendicular angle, it will create a shadow, and that part of the letter will be easily seen from the camera's position.

Figure 4.82 is a good example. The word "hello" is predominantly made with vertical and horizontal letter elements. If the light from a flashlight comes from either the left or right side of the paper, the horizontal elements become almost impossible to see. If the light comes from the top or bottom of the paper, the vertical elements of the letters will be washed out and difficult to see. A compromise is called for. Usually lighting indented writing from a diagonal view, or from one of the corners, will best visualize the letters. Of course, simply moving the flashlight around the paper will show this.

Figure 4.83 reiterates the need to keep the flashlight moving during the exposure to avoid the flashlight streak produced otherwise. An entire sheet of paper can be evenly lit if the flashlight is kept moving.

What is the recommended exposure for indented writing on white paper? 100 ISO film and f/11 are standard elements of an exposure when the evidence may ultimately be used for critical comparisons. The camera is positioned on a tripod over the evidence placed on the floor. The evidence is placed on a clean substrate, like our brown packaging paper, not directly on the floor, where the evidence can be contaminated. In the first photograph we should compose on

Figure 4.82
"Hello" lit from three directions.

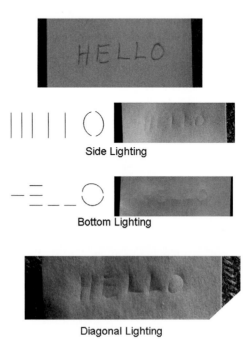

Side Lighting

Bottom Lighting

Diagonal Lighting

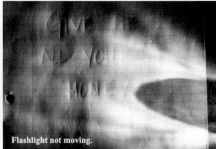

Flashlight not moving.

Flashlight moving.

Figure 4.83

Flashlight: static and moving.

the entire sheet of paper. In this case, it is not the paper that we want to be properly exposed, so the aperture priority mode is not recommended, which would try to properly expose the paper, not knowing it is the indented writing we are truly interested in. With a rechargeable Mag Lite or Streamlight, flashlights frequently issued to crime scene search personnel, a 2-second shutter speed will be set on the camera. Because the paper is white and highly reflective, the flashlight should be moved to about 6′ from the paper. The shutter can be set to a 10-second delay, which will give the photographer ample time to walk 6′ away, get the flashlight almost parallel to the floor, and direct the flashlight from one corner to another corner of the paper while moving the flashlight during the entire 2-second exposure.

Figure 4.84 will be a typical result. Perhaps this suspect was apprehended between robberies, because he had another similar note to the one left at one bank robbery still in his pocket when apprehended within an hour of the first robbery. A pad of paper in his apartment had indented writing on it that showed aspects of both notes.

It is recommended a scale be laid alongside the full sheet of paper for the first photograph, so it does not cover any of the paper. After that, if you would like close-ups of particular words, additional photographs can be taken.

The entire sequence of close-up photographs is therefore:

- An image of the full sheet of paper, with a scale alongside; 100 ISO film; f/11; flashlight 6′ away from the paper, aimed diagonally from one corner to another

Figure 4.84

Indented writing example: bank robbery (Courtesy of the Arlington County Police Department, Arlington VA).

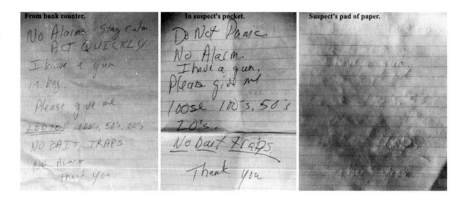

corner, almost parallel to the floor (so the indentations are clearly "shadowed"); 2-second shutter speed.

- +1 bracket: 4-second shutter speed.
- −1 bracket: 1-second bracket
- Another series of brackets (0, +1, −1) with the flashlight coming from a different corner.
- Selected partial phrases or individual words as desired. All with scale.

DUST PRINT PHOTOGRAPHS

Perhaps the shallowest type of 3D evidence a crime scene photographer will be required to photograph is a dust print lift. Dust sitting on top of a non-porous surface is, indeed, a 3D type of evidence. In this case, however, it is not shadowing that will reveal the dust print. Oblique light must highlight the dust itself. It is necessary to enhance the contrast between the dust and the substrate it is currently on, which can be the floor at the crime scene or the black Mylar film of a dust print lifter used to attract the dust from the original surface.

If the dust prints can be visualized "as is," they should be photographed "as is" before lifting them. Many times, however, a dust print is located after a blind search for shoe prints in an area suspected of having shoe prints, but where none can originally be found. Several electrostatic dust print lifters are available on the market, and most seem to recover such dusty prints pretty well. Once the dust print is on the black Mylar film, perhaps the first thing that should be noted is that there has been a left-to-right reversal. A dust print appearing to be a right shoe was made by a left shoe, and vice versa. Figure 4.85 shows a dust print and the shoe that made it.

This must be mentioned in your notes, because there will be times when the photograph of a dust print is not obviously a dust print lift, and it is essential the shoe print examiner compare the photograph to the correct shoe.

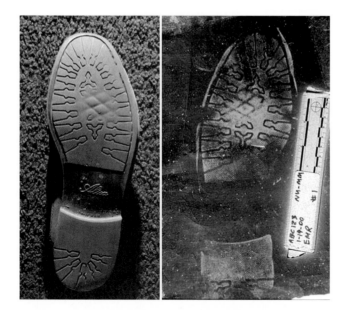

Figure 4.85
Dust print lift: left to right reversal.

Another interesting aspect of dust prints is that they may be a **positive dust print** or a **negative dust print**.

The positive dust print is made when a dusty shoe leaves its residue on a relatively clean surface. The shoe print is the dust. A negative dust print is made when a dirty surface is walked on and the shoe or foot removes dust in the shape of the shoe or foot. Either a positive or negative dust print can be identified. See Figure 4.86.

Here, again, some have switched from using an electronic flash to using a flashlight to visualize a dust print to be photographed. This author was originally taught to use an electronic flash, and many of my images came out fine. Too many times, however, the flash improperly lit the entire shoe print. Because it was possible to light the dust print with a flashlight in a way that lit the entire shoe print well along its entire length, it made sense to consider this lighting technique to capture the image in a photograph as well. It just became necessary to determine the shutter speed necessary for a proper exposure. The film would be ISO 100; and the f/stop was going to be f/11 to avoid any diffraction problems. A Mag lite or Streamlight flashlight was the light of choice. If you use a different flashlight, this shutter speed recommendation will have to be adjusted for your flashlight. It was known that the flashlight would have to be moving to avoid the streaky look of a nonmoving flashlight.

Fanning the flashlight by moving the flashlight in an arc shape was first tried, but soon it was noticed that this technique usually resulted in the part of the dust print closest to the flashlight being overexposed, because it was receiving light the full

Positive dust print: A dust print where the details of the shoe print have been deposited on a surface with dust.

Negative dust print: A dust print where the shoe or foot walking across a dusty surface removed dust, and now the shoe or foot print is a clean area on a dusty floor.

Figure 4.86

Positive and negative dust prints.

duration during a long exposure, while the opposite end of the dust print was only partly lit by the fanning action and tended to be underexposed. So it was necessary to have the flashlight moving while it remained parallel to the long sides of the shoe print. Figure 4.87 shows both the correct and incorrect methods of moving the flashlight during the exposure. Trial and error was used to determine the proper exposure for a dust print lift; it turned out to be 6 seconds. A +1 bracket, then, would be 12 seconds; and a −1 bracket would be 3 seconds. These are all with the flashlight being about 6″ to 1′ from the heel or toe of the dust print.

One other aspect of these long exposures became apparent. Because the labels on the scale, placed parallel to one of the long sides of the dust print, are usually white, these frequently became quickly overexposed. Two preventive cures were developed. As the flashlight is being moved over the dust print, parallel to the right and left sides of the print, try to avoid it coming to the point where it actually directly lights the scale. Try to avoid having the strongest part of the flashlight beam light the scale. A second method to keep the scale from becoming overexposed is to write the identifying information usually put on the scale in bolder black lines. If a bold Sharpie pen, or the equivalent, is available, use it. Then, wherever the label is pure white, cover most of it with black ink.

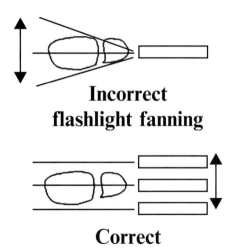

**Incorrect
flashlight fanning**

Correct

Figure 4.87
Incorrect and correct flashlight
movement over dust prints.

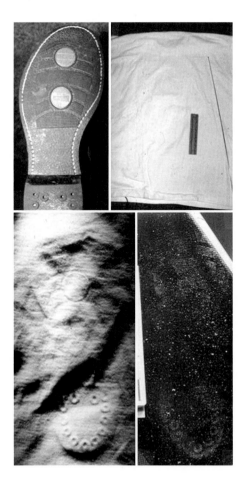

Figure 4.88
Dust print from T-shirt.

The recommended sequence for a dust print lift, then, is:

- Without a scale in place, take the first image with 100 ISO film, f/11, and the flashlight panning the dust print for 6 seconds.
- Add a scale and take another "0" exposure.
- Bracket +1: 12 seconds
- Bracket −1: 3 seconds

Take a "0" exposure from the left side if the first series of four images was with the flashlight coming from the right of the dust print.

Bracket +1 from the left side.

Bracket −1 from the left side.

During the exposures, ensure the flashlight is uniformly lighting both the heel and toe of the dust print at the same time. With just a little practice, your success rate of well-exposed dust print images will rise drastically.

Dust prints can be lifted from hard surfaces like linoleum, parquet wood, oak strip floors, tile, and cement. They have also been lifted from fabric when clothing has been stepped on, or when a victim has been stomped on (see Figure 4.88). In a test, dust prints have also been successfully lifted from skin after a recent "stomping" incident. See Figure 4.89.

A nonelectric technique is required for live skin. The Lynn Peavey Company, a crime scene product distributor, makes Stati-Lift, which is perfect for this application. It is made up of two layers of plastic, one clear and one almost as opaque as the Mylar film used with other electric dust print lifters. Just by separating the two sheets of plastic with the tabs provided, a static charge is applied to both sheets of plastic. Squeegee the darker layer over the dusty impression with one hand, and the dust transfers to the film. See Figure 4.89.

As the surface becomes more and more dusty, successfully lifting a positive dust print becomes more difficult or impossible. Walking across a carpet that is very dusty will not produce dust prints of value on adjacent portions of the carpeting. Stepping from a carpeted floor onto upholstered furniture will also not usually transfer a dust print. Walking across that same carpet, and then stepping onto the linoleum kitchen floor, will usually leave a great dust print.

One final aspect of dust print lifters must be mentioned. Students have been able to retrieve dust prints with detail after the dust prints have been made, and then left several days to simulate a time period between a crime being committed and the crime being discovered. Subsequent falling dust may cover dust prints to the point they are no longer visible. In these situations, a first or second dust print lift may remove the extraneous dust covering the dust print of interest. Recognizable shoe print patterns have been recovered up to 43 days after the original dust print was left on a surface. Additional snow falling on top of a shoe print in snow will effectively fill in the impression, making it impossible to recover the original shoe print. Additional dust falling over a dust print does not eliminate the

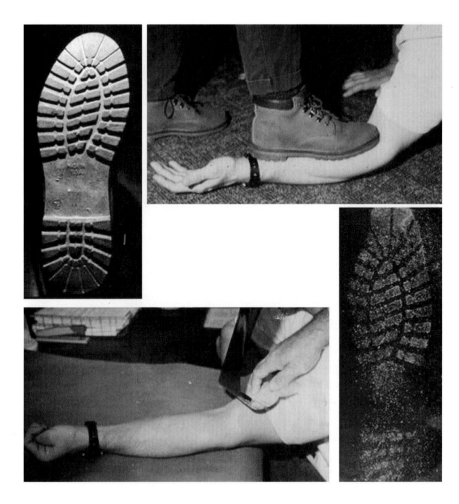

Figure 4.89
Dust print from skin.

possibility of recovering some shoe print detail with multiple dust print lifts. Figure 4.90 shows this effect.

BOUNCE FLASH

WHEN TO USE BOUNCE FLASH

Using an electronic flash so that it is aimed directly toward the scene of interest is usually a very efficient way to light the crime scene. Several situations exist in which doing what is "normal" may actually produce unacceptable results. If a mirror is in the scene, aiming an electronic flash toward it may cause the reflection of the mirror to be seen in another area of the scene. This mirror reflection may be so distracting that it overwhelms the intent of the photograph. A mirror can be a problem, even if it does not throw its own reflection elsewhere into the scene. When direct flash strikes many objects, a shadow is produced behind that object. Our eyes are so used to this that we hardly notice it. If a mirror is in the background, the light reflecting from the mirror can now create a shadow in front of

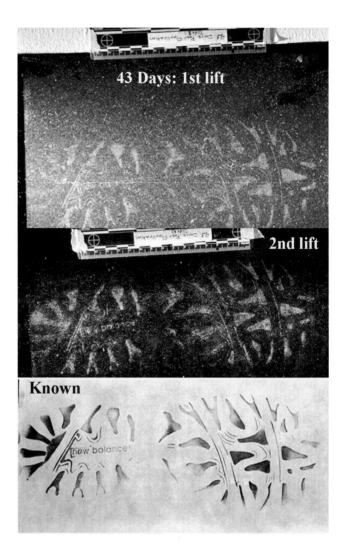

the object as well. When the same object has more than one shadow, it looks a bit strange and becomes distracting. Figures 4.91 and 4.92 show both of these effects, with bounce flash used to eliminate the distracting elements of each.

Mirrors are not the only causes of these types of problems. Many highly reflective surfaces are present in a crime scene. The crime scene photographer will eventually be able to anticipate these problems and use bounce flash to overcome the problems that would be created by these reflective surfaces. Metallic surfaces are highly reflective. When struck by direct flash, they can develop a "hot spot" or glare from reflection. The glare can be eliminated if light softly washes over the same area from the ceiling. Figure 4.93 shows this effect and the cure: bounce flash.

Figure 4.94 shows the effect of a direct flash aimed at "evidence" on a glass-topped table. The reflection of the tabletop is thrown onto the wall behind the

 Direct Flash

Bounce Flash

Figure 4.91
Mirror reflection appearing in background.

Direct Flash

Bounce Flash

Figure 4.92
Mirror creating multiple shadows.

table. Bounce flash eliminates this distracting reflection. Whenever glass is in the field of view, it can behave similar to a mirror and reflect light elsewhere in the scene. Be cautious with furniture covered with glass anywhere in the scene.

Liquids are also highly reflective and will bounce the light of a direct flash unpredictably around a scene. One example ought to suffice for this concept to be remembered. An unhappy non-US citizen learned he was going to be deported and entered a U.S. Immigration Office lobby prepared to negotiate his case. He carried a 5-gallon gas can, a Bic lighter, and a razor blade. As he began dousing the lobby with the gasoline, the lobby cleared quickly. He then slashed one wrist with the razor blade and lit the lighter. He said he wanted officials to reconsider his deportation quickly, before he lost enough blood to faint and light the gaso-

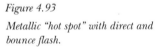

Figure 4.93
Metallic "hot spot" with direct and
bounce flash.

Figure 4.94
Glass-top table with direct and bounce flash.

line. Because the cut wrist did not appear to be cut badly, time was on our side, and a negotiator was called in to deal with the subject. The fire department eventually staged outside a wooden lobby door with a high-pressure, wide-spray water hose; on cue, they opened the door, doused the subject and his lighter, and the subject was subdued. The lobby was now wet and bloody, and an interesting rainbow of gasoline floated on all the other liquids. To illuminate that crime scene with a direct flash would have produced many interesting, though distracting, reflections. Bounce flash was the only way to photograph this scene. Unfortunately, these photographs have not been located at this time.

BOUNCE FLASH IS DIM LIGHT

To light a crime scene with a flash bounced off the ceiling, the light must travel farther than normal. It must travel from the flash head to the ceiling and then reflect down onto the surface of the scene of interest. The inverse square law taught us that when light goes farther to get to its destination, it is dimmer when it gets there, which is, unfortunately, just one cause of the dimness of bounce flash.

A difference exists between direct flash lighting an object at *x* distance, and bounce flash lighting the same object at that same *x* distance. Would you not expect that direct flash would be brighter than bounce flash at the same distance?

In other words, even if the distance the light has to travel to reach the subject to be photographed is 20′, direct flash at 20′ is brighter than bounce flash at 20′. Why?

Direct flash is very effective at lighting its subject matter. The light proceeds directly from the flash head to the subject matter. When the flash is aimed at the ceiling and bounces from it downward toward the scene of interest, it is diffused wider than the direct flash beam of light. Some of this light striking the ceiling bounces away from our intended target. Some of the light is lost to the surroundings that may be outside the area of the particular image that has been composed in the viewfinder. Because this light misses our subject altogether, the light that does fall on our subject is dimmer still.

Many ceilings are not just solid surfaces painted white. Many ceilings, particularly in offices, are made with acoustical tiles. Being porous so they will absorb sound, they can also absorb light. Some of the light striking acoustical tiles will be absorbed and never reflect from the ceiling. This also diminishes the light eventually reaching our subject matter. Bounce flash is dim light.

COMPENSATING FOR DIM BOUNCE FLASH

With less light reaching our crime scene, an exposure compensation must be made to ensure there will be proper exposures. Because these photographs are being taken inside, we have probably already loaded our cameras with 400 ISO film. It is usually impractical to reload the camera with different film for just a few shots. If using a digital camera, however, it would be possible to bump up the ISO film rating to ISO 800 or so for the bounce flash shots to help compensate for the lost light. Remember to reset the camera for the normal ISO 400 film speed after the bounce flash shots are over.

Our shutter speed is locked to our sync speed, so the exposure compensation necessary must come from somewhere else.

Set the flash head for the manual flash mode. The automatic exposure flash mode and the dedicated exposure flash mode allow for the possibility for less light than the manual flash mode, but because the light loss can already be significant with bounce flash, avoid any further light loss and do not use these other flash exposure modes.

Only one camera variable is left to compensate for the light lost because of the use of bounce flash. It will be necessary to open the aperture to let in more light. To use a wider aperture, it must first be established what the "starting" aperture is. Because it is necessary to know the distance the light will have to travel, the first step is to calculate the flash-to-bounce surface-to-subject distance.

A couple of ideas must be dealt with here, and both deal with minimizing the distance the light has to travel.

First, take the flash off of the camera's hot-shoe. If we use the flash connected to the camera with a PC or remote flash cord, we can hold the flash at arm's reach aimed at the ceiling, and this can help minimize the distance the light has to travel. The closer the flash unit is to the ceiling, the shorter the distance it has to travel, and the brighter it will be when it finally gets to where it is needed.

Now, where do we aim the flash? The optimal location on the ceiling to aim the flash, to minimize the total distance the light has to travel, is at a point that is midway between the photographer and the subject matter. If you had a rubber ball and threw it at the ceiling to strike this location, it would bounce down and hit the subject directly. At times, however, this will not work. There may be a concave light fixture there, or there may be a hanging fixture just beyond this halfway point that would block the light striking the midpoint on the ceiling. Nevertheless, if the midpoint is not obstructed in any manner, it is always the most efficient direction to aim the flash head. If some obstruction makes the midpoint unusable, we should consider other options, the next being a point on the ceiling one third the distance between the photographer and subject. If that will not work, we should consider a point directly over the head of the photographer. If neither of these will work, we will be forced to consider aiming the flash at a wall either at one side of the subject or behind the photographer so the light can bounce directly on the subject matter without the photographer casting a shadow over the subject matter. See Figure 4.95.

Figure 4.95
The optimal bounce angle.

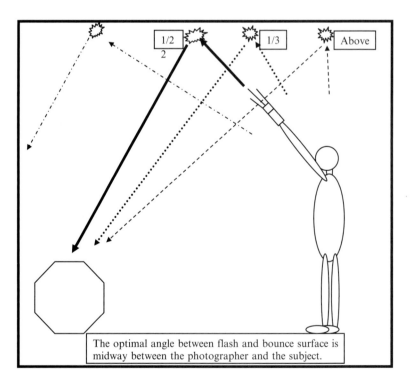

The optimal angle between flash and bounce surface is midway between the photographer and the subject.

Where should the flash not be aimed? The flash head should not be positioned to bounce the light beyond the halfway point between the photographer and the area of interest. This would cause the light to bounce down from the ceiling behind our subject of interest and may either backlight our subject and throw him or her into shadow, or the light may not light them at all.

With the flash held at the optimal angle toward the ceiling, mentally calculate the flash- to bounce surface- to subject distance. Make note of the flash calculator dial's f/stop recommendation for this distance, which is merely the beginning point, not the f/stop that will actually be used. Remember, several dimming factors are present in addition to this new distance the flash will be traveling.

With this f/stop in mind, set the camera for 2 stops wider than what is recommended. This will compensate for the less intense bounced light. Why two stops? Trial and error has determined this is a good starting point.

When taught bounce flash in the late 1970s, we were advised to bracket both a +1 and −1 from this point. However, the −1 bracket was never to be the best of the three shots, so it is no longer recommended. Just do a +1 bracket. That means, in effect, to determine the flash- to bounce surface-to-subject distance, see the f/stop necessary for that distance, and then open 3 stops. This will be a +1 bracket.

Let us return to the calculator dial for an example.

Assume the photographer is standing 10′ from the object to be photographed and must use bounce flash because of various circumstances. The photographer is using 400 ISO film, and his camera's sync speed is 1/60th of a second. A direct flash photograph would require an f/22 according to the calculator dial.

Figure 4.96
The flash calculator dial.

Aiming the flash to a position on the ceiling midway between photographer and the subject, the distance to the ceiling is 7′. From that point on the ceiling to the subject on the floor is 13′, for a total of 20′ the light of the flash has to travel. The calculator dial indicates an f/11 is the appropriate f/stop for a 20′ distance, but that really means a 20′ distance of direct flash, not bounce flash.

So, we open up 2 stops from the recommended f/11 and take the first photograph with an f/5.6. We then bracket +1 and take another photograph at f/4. Many readers may now be wincing. Was it just recommended to take a photograph with an f/5.6 and f/4? Yes, and it hurt just touching the keyboard to type "f/5.6" and "f/4." Is a new Rule of Thumb warranted?

 Rule of Thumb 4.5: When using bounce flash, if the subject of the photograph is 10′ or farther from the photographer, the use of an f/5.6 or a wider aperture will be needed, which will diminish the DOF. Consider ways to get closer to the subject to avoid having to use wide apertures.

Because of the light loss when using this technique, bounce flash is not recommended for long or large scenes. Bounce flash requires the use of much wider apertures, and the resulting DOF reduction may be too great for large scenes. If faced with this circumstance, consider segmenting the large scene into more manageable smaller ones. This makes more sense than trying to photograph long scenes with wide apertures, knowing that the entire scene will not be in focus. Relatively small areas, however, such as tabletops and most kitchens and bathrooms can often benefit from the softer diffused light produced with bounce flash.

PAINTING WITH LIGHT[*]

WHEN TO USE PAINTING WITH LIGHT

Painting with light:
Painting with light (PWL) is the photographic technique to be used whenever a single flash would be inadequate for the size of a large dimly lit scene. The PWL technique properly lights a large scene as if it were daylight.

Painting with light is the photographic technique to be used whenever a single flash would be inadequate for the size of a large dimly lit scene. Whereas the flash calculator dial may indicate an f/2 can light a single object 60′ away, it should not be used for a scene 60′ long, because an f/2 will also ensure minimal DOF. In addition to outdoor scenes during the night, painting with light can also be used in large interior areas, such as parking garages, gymnasiums, theaters, and indoor concert venues.

Many law enforcement agencies have access to light trucks that can bring in auxiliary lights. However, as was stated earlier, it is not just the light volume that is an issue. Daylight color film will record the proper colors of objects within the

*Most of this section was previously published in *Evidence Technology Magazine*, vol. 2, No.5, Sept./Oct. 2004.

scene only if lit by either midday sunlight or electronic flash. Other lights will usually introduce colored tints into the scene, and, at times, these tints will be unpredictable and objectionable. Using black-and-white film in these situations would certainly avoid the color tint issue.

The better choice is to use the painting with light technique. It allows for the scene to be properly colored, but when practiced so that it can be efficiently done, it is even more convenient than having to wait for light trucks to respond and set up.

Figure 4.97 shows a scene lit by a single flash, showing how inadequate it is for the size of the scene. The same scene is then lit with the painting with light technique. When done properly, it is as if God turned on the light so the entire scene could be properly exposed. The painting with light technique properly lights a large scene as if it were daylight. Multiple flashes are used to light one negative so that the accumulated light is adequate for the size of the scene.

PAINTING WITH LIGHT FROM TWO SIDES

With a large outdoor nighttime crime/accident scene, painting with light provides adequate light to properly expose the entire scene. Basically, painting with light entails the camera being mounted on a tripod, and the shutter being opened for as long as it takes to have someone circle the entire scene, firing a hand-held flash unit multiple times to ensure the entire scene receives enough light for a proper exposure.

Twenty variables have been determined that apply to the basic PWL situation. Perhaps that sounds intimidating. However, once you think about it, because all the variables are described in the following paragraphs, there will be nothing for the individual to have to "invent" for themselves. All the hard work has been done. The reader now only has to understand the variables and then apply them when the need arises, which is far less intimidating than a challenge from a photography instructor to you to determine all the variables yourself. Hopefully, as the variables are described, your reaction to most of them will be, "Of course, why would it be done any other way."

Figure 4.97

A large scene lit with a single flash and by PWL.

The variables seem to logically fit into three areas nicely:

- Tripod variables
- Camera variables
- Flash variables

The variables make perfect sense once they have all been explained. Many are interrelated and will make sense only when their counterparts have been explained. Because of this interrelationship, a sequence cannot be assigned to individual variables that will make each totally understandable as they are presented. Repeating the phrase, "this will make sense soon, as soon as 'x' has also been discussed" has been avoided. Give all of them a first reading, and when you begin the second reading, you will find yourself nodding in agreement as each variable is explained.

Tripod Variables

Assume a large night-time outdoor crime scene has to be photographed.

1. The first thing that must be done is to establish a viewpoint of the scene. Some scenes would fit beneath a large circle, because they are as long as they are wide. Many scenes, however, are wider from one point of view and narrower from another point of view. If this is the case, select the point of view that has the scene arranged so that it is at its narrowest orientation as you view into it. The painting with light technique has the virtue of being able to light extremely long scenes; wide scenes will be problematic. It may be necessary to walk around the scene so that this narrow alignment of the evidence can be established. Once this has been done, other variables will make more sense.

2. Now that the viewpoint has been established, the tripod eventually will have to be set up somewhere. Set it up 21′ from the closest item of evidence. Many long strides are approximately 3′ in length. Stand over the closest item of evidence and take seven giant steps away from it, and set the tripod there. From the tripod to the first item of evidence forms the line that, if continued, has the evidence laid out long (away from the camera) and narrow.

   This variable gets questioned perhaps more than any other. Why 21′? Eventually, you will see that the bottom of the camera's composition will be set to a 15′ distance from the camera. Having the closest item of evidence at 21′ ensures it is near the camera without risking it getting too close to the bottom edge of the photograph to possibly be cut off. Could 18″ have worked just as well? Yes. If you are not happy until you force a concession from everyone you meet, you may select your own distance for this closest item of evidence. Just have it be far enough beyond the 15″ point distance so it does not risk getting cut off.

3. Mount the camera on the tripod. Raise it to your eye height, so that viewing the scene through the viewfinder is similar to viewing the scene without a camera; this is called your natural perspective. Anyone viewing your photograph of the scene would see the scene as if they were standing at your side as you viewed it without a camera.

4. The camera will have to be tilted up or down on the tripod some specific angle. When you are determining the tripod location by pacing off 21″ from the nearest item of evidence, after just two paces, 6″ from the evidence, temporarily set down an evidence marker. That marker will now be 15′ from the tripod. While looking through the viewfinder, raise and lower the camera until the bottom edge of the viewfinder aligns with the marker 15′ from the camera. Your view of the scene is now 15′ and beyond. The marker at 15′ can now be picked up.

Camera Variables

5. Because it is a dark scene, load the camera with ISO 400 speed film. Tests have shown that many ISO 800 films have excellent resolution and do not suffer from the graininess that 800 ISO films once had. However, they are more expensive, and with the number of film manufacturers dwindling, 400 ISO film may be more accessible. Digital cameras will easily be able to have different ISO equivalents set as desired. However, until digital noise with ISO 800 settings is as low as the digital noise level of ISO 400 settings, the latter is still the film speed preferred. Also, some older flash units do not have an ISO 800 setting, restricting the photographer to its fastest ISO setting of 400 ISO.

6. Use a 50-mm lens focal length. This photograph is an example of an exterior overall photograph, and the 50-mm lens shows the scene as the eye saw it. Evidence within the scene seems neither nearer to nor farther from the camera when viewed through a 50-mm lens.

Using a 50-mm lens is also the reason we want to establish a view of the evidence so that it is laid out narrowly rather than wide. The 50-mm lens can encompass a view of just 46° of the real world. Were the scene aligned so the evidence was wide, rather than narrow, the use of a 50-mm lens would force us to back away from the evidence to include it all in our composition. Because the evidence is now viewed so it falls into a relatively narrow alignment, we can get closer to the first items of evidence.

7. A camera exposure mode must be selected. Choose the manual exposure mode, so that you can select a specific shutter speed and a specific f/stop.

8. Select the "bulb" shutter speed, which is the shutter speed that has the shutter remain open for as long as the shutter button is depressed. Because the shutter has to be open for a long time, also use a remote shutter release cable that is capable of locking the shutter open. This can be a manual remote shutter release cable that screws into a hole in the shutter button, or it can be an

electronic shutter release cable. Some cameras have a shutter setting that will lock the shutter open the first time the shutter is depressed and then close the shutter the second time the shutter is depressed. This will also work.

9. The f/stop selection will be based on half the width of the back of the scene. While looking through the viewfinder, have an assistant walk to the back center of the scene, at the distance from the camera of the last item of evidence. Establish an imaginary line between the center of the back of the scene and the camera. Perpendicular to this line, have the assistant begin walking away from the center of the scene until they just barely step out of your field of view as you view the scene through the viewfinder. Yell "Stop" to them when they step outside the camera's viewpoint. Have them place a marker at this location. Repeat this on the other side of the scene and have a marker placed there also. Both markers should be just out of the field of view of the viewfinder. Have your assistant pace the distance between the two markers or measure this distance with a tape measure. If their pace is also approximately a 3′ stride, multiply their number of paces by 3 to determine the width of the back of the scene. Divide the total width of the back of the scene by two. An assistant will eventually hold a flash unit at each marker location, aimed toward the center of the scene. It will only be necessary for each flash unit to light just half the width of the back of the scene. For example, if the total width of the back of the scene is 60′, half of that is 30′. Each of the multiple flash firings will have to light a 30′ long area.

10. Now that the f/stop has been determined, it will be possible to focus the camera by the hyperfocal focusing technique. Recall that . . .

- f/22 had a DOF range of 6′ to ∞, when focused at 12′.
- f/16 had a DOF range of 8′ to ∞, when focused at 16′.
- f/11 had a DOF range of 12′ to ∞, when focused at 24′.
- f/8 had a DOF range of 15′ to ∞, when focused at 30′.

Of course, this presumes a flash with a guide number of 120. Should you be using a flash with a different guide number, your own calculations will have to be determined.

In our hypothetical case with a 60′ wide scene, the flash will have to light 30′. With a 120 GN flash, the f/stop to light a 30′ distance is an f/8, according to the flash calculator dial. To hyperfocal focus with an f/8, focus on something 30′ away from the camera. If your lens has a distance scale, just align the 30′ distance over the focus point. Without a distance scale, have an assistant stand 30′ from the camera with a flashlight held to light their face, and focus on their face.

This will change with the width of the back of each scene. However, for some reason, most times this author had to use the PWL technique to light the scene, f/8 turned out to be the most frequently used f/stop.

11. If the crime scene is in an area with the ambient lights of buildings located behind the camera, or if roads are nearby where cars can swing around and ultimately hit the back of the camera with their headlights, then the camera's viewfinder will eventually have to be covered. Extra light coming into the camera through the viewfinder can ruin a painting with light photograph. Many, if not most, camera straps have a viewfinder cover attached to them for this very purpose. If your camera does not have a viewfinder cover, throw a roll of black electric tape into your camera bag, and just before the flashes begin lighting the scene, cover the viewfinder with black tape.

Flash Variables

12. Use a 120 GN flash. The settings for this hypothetical case are based on the use of a 120 GN flash. A weaker flash will necessitate the use of wider apertures, which will negatively affect the DOF range. A flash with a stronger GN will, of course, be able to be used, because it would ensure smaller apertures are used for the same sized scenes, thereby ensuring even better DOF is obtained.

13. The flash exposure mode should be set to manual. This large dimly lit scene requires all the light the flash is capable of. The benefits of the automatic or dedicated flash modes revolve around their ability to cut the flash power off early, which is not applicable in this situation.

14. Because a 50-mm lens is being used, the flash head has to be set to "normal." The flash head will be used detached from the camera's hot-shoe, so this will have to be done manually. The camera cannot "tell" the flash which focal length of lens is being used.

15. Markers have already been placed where the flash operator can stand safely and not be in the field of view. In this case, the "safe spot" is 30′ from the middle of the crime scene. It is important that the flash not be in the field of view when it is fired. If there has been a miscalculation about the "safe spot" for the flash and the camera can "see" the flash go off, the results are interesting, but not conducive to a successful photograph. If the camera can "see" the flash pop, the result is a variation of the Star of Bethlehem.

Figure 4.98 shows the painting with light setup, except the focal length of the lens was intentionally widened to include the two markers that would have been "safe spots" had a 50-mm lens been used. Now, three flash bursts can be seen, one at the left rear and two at the right rear. The flash bursts do not contribute to the success of final image. Figure 4.99 shows what many would imagine the true Star of Bethlehem would look like. Interesting? Yes. A beneficial addition to the photograph? No.

From these flash positions at the rear of the scene, the flash operator needs to work around the scene until it is completely lit. When doing so, the flash

Figure 4.98

PWL: intentional use of wide-angle lens to show the "safe spot" markers and flash bursts.

Figure 4.99

PWL: flash burst visible in left rear.

operator must remain 30′ from the imaginary midline of the crime scene. The best way to ensure the flash operator does not accidentally begin to get closer to this imaginary midline is to place two more markers 30′ from the camera's position, one on the left and one on the right. Then, the flash operator continues on a line from the marker at the rear of the scene to the marker opposite the camera's position.

16. The flash operator will be manually firing the flash by pushing the appropriate button on the flash. The flash, however, is in full manual mode, and with the camera set at f/8, will light a distance of 30′. If the flash is just held at chest height, it may overexpose the area nearest the flash operator. To help prevent this overexposure possibility, the flash is held at arm's length directly above the photographer's head. This will help prevent burning out the scene just beyond the flash operator's position.

17. Now that the flash is as high as the flash operator can get it, where should it be aimed? The first inclination is to aim the flash at the midpoint of the scene. However, the midpoint will actually receive some light from both sides of the scene. Because of this, to avoid overexposing the midline, it is suggested to aim the flash so it strikes a point about 4/5th of the way into the midline. Then, the midline overlaps of light will blend well.

This is easier said than done and is one of the critical aspects of the painting with light technique. Should the flash operator accidentally hold the flash so it projects its light parallel to the scene or above the scene, the scene will be very underexposed. It is essential that the flash be aimed slightly down from its position held above the photographer's head. Standing directly under the flash head, it is very hard to determine the angle it is aimed into the scene. For this reason, it is recommended that the flash operator stand to either side of the flash unit.

Figure 4.100 shows how the angle of the flash head can be determined when standing at either side of the flash. From beneath or from behind the flash unit, the angle it is aimed into the scene cannot be determined.

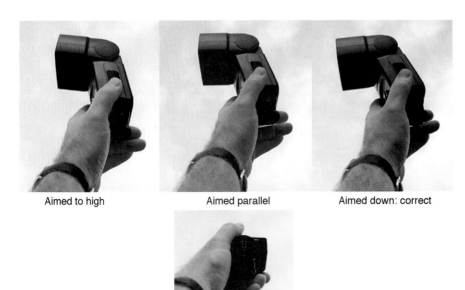

Aimed to high   Aimed parallel   Aimed down: correct

From underneath, the angle is impossible to determine

Figure 4.100
PWL: flash angles into the scene.

18. Once the first flash has been fired, the flash operator must walk to a new position to fire it again. As mentioned previously, they should walk toward the marker 30′ from the camera position. Because the flash throws light out in the shape of a bubble, it will be necessary to go to a position that will enable the sides of the flash bubbles to overlap so no gaps are present in the overall lighting. From trial and error, it has been determined the correct distance to walk to arrive at this new position is "5 casual paces." Not giant steps! Not baby steps! Walk five of your normal paces down the line. Photographers are short-legged and long-legged. When tall and short people walk alongside each other with "normal paces," they will stay together without much effort. During this 5-step walk, the flash usually will have a chance to recharge and be ready to fire again, if not immediately, very shortly thereafter.

This 5-step interval between flashes into the scene only needs to continue until the area 15′ from the photographer has been lit. Remember, the first 15′ between the camera and the scene will not be in the field of view and, therefore, does not need to be lit.

Walking around the camera's position to get to the opposite side of the scene provides the flash operator the perfect opportunity to fire one flash into the scene from the camera's position. Lighting the scene from both sides does not light the front of the evidence close to the camera. The scene looks more normal if the closer items of evidence get some front lighting. This flash does not overexpose the front of the scene at all. Just remember to aim the flash so it will strike the ground about 25′ from the camera.

Beginning at the point 15′ from the camera, a series of flashes are then popped into the scene from the opposite side as the first flashes. Each side will get five to eight flashes, and one flash will come from the camera's position.

Coordination between the camera operator and the flash operator:

19a. To get ready for the actual photograph, the camera operator will cover the viewfinder with the viewfinder cover. He will hold up a piece of black construction paper over the end of the lens, trip the shutter with the shutter release cable, and lock it open. He will then await the flash operator's signal that the flash has charged and he is in position. Figure 4.101 shows the camera operator awaiting this signal.

In this position, it is critical that the camera operator concentrate on the position of the black paper and the lens. It is imperative the tripod not be moved and that the paper not strike the lens and move the camera. Doing either will introduce an offset into the sequential exposures on one negative, which will look like blur and ruin the image. The camera operator will be able to see when the flash goes off with his or her peripheral vision: it is not necessary for the camera operator to be looking down into the scene. His entire job is to cover and uncover the lens during the flashes without bumping either the camera or tripod.

19b. When the flash operator notices the flash has fully charged, it is held overhead, aimed to strike the ground about 4/5 of the distance into the midline, and he positions his finger over the manual flash button. He then yells, "ready!" This is not a polite whisper or said in normal conversational tones. The flash operator may be 60′ or so from the camera operator, and there may be ambient noise in the area.

Figure 4.101

Camera operator's position waiting for "ready."

Listening for "Ready!"

20. On hearing the flash operator yell "ready," the camera operator raises the black paper away from the lens, and when it reaches the apex of a 90° rotation, the camera operator yells, "go!" (Figure 4.102). This is not a polite whisper. It is not said in a conversational tone. There may be distracting ambient noise in the area. The camera operator yells, "go" when the black card has reached the top of the arm swing. Without waiting for the flash to go off, the camera operator immediately swings the black paper back to a position in front of the lens. If the flash operator did not get the flash to go off in that brief time period, the process will have to be repeated. After all, why would the flash operator have said "ready!" if they were not ready.

This process is repeated as the flash goes off multiple times from one side, once from the camera position, and then from multiple positions from the other side. When all the flashes have been fired, the shutter release cable is unlocked, the shutter closes, and the black paper can be removed from in front of the lens.

Painting with Light Tips

1. It was mentioned that black construction paper be used to cover the lens during the painting with light technique. When black construction paper cannot be found, a hat or clipboard has been used instead. If the clipboard is light colored, it should not be used. If it is glossy or will reflect light striking it into the lens, it should not be used. The key here is to use a dark opaque object to cover the lens to prevent stray light entering the camera between flashes. Black construction paper is ideally suited for this job. Put some in your camera bag.

2. The manual flash mode eats batteries. Do not attempt to try this process with used batteries in the flash unit. It is very frustrating to have pedestrians and traffic stopped from entering the scene, have support personnel staking out

"Go!"

Figure 4.102

Camera operator's position for "go."

the perimeter to prevent strays from entering the scene from various directions, and then have to wait for used batteries to charge the flash between flash cycles. Put new batteries into the flash unit.

3. Better yet, use two flash units, both with new batteries. Two people are acting as camera and flash operators, and often that means two camera kits are available. When the flash operator has two flashes, both with new batteries, the actual process of painting with light can be just a few minutes long. With two flash units, the flash operator can walk around a large scene hardly needing to stop at all, because he keeps alternating from one flash to the other so one is always ready to go. Both flash units, of course, have to have the same guide number.

4. It would be possible for multiple flash operators to stand around the scene and pop off multiple flash units with one ready/go sequence and then all move to another position. This can be tricky if one does not notice that one or the other flash units did not fire in the proper time frame. Unlit gaps in the scene will be the result. If the use of multiple flash operators is considered, it is also imperative to ensure all the flash units have the same guide number. If not, there can be a noticeable difference in flash bubbles.

Figure 4.103 shows most of the painting with light process. A 50-mm lens encloses eight items of evidence in its field of view. "Safe spots" for the flash operators have been determined, and multiple flash bursts come from the same distance from the midline. Flash bursts overlap in the middle of the scene and between adjacent bursts. One flash is directed into the scene from the camera's position. A line at 15′ represents the bottom of the camera's composition as it is tilted on the tripod. The first item of evidence is 21′ from the camera. "A-frame" markers have been laid on their sides at the 15′ and 30′ points: at 15′ for the camera tilt and at 30′ for hyperfocal focusing. Both of these will be picked up before the shooting begins.

5. The painting with light technique can be done with only one person if it is dark enough to leave the shutter open without being blocked between flashes. Then, a single person can run around the perimeter of the scene firing off the required number of flashes.

6. At times, there will be a relatively large item in the crime scene, on one side or the other (like a car). If it were in the middle of the scene, it would be easy to get the prescribed distance from it with the flash so it would be properly exposed. In this case the side visible to the camera is too far away from the flash operator on the opposite side of the scene for a normal flash burst to fully illuminate it. Because it will be captured by the camera, you will want it properly exposed.

The trick is to turn your flash operator into a "snowflake." Recall how falling snow can be eliminated from an image by determining an exposure for a 2-second

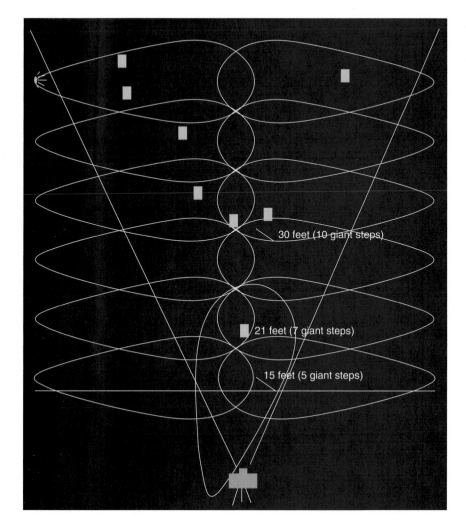

Figure 4.103
The PWL process with multiple flashes.

30 feet (10 giant steps)

21 feet (7 giant steps)

15 feet (5 giant steps)

shutter speed? Because the snowflakes are never in the same spot long enough to be properly exposed, they are blurred sufficiently to appear to disappear. It may be necessary to have the flash operator walk into the scene to properly expose the side of the object toward the midline of the scene. If we just ensure the flash operator keeps moving and is never directly lit by any direct light source, they may not appear in the photograph.

This solution, however, creates new problems. If the flash operator walks into the crime scene area holding the flash unit above his head, the flash burst will be recorded on the image even if the flash operator is not. Correct. So the flash operator must bring down the flash unit so that the flash operator's body blocks the flash from being seen by the camera when the flash goes off. This, however, gets the flash, strong enough to light a distance of 30', close to the ground where it will likely overexpose the area near the flash. This necessitates a manual flash intensity reduction. The calculator dial indicates f/8 is proper for 30′ at full manual power, f/8 is proper for 20′ at ½ power, and f/8 is proper for 15′ at ¼ power. It is recommended to use the ¼ power setting and get 15′ away from the object of interest. Trying to light an object the size of a car with a flash head set to "normal" from just 15′ may not light the full width of the car. Change the flash head to "wide" angle to force the light wider than normal. True, this may diminish the flash output straight ahead somewhat, but we can live with that. Remember, the side of the car closest to the midline did receive some light from the flashes on the opposite side of the scene but not enough for a proper exposure. Having the flash held by the flash operator slightly dimmer will augment this other light nicely.

Figure 4.104 demonstrates the movement of the flash operator. Without ever stopping, when they approach the 15′ distance from the car, they yell "ready," the camera operator yells "go," and the flash is popped. Continuing across the scene, the flash operator exits stage left. Many examples of this technique have been successful. There can be no bright ambient light around the scene to light the flash operator. It is also critical that the flash operator light the car from the side, and that the flash operator avoid lighting the rear of the scene with the flash burst. If this is done, the flash operator may be noticed as a dark silhouette.

The flash operator should be wearing dark clothing if we want them to "disappear" from the scene. Light clothing may reflect some stray ambient light into the camera.

If the midline side of a large object in the crime scene needs to be lit, it is suggested that this flash be the first one. All the other flashes from around the scene perimeter should follow this flash burst. The darker the scene, the easier for the flash operator wearing dark clothing to become invisible.

Figure 4.104

PWL: flash operator turned into a snowflake.

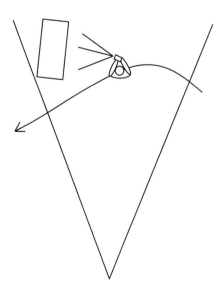

PAINTING WITH LIGHT FROM ONE SIDE

There will be times when it is impossible to flash into the scene from two sides. A riverbank or cliff may be on one of the sides of the scene or a building façade may be on one side of the scene. Many things can prevent a flash operator from getting to both sides of a scene. In these cases, only a slight modification of the painting with light technique as already explained is required to deal with these possibilities.

The vertical line on the far right side of Figure 4.105 can be imagined to be a building façade, cliff, or riverbank. In this situation, the flash operator can only stand on the left side of the scene. It will now be necessary for multiple flashes, all on the left side of the scene, to light the full width of the scene. With a 120 guide number flash unit, these will be the choices.

- f/22 can light 10'.
- f/16 can light 15'.
- f/11 can light 20'.
- f/8 can light 30'.
- f/5.6 can light 40'.

First, determine the distance from the wall/river/cliff that the farthest item of evidence is at, which is the width of the crime scene that will have to be properly exposed. That will decide the necessary f/stop. It will be necessary to light the wall and this item of evidence with a flash. For this hypothetical exercise, we will presume the item farthest from the wall is 28' away. This means that an f/8 will adequately light the scene 30' wide.

Establish a point that is 30' from the wall by pacing the distance or using a tape measure. Place a marker just outside this distance at about 30.5'. At the other end of the scene, place another marker at the same distance. This will establish the "safe" line. Flash operators can stand on this line, and they will not appear in the photograph.

The camera on a tripod can be placed so the camera is 30' from the wall. Now the camera on the tripod can be rotated to the left or right so that either the right side of the field of view or the left side of the field of view aligns with an imaginary line 30' from and parallel to the building façade/river bank/cliff. Notice, in the case of the graphic, that the left side of the field of view, as seen through the camera viewfinder, has been positioned to be parallel to and 30' from the building façade. The camera on the tripod can be moved along this imaginary line 30' from the wall to a point at which the tripod is also 21' from the closest item of evidence.

The camera is tilted so that the bottom of the viewfinder hits the 15' location. An item 30' from the camera is focused on to hyperfocal focus. Now, everything from 15' to infinity will be in focus.

Figure 4.105
PWL from one side.

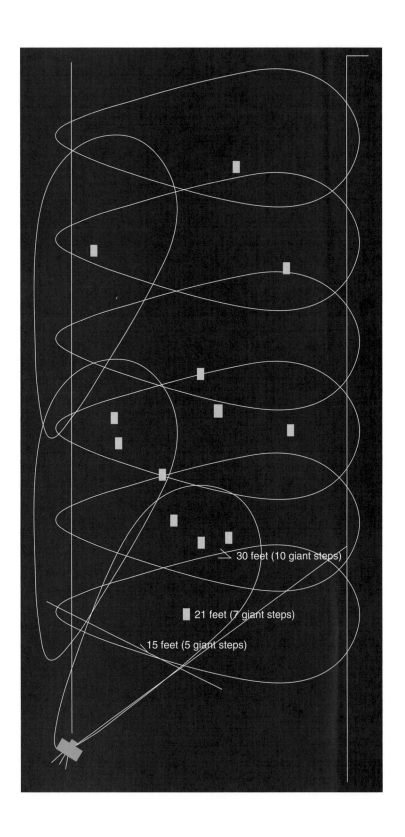

30 feet (10 giant steps)

21 feet (7 giant steps)

15 feet (5 giant steps)

Nine different flashes were needed to completely cover this scene. Six were directed into the scene from the "safe" line, one was aimed into the scene from the camera's position, and two more were aimed along the "safe" line. Notice that the flashes on the line 30′ from the wall created flash bubble gaps that had to be covered by the second line of flashes that are aimed down the "safe" line. If this is not done, a zigzag of dark unexposed triangles will be seen at the left side of the image.

Figure 4.106 is an example of painting with light from one side created as part of a training exercise at a local police academy. In this case, the "safe" line was a sidewalk on the right side of the image that was parallel to the building wall.

The painting with light technique certainly works well once it has been practiced a bit. Other agencies may prefer to use either of two alternate lighting techniques for a large dimly lit scene: time exposures and taking the photographs with the aperture priority technique. These techniques are explained in Chapter 9, under the discussion of traffic accident photography.

Before this chapter on flash concludes, and now that the painting with light technique has been discussed, it is a good time to discuss a potential problem that will occasionally occur when doing flash photography. Imagine the need to photograph shoe prints in dirt at midnight. When you open your camera bag, you find the PC or remote flash cord is either missing or defective. If it is impractical to call for another cord, or none is available, how can you proceed with your photography?

One solution may be to use a flashlight as your light source. Take a meter reading of the flashlight beam as it lights the shoe print in the manual exposure mode with the f/stop set to f/11. The meter reading will suggest the appropriate shutter speed. Because the camera will be on the tripod, slow shutter speeds are OK. Just remember to keep the flashlight moving to avoid the flashlight streaky look. Or, you can set the camera to aperture priority, using an f/11, and the camera will automatically select the shutter speed for the proper exposure.

Another solution would be to use the flash off-camera without a PC cord, and then set the shutter speed to 2 seconds and set a 10-second delayed shutter. After the shutter is depressed, you will have 10 seconds to get into position with the flash. When the shutter opens, manually fire the flash during the 2 seconds the

Figure 4.106
PWL from one side.

shutter will be open. As long as no strong ambient lights are in the area, having the shutter open for 2 seconds will not lead to overexposure.

SUMMARY

An electronic flash is designated by its guide number, a method of rating the relative strength of the flash unit. Ensuring the flash output matched both the camera viewpoint and the focal length of the lens was emphasized.

Many flash units allow the crime scene photographer to choose from the manual, automatic, or dedicated flash exposure modes. These flash exposure modes were explained, along with their relative strengths and weaknesses.

The need to use the camera's designated sync shutter speed was explained.

The various techniques by which the photographer can control the flash output were also explained. The more controls that any photographer has at his or her disposal, the more precision that can be expected in the final product: a well-exposed photograph. Some techniques may work under certain circumstances, and others may work better in other conditions.

The use of the flash calculator dial to determine the proper f/stop for the flash-to-subject distance was explained. The inverse square law, as the foundation for determining the proper exposure when using electronic flash, was explained. The common exposure errors that may be encountered were pointed out.

Bracketing with flash was also explained.

When wide extremes of light are in the scene, ranging from extremely bright areas to very dark areas, the film, by itself, will not be able to capture details in both of these extremes. Details will be lost either in the highlight areas or in the shadow areas. Some areas will be either overexposed or badly underexposed. Rather than lose potentially important information in either area, the use of fill-in flash can ensure details in both exposure extremes are captured.

Certain types of evidence are best lit with the light coming from the side. Whenever 3D texture, pattern, and similar details are present, side/oblique lighting often improves the visibility of these qualities. This chapter explained how to photograph such objects. Reflectors can also be used to soften the hard shadows cast by oblique flash and illuminate evidence hidden in the shadow area.

The benefits of bounce flash and how to use this technique was explained.

Finally, the painting with light technique was explained. How to use this technique with the flash directed in toward the scene from two sides and from only one side were explained.

DISCUSSION QUESTIONS

1. Explain the meaning of flash guide numbers. How can it be determined which guide numbers of flashes are best suited for crime scene work?

2. Is there just one sync speed a camera can use with a particular flash unit? Although most cameras designate one shutter speed as that camera's sync speed, can "faster" or "slower" shutter speeds ever be used effectively at crime scenes?

3. Discuss the nuances of the manual flash exposure mode, including the pre-supposition that the flash is used in a "normal" room and the flash is used as direct flash rather than bounce or oblique flash.

4. Indicate one circumstance when it would be better to choose the manual flash mode over either the automatic or dedicated flash mode. Why?

5. Indicate one circumstance when it would be better to choose the automatic flash mode over either the manual or dedicated flash mode. Why?

6. Indicate one circumstance when it would be better to choose the dedicated flash mode over either the manual or automatic flash mode. Why?

7. In what circumstances does fill-in flash result in distinct benefits to exposure?

8. Explain all the concerns related to tire track photography.

9. Bounce flash produces dim light on the subject. Explain why this is so and the method to determine a proper exposure when using bounce flash.

10. Explain the differences between the PWL technique from one side, and the PWL technique from two sides.

EXERCISES

1. Take a flash picture of any object 15′ from the camera using the camera's sync shutter speed. If your camera will allow it, photograph the same object, from the same position, at 1 stop faster than sync and at 2 stops faster than sync. If your camera allows these shots, you should be able to anticipate the effects of all three different shutter speeds.

2. With the flash mounted on the hot-shoe, take a photograph of any object 10′ from the camera. Keep the flash on the hot-shoe and repeat at distances of 5′, 4′, and 3′. At what distance is it imperative to remove the flash from the hot-shoe for your equipment?

3. Standing 15′ from a white (or, just slightly off-white) wall, determine a proper exposure for the manual flash exposure mode, and take the picture. Using either automatic or dedicated flash exposure modes, from the same position, photograph the same wall without making any exposure compensations. For the third shot, in either the automatic or dedicated flash exposure mode, make an exposure compensation to ensure the wall is not underexposed.

4. Fill the frame with a large black or deep navy blue object (do not use the black or blue painted sheet metal of a car). First, photograph with the manual flash mode. Second, use either the automatic or dedicated flash mode without any exposure compensations. Third, use either the automatic or dedicated flash mode with an exposure compensation to prevent overexposures.

5. On a day with bright sunlight producing deep shadows (an f/16 kind of day), place an object under a car by one of the tires so it is in deep shadow. Meter the sunny part of the scene, set the camera for that exposure, and photograph the object. Next, determine the correct fill-in flash exposure for the same object from the same position. Take that photograph also and compare it to the first.

6. Sometime after 10 PM, find or create a shoe print in dirt, place the camera on a tripod, and photograph it with the flash as direct as possible. Next, determine a proper oblique flash exposure, and take a second image. Compare the two.

7. In the bathroom, place several items around the sink and take a direct flash photograph. Next, determine the correct bounce flash exposure for the same scene, from the same position, and take a second photograph. Compare.

Compare your images with those in this chapter and with other images on the companion website, referring particularly to the following image folders: bite marks, bounce flash, dirt, dust prints, fill flash, indented writing, manual flash, and sync.

FURTHER READING

The Amphoto Editorial Board. (1974). "Night Photography Simplified." Prentice-Hall, Englewood Cliffs, NJ.

Child, J., and Galer, M. (2000). "Photographic Lighting." Focal Press, Oxford.

Cornfield, J. (1976). "Electronic Flash Photography." Peterson Publishing Co., Los Angeles, CA.

Davis, P. (1995). "Photography." 7th Ed. McGraw-Hill, Boston, MA.

Frost, L. (1999). "The Complete Guide to Night & Low-Light Photography." Amphoto Books, New York, NY.

Hedgecoe, J. (2003). "The New Manual of Photography." DK Publishing, Inc., New York, NY.

Lefkowitz, L. (1986). "Electronic Flash." The Kodak Workshop Series. Eastman Kodak Company, Rochester, NY.

McCartney, S. (1997). "Mastering Flash Photography." Amphoto Books, New York, NY.

Neubart, J. (1988). "The Photographer's Guide to Exposure." Amphoto Books, New York, NY.

CHAPTER 5

CRIME SCENE PHOTOGRAPHY

LEARNING OBJECTIVES

On completion of this lesson, you will be able to . . .

1. Explain why the carefully written documentation of each photograph is necessary.
2. Explain three techniques by which photographs are documented.
3. Explain how overall photographs relate to the general crime scene area.
4. Explain how exterior overall photographs are to be taken.
5. Explain how interior overall photographs are to be taken.
6. Explain how midrange photographs are best taken.
7. Explain the four types of close-up photographs.
8. Explain additional types of photographic concerns related to documenting the wounds of suspects and victims.
9. Explain the complete photographic documentation of a homicide victim.

KEY TERMS

Body panorama
Close-up photographs
Exterior overall
  photographs

Interior overall
  photographs
Labeled scale
Midrange photographs

Natural perspective
Overall photographs
Photo identifier
Photo memo sheet

PHOTO DOCUMENTATION FORMS

A crime scene photograph can eventually be offered as an item of evidence in court. All evidence needs to have a chain of custody established. How can the court be assured that the photograph being offered in court as evidence actually is an image from the incident in question? Could it be a stock image used as an example of what things should look like but actually produced as part of a training exercise? Could it have come from another crime scene?

One way an image can be established as being a photograph from the incident in question is to have the photographer who took the image at the crime scene testify in court. Photographers can state that they took the photograph, what the subject matter of the image was supposed to be, where it was taken, and when it was taken. Photographers can also testify about the camera and flash variables used to capture the image and defend these choices if necessary. However, it is not absolutely necessary to have the photographer who took the photograph introduce an individual image in court. Courts will sometimes allow anyone who was present at the crime scene and is familiar with the subject matter of the photograph to testify that an image is "a fair and accurate representation of the scene" as they saw it at the crime scene in question.

Because crime scene images are frequently critical to a case, they are frequently challenged. An entire chapter will be included later in this text to directly address many of the issues related to the legal admissibility of photographs and images of the crime scene. In this chapter, three standard methods used to document the photographs at a crime scene will be discussed; not only because it is a chain of custody issue, but also because if we can somehow have the images themselves be self-documenting, they will more easily be accepted in court as evidence.

THE PHOTO IDENTIFIER

Photo identifier:
Information that should be recorded on the first frame of every roll of film: (1) case number, (2) date and time, (3) address or location, (4) photographer's name, (5) whether it is the first, second, or third roll of film for this incident.

Every roll of film should begin with the first frame of film being a photograph of the **photo identifier**.

Fill the frame with the identifier when taking this photograph. Although variations of identifiers are used by different law enforcement agencies, some common elements exist for most photo identifiers. These can be preprinted sheets that the photographer merely fills out, or they can be hand-written as needed. It is the information that is important. The photo identifier enables every roll of film to be associated back to the photographer and the specific crime scene in question.

The basic information that should be included in the photo identifier includes the following:

- The case number: The case number is preferred to the crime type, because the crime type may change between the time of the initial call that reported the incident and the type of case that is going to court. Incident types may change for many reasons. The seriousness of the case may go up in severity. A case originally reported as "shots fired" may subsequently end up in court as a homicide prosecution. The seriousness of the case may be revised downward. The initial report of a "rape" may end up in court as a simple assault because of plea bargaining. The case number does not change.

- The date when the first image was taken: many times the time of the first image is also indicated.
- The address/location of the photographs: Most frequently this is a street address. Sometimes a name of the business is also included. First Virginia Bank, 123 Bank St. If a room or suite number applies, add this also.
- The name of the photographer: Badge numbers or employee identification numbers have also been used.
- The roll number of the film currently being exposed (1, 2, etc.) by that photographer, on that date, for that case number, at that location.

A Rule of Thumb is associated with the photo identifier.

Rule of Thumb 5.1: Any time one of the variables on the photo identifier changes, a new photo identifier needs to be photographed.

All the images following any photo identifier are directly linked to it. Therefore, if one of the elements on the photo identifier changes, a new one is needed. For example, a bank robbery certainly will have the bank itself be one location photographs are captured. The scene of the recovery of the getaway car can be a different location and should have a different photo identifier. If the suspect is arrested with the money, a mask, and a weapon, hiding in an alley two blocks away, photographs taken there will require another photo identifier. When a search warrant is served on the suspect's residence, another photo identifier will have to be used. All share the same case number, but the locations differ. If any particular image is challenged in court, it should eventually be able to be linked to the photo identifier immediately preceding it on the roll of film it is on.

With a digital camera, a similar concept is used, although no roll of film is used. The photo identifier should be photographed at the beginning of every new incident or when any of the variables related to the photo identifier changes during one incident.

Examine Figure 5.1 used at the agency this author retired from. If this image is not printed in color, the reader should know the paper might look dark, because it is a shade close to 18% gray. When this author first began crime scene work in the late 1970s, the identifier was the same, except it was on white paper. Why the change? Many automatic printers determined the correct exposure values to print the entire roll of film by the density of the first negative. Because we had used identifiers on white paper, we intentionally overexposed the identifier to keep it from looking underexposed as the camera meter was prone to do. Some exposure errors were generated by the use of the white paper as the first frame of every roll of film. We switched to gray identifiers, and our exposure errors were reduced.

Figure 5.1

Photo identifier (Courtesy of the Arlington County Police Department, Virginia).

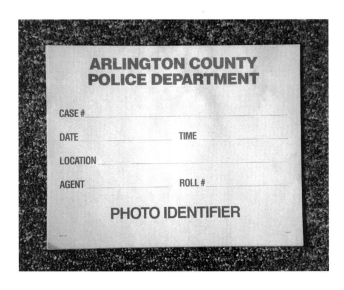

If one case does not require an entire roll of film, the film can be left in the camera; then, at the beginning of the next case, a new photo identifier is used to begin the series of photographs for that case. Rolls of film can come in 8 exposures, 12 exposures, 24 exposures, and 36 exposures. If an entire roll of film is not used at one incident, this enables the judicial use of film without wasting too much. It is possible some agencies have standard operating procedures regarding this. If your agency's policy is to put only one incident on each roll of film, comply with your agency's policy.

THE PHOTO MEMO SHEET

Photo memo sheet: A form to log all the specific data related to the camera, film, and specific variables used to capture each individual photograph.

Every individual photograph should be logged on a **photo memo sheet**, which is a form to log all the specific data related to the camera, film, and specific variables used to capture each individual photograph.

Different agencies have variations of the example photo memo sheet shown below. Figure 5.2 is actually a montage of photo memo forms from several different agencies. Good ideas from several agencies have been incorporated into this form, so that this form will tell a photography instructor just how the student created each image. This form probably has more than a working law enforcement agency might expect from crime scene photographers at actual crime scenes, but if you like any of the ideas on this form, feel free to use them.

Two reasons exist for documenting all the data regarding each image.

First, when the case eventually goes to court, several months or years after the images were originally captured, a properly filled out photo memo sheet can be used to "refresh the memory" of the photographer about the specifics of each image.

Case Number	Date/ Time	Address/ Location	Lens: Indicate mm used, or Macro = M

| Photographer | Roll # | Camera | Film ISO | Light: Available: A; Flashlight: FL M Flash: M, M/2, M/4, M/16; Auto: P, B or R; orTTL |

Figure 5.2

Photo memo sheet.

(With Close-Up filters, indicate either M+1 through M+7 in the lens column.) (Indicate GuideNumber of Flash) _____

#	Lens	Light	SS	F-#	Description
1					
2					
3					
4					
5					
6					
7					
8					
9					
10					
11					
12					
13					
14					
15					
16					
17					
18					
19					
20					
21					
22					
23					
24					

Notes:

Second, if any particular image did not come out quite as expected, the details in the photo memo sheet can possibly be used to figure out what went wrong and can be used to ensure images taken under similar circumstances in the future are improved. The photo memo sheet is a means by which past mistakes can be corrected. If one particular photograph is not as successful as it might have been, knowing how it was taken may suggest why the photograph looks the way it does.

It might then suggest a different technique to make the next attempt in similar circumstances more likely to be successful. With good notes, we can learn by our previous mistakes.

The best time to log a photograph on the photo memo sheet is immediately after taking the photograph. Trust nothing to memory. Immediately log each photograph as it is taken. If the luxury of having an assistant act as your scribe is a possibility, take advantage of it. Otherwise, this job falls on the shoulders of the individual photographer. It may take a bit of juggling to manage a camera, a flash on a PC cord, and a clipboard with the photo memo sheet on it, but one soon gets used to holding the clipboard with one's knees while taking an individual photograph.

Invariably, line 1 on the photo memo sheet should be "photo identifier" as its "description."

THE LABELED SCALE

Labeled scale: A scale that has more than dimension information on it. It includes similar information as the photo identifier: (1) case number, (2) date and time, (3) address or location, (4) photographer's name. In addition, it frequently includes (5) the evidence number that is being assigned to a particular item of evidence. Also, (6) the name of the subject being photographed can be added.

Each item of evidence should include several close-up photographs. It is standard practice to include a **labeled scale** in one of the close-up photographs.

Not just a scale; a labeled scale. Why not have the scale indicate more than just dimensions? Remember, to the extent we can have the image "tell" the judge, jury, and both attorneys what they want to know about an image, the easier it will be to get the image admitted into court as evidence. By use of a scale with the proper labeling information on it, the image can approach our goal of having the image be "self-documenting."

A labeled scale can be a preprinted form, or it can be as simple as a 6″ ruler with a removable label on it. Either way, the same information should be included on the scale.

If you will notice, both of the labeled scales in Figures 5.3 and 5.4 are predominantly gray. Scales come in a variety of colors: black, white, gray, fluorescent, and transparent. It is normally recommended you choose the scale so that it is roughly the same reflectance as the item of evidence it is being used with. To the extent possible, it is a good idea to have the scale reflect the same quantity of light; it would be counterproductive to have the light reflected from the scale alter the exposure of the evidence it is being used with. If a light meter or sensor of some type is being used to determine the exposure, having the scale be close in reflectivity to the evidence is important. Alternately, using a gray scale usually will not alter the exposure value required for any item of evidence.

As mentioned in the preceding chapter, it was recommended to use a black scale with the black Mylar film of a dust print lifter, and to use a white scale with indented writing on white paper. The reader may find it useful to keep differently colored scales for different types of evidence.

Many agencies assign an evidence number to the evidence as it is chronologically encountered. As each item of evidence is collected, it is assigned a number.

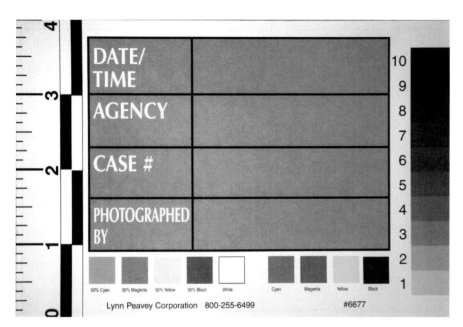

Figure 5.3

A commercially available labeled scale.

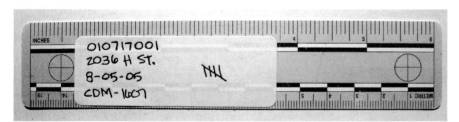

Figure 5.4

A labeled scale.

This number should be indicated on the labeled scale when the close-up is taken. An effective way to accomplish this is to use four slash marks and a diagonal slash for each five numbers, having the numbers be cumulative, rather than having to change or erase each number each time. By doing this, it is not necessary to change the label as new items of evidence are photographed. It is also not necessary to scratch out the preceding number to add a new number. For example, evidence item 7 would be indicated like this: ⅢⅡ

If someone or their injuries are being photographed, indicate their name on the labeled scale. This should be on a separate label so it can be removed after the photographs of that person have been taken.

Remember, the labeled scale needs to be placed on the same plane as the evidence. This Rule of Thumb is critical to good crime scene photography.

Rule of Thumb 5.2: Whenever scales are used, they should be positioned on the same plane as the evidence they are associated with.

If a handgun is being photographed, and you focus on the top of the gun, the labeled scale has to be at the same height as the top of the gun. A variety of props can be used to raise the labeled scale up to the necessary height. Lens caps, coins, and pocket notebooks have all been used frequently. When all else fails and it seems impossible to find props of just the right height, folded paper can be made into small A-frames of the desired height. Forget why the scale should be on the same plane as the evidence? There are two reasons.

1. To have the scale in the composition enable the size of the evidence to be determined, it is necessary to have the scale be on the same plane as the evidence. If the top of the evidence is the point of focus, the scale must be at that same height.
2. With close-up photographs, depth of field is severely reduced. It would be possible to have the top of a weapon be in focus with the scale laying alongside the weapon but with the surrounding surface (foreground and background) slightly out of focus.

These images show two items of evidence and the props used to get the labeled scale as high as the evidence. The trick is to consciously avoid having the props appear in the photograph. With just a little practice, this becomes second nature to the experienced crime scene photographer.

If the evidence is a shoe print in dirt, where the pattern of the shoe print is 1″ deep into the dirt, the labeled scale has to be placed into a depression you have dug for it that places the labeled scale 1″ lower than the surrounding dirt. This ensures that the photograph of the evidence can eventually be enlarged to be life size, so a critical comparison can be made between the evidence and the photograph.

It may be advantageous to "prove" that scales on the wrong plane will vary in their ability to be focused on properly. Crime scene photographers have frequently just placed a scale alongside an item of evidence without much care about the scale being on the same plane as the height or depth of the evidence. When this "error" has been pointed out, the reply has sometimes been that this theory may be appropriate for a lecture in a classroom situation, but out in the "real world" such things are not really important. It is hoped that Figure 5.9 convinces the reader which position is correct. In Figure 5.9, two scales are placed adjacent to each other but at four different heights.

On the top, the two scales are the same height. A stack of nickels was used as a prop for both scales. On the lower left, the top scale of the pair was lowered by one nickel, while the focus was set for the bottom scale. The fine dots making up the background of the label on the top should appear to the reader to begin to blur. When the top labels in the next two pairs of scales have been lowered by ¼″ and ½″, the blur becomes very obvious.

Figure 5.5
Labeled scale on folded paper.

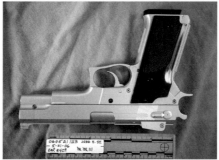

Figure 5.6
Do not let props be seen.

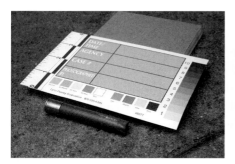

Figure 5.7
Labeled scale on prop.

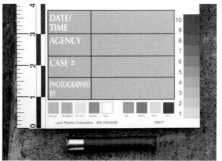

Figure 5.8
Do not let props be seen.

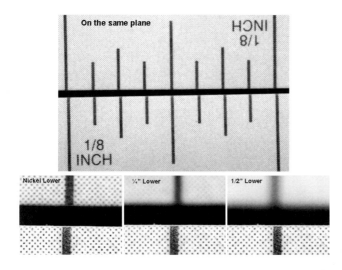

Figure 5.9
Scales on different levels: I.

That is only half of the problem with scales not being on the same plane as the evidence.

If the scale is not on the same plane as the evidence, it is actually a different proportion, and the scale cannot be used to enlarge the evidence to "life size." If the scale is lower than the plane of the evidence that is focused on, the scale will be too small. This would be the case if a gun were being photographed, with the scale laid on the floor next to the gun. If the focus is set for the top surface of the gun, the scale will be too small.

If the scale is higher than the evidence, it will be larger than the evidence. This would be the situation if a shoe print in dirt was being photographed, and a scale was laid on the top surface of the dirt alongside the impression. If the bottom of the shoe print is the point of focus, the scale will be larger than it should be.

At actual crime scenes, this has been pointed out to crime scene photographers, and similar comments about the "real world" working differently than photographic theory would suggest they have been sometimes heard. It is hoped that Figure 5.10 will convince you otherwise. In the center of the image, two scales were laid alongside of each other, and each 1″ length matches up. The left two images show the two scales at different distances from the camera. Both of the left scales are ¼″ higher than the right scale, which is the scale focused on. Not only is the right scale a bit out of focus, but it is also smaller. If the purpose for putting

Figure 5.10
Scales on different levels: II.

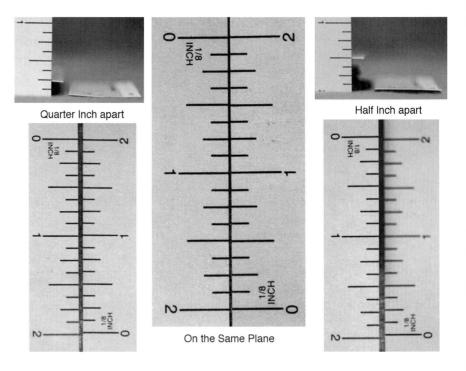

Quarter Inch apart

On the Same Plane

Half Inch apart

the scale into the composition was to enable the image to be blown up to life size because the scale is the wrong size, this cannot be done.

The two right images also show the two scales at different distances from the camera. However, now both of the right scales are $\frac{1}{2}''$ lower than the left scale, which is the scale focused on. Not only is the right scale out of focus, but it is also smaller. Again, if the purpose for putting the scale into the composition was to enable the image to be blown up to life size, because the scale is the wrong size, this cannot be done.

In either case, if the evidence is below grade or laying on a surface, the crime scene photographer has to make the effort to lower or raise the scale to the same plane of the evidence being focused on.

OVERALL PHOTOGRAPHS

Eventually, individual items of evidence will be photographed. For the importance of any individual item of evidence to make sense to the viewer, it will be necessary for that item of evidence to be related to the crime scene. The crime scene itself, in turn, will need to be related to the general surroundings around the crime scene. In this way, the complete story of the crime in question is documented. Every item of evidence needs to be linked to the crime scene. The crime scene itself must be linked to the general surroundings. The **overall photographs** do this job.

NATURAL PERSPECTIVE

For the most part, overall photographs are taken from a **natural perspective**, which is a viewpoint of the scene that has the photographer standing at full height.

A viewer of overall photographs can presume this was the position the photographer was in when the photographs were taken.

One exception to this rule is that at times it may be necessary to take aerial views of the crime scene to show the relationship of parts of the scene to the general neighborhood. Aerial photography, whether taken from a balcony, a rooftop, a cherry picker utility truck, a fire department's ladder truck, or from aircraft will be discussed separately.

Usually, overalls are not taken with the photographer kneeling or standing on a support to get higher. If this is done, notes should be kept to document this fact. Unless the photographer specifically mentions this was done, the viewer of the image can presume the photograph was taken from a natural perspective. There will be times that a particular point of view is being documented, and sight obstructions may need to be included from the viewpoint of the photographer. A witness at a certain location may report having seen something from their

Overall photographs: These document the general conditions of the scene, both with exterior and interior views, and how the specific crime scene relates to the surrounding area.

Natural perspective: A viewpoint of the scene that has the photographer standing at full height.

original position. Photographs will then be taken from their viewpoint. A driver of one of the cars involved in an accident may report not seeing the other vehicle because of a sight obstruction. In this case, it will be necessary to get the camera at the height of the driver's eyes to document obstructions from that perspective. However, these are the recognized exceptions to the general rule that overall photographs are normally taken from a natural perspective.

EXTERIOR OVERALLS

The crime scene photographer will want to create a complete story of the crime with the photographs that are taken. Most of the time, the judge and jury do not return to the scene of the crime. Adequate photographs can do this job for them. **Exterior overall photographs** relate the crime scene to the general surrounding area.

If the crime was committed indoors, they also show external views of the building in which the crime occurred. It will be the job of the crime scene photographer to walk the viewer of the photographs into the crime scene from its outer perimeter.

When possible, the crime scene photographer should begin at the intersection closest to the crime scene. This may not be possible in some rural areas. But the idea is a good one to acclimate any interested parties of the crime to the general surrounding area.

When photographing the street signs, do so in a way that indicates to the viewer which of the two street names is the one on which the crime occurred. Without this consideration, it would be possible to photograph the signs giving equal emphasis to both streets, which can leave the viewer confused about which is more important. When composing this shot, make the street the crime occurred on more predominant in the photograph. Figures 5.11 and 5.12 show this difference in emphasis. Both street names are visible in both photographs,

Exterior overall photographs: These photographs relate the exterior of the crime scene to the general surrounding area. If the crime was committed indoors, they also show external views of the building in which the crime occurred.

Figure 5.11
Street signs: emphasis on Poplar Tree Rd.

Figure 5.12
Street signs: emphasis on Pergate Lane.

but the photographer intentionally directed the viewer of each image to one of the streets by making one of them more prominent.

Many times, this kind of shot will have a large amount of sky as the background, which will tend to cause an underexposure. If this is the case, it may be best to take an exposure reading off a grassy area or well-traveled asphalt in the area to avoid exposure errors. If no green grass or well-traveled asphalt is nearby, open up the aperture 1 to 1½ stops to prevent the underexposure the sky is likely to cause if the meter is relied on totally. At night, hitting the highly reflective sign with flash may wash out the sign entirely. Care must be taken that the flash does not strike the sign directly. Position the flash so the most extreme angle is used. Or reduce the flash intensity drastically to avoid reflection washout.

If a street sign is not handy, some other well-known point of reference can be used to acclimate the viewer of the photographs to the general area of the crime scene. This can be a well-known building, a natural feature of the area, or anything else the photographer believes can be used as a point of reference.

Once the street sign, or point of reference, has been photographed, take a photograph from the street sign, showing the intersection in question, toward the crime scene.

Figure 5.13 shows the two streets indicated in the photograph of the street signs. Just as one of the sign names was prominent when the signs were photographed, that same street is shown more prominently in this image. Moving from the intersection toward the crime scene, the crime scene photographer should take one or more additional images. Like Figure 5.14, these additional images will give the viewers a good feel for the crime scene neighborhood. All these images should be taken with a 50-mm lens. The use of a wide-angle lens or a telephoto lens will elongate or compress the scene, respectively.

Once at the crime scene building, the entire exterior should be photographed. The purpose is to show all the possible ways the suspects could have entered or left the building. Even when it is believed that information is already

Figure 5.13
From intersection toward the crime scene.

Figure 5.14
Mid-block to the crime scene.

known, it is good practice to photograph the entire façade, because new information may be developed during the course of the investigation that contradicts the original hypothesis.

Figure 5.15 shows a simple residence photographed from all sides. In this series, a 50-mm lens was used. Preference was given to views of each side with the film plane parallel to the walls of the building whenever possible. When obstructions require it, angled views were used. If one or more sides of the building are particularly long, a wide-angle lens can be used to minimize the number of overlapping photographs required to capture the entire building. If a wide-angle lens is used, however, particular care should be taken to eliminate anything besides the building walls. If parking lots or lawn details are included in this composition, they will be elongated and incorrectly depict the extent of the surrounding grounds. In this case, the composition of the building should ensure the bottom of the viewfinder ends with the bottom of the wall being photographed.

In Figure 5.16, this was ignored, and the distance between the photographer and the building appears to get longer and longer. The top image was taken with a 50-mm lens, which is the true distance between the photographer and the building. While standing in the same spot, the middle image was taken with a 35 mm lens. There appears to be a greater distance between the photographer and the building. While the photographer stood in the same spot, the lower image was taken with a 24-mm lens. There appears to be an even greater distance between the photographer and the building. The distance between the front sidewalk and the front of the building may become an issue being contested in

Figure 5.15

Exterior overall of a single family home.

court. In this case, only the image taken with a 50-mm lens should be allowed in court, because the other images do not depict a fair and accurate distance between the front sidewalk and the building. Again, wide-angle lenses can be used to photograph the building as long as the image is composed so that the bottom of the building is at the bottom of the composition. Figure 5.17 shows

Figure 5.16
Building: 50-mm, 35-mm,
and 24-mm lenses.

Figure 5.17
Building: 50-mm, 35-mm,
and 24-mm lenses.

the proper composition of a large building with lenses of different focal lengths. In these three images, taken top to bottom with a 50-mm, a 35-mm, and a 24-mm lens, the wider angle of the lens just incorporates more of the building in the field of view, without deceiving the viewer about the size of the lawn in front of the building.

When taking exterior overalls, include a photograph of the building address or the building name. For instance, if the First Virginia Bank at 123 Bank Street is robbed, it would be appropriate to get a photograph of each designation; better yet would be if they could both be composed in the same image.

If it is possible to photograph the exterior of a building with the film plane parallel to the walls of the building, this is preferred. At times, this will be impossible, because the surrounding buildings may make it difficult to get in a position where the camera can be positioned with the film plane parallel to the walls.

Figure 5.18 shows views of the same building. The top two images show the building photographed from diagonal viewpoints. The bottom two images are the left and right halves of the building, with overlap in the middle. If a diagonal view of a wall of a building and having the film plane parallel to the wall are both possible, compose the building with the film plane parallel to the wall.

Figure 5.18

Large building: diagonal views versus film plane parallel.

Five reasons exist why the compositions with the film plane parallel to a wall are preferred over diagonal viewpoints.

- Diagonal viewpoints have the near side of the building appear larger than the far side of the building. It is natural to believe the photographer composed the images for specific reasons and that larger parts of the scene are larger because the photographer wanted to emphasize one part over another part. If the film plane is parallel to the façade, the same emphasis is on the left and right sides of the building. Both sides can be seen equally well. Neither side is diminished in importance by making one side smaller.
- When one side of the building is closer to the camera and the other side of the building is farther away, a depth of field problem is created. The photographer will want to ensure the entire façade is in focus, but it is possible that this may not be able to be done. With the film plane parallel to the façade, it is easy to ensure the left and right sides of the building are in focus at the same time.
- When the lighting is dim, electronic flash will have to be used. Flash cannot light areas near the camera and areas farther from the camera at the same time equally well. However, if the façade is composed so that it is the same distance from the camera, it will be easy to ensure equal lighting has been achieved.
- One of the purposes of these exterior overall photographs is to show all the possible means of ingress and egress. Having the film plane parallel to the façade does this job better. Doors and windows are easier to see when they are larger, and having the film plane parallel to the façade has the camera closer to the façade so that all doors and windows appear larger and easier to see.
- If evidence is outside a building, this evidence will have to be related back to a fixed feature of the building with midrange photographs and measurements. All the fixed features of a building are better seen when the façade is photographed with the film plane parallel to the walls of the building.

Figure 5.19 shows a burglary scene, with many shoe prints in the dirt outside the residence.

Another aspect of exterior overall photographs is to show the immediate area surrounding the crime scene. This can be done two different ways. Views toward the crime scene building can be taken from various directions.

- Views toward the home from both directions of the street.
- Views toward the home from both directions of a rear alley or a street behind the residence.
- Views from the residence outward down the street in both directions.
- Views from the residence showing other possible escape routes.

Figure 5.19

Burglary scene: shoe prints in dirt (Courtesy of Criminalist Rebecca Shaw, Arapahoe County Sheriff's Office, Colorado).

- Vehicles parked nearby. A suspect may have parked nearby but was unable to get to his or her vehicle for a number of reasons and left it behind, intending to return later to get it. If vehicles parked nearby do not belong to a suspect, they may also provide law enforcement agencies the names of potential witnesses.

Taking exterior overall photographs at the beginning of the crime scene documentation is also productive, because these locations are usually open to the public's view and not usually subject to any possible subject's reasonable expectation of privacy. A search warrant may be necessary for the search and seizure of evidence within buildings and residences. The exteriors of buildings, however, are usually not regarded as restricted areas requiring search warrants. While the search warrants are being secured for interior spaces, much is to be done outside that does not suffer from similar restrictions.

INTERIOR OVERALLS

Once the legalities of entering possibly restricted areas have been addressed, taking interior overall photographs is usually a priority, because the scene will have to be documented "as found" before anything can be moved. Depending on the nature of the crime being investigated, many rooms may need to be photographically documented, or just one room may need to be photographed. **Interior**

overall photographs are the bridge between the exterior overall photographs and the individual items of evidence within the crime scene. After the exterior has been photographed and before the evidence is photographed, the interior is photographed.

At times, walking through an exterior door will put us into the crime scene room. Other times, when the building is initially entered, you will be in a room that is not the crime scene, or you will find yourself in a hallway or lobby. Depending on the nature of the crime, you will have to decide whether each of these other rooms, hallways, or lobbies will also have to be photographed to adequately document the full extent of the crime scene. Preference should be given to the use of a 50-mm lens for most of this photography. It will capture these areas as you saw them. The 50-mm lens will neither compact relative distances nor will it elongate the size of any room.

Should you decide to photograph the entire route to the crime scene room, Figures 5.20 through 5.22 show representative examples. The camera should be turned vertical for subject matter that is taller than it is wide. If doors have distinguishing numbering or lettering, be sure to include this information. If a door is closed, partially open, or wide open, make sure to photograph it "as found" before moving it. In one photography school, it was taught to photograph a door from the outside before entering through it and then to photograph it from the other side after walking through it. The extent of the crime scene will dictate whether this needs to be done.

Once inside the crime scene room, take a complete set of interior overall photographs. The phrase interior overall photographs suggests that a full 360° view of the room will be photographed, which is true, but somewhere along the line, the purpose of the interior overall photographs seems to have gotten lost. Some have considered this task to be a necessity but have tried to get past this requirement as fast as possible, so that other, more important photography can be started.

This mindset usually has the crime scene photographer photographing the room from each of the four corners while facing the opposite corner, which is also the way this author was originally trained. If the room was big enough so that these four shots with a 50-mm lens did not encompass the entire room, it was even suggested that a wide-angle lens be used. Figure 5.23 shows the result. This has become the standard taught by many law enforcement agencies.

Figure 5.23, however, suffers from a fatal flaw. Because a wide-angle lens was used to compose each of the four segments, the end result is that the room looks much bigger than it really is. Granted, an entire 360° view of the room has been achieved, requiring just four images to do so. Are these four images, as a whole, a fair and accurate representation of the scene? I do not believe so. In many crime scenes the room dimensions have proved critical to a reconstruction of the scene. From X position in the room, subject Y could have accessed Z area with a gunshot or thrown knife. Had the room been a different size, this would not have been

Interior overall photographs: These are the bridge between the exterior overall photographs and the individual items of evidence within the crime scene. After the exterior has been photographed and before the evidence is photographed, the interior is photographed.

Figure 5.20
Building exterior door.

Figure 5.21
Hallway to crime scene room.

Figure 5.22
Door to crime scene room.

Figure 5.23

Interior overalls: corner-to-corner.

possible. Would it be sufficient to just mention to the viewers of the photograph, the judge and the jury, that the use of the wide-angle lens tends to make the room appear a bit larger than it really is? I think not. Photographs have always been regarded as such strong types of evidence that the mantra, "Is this a fair and accurate representation of the scene?" is almost as well known to the general public as are the Miranda rights so often repeated on television and in the movies. If the size of the room is critical to an issue before the court, an image depicting the room as larger than it really is can be justification for that image being excluded from being admitted into evidence for that trial.

To be fair, the images in Figure 5.23, taken with a wide-angle lens, do show the relative position of all the furniture, windows, and doors in a minimal number of images. This cannot be denied. If it were possible to separate this convenient documentation of the room arrangements from the distorted suggestion that the room is larger than it really is, these photographs would be an efficient method to take interior overall photographs. Do these images accurately show the relative arrangements of furniture and room features? Yes. Do these images accurately portray the dimensions of the room? No. Can these images be used in court and still be considered "fair and accurate representations of the scene?" If the testimony includes a statement that the wide-angle lens also makes the scene appear deeper than it was in reality, this would take the wind out of the sails of a defense attorney's objection that the images distort and elongate the perceived distance from the photographer's position to the back of the scene.

Another problem with Figure 5.23 is that the diagonal views of the room make the far corner of the room much smaller than necessary. Being smaller, these areas are harder to see. Why make potentially important areas of a crime scene room smaller than necessary? As with exterior overalls, small can be regarded as being less important, because if it were important, the photographer certainly would not have intentionally made that area of the scene small. Should there not be an evenness to the emphasis placed on the interior walls?

Let us return to the purpose of interior overall photographs. Eventually, all the evidence within the crime scene will need to be related to the fixed features of the scene. It is these fixed features that are being photographed with the interior overalls. What is important is not to show the 360° view of the room in as few shots as necessary; what is important is to be able to see the fixed features of the scene as clearly as possible. Composing photographs so that some parts of the wall will be farther from the camera, and therefore smaller than necessary, can be seen as self-defeating.

Is there a better way? Yes. Photographing the four walls with the film plane parallel to them is the answer. In so doing, the wall at the right side of the image is as large as the wall on the left side of the image. Fixed features of the walls, doors, windows, electrical outlets, floor vents etc., are all easier to see when they are closer to the camera. Compare Figure 5.24 to Figure 5.23.

Yes, it now takes six images to capture the entire 360° view of the same room. But notice the emphasis on the left and right sides of each wall is uniform. The entire room is visible, and most importantly, the room does not look differently sized than it really is. There would be no problem answering "Yes" to the question, "Are these fair and accurate representations of the scene?"

Other photography texts suggest the size distortion produced by the corner-to-corner shots with a wide-angle lens is minimal and should not be a problem in court. Your agency will be the final arbiter of this disagreement, and you will be required to take your crime scene photographs as your agency demands. This section has provided an alternate point of view, and if your agency's standard operating procedures related to crime scene photography are not written in stone, you now have a choice.

Whether taken from one corner aimed at the opposite corner or taken with the film plane parallel to the wall, this author has never been pleased with the look of direct flash when taking interior overall photographs. Because flash is so directional, frequent "hot spots" are seen, and the extreme edges of the images are sometimes a bit dimmer than the center of the image. Care must be taken that the flash head is appropriate for the focal length of the lens used. Another frequently encountered problem was that white walls would "fool" the meter into underexposing the image. If the predominant part of the composition is a white wall, light reflected from it will cause the flash sensor or camera meter to shut the flash off a bit early. White walls usually require an exposure compensation of +1 or even +1½.

Figure 5.24

Interior overalls: film plane parallel to walls.

If you choose to compose the scene with the film plane parallel to the opposite wall, another solution is available to these exposure problems. Many are pleased with the look of bounce flash in these circumstances. Because the scene is just one wall rather than half the size of the room, bounce flash can be used without noticing a loss of depth of field. The softer light coming from the ceiling is more pleasing than the look of direct flash.

Figure 5.25 was taken with direct flash. Notice the shadow of the goose in the center of the image and the reflections of the flash off the windows in the front door and the glass over the two prints at the left side of the image. Do you find those four glass reflections annoying? Figure 5.26 is a vast improvement. Bounce flash would not work if attempting to light half of the room. The required large aperture would risk parts of the scene being out of focus. With a smaller scene, however, bounce flash works perfectly.

Most interior overall photographs are also taken from eye level or from a natural perspective.

Figure 5.25
Interior overall: direct flash.

Figure 5.26
Interior overall: bounce flash.

MIDRANGE PHOTOGRAPHS

Midrange photographs:
The purpose of midrange
photographs is to show a
relationship between an
individual item of evidence
and a fixed feature of the
scene previously pho-
tographed in one of the
overalls.

Close-up photographs:
Fill the frame with just the
item of evidence, while
maintaining the film plane
parallel to the evidence.

Overall photographs have been concerned with documenting the crime scene, both indoors and outdoors. With **midrange photographs,** we now begin to document individual items of evidence.

Before a series of **close-up photographs** of the evidence is taken, it is necessary to tie the evidence to the crime scene. Close-up photographs will fill the frame with just the item of evidence and show various views of the evidence.

These will only be informative if it is known where these close-ups are located within the crime scene. It is the purpose of midrange photographs to do this job.

To the extent possible, midrange photographs should also be taken from a natural perspective. At times, this will not be possible. To photograph a gun under a bed, it will be necessary to get the camera low enough to view the gun under the bed. As a general rule, however, the midrange photographs are taken from a natural perspective.

The midrange photograph is also taken immediately before a series of close-up photographs of that particular item of evidence. It is not suggested that all the midrange photographs for all the items of evidence be done first and then the series of close-up photographs be done next. Once an item of evidence has been selected to be documented, its midrange photograph is taken followed by close-up photographs of the same item of evidence.

PROPER VIEWPOINT TO AVOID PERSPECTIVE DISTORTION

When the term "distortion" is mentioned, two different kinds of distortion can come to mind:

- There can be lens distortion. A 50-mm lens is usually used to avoid the distortion related to the use of different focal length lenses. As previously mentioned, a telephoto lens can "distort" the scene by compressing the foreground-to-background distance. A wide-angle lens can "distort" the scene by elongating the foreground-to-background distance.

- "Distortion" is also related to the photographer's point of view. A midrange photograph includes two items: a fixed feature of the scene and an item of evidence. Depending on the photographer's point of view, these two items can be aligned in the viewfinder in one of three different ways:

1. They can be composed so they are parallel to the film plane.
2. They can be composed so they form a straight line from the photographer's point of view.
3. Or, somewhere in between, they can form a diagonal line, with one item closer to the camera and the other item farther from the camera.

Figure 5.27 shows these variations. The top three images have the photographer, the pistol, and the door frame aligned in a linear point of view. The top left image has the photographer standing at his full height or from a natural perspective. The top middle image has the photographer stooping down a bit. The top right image has the photographer kneeling. The pistol appears to be getting closer to the door frame. None of these three images is to be trusted when the issue is to

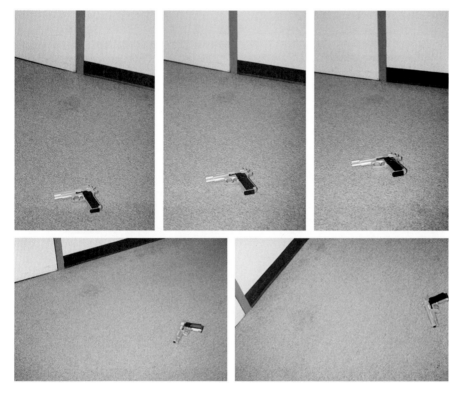

Figure 5.27

Midranges of a gun and door frame.

show the viewer a fair and accurate representation of the distance between the pistol and the door frame. The lower left image shows a diagonal point of view. The pistol is closer to the photographer than the door frame is. This point of view is better than any of the three top images, but it is still not the preferred point of view.

The bottom right image has both the pistol and the door frame an equal distance from the photographer. It can be said that the film plane is parallel to an imaginary line drawn between the pistol and the door frame. Another way to express this point of view is to state that the photographer intentionally formed an isosceles triangle with the camera an equal distance from both the pistol and the door frame. This point of view, and only this point of view, depicts a fair and accurate distance between the pistol and the door frame. The other points of view have the tendency to make the pistol appear somewhat closer to the door frame than it is in reality.

Figure 5.28 shows another variation on this theme. The three top images are linear points of view, with the photographer intentionally changing the height from which the image is captured. The cartridge casing appears to be getting closer to the electric box. The bottom image has the photographer forming an isosceles triangle with both the casing and the electric box equal distances from the camera. Of the four images, only the bottom image shows an accurate distance relationship between the casing and the electric box.

Figure 5.28

Midranges of a casing and electric box.

At times, because of the configuration of the scene, it may prove impossible to assume this "isosceles triangle" position. Obstructions may make it impossible to get an equal distance from both the item of evidence and a fixed feature within the scene. If this occurs, the only alternative is to assume a variation of a diagonal point of view. If this is all that is possible, then so be it. Just remember, the more equidistance achieved between the point of view, the evidence, and the fixed feature, the better.

COMPOSITION: WATCH THE BACKGROUND

When the photographer is so intent to form this "isosceles triangle" point of view with a midrange photograph, other concerns should not be forgotten. In particular, students of photography seem for forget the Cardinal Rule to fill the frame with just the subject matter at hand, eliminating extraneous elements from the composition.

Figure 5.29 shows two examples of the improper and proper compositions for midrange photographs. Do not get tunnel vision. Do not be so concerned that the isosceles triangle point of view is established that you forget other composition ideas and include too much background in your image. The crime scene photographer who becomes known in his or her department as the "go-to" person, if the best photographs of an important crime scene are required, is the one who can juggle more than one idea at the same time.

Figure 5.29

Midrange composition: watch the background.

Figure 5.30 presents one other concern. Sometimes, it is possible to attain the "isosceles triangle" point of view from two different positions. These two photographs show an evidence marker in relationship to a tree trunk, but both are from different points of view. Is one better than the other? Certainly, the first thing that might be mentioned is that had the photographer gotten closer to both the marker and the tree, neither background would be in the field of view, because then the camera would be aimed down more. This is true. However, it may be the case that the suspect or victim is somehow associated with the construction scene across the street. If this is the case, you might be able to do double duty by composing the midrange photograph with the construction scene in the background. This should be a decision you consciously made rather than the result of some serendipity. If the construction scene across the street has no relationship to the crime, it definitely should not be included in the background. Should the viewer of the photograph not be entitled to make the assumption that if it appears in the image, it is there because the photographer chose it to be there? It definitely should not be in the background only because the photographer failed to consider it as a distraction.

If you have been convinced that having the film plane parallel to an imaginary line between the evidence and a fixed feature of the scene is the proper alignment for midrange photos, decisions are still to be made. Those same two items, the evidence and a fixed feature of the scene, can appear in the composition high in the field of view, somewhat centered in the field of view, or low in the field of view. Why choose one or the other? If you are becoming the photographic purist this author hopes you become, you are convinced a particular photograph should tell just one story. If the relationship of the evidence to a fixed feature of the scene is the only story that should be told in this one particular photograph, everything else that does not contribute to this goal should be eliminated from the field of view. This can often simply be done by either raising or lowering the camera. Doing so maintains the same relationship of the evidence to the fixed feature of the scene but allows the photographer to include or exclude elements in the foreground or background as required.

Figure 5.30

Midrange composition: Is one view better?

Figures 5.31 and 5.32 are two sequences of midrange photographs. In each trio of images, the same scissors are being shown in relationship to the same fixed feature. In Figure 5.31, the scissors are being related to a metal fence post on the right side of the image. In Figure 5.32, the scissors are being related to the leg of a bench that has been bolted down at the scene. In each of the trio of images, the same distance relationship between the evidence and the fixed feature is documented. Each of the six images tells the story of this distance relationship. Only two of the six images, however, tell their story "cleanly," without unnecessary or distracting elements. The top image of each sequence is the "best" image. Each of the top images tells its distance relationship story, while intentionally eliminating elements that do not contribute to the intent of the crime scene photographer. It is hoped that none of my readers are saying, "I'd be satisfied with any of them." Once you see a difference between each sequence of three images, choose

Figure 5.31

Three midrange compositions: I.

Figure 5.32

Three midrange compositions: II.

the one that eventually tells the viewer of the image that the photographer is extremely aware of their craft and composes images carefully and purposefully.

One obvious point needs to be made. In many crime scenes, both inside and outside, so much clutter is everywhere that it is impossible to remove all the extraneous items from the field of view. We will all run into these situations frequently. In these situations, you can still use the information presented in this section to choose the composition that may be a bit less cluttered.

LENS DISTORTION, REVISITED

When elements of a scene are aligned vertically, or at varying distances from the photographer, using different focal length lenses will have a dramatic impact on the perceived distances between those elements. Telephoto lenses will tend to compress the scene, making the elements appear closer together. Wide-angle lenses will tend to elongate the scene, making the elements appear to be farther apart. Now, consider if these same lens effects will affect elements of the scene that are the same distance from the photographer. In other words, if a midrange photograph has been composed properly, if the "isosceles triangle" positioning has been established, will different focal length lenses have the same negative distortions? No.

Figure 5.33 shows the effect of using different focal length lenses when properly composing a midrange photograph. Because composing each shot with a different focal length lens will require the photographer to stand at different distances from the pistol and the wall corner, a slight variance in the views exists. Notice as the focal length changes from the top of the sequence to the bottom of the sequence, the view also changes from being able to see more of the top of the gun to seeing the gun from farther away, which shows less of the top of the gun. The main point of this sequence, however, is that the perceived distance between the pistol and the wall corner does not change with the use of different lenses.

How can this be helpful to the crime scene photographer? If crime scene obstructions or if other evidence in the crime scene prevents the crime scene photographer from composing a midrange photograph with a 50-mm lens, the photographer should know that a proper midrange photograph can still be obtained by standing a different distance from the evidence and the fixed feature and composing the scene with a different focal length lens. For instance, if the composition of the pistol and the wall corner with a 50-mm lens would require the photographer to stand in the middle of a large puddle of blood, it is now realized that the photographer can stand closer to the evidence or farther from the evidence to avoid the puddle of blood and still capture a midrange photograph depicting the fair and accurate distance between the evidence and the fixed feature of the scene.

A question always asked when on the topic of taking midrange photographs is, "Can one image be used as a midrange photograph for several items of evidence

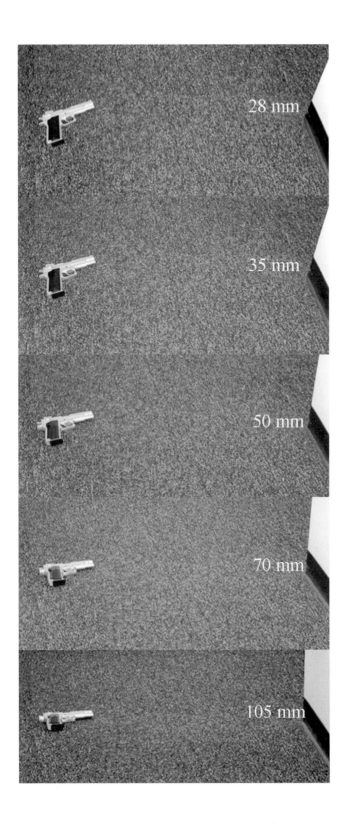

Figure 5.33
Midrange photographs with different focal lengths.

that are grouped relatively close together?" Yes. If the same fixed feature will be used for multiple items of evidence and they are all relatively close together, one well-composed shot can suffice as a midrange photograph for all the items of evidence.

CLOSE-UP PHOTOGRAPHS

To compose a close-up photograph properly, it will frequently require the photographer to abandon the natural perspective point of view. Many times, the photographer will be leaning over the evidence, if not actually kneeling, to get close enough to fill the frame with the evidence.

Several types of close-up photographs exist, all of which attempt to fill the frame with the evidence. This section explains the different types of close-up photographs.

CLOSE UPS: "AS FOUND," "AS IS," "IN SITU"

The first close-up photograph should show the evidence "as found" in the scene before any alteration or movement of anything in the scene. Nothing should be added to the scene, and nothing should be taken away from the scene when the evidence is first photographed. Another description of this is sometimes expressed as photographing the item of evidence "in situ" (Latin for "in place"): in the situation it was originally found before any movement.

Regardless of the size of the evidence, there should be an attempt to fill the frame with the evidence. If the evidence is textbook sized or larger, this can be done with the normal 50-mm lens. If the evidence is smaller than a textbook, magnification will be required. As mentioned earlier, close-up filters or diopters can be used or a macro lens can be used. Some may use extension tubes to achieve magnification. Whatever equipment is used, the goal is the same: enlarge the evidence so it fills the frame. If the evidence is important enough to photograph, why not make it as large as possible, so small aspects of the evidence may better be seen? If the evidence is not made as large as possible, the alternative is to include more and more of the substrate beneath the evidence in the field of view. It is best to fill the frame with the evidence.

When doing close-up photographs, you should invariably be using one of two apertures. If the close-up photograph will be used for comparisons with another item of evidence, then f/11 should be used, and the other camera variables altered to ensure a proper exposure with the f/11. The f/11, as you should recall, still gives a good depth of field, while avoiding image softness that can be caused by diffraction. If the photograph will not be used for critical comparisons, an f/22 should be used. There would be no benefit to using any other aperture in this case.

Figure 5.34 shows variations of a close-up photograph of an expanded Black Talon bullet and its jacketing. If both are possible, why would anyone think the one showing the bullet smaller than it has to be is better? When the evidence is smaller than it can be, the image includes more substrate beneath the evidence. It is best to fill the frame with the evidence. Why have more of your silver halide crystals or your digital pixels assigned to capturing the details of the substrate rather than the evidence?

Figure 5.35 shows a set of close-up filters being used with a column of nickels. The set of three filters come with a +1, +2, and +4 magnification, and they can be stacked differently for different combinations. Figure 5.35 shows each variation from +1 through +7. It should be noted that the manufacturers of these filters usually suggest that all three (the +7) filters should not be used at the same time, because it is possible the image will be degraded with so much glass added to the end of the primary lens. This may be true, but this author has not seen the degradation warned about in any images captured with all three filters.

Many photographs of items of evidence, usually with a scale in them, will eventually be used in court. Some judges and some defense attorneys may question whether the evidence was in fact originally found with a scale laid conveniently alongside the evidence. Just in case this ever comes up in a trial, the "as-is" photograph can save the day. Also, it is sometimes suggested the scale may be hiding or covering some other evidence. The "as-is" photograph can also be used to prove this has not occurred.

Figure 5.34
Close-ups: fill-the-frame.

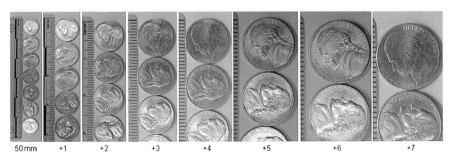

Figure 5.35
Close-up filters and nickels.

50 mm   +1   +2   +3   +4   +5   +6   +7

WITH A FULLY LABELED SCALE, ON THE SAME PLANE

The second close-up photograph that should be taken of every item of evidence is one that includes a fully labeled scale alongside the evidence, which is done for the following three reasons:

1. With a fully labeled scale in the image, the viewer can get some sense of the size of the evidence. The size of some types of evidence cannot be appreciated without a scale. For instance, some shoe treads have the same pattern whether they are child sized or adult sized. A close-up image of one of these shoe impressions in dirt does not indicate which size they are without a scale.
2. With a scale in the image, it will be possible to enlarge the original image to life size, so that is can be compared with a known item of evidence by an examiner.
3. The labeling information on the scale helps document the image.

Figure 5.36 shows two variations of a fully labeled scale being used. Because it is necessary to include the fully labeled scale in the composition, it is necessary to back away a bit from evidence that is relatively small. Paint chips, fingernail fragments, or pubic hairs are other good examples of evidence with dimensions smaller than the scale. This does make the evidence appear smaller than it was in the "as-is" close-up photograph, but this will be "cured" with the next type of close-up photograph. If the evidence is larger than a textbook, the addition of the fully labeled scale does not usually require making the evidence much smaller than it was in the "as-is" close-up.

It must be mentioned, again, that the fully labeled scale must be raised or lowered to the plane of the evidence on which it is being focused. If the evidence is a pistol on the ground, the fully labeled scale must be raised to the height of the top of the pistol. If the evidence is a shoe print in dirt, the fully labeled scale must be lowered to the depth of the shoe impression. Recall Rule of Thumb 5.2. The fully labeled scale should never just be placed on the surface next to the evidence.

Figure 5.36

Fully labeled scale, on the same plane.

A CLOSE-UP PHOTOGRAPH WITH THE EVIDENCE AGAIN FILLING THE FRAME, BUT THIS TIME, WITH JUST A PORTION OF THE SCALE IN VIEW

When photographing a small item of evidence requires backing away from the evidence to include the fully labeled scale, another close-up with just a portion of the scale in view should be taken next, which is a true attempt to fill the frame with the evidence again. This time, however, just a portion of the scale is composed in view. In this close-up photograph, the evidence should be as large as it was in the original "as-is" close-up. Scale increments still need to be seen. Some scales have both inches and metric increments. It would be necessary to distinguish between the two.

Figure 5.37 has the small item of evidence again filling the frame. If this image were eventually going to be used for comparison purposes, the next step would be to bracket the shot with both a +1 and −1 exposure. This gives the examiner the opportunity to use the image that best shows different aspects of the evidence for their comparison. Even if you are using a digital camera and can see that your initial shot was properly exposed, it is still recommended to bracket this shot. Again, sometimes it is not whether the shot was properly exposed that justifies a set of brackets. Sometimes brackets are taken to allow an examiner to choose which image best shows them a particular feature of the evidence. Some aspect of the evidence may best be seen when the overall image is overexposed or underexposed.

If the item of evidence is relatively large, it may not be necessary to take this shot with just a portion of the scale in view. A shotgun, for example, will fill the frame the same regardless if a fully labeled scale were alongside it or if just a portion of the scale were in the image.

Figure 5.37

Close-up, with just a portion of the scale in view.

THE "ALTERED" CLOSE-UP

Many times it will be important to take additional close-up photographs of the evidence at the scene before it is packaged. Once the evidence has been measured in place, it can be carefully picked up and inspected. Other aspects of the evidence may reveal themselves, which were not apparent when the evidence was viewed "as found." For instance, a knife may have a bloody fingerprint on the bottom side, and it was not immediately apparent "as found." In situations like this, additional "altered" close-up photographs should be taken as soon as possible. The newly discovered aspect of the evidence may be transitory or fragile, and it may not survive the packaging and transportation from the scene to the storage facility. Additional close-up photography may be the only opportunity to document a newly found critical aspect of the evidence.

Another possible reason to do an "altered" close-up is that it might provide a better background than the "as is" background. If the original background was multipatterned or multicolored, it may have been difficult to see as it was originally found. If the original background was the same tone or color as the evidence, the evidence may seem to be lost against a similar background. Altering the background the evidence is on may be a better image of the evidence for a variety of reasons.

A new Rule of Thumb presents itself here.

 Rule of Thumb 5.3: "Altered" close-up photographs should be done in a way that makes it immediately obvious to the viewer that this photograph is of evidence that has been moved. The photographer must be assured that no one viewing the "altered" photograph can confuse it with any of the "as-found" photographs.

Do not just roll the evidence over and take another photograph of it against the same background as it was originally photographed against. Make it obvious this next photograph is, in effect, "staged," which is usually best done by changing the background in the photograph. Rather than looking for a clean background that is naturally found in the scene, placing the evidence on plain brown evidence wrapping paper or on a brown paper bag used to package evidence is often the best choice. Packaging materials are always available at crime scenes.

Figure 5.38 shows an example of an altered close-up photograph. This view shows the mushroomed nose of the bullet rather than the base of the bullet. Being on a plain background, it cannot be confused with any of the other images showing the bullet at the crime scene where it was originally found.

A close examination of the evidence at the scene is therefore essential. A close examination of the evidence at the scene may reveal significant identifying features, other trace evidence, or other evidence on the original item of evidence, and these features will eventually affect how the evidence is processed and packaged.

Figure 5.39 shows a grouping of all the different types of close-up photographs that may be taken at a crime scene.

THE PHOTOGRAPHIC DOCUMENTATION OF BODIES AND WOUNDS

BODIES AT THE CRIME SCENE

Like any other item of evidence, a body at the crime scene has to be fully documented photographically. The series of required photographs begins with a midrange photograph relating the body to a fixed feature of the scene. As with

Figure 5.38
The "altered" close-up.

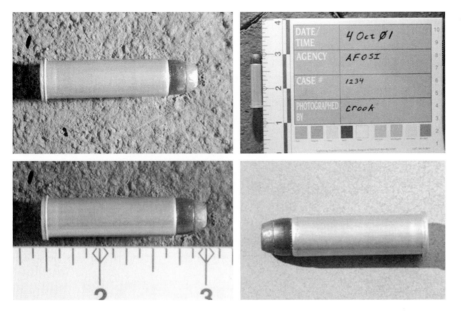

Figure 5.39
A close-up series.

other midrange photographs, the best composition is established by having part of the body and a fixed feature of the scene the same distance from the photographer: the "isosceles triangle." The previous sentence used the phrase "part of the body." Why is this so? The entire body will eventually be documented in the close-up photographs that follow, so it is not necessary to have the entire body be included in the midrange photograph. If it is not a problem to include the entire body in a midrange shot, that is great, but it is not necessary. Figure 5.40 is an example of the proper positioning for a midrange photograph of a body. The fixed feature of the scene and part of the body are equal distances from the camera.

Body panorama: A series of photographs showing the body from all four sides. It includes a full-face shot for identification purposes. It also includes a photograph from directly over the body.

After the midrange photograph of a body, a complete **body panorama** series is photographed. This includes a series of photographs showing the body from all four sides.

These are close-up photographs of the body, attempting to fill the frame with just the body, eliminating everything else. A full-face shot, for identification purposes, should also be part of a full body panorama series. It should also include a photograph from directly over the body.

Figure 5.41 shows the six images normally considered the full body panorama. Both the head-to-toe and the toe-to-head shots have to be taken with a 50-mm lens to ensure the body is not stretched or compressed. However, when taking the left side or right side views, you can use any lens necessary.

Recall the previous shots of the pistol and the wall corner (Figure 5.33). Any focal length of lens can be used as long as the item(s) photographed are running horizontally across the field of view. There will be no distortion of horizontal details. Figure 5.42 shows the same body photographed with different focal lengths. All accomplish the goal of photographing the right side of the body; none distort the perceived body length.

When the body is photographed from the left or right side, the photographer has the option of having the body positioned high in the field of view, aligned in the

Figure 5.40

Midrange of body (Courtesy of GWU MFS Student Saraya Dickson).

Figure 5.41

Full body panorama (Courtesy of GWU MFS Student Saraya Dickson, Victim: GWU MFS Student Bethany Pridgen).

middle of the frame, or positioned low in the field of view, as in Figure 5.43. This choice will depend on whether clutter or elements not necessary for the success of the image are on either side of the body. The body remains the same. Excluding details in the foreground or background are controlled by this technique.

Several comments about the full-face photographs need to be made. This shot is usually of the face from straight on, not a profile shot. Slight angles to the left or right will also work. The purpose of these shots is to produce an image that can be shown to neighbors, co-workers, or anyone else who might know the victim to find someone who can identify the victim. To accomplish this goal, the photograph must actually look like the victim. To do this, we again borrow a technique used by professional portraiture photographers. To fill the frame with a face and still have it look like the person being photographed, professional portraiture photographers use lenses with focal lengths between 100 mm and 120 mm, because if you get close enough to a face to fill the frame with it when using a 50-mm lens, the face gets distorted to varying degrees. The nose, being closer to the camera, usually looks a bit enlarged. See Figure 5.44.

Rule of Thumb 5.4: To fill the frame with a face, while remaining back a distance that does not enlarge the nose, requires a lens in the 100-mm to 120-mm length, which is the perfect portraiture lens focal length range.

It makes the face look like it appeared to the viewer. This becomes a bit of a problem for short photographers. While standing directly over the body, many 100-mm

Figure 5.42

The side of a body: different lenses.

Choose this framing if there are background details that are distracting

If there are no distracting foreground or background details, choose this composition.

Choose this framing if there is foreground clutter that is distracting.

Figure 5.43

The side of a body: high, middle, and low (Courtesy of Donna McLaughlin).

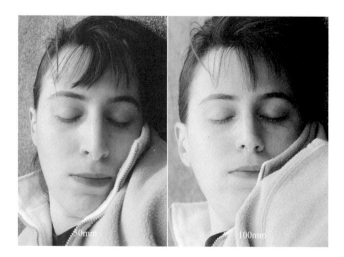

Figure 5.44

Full-face shot: 50-mm and 100-mm lenses (Courtesy of GWU MFS Student Paul Nelson).

lenses have their close-focusing distance too long for a short photographer. This means the short photographer will not be able to focus the camera while standing directly over the victim. The simple solution is to eventually tilt the victim's head to the left or right so the photographer can stand a little farther away and get the camera within its focusing range.

Another issue is the lighting of the face. Outside during the day, the light of the sun may be coming from below the victim's face. This angle of the sun produces distracting shadows on the victim's face. Notice in Figure 5.41 that the full-face photograph is different from the other images. Instead of tilted to the side a bit, it is now positioned so the face is facing straight up. This can be done after all the other "as is" photographs have been taken. The full-face shot does not have to be taken "as found." Many times it cannot be taken "as found." There may be blood or hair or other reasons the face cannot be photographed filling the frame "as found." Besides repositioning the face, the lighting of the face can be "arranged" to eliminate distracting shadows. If the sun is not producing "attractive" shadows, eliminate them altogether by blocking the sun totally. Then, you have the choice to photograph the face in total shade, as in Figure 5.41. Or, you can add light with your electronic flash. If this is your choice, just remember to align the flash so, as you view the face through the viewfinder, the flash is directed across the face from either a 10 to 11 o'clock or 1 to 2 o'clock position. These angles of light across a face are the most natural and pleasing. We are used to seeing faces lit from above. Even if the victim is in a prone position on the ground, we can position the electronic flash as if it is coming from above the head.

When photographing a face indoors, these same issues must be remembered. Try to use at least a 100-mm lens and light the face with the flash positioned high and off to either the left or right a bit.

When a body is on the floor or ground, many times the face is not positioned to be facing the ceiling straight on. The face will frequently have the chin a little higher than the forehead. To take a photograph of a face, you will want to make sure the film plane is parallel to the face, not parallel to the floor or ground. You do not want to emphasize the chin and neck. Try to have the camera an equal distance from the chin as it is from the forehead. See Figure 5.45.

When photographing bodies, on many occasions you cannot manage to get the entire right or left side of the body in your field of view, even with a zoom lens with wide-angle choices. Sometimes, it is just not possible to get far enough away to get the entire body in one shot. So be it. Do not get discouraged. Take two photographs of the side of the body: the top half and the bottom half. These two images will be able to be arranged so they appear as one entire body by either the wet-chemistry darkroom technician or your digital image processing software. See Figure 5.46.

This image shows how to align the two shots. It is not necessary to have them appear to be one original image. It is not necessary to use "stitching" software that

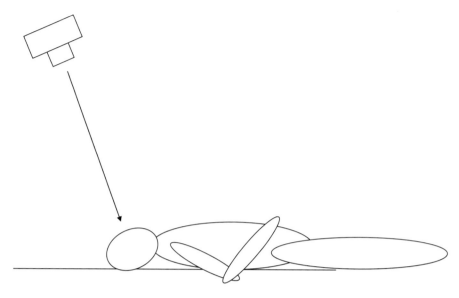

Figure 5.45
Camera with the film plane parallel to the face.

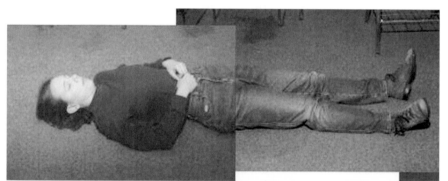

Figure 5.46
Composite body shots.

is available on the market. It is OK to have them obviously be a composite. It is another way to accurately document the scene as it was. If it took two images to capture the entire length of a body, show both of them. A hint may be useful when doing this. Do not stand opposite the waist of the subject and photograph both halves of the body from this position. It is better to stand opposite the shoulders of the victim when photographing the head and waist, and it is better to stand at the knees of the victim when photographing the waist and feet. Otherwise, you will see the head and feet appear a bit smaller than the waist when the composite is aligned.

The body panorama series includes an image with the camera directly overhead. This shot may not be familiar to some crime scene photographers. It is not shot with the photographer attempting to lean directly over the victim. It is not shot while standing on a chair or ladder. The camera is placed on a tripod, with its legs and center post fully extended, and then the camera is positioned as high as possible over the body, sometimes actually touching the ceiling. This viewpoint captures the body as it will be depicted in the crime scene diagram: from a bird's eye view.

Figure 5.47 shows a camera mounted normally on the tripod, the left image; and the camera mounted on the tripod for this technique, the right image. With the camera reversed as shown, the back of the camera can actually be placed against the ceiling. Two reasons exist to get the camera up against the ceiling, if the ceiling is close enough to reach. The closer the camera is to the ceiling, the more of the scene it can capture, even if it is just a short distance farther away. With the camera against the ceiling, it is steadier than just hovering over the body in mid-air. The ceiling can help prevent camera shake and therefore blur.

What are the camera variables for taking this photograph?

Film selection: Indoors or outdoors at night or with dim lighting, use ISO 400 film. Outdoors, with midday sun, use ISO 100 film.
Shutter speed: Use your camera's sync speed, because flash will be used.

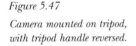

Figure 5.47

Camera mounted on tripod, with tripod handle reversed.

Aperture: Set the camera to f/22. The camera will eventually be no farther than 10′, and f/22 is the recommended aperture for flash images within 10′ with 120 GN or stronger flash units. It may be necessary to adjust the power of the full manual flash to ½ power.

Flash mode: Set the flash to the manual flash mode. If other bright lights are around the scene, or if the scene is outside at midday, flash sensors may cut the flash off early. If any quantity of blood is around or on the body, the reflected glare from the flash striking a liquid will also cause the flash to be cut off quicker. The manual flash mode will not be affected in these circumstances. Use the manual flash mode for both indoor and outdoor scenes. If outside, the flash will also act like fill-flash to lighten any shadows present. The flash duration can also eliminate concerns of blur from camera shake with the camera on a tripod hovering over the body when the camera cannot be placed against the ceiling. It may be necessary to adjust the power of the full manual flash to 1/2 power, depending on the GN of the flash unit and the eventual height the flash can be held above the body.

Focal length: To determine this, stand as far away from the body as the camera will eventually be positioned above the body. With a zoom lens, vary the focal lengths until the body is easily included in the field of view, with a couple of feet in each direction beyond the body. A full body usually requires a 35-mm lens. This can be variable, because different photographers may be able to get the camera higher depending on their own height and arm length.

Focus: You should pre-focus the camera when standing as far away from the body as the camera will eventually be positioned above the body.

Camera alignment: You should ensure the camera has its film plane parallel to the body and the camera is over the midpoint of the body. An assistant can help ensure the camera is properly positioned over the long axis, whereas it is usually easy for the photographer to visualize the mid-point from head to feet.

Figure 5.48 indicates the wording the assistant standing on the long axis of the body will use to correct the original position of the camera. The photographer cannot tell whether the camera is positioned exactly above the body. The assistant can help get the camera over the body and ensure the camera is aimed straight down over the body.

Delayed shutter release: A 10-second delayed shutter is set to provide time to properly position the camera above the body.

Figure 5.49 is the result of this effort, a great addition to the body panorama series. Another use for this overhead shot is to help produce a more realistic crime scene diagram. Instead of using the bodies provided by some crime scene diagramming (CAD) programs, some software programs, like Crime Zone, allow you to trace the body of the victim and place the tracing into the crime scene diagram. This makes the victim more real than just using stick figures in a diagram.

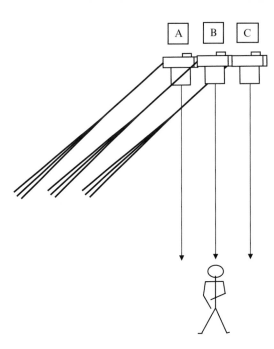

Figure 5.48

Ensuring the camera is directly above the body.

A) Not far enough: therefore needs to go "Farther."

B) Over the body, therefore "OK."

C) Too far beyond the body, therefore needs to go "Back."

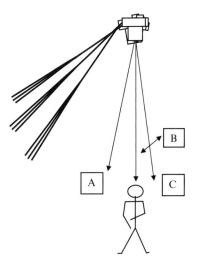

A) Aimed a bit towards the photographer's side, therefore, "Lower" the tripod end.

B) Aimed over the body, therefore, "OK."

C) Aimed a bit beyond the body, therefore, "Raise" the tripod end.

Figure 5.49

Overhead image (Courtesy of Victim, GWU MFS Student Katherine Walters).

Figure 5.50 shows the positioning of the photographer to take overhead images. With the flash mounted on the camera's hot-shoe, the camera/flash/tripod combination can become heavy enough to give some doubts about being able to hold the camera over the body for 10 seconds until the shutter trips. Finger strength is not crucial to the success of this type of photograph. The top left image shows my able assistant holding the tripod with just finger strength. Do not rely on finger strength. The next image to the right shows my assistant with her right elbow on the lower end of the tripod legs. The tripod is now supported by bones, not finger muscles. My assistant could open both hands, and the tripod remains where it is because bones are supporting it. Of course, it is suggested you still wrap your fingers around the tripod legs, but it is not finger muscles that are supporting the tripod. The image on the far right shows this same tripod support technique being used. The image at the low left is an example of the kind of image that can be produced using this technique, quite a nice addition to your portfolio of body shots.

After the photographic documentation of the entire body, documenting the wounds to the body is next in order. As with all types of evidence, the complete documentation of a wound begins with taking a midrange photograph. The body, however, has already been photographed in its relationship to the crime scene. With a wound, it is necessary to take a midrange photograph of it showing the wound's relation to a fixed feature of the body. This will normally be a joint, like a wrist, elbow, knee, or shoulder. Once the close-ups of the wound begin, it is essential that the location of the wound's position on the body has been documented. The midrange photograph does this.

A close-up "as is" is followed by a photograph with a fully labeled scale. Then, a portion of the scale is included in a series of exposures, bracketed 0, +1, and −1.

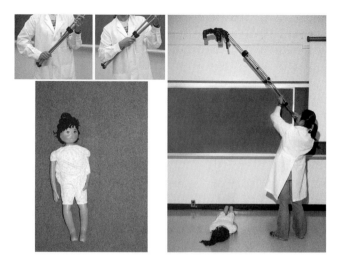

Figure 5.50

Overhead collage (Thanks to Melinda Hung, GWU MFS Student and Departmental Teaching Assistant).

Figure 5.51 shows a grouping of this sequence. Again, even if you are using a digital camera and can see that the first exposure was "correct," it is still recommended that you complete the series of photographs with brackets.

When taking photographs of wounds, several exposure issues must be kept in the back of your mind.

If you fill the frame with Caucasian skin, this is quite a bit lighter than the 18% gray scene the camera's meter is expecting. The same applies for the use of automatic and dedicated flash modes. These also expect a scene that reflects 18% of the light that strikes it. The usual result will be an underexposed image, which is the result of the "dirty snow" phenomenon. The lighter scene reflects more light toward the relevant meter/sensor. The meter/sensor wants to provide an exposure for 18%. Receiving more light than that, the meter/sensor will force the camera operator to set an exposure setting that will result in an underexposure, unless corrected by a smart camera operator. Filling the frame with Caucasian skin normally requires an exposure correction of +1.

Skin much darker than 18% gray will require exposure compensations on the negative side. When in doubt, bracket. Very dark skin should be exposed with a −½ and −1.

Wounds are frequently bloody. If blood is in the scene, it will reflect more light than normal scenes. It is not the red color of the blood that is the issue. It is the fact that the blood is a liquid, and all liquids reflect much more light than normal nonbloody items. Think of blood as being similar to a chrome bumper. When light strikes it, there will be a hot spot of glare.

Figure 5.51

Animal bites to a hand: the full wound series.

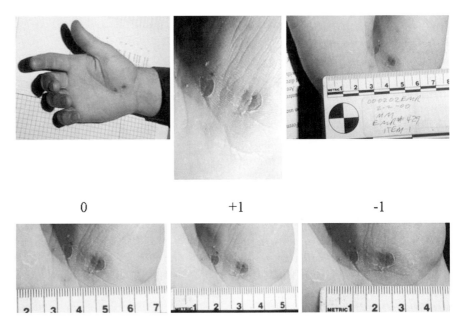

0   +1   -1

The result is underexposure, as in the left side of Figure 5.52. Select an exposure compensation toward overexposure. The right side of Figure 5.52 was taken with a +1 exposure compensation. If using flash, consider using the manual flash mode, which will not turn the flash off early because of higher amounts of reflected light.

Another trick to help prevent the strong direct flash from causing underexposures is to soften the flash somehow. Two methods are readily available. Adjust the flash head to a wider flash throw. If using a 50-mm lens, consider setting the flash head to 35 mm or even 28 mm. Or consider the use of the wide-angle flash diffuser that clips on over the flash head. Either, or both, of these techniques may help prevent underexposures. This may seem odd, because the use of a wide-angle flash head and/or the diffuser reduces the flash intensity. The point is that if they also reduce liquid glares and hot spots, the metering system may not be "fooled" into underexposing the image. When in doubt, bracket.

This issue does not just apply to blood. If the skin is very sweaty or oily with lotion, these same tendencies toward underexposure must be dealt with.

Besides underexposure, the issue of "hot spots" reflecting from wet skin is another issue. A trick to consider in the situation where all previous images have captured many "hot spots" is to use bounce flash. Many times, a soft reflected light coming from the ceiling is less likely to create the glare associated with direct flash.

The opposite extreme of this reflection issue is a body that absorbs light because it is covered in soot or is itself charred black. Victims of fatal fires require extra consideration for proper exposures. If an exposure meter or flash sensor is "fooled" by the amount of reflected light returning to the camera's position, the result will frequently be overexposures. If less than the normal 18% of light is reflected from the scene, the meter/flash sensor wants to ensure the photographer gets the 18% expected and overexposes a dark scene. Again, the cure is to use the flash in the manual exposure mode, which is not "fooled" by reflected light. Always bracket in tricky lighting situations. Figure 5.53 shows the victim of a fatal fire in the classic pugilistic attitude, with elbows and wrists bent.

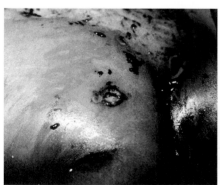 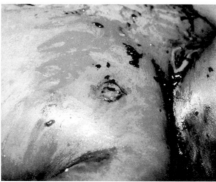

Figure 5.52

Bloody gunshot wound, original and +1 exposure.

Figure 5.53

Fire victim in classic pugilistic attitude (Courtesy of Arlington County Police Department, Virginia).

After all the wounds have been documented on the body "as found," two more general categories of photographs usually must be taken before the body is moved. Look for anything that can be used to identify the body. This can be a wallet's contents or papers in pockets. It can also be engraved jewelry. This can include scars, marks, and tattoos. This text cannot include all the means by which bodies have been identified, but when they relate to a body, it is best to document them photographically.

At many death scenes, the postmortem interval is frequently critical to the successful resolution of the incident. Document all of the clues to the postmortem interval. This may include the positioning of the body, if rigor mortis is not consistent with the body's current position. The same applies to lividity. Is it consistent with the body's current position? What color is the livor mortis? The color of lividity may indicate the manner of death. Document the body's decomposition condition. At what stage has it proceeded to? Have insects gained access to the body? Document their variety, the extent of their infestation, and signs of their long-term presence at the scene. Insects can be the cause of artifacts that can be confused with wounds, as in Figure 5.54. Are these bullet holes, stab wounds, or holes totally or partially caused by larval activity? In either case, they should be documented with photographs. Not until the autopsy will the answers be certain.

Figure 5.54

Insect infestation: True wounds or artifacts?

Are there other signs at the scene that may indicate how long the victim has been dead? Is the victim in pajamas, work clothing, or casual clothing? What looks like the last meal that was prepared? Are letters or newspapers dated? Again, it is not the purpose of this text to cover all postmortem interval signs. As your particular scene evolves, make sure to document all these issues with proper photographs.

Obviously, if weapons are near the body, they will not only be photographed in relation to fixed features of the scene but also in relationship to the body. Figure 5.55 documents that this suicide victim was still clutching the revolver used in the incident.

Cadaveric spasm (as seen in this Figure) is somtimes confused with rigor mortis, where selected muscles immediately stiffen at death.

When the top side of the body has been adequately documented with photographs, a clean sheet is usually laid alongside the body and the body is rolled onto it. The series of earlier photographs are repeated with the back of the body and the wounds that now become visible. In particular, the area that had previously been under the body is itself searched and will usually be photographically documented.

As the body is prepared for removal from the crime scene, it would be a good addition to your collection of crime scene images to photograph the process of putting the body in the body bag. Also photograph the body bag tag or label.

Figure 5.55

Cadaveric spasm (Courtesy of Arlington County Police Department, Virginia).

BODIES AT AUTOPSY

Sometimes the law enforcement agency tasked with working the crime scene can manage to have one of the original crime scene photographers also attend the autopsy. Many times the sheer volume of cases may make this impractical. When it can be managed, however, it is more efficient for the autopsy photographer to have some knowledge of the initial crime scene.

If possible, this series of photographs should begin with a shot of the same sealed body bag and tag/label as was photographed at the original crime scene. When the body bag has been opened, the body and the sheet the body is frequently wrapped in is removed, a full-face shot should be taken early. Sometimes this may require an additional photograph of the face after it has been washed, because blood, debris, and hair may initially mask a full view of the face.

Next, full views of the body as originally clothed should be taken. Wounds that penetrate clothing are photographed before the clothes are removed. When the clothing has been removed, the body is again photographed, topside, then rolled over and photographed again. Midranges and close-ups of newly discovered wounds, bruises, and injuries are collected. Figure 5.56 shows a defense wound to the hand of a stabbing victim. Figure 5.57 shows the inside and outside of a skull. The amount of soot and gunshot residue on the skull help corroborate this was a contact gunshot wound, rather than a shot fired at a distance. Figure 5.58 shows multiple stab wounds to the back.

If wound paths are determined by the insertion of probes, these images are critical to the investigation. The angle and directionality of wounds are very important to any reconstruction of a crime.

Frequently, X-rays are used to show the position of broken knife blades, projectiles, and foreign objects in a body. X-rays can be photographed without a flash. Usually ISO 400 film, 1/60th shutter speed and f/8 will produce good results. Figure 5.59 shows a close range shotgun wound to the torso and the X-ray showing the pellets within the body.

Figure 5.60 shows a letter opener in a skull and the X-ray.

When these objects are removed from the body, they should be photographed again, with and without a scale.

SENSITIVE PHOTOGRAPHS

There will be times when wounds or injuries are located on parts of the body usually covered by a swimsuit. If the victim is still alive and will permit photographs of these wounds and injuries, several important concerns are related to the successful capture of this kind of evidence.

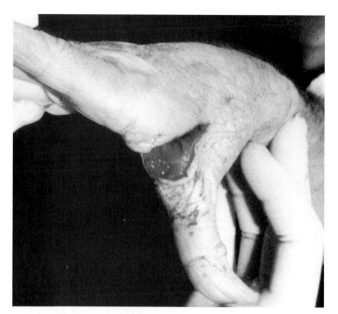

Figure 5.56
Defense wound (Courtesy of
Arlington County Police
Department, Virginia).

 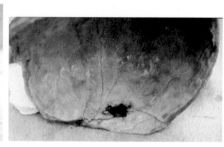

Figure 5.57

Contact gunshot wound to
skull.

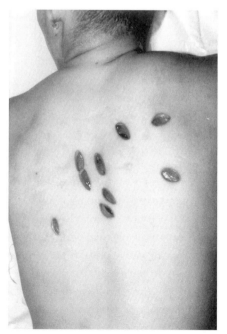

Figure 5.58

Multiple stab wounds to the
back.

Figure 5.59

Shotgun wound and the related X-ray.

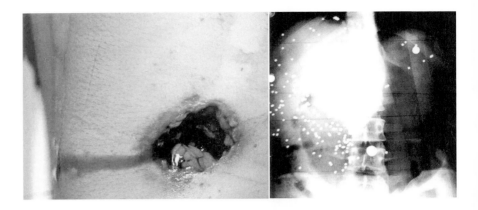

Figure 5.60

Letter opener in a skull and the related X-ray.

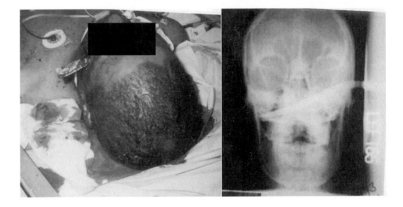

Of course, the first issue is obtaining the victim's permission to take photographs of these areas. Many may be reluctant to allow photography of their "private parts." It must be remembered that the victim cannot be forced to permit such photographs. They must willingly allow it. Obtaining the victim's permission will often depend on the crime scene photographer's professionalism and ability to explain the need for such photographs.

One issue is whether female photographers are available to photograph female victims, or male photographers are available to photograph male victims. When possible, having the same-sex photographer will often put the victim more at ease. However, in many smaller jurisdictions providing multiple different-sexed photographers during a particular shift is not possible. Because of the workload, at times the only photographer currently available is an opposite-sex photographer. This will have to be explained to the victim with a sensitive yet professional approach.

Explain that you understand their possible reluctance to allow you to photograph their wounds. Quickly tell them that you are a professional photographer, and you understand their concerns. Explain that you have had to take these kinds of photographs many times, and one of your primary concerns is that you obtain the photographs that are required to assist in the prosecution of the crime that has been committed against them. Explain that you know they have already been victimized once, and the last thing you want to do is to victimize them again.

To accomplish this goal, you want them to feel as comfortable as possible during the process, and you want them to have someone they feel comfortable with in the room while these photographs will be taken. This can be a spouse, family member, clergy, or other available friend. If none of these are available, it can be hospital staff if the victim is currently in the hospital. Ask them who they would like to have with them during this photography. Even if they do not seem to care about this issue, strongly insist on another person being in the room when the photographs are taken. You do not have to tell them this, but having someone else in the room is also for your protection. You certainly do not want a confused or vindictive victim accusing you of inappropriate behavior while they are partially undressed and alone with you in a room.

Explain one of your utmost concerns is protecting their modesty during the process of photographing their wounds. Although this should be your primary concern, the modesty of others is also one of your concerns. The judge, jury, and attorneys will also be looking at these images, and you do not want to unnecessarily offend them by showing too much skin if it can be avoided. Even if the victim does not express a concern for their own modesty and is very willing to allow you to photograph anything you please, keep these other viewers in the back of your mind and take your photographs with them all in mind.

Also, these photographs represent your professional abilities. Anybody can take a full-frontal image of a nude person, but the professional will know how to photograph what is necessary, while being careful to exclude from their composition any unnecessary areas of skin.

Several investigators have expressed a wish for a single photograph of the entire area showing wounds or bruises. Usually, a series of well-placed photographs of "parts" will accomplish the same goal. If the victim has been cut and bruised between the neck and knees, it is not necessary to take one photograph of the body between the neck and knees. It is possible to show the entire extent of these injuries by showing a montage of several images.

Figure 5.61 is a perfect example of several images being able to substitute for a single full-frontal image. This rape victim was also tortured by having cuts applied to her body between the neck and knees. By carefully draping her body, the photographer has ensured that the viewer of this series of images gets the full impression of the extent of the injuries to the victim. These images did not happen by accident. A very experienced photographer knows what is necessary to tell the story to the court and drapes the victim to accomplish that goal without compromising the victim's modesty or the photographer's professionalism. As a juror, would you also not appreciate this draping of this

Figure 5.61
Injuries to rape victim.

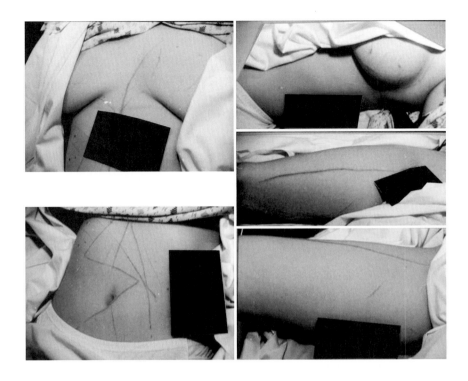

victim, while still enabling you to see the evidence necessary to establish an element of the crime?

Figure 5.62 represents another method used to accomplish our goals, besides judicious draping, which is the cropping of images to show only what is needed. Actually, Figure 5.62 corrects what can be regarded as a mistake. In the original photograph, the victim's face was included. In that shot, she was very upset and

Figure 5.62
Sexual assault victim.

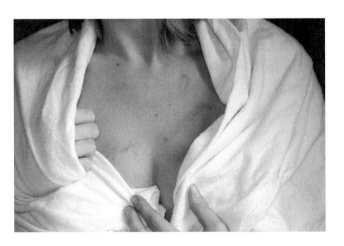

crying. Usually, it is recommended not to include the victim's face in any photograph of sensitive body areas. For two reasons: First, if the victim is told you are intentionally not including the face in photographs showing sensitive parts of their body, they may be more willing to allow the photograph be taken in the first place. Second, if the victim is noticeably upset, the judge may not allow the photograph to be admitted into court as evidence if the photograph unduly prejudices the jury. A crying victim may be regarded as being overly solicitous of sympathy for the victim.

It is true that the victim's face should be photographed, but it should not be in the same photograph as sensitive body parts. How would it be established that the face belongs to the other photographs of the wounds and bruises? Try to have some common element in the face shot and in the body shots. Shirt or blouse fabric can be common to all the photographs. If the victim is wrapped in blankets or bed sheets, ensure a prominent fold in the material is noticeable in the face shot and the other body shots.

Figure 5.63 includes two images. The left image is the way the original image was captured, with blackout areas placed as necessary. The right image is the photograph that should have been taken. The incident is a rape, where the victim also sustained bruising to both breasts. The only purpose for the photograph is to document the two bruises on the breasts. In this particular case, the victim reacted spontaneously to a request to photograph the bruises by pulling up her hospital gown. The photographer, unfortunately, also reacted spontaneously and took a quick photograph without carefully arranging the photograph to show only what was necessary.

What is wrong with the left image?

1. The victim's face is in the same photograph as sensitive body parts.
2. Much more than the bruising is included in the composition. Would you like to have your professionalism judged by this photograph?

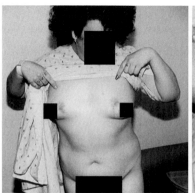 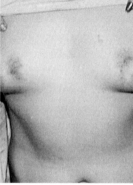

Figure 5.63
Composition issue: rape victim.

3. If an image can be excluded from court because it may prejudice the jury against the defendant, it can and should also be excluded from court if it may prejudice the jury against the victim. In this case, also visible in the image are tattoos on the victim's fingers. In today's climate, many tattoos that women adorn their bodies with have become socially acceptable: roses, Celtic symbols, college initials, and a myriad of other designs that are so commonplace they hardly attract any attention. However, at one time, tattooed women were mostly found on the back of Harley motorcycles. The letters, "Love" on her right fingers belongs to this latter category of tattoos. It would be possible for someone to see these tattoos and think less of the victim for it. Because of this possibility, they should have been excluded from the photograph. Actually, it is doubtful that the photographer even saw these tattoos when the photograph was taken.

The image on the right is what should have been composed in the viewfinder before the shutter button was depressed.

A caution is warranted here. Do not get the impression that carelessly composed photographs can always later be cropped either in the wet-chemistry darkroom or with digital imaging software. The defense attorney has a right to request copies of all the images related to a particular incident. That includes all the original images. A defense attorney may challenge your images, not on their factual content but on your professionalism. Whether this challenge is successful or not is moot. Having your professionalism questioned in court may be remembered by other attorneys in the court and by the judge. Do not give anyone the opportunity to do this. Far better is to develop a reputation in your agency as the person who can get a photographic job done right the first time.

Figure 5.64 is a well-composed and intentionally thought-out image of bruising to the buttocks. It demonstrates a careful composition designed to show what the photographer wanted you to see while excluding from view any elements not essential to tell this one particular story.

It is a "great shot" for another reason. When these kinds of images are required to show injuries, wounds, or bruising, they can sometimes be done in a way that is "sexless." With this particular image, it is impossible to tell if the victim is a man or a woman or how old he or she is. Imagine the need to have one of your loved ones requiring such photographs. If absolutely necessary that such images be taken, would you not prefer the crime scene photographer "isolate" the area requiring photographic documentation. Figure 5.64 does this perfectly. The bruise to the buttocks is documented, but all four sides of the bruise have been cropped off or draped in some way. The viewer of the photograph must be struck with the impression that this image did not just happen accidentally; it was composed by a professional crime scene photographer.

In one well-known situation, larger views of the body may have to be shown. When a weapon, tool, or other object was used that leaves a distinctive pattern on

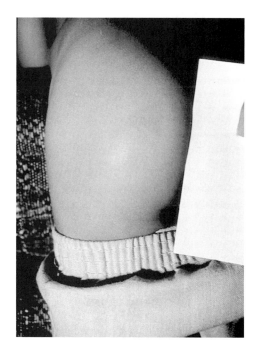

Figure 5.64
Bruise to buttocks.

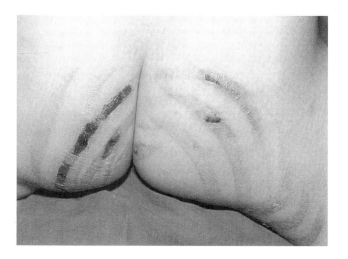

Figure 5.65

Stove top element wound (Courtesy Of Joanna Collins, GWU MFS Student and AFOSI Special Agent).

the body, the entire pattern needs to be included in one photograph, which is the only way an examiner may be able to match a known weapon, tool, or implement back to the image. This can occur with weapons like crescent wrenches, bite marks, shoe treads, or a variety of other things. Figure 5.65 is an example of a child punished by being placed on a hot stovetop. Sometimes the injury, wound, or bruise is nondistinct. In these cases, imaging parts and pieces of the entire wound can tell the story that needs telling. If a pattern is present, however, and this pattern can be linked to a specific item that made it, the entire pattern should be captured in the same image. In these cases, at least one image of the series taken must have a scale included alongside the pattern. This will have to be explained to the victim so they will give you permission to photograph larger areas than normal.

Many times, fresh wounds do not show the pattern that would be useful to do a comparison to a known object. The wound may be swollen. The wound may have irregular edges. Bloody areas may make it difficult to see the wound with all its definition. At times it is wise to tell the victim that there will need to be some follow-up photography at a later time. Coming back in a day or two, or maybe even after a week or so, may be required. Sometimes the best images of the wound or the pattern left by the weapon are most efficiently imaged on a later day. Certainly photograph the wounds as soon as possible, but follow-up photography may be required.

SUMMARY

Photographs can become critical items of evidence in court. As such, they must be able to withstand challenges from the defense. The careful documentation of each photograph is the best way to ensure that a photograph may survive some potential challenges. This chapter explained the three primary ways that photographs are documented.

Three basic types of crime/accident scene photographs exist: overalls, midranges, and close-ups. The overall photographs document the general conditions of the scene, both with exterior and interior views, and how the specific crime scene relates to the surrounding area. When taking overall photographs, the photographer temporarily ignores the specific items of evidence within the scene, and concentrates on documenting the general scene conditions.

The purpose of midrange photographs is to show a relationship between an individual item of evidence and a fixed feature of the scene previously photographed in one of the overalls. "Fixed features" are truly fixed at the scene. Inside, they are wall corners, doors, windows, and electrical outlets. They are not items of furniture that can easily be moved. A car parked in a garage is not a fixed feature because it can be moved. Outside, fixed features are streetlights, manhole covers, building facades, sidewalks, curbs, and trees. They are not vegetation that is likely to change or disappear with the seasons. Midrange photographs are designed to include both an item of evidence and a fixed feature of the scene in the same photograph. It is critical to have both of these elements properly composed, exposed, and in focus. This chapter recommended a technique to ensure successful midrange photographs.

A close-up photograph is an attempt to fill-the-frame with the evidence, with the film plane parallel to the evidence. Diagonal viewpoints are to be avoided.

Critical comparison photographs will only be comparable to a known item of evidence if the film plane was parallel to the evidence when the close-up photograph was taken. The photograph should also indicate to the viewer that the photographer carefully chose the viewpoint, and diagonal viewpoints do not show this level of care.

The photographic documentation of wounds requires additional considerations not always relevant to other types of evidence. The complete photographic documentation of a homicide victim, both at the scene and later at the autopsy, was explained. Photographing skin and bodies presents special challenges to the photographer. This chapter explained some of these processes and concerns.

DISCUSSION QUESTIONS

1. Briefly explain the written documentation that should accompany each photograph. When scales are used, explain why they should also be labeled.
2. Briefly explain the types of exterior overall photographs. Explain the issues related to the different lenses that can be used and the perspective that is suggested.
3. Briefly explain interior overall photographs. Explain the issues related to the different lenses that can be used and the perspective that is suggested.
4. Midrange photographs have a specific purpose. Explain. They, too, have lens and perspective aspects. Explain them.
5. Explain the four different types of close-up photographs that can be taken of evidence.
6. Explain the "full body panorama" series of photographs.
7. Explain the advice to "zoom in, crop, and drape" when taking photographs of "sensitive" areas of the body.

EXERCISES

1. Fill-the-frame with a photo identifier and photograph it both with natural lighting outside and with electronic flash inside.
2. Take a series of photographs completely documenting two sides of a building.
3. Take a photograph relating the building in which a crime occurred to the surrounding area or neighborhood.
4. Take a series of photographs completely documenting two walls of a room.
5. With an item of evidence outside, take a midrange photograph and four close-up photographs of it.
6. With an item of evidence inside, take a midrange photograph and four close-up photographs of it.
7. With a body outside, take a midrange photograph and a complete body panorama of it.

8. On this body, draw a 1″ "wound" and take a midrange photograph and a series of close-ups of it.

9. With a body inside, take a midrange photograph and a complete body panorama of it.

10. On this body, draw a 1″ "wound" and take a midrange photograph and a series of close-ups of it.

Compare your images with those in this chapter and with other images on the supplemental website of images, referring particularly to the following image folders: body, close-ups, crime scenes, exterior overalls, interior overalls, and midrange.

FURTHER READING

Duckworth, J. E. (1983). "Forensic Photography." Charles C Thomas, Springfield, IL.

Editors of Eastman Kodak Company. (1976). "Using Photography to Preserve Evidence." Eastman Kodak Company, Rochester, NY.

Lester, D. (1995). "Crime Photographer's Handbook." Paladin Press, Boulder, Colorado.

McDonald, J. A. (1992). "Close-Up & Macro Photography for Evidence Technicians." Photo Text Books, Arlington Heights, IL.

McDonald, J. A. (1992). "The Police Photographer's Guide." Photo Text Books, Arlington Heights, IL.

Miller, L. S. (1998). "Police Photography." 4th Ed. Anderson Publishing Company, Cincinnati, OH.

Redsicker, D. R. (1994). "The Practical Methodology of Forensic Photography." CRC Press, Boca Raton, FL.

Siljander, R. P., and Fredrickson, D. D. (1997). "Applied Police and Fire Photography." 2nd Ed. Charles C Thomas, Springfield, IL.

Staggs, S. (1997). "Crime Scene and Evidence Photographer's Guide." Staggs Publishing, Temecula, CA.

CHAPTER 6

ULTRAVIOLET, INFRARED, AND FLUORESCENCE

LEARNING OBJECTIVES

On completion of this chapter, you will be able to . . .

1. Explain the various results of light striking different surfaces.
2. Explain where on the electromagnetic spectrum the UV range is located.
3. Explain various uses of UV light to visualize otherwise "invisible" evidence.
4. Explain where on the electromagnetic spectrum the visible light range is located.
5. Explain the Stokes shift.
6. Explain some of the different types of evidence that can be made to fluoresce so they can more easily be located and collected.
7. Explain where on the electromagnetic spectrum the IR wavelengths are located.
8. Explain several types of evidence that can be visualized in the IR range of the electromagnetic spectrum.

KEY TERMS

Electromagnetic
   spectrum
Fluorescence
Infrared light

Luminescence
Phosphorescence
Photographic infrared
   range

Ultraviolet light
Visible light

THE ELECTROMAGNETIC SPECTRUM (EMS)

Figure 6.1 shows the typical wavelength design. The wavelengths of ultraviolet (UV), visible light, and infrared (IR) light are expressed in terms of nanometers (1 nm = one billionth of a meter from peak to peak).

LIGHT ON THE ELECTROMAGNETIC SPECTRUM

Electromagnetic spectrum: This spectrum depicts light as a wave of radiation, which has both a peak height (amplitude) and a distance between peaks (wavelength).

Other forms of radiation on the **electromagnetic spectrum** include those with much shorter wavelengths (gamma rays and X-rays), as well as those with much longer wavelengths (microwave and radio waves).

Most have heard the phrase "the speed of light." Sometimes this is explained with a reference to lightning. Because the speed of light (generally considered roughly 300,000,000 meters per second or 186,000 miles per second) is much faster than the speed of sound (generally considered 1120 feet per second or 761 miles per hour), we will see a lightning flash first; and then, depending on the distance of the lightning from the human sensor, the sound of the lightning can arrive several seconds later. Because the speed of light is so fast, and the distances we can see on earth relatively short, the "seeing" of light coming from any object on the earth appears "instantaneous."

When we look toward the heavens, things begin to differ. Because the moon is approximately 239,000 miles from the earth, it takes light approximately 1.28 seconds to travel from the moon to the earth. Because the sun is approximately 93,000,000 miles from the earth, it takes light approximately 8 minutes and 19 seconds to travel from the sun to the earth. That means that if the sun were immediately snuffed out and ceased to exist, apart from other terrible physical effects, we would still see the light of the sun for another 8 minutes and 19 seconds, because we would still be seeing the light that is traveling between the sun and the earth until it all reaches the earth.

Figure 6.1
Wavelength and amplitude.

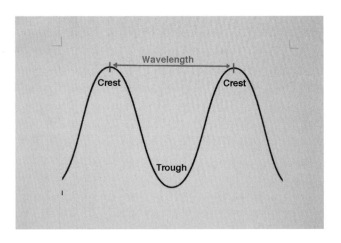

Why mention these interesting facts about the speed of light? It is to get the reader used to thinking in terms of our seeing things some "time" after light left the source of the light. A time and a distance exist between a person's seeing something and the cause of those light waves. When writing about light, some foundation is almost always warranted.

Figure 6.2 is a representation of the entire electromagnetic spectrum. On it you can see the small area designated for **visible light** (Figure 6.3).

Visible light: The range on the electromagnetic spectrum between 400 and 700 nanometers (nm). A nanometer is one billionth of a meter.

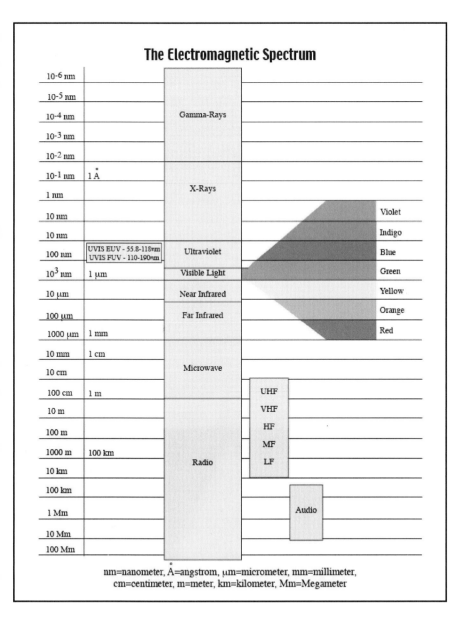

Figure 6.2

The electromagnetic spectrum. (http://lasp.colorado.edu/cassini/education/Electromagnetic%20Spectrum.htm)

Figure 6.3

The visible light range (Courtesy of Jeff Robinson, scamper.com).

The Visible Spectrum

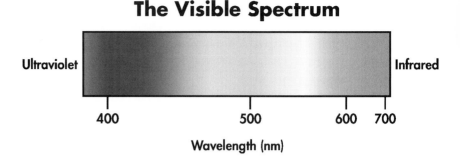

Ultraviolet

Infrared

400   500   600   700

Wavelength (nm)

Visible light is generally regarded to be the range on the electromagnetic spectrum between 400 and 700 nm. Below the 400 nm range is the region of **ultraviolet light**, which is 100 to 400 nm.

Above the 700 nm range is the region of **infrared light**, with the region of infrared applicable to photography being approximately 700 to 1100 nm.

Ultraviolet light: The region on the electromagnetic spectrum between 100 and 400 nm. It is also sometimes divided into long, medium, and short wave ultraviolet light (UV). At times, it is thought of as UVA, UVB, or UVC.

Infrared light: The photographic area of the infrared (IR) range is in the near IR part of the electromagnetic spectrum, 700 to 1100 nm.

LIGHT ENERGY AND DIFFERENT SURFACES

Light, as a region of waves on the electromagnetic spectrum, reacts different ways when it strikes different surfaces. As manipulators of light, photographers should understand this interaction. Sometimes, we just need to be aware of these processes. Sometimes, we must control or manipulate the photographic variables to achieve different effects. The main reactions of light with the different substrates it strikes are (a) reflection, (b) absorption, (c) transmission, and (d) the conversion of light from one state to another, usually regarded as luminescence.

White light is the combination of various colors. Most are aware of the ability of a prism to "split" white light up into its component colors, the rainbow as shown in Figure 6.4.

When white light strikes some surfaces, there can be a total reflection of all the light. When all the light is being reflected, the result is an impression that the

Figure 6.4

A prism splitting up white light (Courtesy of Jeff Robinson, scamper.com). Please visit the book's companion website (http://books.elsevier.com/companions/9780123693839) for a color version of this figure.

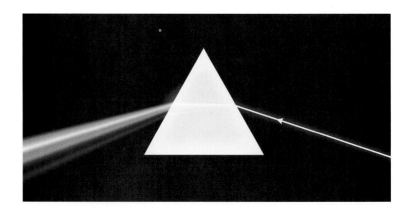

surface itself is white. Actually, our interpretation of paper being lit by white light as being white paper really means our eyes are being stimulated by all the reflected light coming off the paper. The paper is reflecting all the light that was present in the light that struck the paper. The eyes sense all the wavelengths of all the colors in the reflected light, and the brain interprets this combination of stimuli as a surface that is itself white. Technically speaking, the paper itself is not white; because it reflects all the light that strikes it, the brain "assigns" the color white to it.

When white light strikes some other surfaces, all the colors that make up white light can be absorbed by the surface. When no colors are reflected by a surface, the brain interprets this surface as being black. Technically speaking, this surface itself is not black; because it absorbs all the light that strikes it, the brain "assigns" the color black to it. Actually, black is the absence of color. It is really not a color at all.

Most surfaces partially absorb some parts of white light, and partially reflect some parts of white light. The results are surfaces that appear to be colored or manifesting varying shades of gray.

Why do we say a shirt is red? It is "red" because when all the colors of the rainbow, white light, strike it, the material and dyes in the fabric absorb the violet, blue, green, yellow, and orange of the white light, and the shirt reflects only the red part of the white light. Why do we say a dress is blue? It is "blue" because when all the colors of the rainbow, white light, strike it, the material and dyes in the fabric absorb the violet, green, yellow, orange, and red of the white light, and the dress reflects only the blue part of the white light. Why do we say something is gray? An object is considered gray because only a portion of the white light striking it is being reflected, and some of the white light striking it is being absorbed. Being gray is a result of a surface partially reflecting white light and a partially absorbing white light.

Besides being able to be reflected and absorbed by different surfaces, light can also be transmitted through some surfaces or materials. Obviously, we think of glass and water right away. But, remember the result of putting one layer of handkerchief over an electronic flash head? The light still went through the handkerchief. It was just 1-stop dimmer, because some was absorbed by the cloth. Most is transmitted through the cloth.

Some materials react to light differently. Some materials, certainly not all, will convert the light. First pointed out by Irish physicist George G. Stokes in 1852, some materials will absorb the light that strikes them and convert that light into a light of longer wavelength and lower intensity, which is normally understood as **fluorescence.**

A stimulating light, usually from a laser or alternate light source, which emits a light of a known wavelength and frequency, is directed onto a surface. That surface may sometimes totally absorb the stimulating light. Unlike what happens when no light is reflected from the surface, and the result is a perception of a black surface, molecules in the surface will become excited, and some of their

Fluorescence: A stimulating light, usually from a laser or alternate light source, is directed onto a surface. Molecules in the surface will become excited, and some of their electrons will rise to a higher electronic state. As they return to their previous state, energy is emitted. The light subsequently emitted has a lower intensity than the original stimulating light. Light of a different wavelength and frequency also has a different color.

electrons will rise to a higher electronic state. As they return to their previous state, energy is emitted. Because a loss of energy occurs during this process (actually, the Conservation of Energy Principle excludes the possibility of energy being destroyed or lost; in reality, some of the light energy is converted into heat energy that is dissipated into the material, which does reduce the net light energy remaining), the light subsequently emitted has a longer wavelength and lower intensity than the original stimulating light. Light of a longer wavelength and lower intensity also has a different color.

A classic example of the conversion of one light into fluorescent light is the use of an argon ion laser, emitting what is perceived to be a light blue light. Figure 6.5 shows the laser emitting its typical light blue light.

Two sections of handkerchief stained with urine and lit with room lighting are shown in the middle of Figure 6.5. On the right are the same urine stains lit with the laser and photographed with an orange filter over the camera lens. Why the use of an orange filter? As stated previously, the fluorescence stimulated by the laser is weaker in intensity than the original laser light. The stronger light of the laser overwhelms the fluorescence it creates. To see and photograph the fluorescent urine stains, it is necessary to eliminate the laser light without turning it off, which is done with a filter. It is known that the fluorescence created by an argon ion laser usually occurs in the orange range of the visible light spectrum. An orange filter appears to be orange to our eyes because it absorbs all the other colors of the visible light range and reflects and transmits orange. By viewing the urine stain wearing orange goggles, one can see the fluorescence created by the laser without turning it off. The orange filter will absorb the blue light of the laser, while allowing the orange fluorescence caused by the laser to be seen. Put an orange filter on the camera lens, and the same view can be photographed.

Figure 6.5

The argon ion laser and urine stains. Please visit the book's companion website (http://books.elsevier.com/companions/9780123693839) for a color version of this figure.

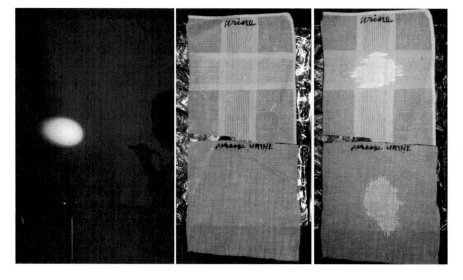

To photograph this fluorescence, aperture priority exposure mode can be used with an f/11 with the camera loaded with 400 ISO film or digital cameras set to 400 ISO. Because the camera meter is designed to provide proper exposures for objects lit with visible light, it makes sense that the camera light meter may not be as successful with fluorescent objects. In fact, the result is usually overexposure. When using the aperture priority exposure mode, begin with the exposure compensation set to −2. You may bracket around that.

Having used the term "fluorescence" a few times, it is necessary to clear up the confusion that frequently is associated with three terms: luminescence, fluorescence, and phosphorescence. **Luminescence** is the general term that includes both fluorescence and phosphorescence. Luminescence means a molecule's ability to "glow" from causes other than heat.

Incandescence is an example of "glowing" caused by heat. So, luminescence and phosphorescence are both types of luminescence. What differentiates them? Fluorescent materials emit a "glowing" light only while they are currently being stimulated by a stronger, higher-intensity light. Turn off the stimulating light and the fluorescence also immediately ends. **Phosphorescence** is the ability of some materials to retain some of the radiation they have absorbed after the stimulating light has ceased.

This energy can then be released over long durations. Some children's toys designated as "glow in the dark" toys are phosphorescent. Expose them to light once, and they can continue to glow long after the lights have been turned off.

When searching a crime scene for evidence that is currently not visible with the normal room lights on, UV lights, forensic light sources (usually understood as a non-laser light source emitting only one color of light), lasers, and alternate light sources (usually understood as a non-laser light source capable of emitting a variety of different lights) are used to stimulate potential evidence so it will fluoresce and become visible.

Some standard combinations of lights and filters are well known:

- If a UV light is being used to stimulate fluorescence, the stronger UV light can be absorbed by either a UV filter or a yellow filter. Doing so allows the weaker fluorescence to be seen. The top part of Figure 6.6 is a semen stain fluoresced with a UV light and filtered with a yellow filter.

Rule of Thumb 6.1: When searching a crime scene with a forensic light source, laser, or an alternate light source emitting a UV light, use a UV filter or a yellow filter to see and photograph the fluorescence generated.

- If a blue light (which can be emitted by a laser, forensic light source, or an alternate light source) is used to stimulate fluorescence, the stronger blue light can be absorbed by using an orange filter. Doing so allows the weaker fluorescence to be seen. The bottom part of Figure 6.6 is the same semen stain fluoresced with a blue light and filtered with an orange filter.

Luminescence:
Luminescence means a molecule's ability to "glow" from causes other than heat, and luminescence and phosphorescence are both types of luminescence.

Phosphorescence:
Phosphorescence is the ability of some materials to retain some of the radiation they have absorbed after the stimulating light has ceased. This energy can then be released over long durations: "glow in the dark" toys are an example.

Figure 6.6

Semen stain fluoresced with UV and blue light.

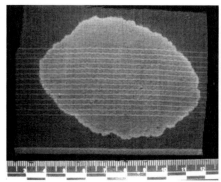 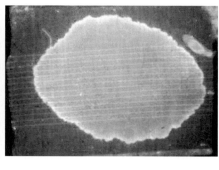

 Rule of Thumb 6.2: When searching a crime scene with a forensic light source, laser, or an alternate light source emitting a blue light, use an orange filter to see and photograph the fluorescence generated.

- If a green light is used to stimulate fluorescence, the stronger green light can be absorbed by use of a red filter. Doing so allows the weaker fluorescence to be seen. Figure 6.7 shows a shoe used to create a bloody shoe print, a white gel lift/transfer of the bloody pattern, and the fluorescence of the shoe print created by use of a green light and filtering this light with a red filter.

 Rule of Thumb 6.3: When searching a crime scene with a forensic light source, laser, or an alternate light source emitting a green light, use a red filter to see and photograph the fluorescence generated.

Because some materials only fluoresce when stimulated by one particular light, having the ability to search a crime scene with a variety of different lights will typically yield the recovery of more evidence. Not all evidence will fluoresce when just stimulated with a blue light.

Figure 6.7

Hungarian red fluoresced with green light and a red filter.

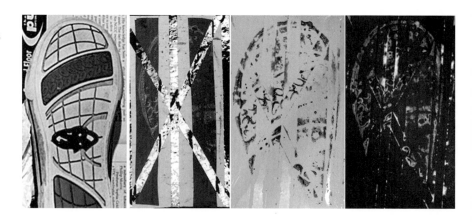

ULTRAVIOLET LIGHT (UV)

As mentioned previously, UV light is located on the electromagnetic spectrum between 100 and 400 nm. UV light can be characterized as being long wave UV (315–400 nm), medium wave UV (280–315 nm), or short wave UV (200–280 nm). Long wave UV is the region sometimes described as being "black light" UV, or sometimes it is called the "disco light" range. In this range, clothing sometimes appears to glow brightly. This would suggest long wave UV light might be useful in locating difficult-to-find trace fibers. Long wave UV light was the first crime scene light used to search for blood, non-blood body fluids like semen at rape scenes, and fluorescent fibers. Before lasers and alternate light sources became the most frequently used light sources at crimes scenes, long wave UV light was the only light able to visualize many types of "invisible" evidence. It still does these jobs well, and many agencies with limited budgets can still locate and collect many kinds of evidence with just a long wave UV light. A long wave UV light is normally a light emitting approximately 365 nm.

Short wave UV light did not have a use at crime scenes until the advent of Reflected Ultra-Violet Imaging System (RUVIS). This equipment uses high-intensity short wave UV light to visualize untreated latent fingerprints. It apparently does this job extremely well. Subsequent use in the field and research indicates that it should not be used at a crime scene until all samples of body fluids have first been collected. Short exposures of DNA to short wave UV light can preclude the eventual successful typing of DNA. The tool is still very useful. It must be used judiciously, however, and not until all DNA samples have been collected and removed from the crime scene.

UV light is also sometimes divided into UVA (315–400 nm), UVB (280–315 nm), and UVC (<280 nm). UVA is considered the "tanning" region of UV light. UVB is considered the region of UV light that is responsible for sunburn and sun damage to skin. UVC is considered "germicidal." UVC is sometimes used as a non-chemical disinfectant, because it can kill germs effectively, which is also the reason that UVC, or short wave UV light, should not be used around DNA samples. It can also prevent the typing of DNA samples.

One other general comment should be made about UV photography before specific examples of its application are addressed. It was previously discussed in Chapter 2 that attempting to capture images in the infrared region required a focus adjustment. A camera's focusing system is designed for visible light, not infrared light. Attempting to capture detail in the ultraviolet will also require a focus adjustment. The focusing adjustment for ultraviolet photography, however, is the opposite of that required for infrared photography. It would, therefore, be good to review the focusing adjustment required for infrared photography.

Figure 6.8 shows the distance ring and focusing point for one particular camera. The dark numbers indicate the distance in feet, and the light numbers indicate the distance in meters. The vertical light line is the focusing point. The dark

Figure 6.8

The IR focusing adjustment.

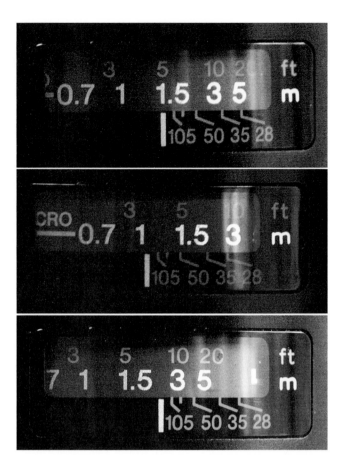

numbers to the right of the focusing point indicate this is a zoom lens, and they are the various focusing adjustments necessary, depending on which focal length you have set the lens to. Let us presume a 50-mm lens focal length is being used in this example. The top image shows the 5′ distance has been aligned with the focusing point, indicating some subject of interest is 5′ away from the camera, and the camera has been focused normally. The middle image shows the focusing adjustment required for infrared imaging. The 5′ distance, which was in focus in visible light, has been repositioned so it is now over the infrared focus adjustment point. This sets the focus for infrared light.

The focus correction for ultraviolet lighting is just the opposite of the focus adjustment for infrared lighting. Because IR lighting required the 5′ distance be rotated to the right until it aligned with the IR focus adjustment point, for UV lighting the 5′ distance should be rotated to the left the same distance as it had been rotated to the right for IR lighting. Approximate the distance between the visible light focusing line and the 5′ number once it is aligned with the IR focus adjustment point, and set the 5′ number that same distance to the left of the white focusing line. The bottom image of Figure 6.8 has the 5′ distance correctly adjusted for UV lighting.

One important caveat needs to be mentioned when trying to determine the correct focusing point with UV light. At times, this full UV focus adjustment will be needed. Other times, UV light will need little or no focus adjustment. With a film camera, you cannot immediately see the results of your first shot; it is, therefore, recommended that you bracket with your focus positions, as well as with exposure settings. Take one photo at the full recommended UV focus adjustment. Then, take two or three more brackets, working your way back to the "normal" visible light focusing point. Finally, take one image at the visible light focusing point. With a digital camera, take one image at the UV focus adjustment position and another image at the visible light focusing position. See which image is correct, or nearest to correct focus, and make focusing adjustments as appropriate.

Cameras specifically designed for work in the IR or UV range will not need these focusing adjustments, whether they are film cameras or digital cameras.

Long wave UV light is the light of choice for doing both reflected ultraviolet photography and for creating ultraviolet fluorescence.

REFLECTED ULTRAVIOLET PHOTOGRAPHY

Reflected UV photography involves capturing images with only long wave UV light being allowed to strike the film or digital sensor, which is easier said than done. Reflected ultraviolet is perhaps one of the most difficult types of photography. The following are several reasons for this:

1. A UV-sensitive camera sensor is required. Not all types of film are sensitive to UV light. Most film has a UV filter layer, which filters out UV light, so when doing UV imaging, you have to use a film that does record the UV area. Black-and-white film is best suited to reflective ultraviolet photography. Not all digital sensors are sensitive enough to UV light to be able to have nothing but UV light create the image.
2. It is necessary to have a lens that transmits UV light. Most lenses are designed to block UV light, because UV light can cause exposure errors for normal visible light photography. The lens elements themselves frequently block UV light, and in addition, most lenses have special coatings on them to block UV light.
3. If the image will be made with nothing but UV light, it will be necessary to block all visible light from entering the lens. One way to accomplish this is to simply turn off the lights and make sure no stray visible light is entering the room through door cracks or window edges. Working in the dark is very troubling for most people, with the need to turn the lights off and on many times required. Perhaps the best way to eliminate visible light is to use a filter that blocks all visible light while allowing UV light to be transmitted through it. An 18A filter does just this. This filter is not to be confused with the UV filter

or 1A filters previously mentioned as lens protection. Both of these filters block UV light while transmitting visible light. Figure 6.9 shows both a UV filter, which blocks UV light while transmitting visible light, and an 18A filter, which blocks visible light while permitting the transmission of UV light through it. Notice that the 18A filter appears opaque because it blocks all visible light.

4. Reflective ultraviolet photography requires the UV focus adjustment.
5. A long wave UV light source is required.

What are the applications best suited for reflective ultraviolet photography? Predominantly this type of photography has two main uses: questioned document examiners will use this type of imaging to help them differentiate between inks that may look similar when viewed in visible light. A document may have been altered with a different pen than the one that wrote the original words. Words, letters, or numbers may have been added to the original writing to change the meaning of the words or to change the value of checks or other monetary instruments. Even though many pens have similar black inks when viewed with visible light, sometimes these inks can be distinguished when viewed under different light sources. The different inks may show a varying tendency to absorb, reflect, transmit, or fluoresce when viewed under different lights.

Figure 6.9

UV-blocking and UV-transmitting filters.

Reflective ultraviolet light can also sometimes visualize deep muscle bruising that has healed and is no longer visible with room lights. Every text dealing with forensic photography sooner or later refers to the article in the November, 1994 issue of the *Journal of Forensic Science*, in which T. J. David and M. N. Sobel reported a case study in which they were able to photograph 5-month old bite marks on the shoulder of a rape victim with reflective ultraviolet photography. They compared the photographed bite mark to the known dentition of a suspect, and their identification of the two was responsible for the conviction of the suspect. The photographs published in the case study should make everyone grateful for the experienced eyes of experts, because those photos did not reach out and shake this author with their obvious similarities.

The foundation for the use of reflective ultraviolet photography in circumstances of deep muscle bruising is based on the migration of melanin to the wound's perimeters beneath the skin during the healing process. Melanin absorbs UV light, and even if the wound can no longer be seen in normal lighting, patterned wounds or injuries can sometimes be detected with UV light long after the initial injury. Another article published by the same authors relates to the observation of subcutaneous wound patterns 7 months after the original incident. Apparently, the optimal time period for the visualization of deep muscle bruising with reflective ultraviolet photography is between 3 weeks and 5 months. Before that, visible light photography and IR photography should be attempted.

What camera settings are required for reflective ultraviolet photography? The biggest problem is that camera meters are not designed to provide proper exposures for UV lighting. Because different subject matter may react differently to UV light, no standard exposure combination can be recommended. Each time reflective ultraviolet photography is to be done, it is wisest to conduct a series of tests on the subject matter of interest. The two most common situations are provided.

1. Reflective ultraviolet photography using a "black light" as the light source:
   - Use an 18A filter over the lens.
   - Use ISO 400 film "pushed" to ISO 3200. That means telling the darkroom technician that the film needs to be "pushed" 3 stops. The film will be processed with the film in contact with the developing chemistry for a longer period. Different increments of time are used for a 1-stop "push," a 2-stop "push," and a 3-stop "push." Tests have shown that "pushing" ISO 400 film 3 stops yields better results than just using an ISO 3200 film processed normally.
   - An f/8 is the recommended aperture. Smaller apertures are recommended to help prevent focus errors because of the necessary focus-shift required with UV light. Using a smaller aperture increases the depth of field range.
   - The tests vary the shutter speeds to determine the best results. Results have varied between 1/8th second to several minutes.

2. Reflective ultraviolet photography using an electronic flash as the light source:
   - Use an 18A filter over the lens.
   - Use ISO 400 film.
   - Use the sync shutter speed.
   - The exposure tests are done by varying the f/stops through the full range of f/stops offered by the lens.

ABSORBED UV LIGHT IN THE VISIBLE LIGHT RANGE

As mentioned previously, UV light was one of the first search lights for non-blood body fluids, particularly semen at rape scenes. It is still an excellent choice for this task. But, it is also an ideal light to search for faint blood stains. Blood on many surfaces will be easy to see. Blood on substrates that are similar in color to blood can be very difficult to see. That is good for crime scene photographers. If the suspect does not see it, he may not try to wipe it away or wash it from the surface. Blood on multicolored substrates can also be hidden. In these cases, the use of a long wave UV to search for and find "invisible" bloody residues can have remarkable effects. Even though UV light is being used, the result can be seen in the visible light range. Figure 6.10 shows a piece of multicolored cloth as seen in visible light, which does not show any stain pattern. Light the same cloth with UV light, and the blood absorbs the light and appears to get darker, so it can easily be seen.

The literature indicates that an alternate light source emitting a blue light will also act much the same way. This author's experience is that UV light does a better job in visualizing faint blood stains.

Because this is visible light photography, normal exposures can be used: ISO 100 film, f/11, sync shutter speed, and the flash set for TTL/dedicated flash.

Figure 6.10
UV absorbed by blood.

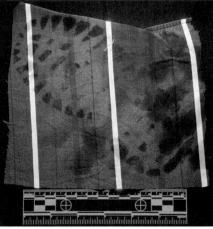

ULTRAVIOLET FLUORESCENCE PHOTOGRAPHY

Ultraviolet light can also stimulate fluorescence in the visible light range, which is one example of the Stokes shift. The UV light is absorbed by various substrates and is changed. Some of the light energy is absorbed by the substrate as heat energy. What is left is emitted as a longer wavelength of light that is also less intense. To view and photograph the fluorescence caused by the UV light, a UV-absorbing filter (camera filters called UV, 1A, 2A, or 2B) can be used. It would also be possible to see and photograph the fluorescence through a yellow filter.

Figure 6.11 shows UV light creating fluorescence when certain fibers and fingerprints are lit. The fibers came from a dryer lint trap and were not visible until lit with the UV light. The fingerprints were processed with a fluorescent powder and lit by UV light. Because this is visible light photography, daylight color film can be used. However, camera exposure meters are designed to provide proper exposures for normally lit subject matter. Fluorescence is definitely not a normal situation. Experience has shown that better exposures are obtained with exposure settings dimmer than the meter may indicate.

For instance, when the fibers in Figure 6.11 were first taken, the camera was set to aperture priority exposure mode with an f/11 with the ISO set to 400 on a digital camera. The resulting shutter speed was 15 seconds, and the image was very overexposed. After a bit of bracketing toward underexposures, Figure 6.11 was eventually captured at 4 seconds, about 2 stops darker than 15 seconds. When in doubt, bracket.

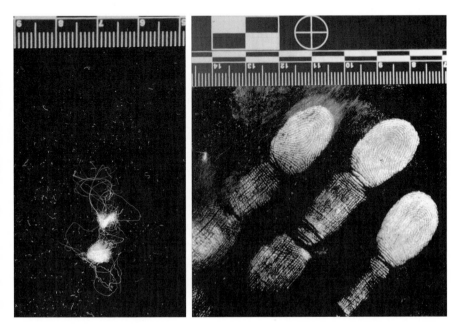

Figure 6.11
UV fluorescence of fibers and fingerprints.

It must be mentioned that any time you are using UV light, you should also be wearing protective goggles. The protective goggles can be UV goggles, which appear clear, or yellow goggles. Sunglasses not only eliminate the glare from the bright sun, they frequently block the UV light coming from the sun. Too much UV light cannot only cause sunburn to the skin, it can also damage your eyes.

Photographing fluorescence in the visible light range can be done with 100 ISO film, f/11, aperture priority exposure mode with an exposure compensation set to −2.

INFRARED LIGHT (IR) ON THE ELECTROMAGNETIC SPECTRUM

Photographic infrared range: The range on the electromagnetic spectrum between approximately 700 and 1100 nm.

The **photographic infrared range** is in the near IR part of the electromagnetic spectrum, 700 to 1100 nm.

The human eye and normal daylight films cannot sense infrared light, so it will be necessary to have a camera sensor that is sensitive to IR light. Infrared films are made, but the use of digital cameras, which are sensitive to IR light, are quickly eliminating the need to use IR film. Not all digital cameras are sensitive to IR light. Most digital cameras have filters over the digital sensor that filters out IR light. Digital cameras can have these filters removed to make the camera sensitive to IR light. This usually means that camera cannot be used for normal daylight photography.

One important benefit of digital cameras sensitive to IR light is that, because the camera is sensitive to IR light, the camera can also focus on IR light, and the otherwise required IR focus adjustment is no longer necessary. Digital cameras can also properly expose for IR light, eliminating much of the guesswork related to IR exposures when film was used.

Obtaining a digital camera sensitive to the IR range of the electromagnetic spectrum is much easier than you might think. Of course, some digital camera manufacturers will modify one of their current SLR cameras so they can be used for IR imaging. This will provide the highest resolution of digital image. This can be costly. Another point of view is to use a digital camera with IR capability as a crime scene search tool. In this situation, it is just necessary to be able to visualize the currently "invisible" evidence. One camera manufacturer makes camcorders and still point-and-shoot digital cameras with IR capability. Sony makes cameras with "Night shot" capability, which is near-IR capability. All that would be necessary to convert any of these "Night shot"-capable cameras to full IR capability is to obtain the normal IR filter, which is a Wrattan #87 filter. This filter blocks all visible light while transmitting IR light through it. Therefore, to the eye, it appears opaque.

Figure 6.12 shows a Sony 717, 5 megapixel, digital camera. It is also shown with a #87 filter over the lens. The key is to use any of the Sony cameras that have "Night shot" capability, also shown in Figure 6.12. This camera has taken many of the images that follow.

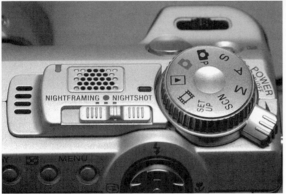

Figure 6.12
Sony 717 with night shot/IR capability.

Sony is not the only digital camera manufacturer making cameras with IR sensitivity. Canon also has made a digital SLR modified for IR work. See Figure 6.13.

Fuji has also recently (summer of 2006) announced a digital SLR specifically modified to be used in the UV/IR ranges.

Why is having the capability to immediately "see" the crime scene with IR light important? In the past, some agencies had the capability to capture IR images at

Figure 6.13

Canon digital camera modified for IR (Courtesy of Tom Beecher, Photografix, Richmond, Virginia).

crime scenes with IR film. This meant that these images would have to be processed before they could be seen. Only then would the investigator have the information revealed by the IR images. Now, with an inexpensive digital point-and-shoot digital camera, or an equally inexpensive digital camcorder, evidence not previously seen can now be inspected and reviewed for information that can provide immediate investigative leads or information that can be used as probable cause in obtaining a search warrant or may even be probable cause in making an arrest. This cannot be overemphasized.

What are the types of evidence IR can reveal at a crime scene? This usually falls into three categories:

- Ink differentiations
- Visualizing gunshot residue
- Visualizing the writing on burned documents

What are the effects of IR light on different substrates?

- The substance may absorb the IR light. If this occurs, the substance absorbing the IR light will remain dark, or if it currently is not dark, it will appear to darken.
- The substance may reflect the IR light. If this occurs, the substance reflecting the IR light will appear to lighten in color or tone. If it currently is black, it will appear to turn a lighter shade of gray, or it may appear to turn white.
- The substance may transmit the IR light. If this occurs, the substance transmitting the IR light may appear to disappear, revealing whatever is beneath it.
- The substance may convert the IR light. If this occurs, the substance converting the IR light will convert some of the energy present in the IR light into heat. The weaker energy remaining will then be emitted by the substance as light of a longer wavelength, which is also a weaker intensity. This light is fluorescent.

Regarding ink differentiations, the dyes and pigments in the ink will show these various effects, and even inks that appear a similar shade of black can show some very drastic changes.

Figures 6.14 and 6.15 show all these effects. In the top left of Figure 6.14, all four inks are being stimulated by visible light, and the result is being visualized in the visible light spectrum. All the inks appear to be dark. In the bottom left, two of the inks have become fluorescent in the IR range, and the fourth ink has gone transparent. In the bottom right, three of the same inks have reacted to the IR light by going fluorescent. Figure 6.15 shows that many inks become transparent when lit by IR light.

These effects become more interesting when they are associated with casework. Usually, it is just important to establish that some portion of a document

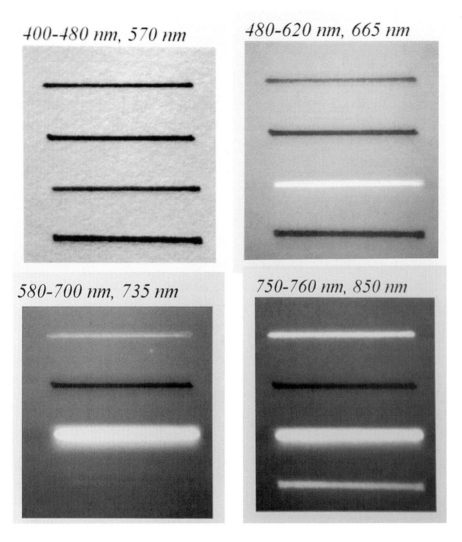

Figure 6.14

Inks reacting to various lights (Courtesy of Jeff Wilson, GWU MFS student).

was written at a different time with a different pen. If a letter, number, or word can be made to react to light differently, this has been established.

Figure 6.16 is an example of an altered check, where the amount of $100 had been altered to read $5100. If the inserted number "5" and the inserted word "fifty" can be shown to be different than the original ink, we will be successful. In this case, it has been done two different ways. In the IR range, both the word "fifty" and the number "5" are transparent to IR light, and they become invisible. In the visible light range, both fluoresce while the original writing remains the same. Had there been any allegation that the original document was written for $5100, either of these images would disprove any such claims. Figure 6.17 is an obliteration of information a subject did not want falling into the hands of the

Figure 6.15

Inks becoming transparent in IR.

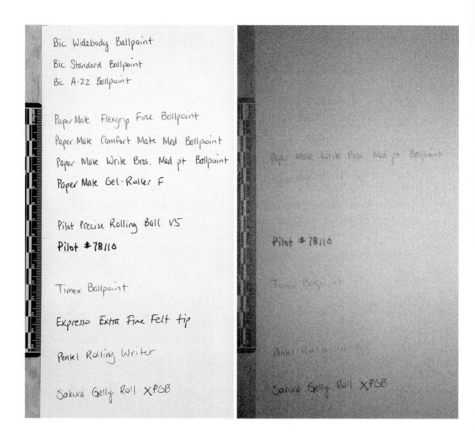

Figure 6.16

Ink fluorescing and transmitting IR.

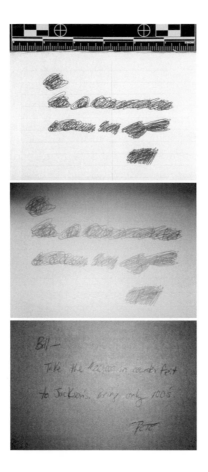

Figure 6.17

Obliterated note visualized by
IR.

law enforcement officials approaching with a search warrant. In this case, it was fortunate that the original note was written with an ink that absorbs IR light, and the obliteration was written with an ink that goes transparent to IR. The middle image of the three shows the Sony camera's near-IR ability used as is, without the # 87 filter. The original writing can be made out, but it is still a bit difficult to see. The bottom image shows the "Night shot" capability of the Sony camera with the # 87 filter. Impressive.

Actually, IR fluorescence can be created two ways: The more normal technique is to stimulate the inks with a blue/green light. Sometimes that may cause fluorescence in the IR range. More unusual, and less likely, is to stimulate the inks with true IR light and examine the inks in an area further down into the IR range. Figure 6.14 shows this result.

How is gunshot residue (GSR) affected by IR light? GSR will absorb IR light, so it will remain dark or turn a bit darker than it originally was, which is unimportant if the surface around the bullet hole is light colored, because GSR itself is dark, and would be easily seen in this case. If the clothing that has a bullet hole in it is very dark or black itself, it will be very difficult or impossible to see the GSR on that type of fabric. The GSR will be darkened by the IR light. Fortunately, however, many black dyes in fabrics reflect IR light. In this case, then, the fabric would lighten to a lighter shade of gray or even white, while the GSR would remain black. See Figure 6.18.

It would later be easy for the firearms examiner to determine the range of fire with test shots from the original weapon and with the original ammunition. However, at the crime scene on the night of the incident, this immediately visualized GSR pattern may contradict an alibi being offered by a suspect or a version of the incident being offered by a witness. This would be very important evidence to be aware of. It could help with interrogations and interviews. It could amount to sufficient probable cause to make an arrest. Or, being offered various accounts of what occurred at the crime scene, it could help you determine which version

Figure 6.18
IR and GSR.

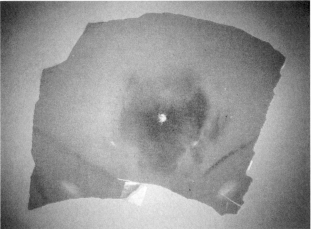

of the incident you will believe. If your agency works a substantial number of shooting incidents, having this immediate GSR visualizing capability is invaluable on the day the incident is first reported.

Figure 6.19 is the unanticipated result of a GSR transfer to a shirt. The incident in question was a shotgun shooting, where the suspect ran off to the rear of a residential neighborhood. A canine was called to the scene to begin a track of the suspect. After tracking just a few houses away, the canine "hit" on a dark blue jersey lying in the grass. We guessed the shirt had been used to wrap the sawed-off shotgun as the suspect fled. To confirm this suspicion, it was decided to examine the shirt with IR imaging to see if we were lucky enough to have the shape of a sawed-off shotgun transferred to the shirt. When this image was revealed, multiple jaws hit the ground at the same time. The subsequent investigation linked the suspect to another previous shooting with a .45-caliber semiautomatic pistol. Sometimes "blind" searches reveal strange types of evidence.

If a suspect does not obliterate an incriminating document with another pen and ink, another way to eliminate the evidence is to burn it. However, many burned documents are reduced to ash, and if this crumbled ash is the evidence you must deal with, not much can be done. Some papers, however, do not disintegrate into ash when burned. Some paper may just darken and turn black. If this happens, it is similar to the GSR on black fabric. The charred paper may reflect IR light and lighten or turn white. If the ink on the paper is an ink that absorbs IR light, the result will be ink that remains visible because it remains dark. Figure 6.20 is an example.

Determining the exposure and the proper focusing point for these infrared images was simplified because the camera was sensitive to IR light. Its auto-focus properly determined the focus. The camera's meter could determine a proper exposure for IR. Aperture priority exposure mode with an f/8 or f/11 produced acceptable results. Because all the visible light was filtered out with a #87 IR filter, a faster 400 ISO film was used.

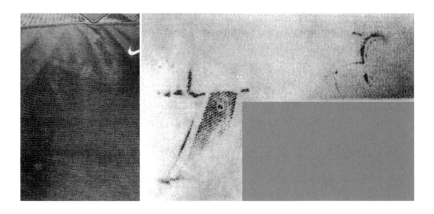

Figure 6.19

IR visualization of a pistol (Courtesy Dave Knoerlein, "Digital Dave").

Figure 6.20

IR visualizing burned writing (Courtesy of Heather Butters, GWU MFS student).

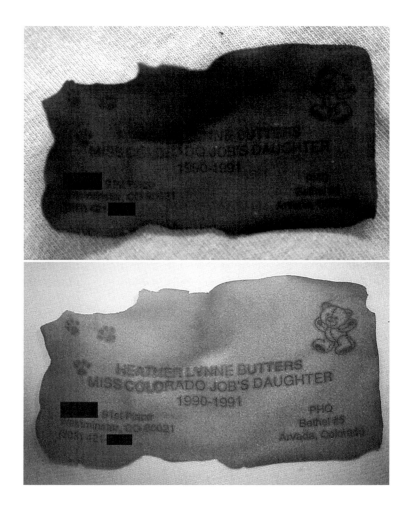

VISIBLE LIGHT FLUORESCENCE

It is possible to create fluorescence in the visible light spectrum several ways.

- UV light can cause fluorescence in the visible light range. To see or photograph the UV fluorescence, use a UV filter or a yellow filter.
- Emitting a strong monochromatic blue light, from either a laser, forensic light source or an alternate light source, will create fluorescence in the orange part of visible light. The best viewing and photography will be done when using an orange filter.
- Emitting a strong monochromatic green light, from an alternate light source, will create fluorescence in the red part of visible light. The best viewing and photography will be done when using a red filter.

Many true alternate light sources will offer more options than these. Some evidence likes to be fluoresced with one light and viewed through a particular filter. The more options you have and use, the more evidence will be found.

Again, it needs to be repeated that fluorescent evidence is not the normal scene that a camera's meter is designed to expose properly. With fluorescence, much of the field of view will be dark, and one item of evidence will be fluorescing. The camera's meter wants to expose the entire scene properly. If the aperture priority exposure mode is used, it is suggested to use an exposure compensation of −2 and bracket. Even when exposing with the manual exposure mode, this can be regarded as a Rule of Thumb.

Rule of Thumb 6.4: When photographing fluorescent evidence, the proper exposure is frequently 2 stops less than what the camera's exposure meter recommends. At a −2 exposure compensation, it would then be best to bracket. The camera's exposure meter is designed to provide proper exposures for "normal" scenes, reflecting 18% of the light that strikes them. The fluorescent situation is far from "normal."

In many situations, fluorescent photography will be necessary at crime scenes.

FLUORESCENT POWDERS AND CHEMICALS

Some of the most frequently encountered types of fluorescent evidence are fingerprints that have been enhanced and treated with fluorescent powders and chemicals.

Figure 6.21 shows a variety of fluorescent fingerprint powders, and one of the most common fluorescent dye stains used to treat super glued prints, Rhodamine 6G. Figure 6.22 is a collection of fluorescent fingerprints. Fingerprint residues are composed of some chemicals, like riboflavin, which are naturally fluorescent, so it is also possible to notice inherent fluorescence before any treatment. Most fingerprints, however, have to be treated with fluorescent fingerprint powders or dye stains to fluoresce.

VISUALIZING NON-BLOOD BODY FLUIDS

Another staple of crime scene searches with lasers and alternate light sources is the search for non-blood body fluids. Semen, saliva, urine, vaginal fluid, and perspiration can all be fluoresced with lasers and alternate light sources; this is most efficiently done with a blue stimulating light used with an orange filter. Figure 6.23 represents some of these fluorescing fluids.

FLUORESCENT FIBERS

No crime scene search would be complete without a search for fibers and trace evidence, many of which fluoresce. At times, we will search with the lights on,

Figure 6.21

Fluorescent fingerprint powders and Rhodamine 6G.

Figure 6.22

Fluorescent fingerprints.

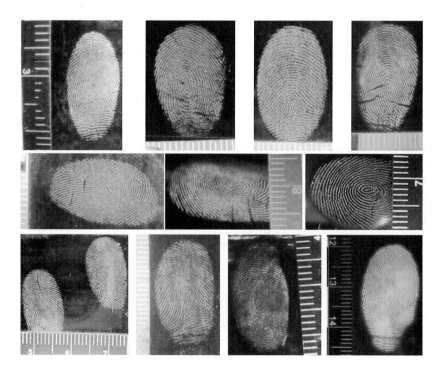

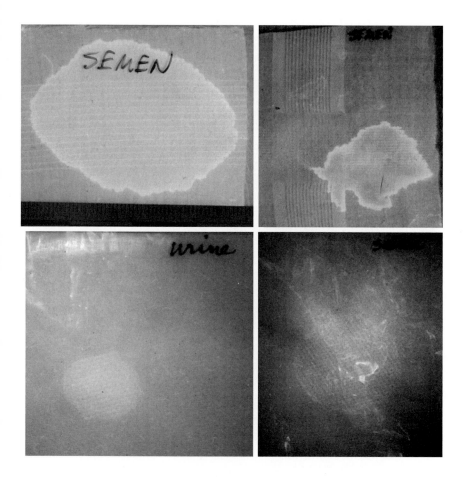

Figure 6.23
Fluorescing non-blood body fluids (Courtesy of Tom Beecher, Photografix, Richmond Virginia).

then with a flashlight, then with a UV light, and finally with a laser or alternate light source. Many times, different fibers and trace evidence will be found under each of these different lighting conditions. Figures 6.24 and 6.25 show fluorescent fibers.

DRUG RESIDUES

Some drugs and the fillers used to increase a drug's volume will fluoresce. Certainly not a positive test for drugs, noticing fluorescence at drug scenes can be helpful. If five guys are sitting around a table with a variety of pills and powders scattered about, it is possible these drugs came from one of their pockets. Hit all their pocket areas with a blue light and view the results with orange goggles on, and it is possible the origin of all the drugs may become apparent. Figure 6.26 shows both a fluorescent crack "sugar cookie" and a homemade pill.

Figure 6.24
Lint in natural light and while
fluorescing.

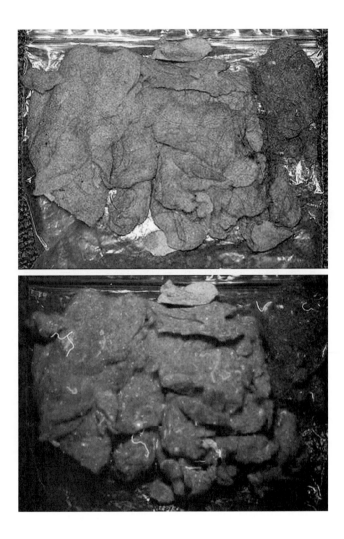

Figure 6.25
Fluorescent fibers.

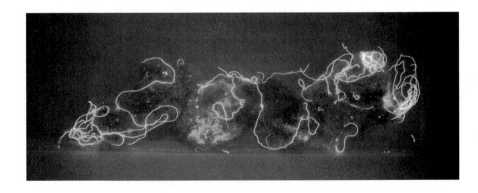

 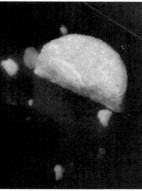

Figure 6.26

A fluorescent crack sugar cookie and pill (Courtesy of Tom Beecher, Photografix, Richmond Virginia).

BLOOD: LUMINOL, BLUESTAR, AND LEUCO FLUORESCEIN

Blood absorbs UV light and gets darker. When chemically treated, blood can also fluoresce. Commonly used to create this chemiluminescence is luminol, a reagent that reacts with the heme in the hemoglobin of blood. The literature suggests luminol is extremely sensitive to blood, able to react with blood in dilutions as weak as 1: 5,000,000. Figure 6.27 shows luminol easily fluorescing under various dilutions.

This makes luminol very useful when trying to detect areas where blood has been washed away in an attempt to hide a crime. Figures 6.28 and 6.29 show two unsuccessful attempts to hide blood by washing it away. A shoe was washed after making a bloody shoe print, but it easily fluoresced when sprayed with luminol. Also, a portion of a towel was stained with two bloody handprints and then washed in a washing machine. Although it appeared to be clean after being washed, luminol revealed faint handprints remaining on the fabric. Another method of hiding blood might be to paint over the stains. Regular household latex paints will not be successful in hiding blood. Figure 6.30 clearly shows fluorescence, where blood had been painted over by six layers of paint, after the application of luminol. Each square, after the top left square which is whole blood, represents an additional layer of paint over the blood.

The standard recommended exposure for luminol is 400 ISO film, f/5.6, and a 90-second shutter speed. Two problems exist when fluorescing blood with this exposure, both related to the 90-second exposure.

First, the 90-second exposure did permit the fluorescence to be photographed, but the rest of the scene appeared to be severely underexposed. Originally, to overcome this known limitation to luminol photography, one photograph was also made while using an electronic flash. Then, the two images could be viewed at the same time so the orientation of the fluorescent image could be seen in its context. This need to take two images to allow the fluores-

Figure 6.27

Luminol fluorescing various dilutions of blood (Courtesy of Tahnee Nelson, GWU MFS Student).

Figure 6.28

Luminol fluorescing a "cleaned" shoe.

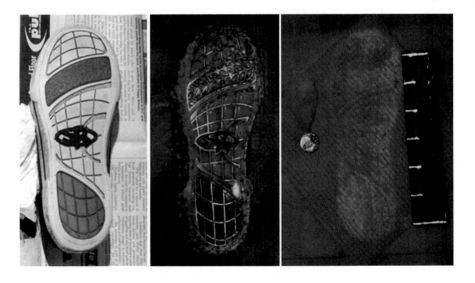

Figure 6.29

Luminol visualizing blood from machine-washed fabric (Courtesy of GWU MFS Student Christine Perletti).

cence to be seen in context has been eliminated by recent recommendations to introduce some form of bounce light during the 90-second exposure. For example, it has been recommended to take a Mag-light flashlight or Streamlight flashlight and turn it on and off as fast as you can while aiming it at the ceiling above the suspected bloody print. This was done approximately 80 seconds into the 90-second exposure. This faint bounce light enabled the surrounding area around the bloody fluorescence to also be captured in the same photo.

Figure 6.30

*Luminol visualizing blood
under multiple layers of paint.*

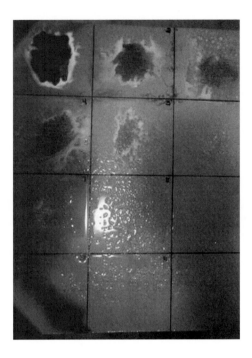

Figure 6.31 shows examples of luminol photography without (left) and with (right) a bounce light off the ceiling during the 90-second exposure. Now, a single image can show both the fluorescence of blood stains and the area in which it is found.

If the exposure takes 90 seconds, more liquid will have to be sprayed over that time frame to sustain the fluorescence. That may be OK with some porous surfaces, but if the surface is nonporous, such an accumulation of liquid may cause

Figure 6.31

*Traditional luminol photo and
bounce light photograph
(Courtesy of Carla Paintner
and Karin Athanas, GWU
MFS Students).*

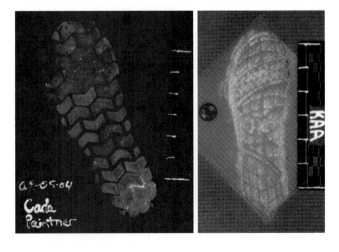

the pattern to run and become indistinct. Worse yet, if the surface is both non-porous and vertical, like a wall, the bloody pattern will run. Three solutions have been offered to counter this problem.

- Shorten the shutter speed, to reduce the volume of liquid used.
- Thicken the luminol reagent.
- Fix/stabilize the blood stain.

To shorten the shutter speed, the concept of reciprocity was used. If one exposure variable is adjusted to let in less light, another variable must be adjusted to let in more light. Because exposures revolve around halves and doubles, it would be possible to cut the shutter speed in half if the light attributable to the aperture was doubled, and so on. These, then, are reciprocal exposures:

- 90 seconds and f/5.6
- 45 seconds and f/4
- 22 seconds and f/2.8
- 11 seconds and f/2

These will certainly reduce the length of time the luminol has to be sprayed and the volume of liquid that will accumulate.

A second thought, offered by west coast criminalist Robert Cheeseman, has been to thicken the luminol with normal food thickeners like xanthum gum or guar gum. These can make the luminol so thick it will not run on vertical surfaces. The problem is that the solution becomes so thick it has to be sprayed by a high-pressure sprayer rather than just a hand pump sprayer. But, it works. One enterprising student applied the thickened luminol with a paint roller to a bloody handprint in their shower stall. The luminol nicely fluoresced the bloody handprint, and the pattern never ran down the wall.

A third solution to the problem of runny prints on nonporous surfaces is to "fix" the blood with a standard blood fixer like sulfosalicylic acid, which is the common blood fixer used in products like Amido Black and Hungarian Red. The obvious problem is that until the faint blood stain has been made to fluoresce, we cannot see it, so where should the sulfosalicylic acid be sprayed? Research is currently being done to determine the best way to combine luminol and sulfosalicylic acid into one reagent, which should solve all the problems.

A few alternatives to luminol exist. Both BlueStar and leuco fluorescein are relatively new chemistries now used to cause faint blood stains to fluoresce. Their advantage is that they create a brighter fluorescence than luminol and the fluorescence lasts longer than the brief luminol fluorescence. Their disadvantage is they also require an alternate light source to stimulate the fluorescence; the fluorescence is not purely chemical. The advantage of having a brighter, longer-lasting fluorescence is a great one. Leuco fluorescein can cause the same item previousl treated with the reagent to fluoresce several months later when lit with an alternate light source again. Figure 6.32 shows the fluorescence caused by leuco fluorescein.

FLUORESCENCE OF BONE

Although this need does not arise frequently, it should be known that bone fragments can fluoresce, and if the issue is a search of a wide area for bone fragments, the use of an alternate light source will simplify the search. Figure 6.33 shows a fluorescent bone fragment.

Because of latent fingerprints, we are all aware evidence may be present at a crime scene that is currently not visible, which is not just limited to fingerprints. Many other types of evidence are currently not visible but can be made visible by use of the proper techniques. With the use of lighting other than the visible light range, more of what is available to be collected as evidence can be seen and collected. We just have to learn to "see" in different ways.

Figure 6.32

Leuco fluorescein visualizing faint bloody shoe print (Courtesy of Tom Beecher, Photografix, Richmond Virginia).

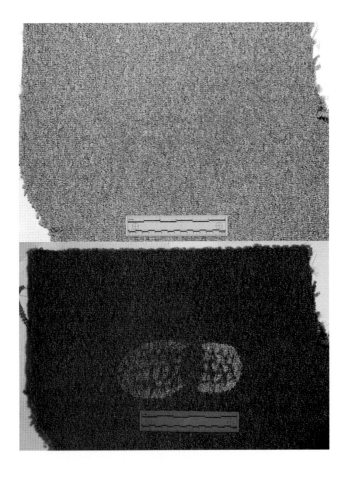

Figure 6.33

Fluorescent bone fragment (Courtesy of Tom Beecher, Photografix, Richmond Virginia).

SUMMARY

The light that is the foundation of photography is related to the electromagnetic spectrum (EMS). This spectrum depicts light as a wave of radiation, which has a peak height (amplitude), a distance between peaks (wavelength), and the number of complete waves passing a point each second (frequency). When light strikes various surfaces, it can interact with those surfaces differently, and crime scene photographers can use these differences to visualize otherwise "invisible" evidence.

UV light can be used to search for faint blood stains, tracks, or trails. UV light can also cause non-blood body fluids, some fibers, and some fingerprints to fluoresce. The evidence can then be photographed while fluorescing and collected.

IR light can also react differently to different substances and substrates, making it possible to visualize various types of evidence not noticed in the visible light range. Inks with different chemical makeup can frequently be distinguished, and gunshot residues on dark fabrics can more easily be seen. Printing or writing on charred documents can sometimes be visualized by use of IR lighting techniques. Some inks can be made to fluoresce in the IR range when stimulated with blue-green light.

Many types of evidence can be made to fluoresce in the visible light range of the electromagnetic spectrum, enabling the evidence to be seen, photographed, and then collected.

DISCUSSION QUESTIONS

1. Explain the relative location on the Electromagnetic Spectrum of UV light, Visible light, and IR light.
2. How is fluorescence related to the Stokes Shift?
3. Light can behave differently when striking various surfaces and substrates. What are these different effects?
4. What are some positive and negative aspects of using UV light at crime scenes to locate and visualize evidence?

5. Discuss some of the advantages of having IR imaging capabilities at major crime scenes.

6. Discuss some of the types of evidence that can be visualized with the use of Alternate Light Sources.

7. Using Alternate Light Sources, UV light, and IR light at crime scene also requires using the appropriate filters or goggles as well. Which filters or goggles are required for these different light sources?

EXERCISES

Obviously, having access to an Alternate Light Source, UV light, and IR light, along with the appropriate filters and goggles is essential for the following exercises. Protective eye wear should be used at all times when working with these light sources. Working with blood and body fluids poses its own health risks. Proper precautions should be exercised when working with blood and body fluids.

1. Photograph a Close-up of a fluorescing fingerprint. Dust a fingerprint with fluorescent powder and a feather duster. Be careful not to over-powder the print. With the camera on a tripod, use Aperture Priority, f/11, the Alternate Light Source emitting blue light, and an orange filter on the lens. Include a partial fluorescent scale with your initials on it. Bracket with a −1 and −2.

2. Photograph a Close-up of a fluorescing fiber. You may be successful finding a fluorescent fiber by collecting lint from a dryer and examining it with a UV or ALS. With the camera on a tripod, use Aperture Priority, f/11, the Alternate Light Source emitting blue light, and an orange filter on the lens. Alternatively, a long-wave UV light and yellow filter can be used. Include a partial fluorescent scale with your initials on it. Bracket with a −1 and −2.

3. Photograph a fluorescing non-blood body fluid. This may be semen or saliva. Not all saliva stains fluoresce readily, so several may have to be checked. Semen readily fluoresces. Use the camera on a tripod, Aperture Priority, f/11, the Alternate Light Source emitting blue light, and an orange filter on the lens. Alternatively, a long-wave UV light and yellow filter can be used. Include a partial fluorescent scale with your initials on it. Bracket with a −1 and −2.

4. Photography a bloody shoe print or handprint with Luminol. This can be animal blood. Bleach may be used as a blood substitute. Photograph in complete darkness, with a flashlight briefly bounced off the ceiling during the exposure. A phosphorescent scale with phosphorescent initials can be included, as well as a penny and positive blood control. Exposure: Use the camera on a tripod, ISO 400, F-5.6, 90 second SS (45 seconds for bleach as a blood substitute); or a reciprocal exposure.

5. View various black inks in IR light through an IR filter (#87), to determine those that absorb IR and remain dark, and those that transmit IR and go invisible.

With an ink that absorbs IR light, write your name. With an ink that transmits IR light, obliterate your name by scribbling over it until it cannot be read in visible light. Photograph the obliteration in ambient light, or with a flash.

6. Using a digital camera with IR sensitivity, and an IR filter on the lens, project IR light on the obliteration, and photograph your name "through" the obliteration. Include a labeled scale, ensuring the ink you use on the label remains visible under IR light.

7. Photograph a piece of black cloth that has been shot at between a 3″ and 8″ range. Use ambient light or a flash.

8. Photograph the same piece of cloth, but image with IR light, and an IR filter over the lens. Include a labeled scale, ensuring the ink you use remains visible under IR light.

9. Create a diluted animal blood stain on a fabric with a busy background pattern. Photograph in ambient light or with an electronic flash.

10. Using a long-wave UV light, the stain should absorb the UV light and appear to turn darker when viewed in the visible light range without any filters. Photograph with the camera on a tripod, Aperture Priority mode, f/11.

11. Compare your images with those in this chapter and with other images on the companion website (http://books.elsevier.com/companions/9780123693839), referring particularly to the following image folders: Fluorescent, Infrared, Luminol.

FURTHER READING

"Applied Infrared Photography." (1981). Kodak Publication No. M-28. Eastman Kodak Company, Rochester, NY.

Baldwin, H. B. "Photographic Techniques for the Laser or Alternate Light Source." http://police2.ucr.edu/laser.html.

Chowdhry, R., Gupta, S. K., and Bami, H. L. (1973). Ink differentiation with infrared techniques. *J. Forensic Sci.* **18,** 418–433.

Cochran, P. "Use of Reflective Ultraviolet Photography to Photo-Document Bruising to Children." http://www.crime-scene-investigator.net/uvchildphoto.html.

Craig, E. A., and Vezaro, N. (1998). Use of an alternate light source to locate bone and tooth fragments. *J. Forensic Sci.* **48,** 451–457.

Dalrymple, B. E. (1983). Visible and infrared luminescence in documents: Excitation by laser. *J. Forensic Sci.* **28,** 692–696.

David, T. J., and Sobel, M. N. (1994). Recapturing a five-month-old bite mark by means of reflective ultraviolet photography. *J. Forensic Sci.* **39,** 1560–1567.

Davidhazy, A. "Ultraviolet and Infrared Photography Summarized." Eastman Kodak Company, Rochester, NY: http://www.rit.edu/~andpph/text-infrared-ultraviolet.html.

Hilton, O. (1981). New dimensions in infrared luminescence photography. *J. Forensic Sci.* **26,** 319–324.

Keith, L. V., and Runion, W. (1998). Short-wave UV imaging casework applications. *J. Forensic Ident.* **48,** 563–566.

Krauss, T. C. (1985). The forensic science use of reflective ultraviolet photography. *J. Forensic Sci.* **30,** 262–268.

Mitchel, E. N. (1984). "Photographic Science." John Wiley & Sons, New York, NY.

Moon, H. W. (1984). Identification of wrinkled and charred counterfeit currency offset printing plate by infrared examination. *J. Forensic Sci.* **29,** 644–650.

Nieuwenhuis, G. (1991). Lens focus shift required for reflective ultraviolet and infrared photography. *J. Biol. Photog.* **59,** 17–20.

Rademacher, G. P. (1988). An accurate method of focusing for infrared photography. *J. Forensic Sci.* **33,** 764–766.

"Ultraviolet & Fluorescence Photography." (1968). A Kodak Technical Publication, M-27. Eastman Kodak Company, Rochester, NY.

West, M. H., Barsley, R. E., Hall, J. E., Hayne, S., and Cimrmancic, M. (1992). The detection and documentation of trace wound patterns by use of an alternate light source. *J. Forensic Sci.* **37,** 1480–1488.

Williams, A. R., and Williams, G. F. (1993). The invisible image? Part 1: Introduction and reflected ultraviolet techniques. *J. Biol. Photog.* **61,** 115–132.

Williams, A. R., and Williams, G. F. (1994). The invisible image? Part 3: Reflected infrared photography. *J. Biol. Photog.* **62,** 51–68.

Zweidinger, R. A. (1994). Use of alternative light source illumination in bite mark photography. *J. Forensic Sci.* **39,** 815–823.

CHAPTER 7

PHOTOGRAMMETRY

LEARNING OBJECTIVES

On completion of this chapter, you will be able to . . .

1. Explain the meaning of the general term photogrammetry.
2. Explain the methods used to extend a perspective grid over the evidence within a crime scene. Then, once the grid has been extended over the crime scene, explain how the grid is reduced around each item of evidence until the precise location of the evidence can be determined in relation to the grid itself.
3. Explain how the perspective grid itself is related to the crime scene.
4. Describe the perspective disc variation to the perspective grid technique.
5. Explain when neither grids nor discs are necessary to do photogrammetry. Explain when the natural grid variation will be able to be used to determine the precise measurements of evidence.
6. Explain the reverse projection variation.
7. Explain the basics of Rhino photogrammetry.

KEY TERMS

Natural grid   Reverse projection
Perspective grid   Rhino photogrammetry
Photogrammetry

INTRODUCTION TO PHOTOGRAMMETRY

To adequately document a crime scene, three activities are normally required. Of course, the crime scene must be photographed. The crime scene and the evidence within it also have to be measured, and sketches and finished crime scene diagrams have to be drawn. This can be done by hand, the old-fashioned way, or a variety of computer aided diagramming (CAD) software programs can be used. Finally, taking meticulous notes is required. All three crime scene documentation methods complement each other; with all three, every aspect of the crime scene has been fully documented. No one method of crime scene documentation can suffice. This is a very time-consuming, but necessary, process.

Would it not be nice if one of these "necessary" steps could be eliminated? Can a way to streamline the process be found? If nothing is lost along the way, an increase in efficiency is always desirable, is it not? This is a pretty strong suggestion: suggesting eliminating the process of taking crime scene measurements. That is not this author's intent. Confident that the techniques in this chapter produce measurements just as accurate as can be derived by use of traditional measuring of the crime scene with a tape measure, this author still regards photogrammetry as a supplement to the baseline coordination and triangulation measurements techniques.

This author has taught photogrammetry at the Virginia Forensic Science Academy, the Northern Virginia Criminal Justice Training Academy, and on campus at The George Washington University. During their first exposure to photogrammetry, all three types of student are usually quibbling about fractions of an inch when the known answers are provided to the problems they have worked. To be sure, there are some very sophisticated variations of photogrammetry: aerial and satellite photogrammetry applications to mapping and image analysis. The FBI uses complex software to extrapolate the height of bank robbers and the lengths of sawed-off shotguns used in the commission of crimes. On the other end of the continuum, however, there are some very basic applications of photogrammetry that can easily be understood and applied to crimes scenes. This marriage of crime scene photography and the need to acquire accurate measurements of the evidence within crime scenes is the essence of photogrammetry.

Photogrammetry: Refers to the activities of "(1) photographing an object; (2) measuring the image of the object on the processed photograph; and (3) reducing the measurements to some form such as a topographic map" or crime scene diagram.

The term photogrammetry refers to the activities of "(1) photographing an object; (2) measuring the image of the object on the processed photograph; and (3) reducing the measurements to some form such as a topographic map"* or a scale crime scene diagram.

More simply, we might say that photogrammetry is "the science or art of obtaining reliable measurements by means of photographs."† Photogrammetry is the method, for example, by which the moon has been mapped. Surveyors did not scout the surface of the moon. The measurements of the moon's surface were derived by the analysis of photographs of the moon.

---

*Moffett, F. H., and Mikhail, E. M. (1980). "Photogrammetry." 3rd Ed. p. 1. Harper and Row, New York.
†Slama, C. C., Editor. (1980). "Manual of Photogrammetry." 4th Ed. p. 1. American Society of Photogrammetry, Falls Church, VA.

We are all familiar with a checker board or a chess board: a neat 8 square × 8 square combination of rows and columns. Many floors are made up of $1' \times 1'$ floor tiles. If a crime were to occur on a floor with these floor tiles under all the evidence, determining the location of all the evidence would be extremely easy. If the crime scene currently being worked is not conveniently located on such a floor, would it not be nice if we could superimpose a uniform checkerboard grid system, like $1' \times 1'$ squares, over the entire crime scene? If this can be done, extrapolating the measurements of all the evidence within the crime scene would be vastly simplified. This is what photogrammetry is all about. "Photographs can be used to make up for a lack of suitable measurements . . . , to supplement inadequate measurements, and, in some cases, substitute for measurements."[*]

Here are just a few circumstances when the proper use of photogrammetry could be used as a back-up to traditional measurements or even as a substitute for measurements at a crime or accident scene. You may be able to think of others.

- Evidence at the scene is so numerous or so intricate that locating all the items of evidence within the scene would require an inordinate number of measurements.[†] Consider a complex accident scene, with skid and scuff marks, scattered debris, liquid run-off, gouge marks, vehicle damage, sight-line questions, etc.
- When the on-scene investigator or the evidence technician has no way of knowing what may later be important to the reconstructionist subsequently called in to analyze the scene.[‡] Without advanced training in accident reconstruction, an evidence technician may not know just what is important for the reconstruction of the accident.
- With unfavorable weather conditions, traditional methods of measuring items of evidence may not be possible. Evidence may be blown away by high winds, washed away by heavy rains, or covered by a snowstorm. Frozen surfaces may make walking around the scene difficult or impossible.[§] Yes, it would be preferable to pull out the tape measure and do it all the old-fashioned way, but sometimes the evidence just might not wait for this process.
- When the investigator must work alone or without the normally available equipment. Are we not always blessed with multiple assistants to help with every request, and we certainly have all the equipment available on the TV show CSI?

[*]Baker, J. S., and Fricke, L. B. (1986). "The Traffic-Accident Investigation Manual." pp. 30–33. Northwestern University Traffic Institute, Evanston, IL.
[†]Baker, J. S., and Fricke, L. B. (1986). "The Traffic-Accident Investigation Manual." pp. 30–33. Northwestern University Traffic Institute, Evanston, IL.
[‡]Repeating history by 'reverse projection' reconstruction. (1988). *Law Enforcement Technol.* **May/June,** 26.
[§]"Perspective Grid for Photographic Mapping of Evidence." 2nd Ed. (1983). p. 2. The Traffic Institute, Northwestern University, Evanston IL.

TV is entertainment; real world crime and accident scenes are often quite a challenge.[1]

- Every outdoor crime scene and every accident scene can easily be roped off, correct? At the peak of rush hour, diverting traffic from the main road, for as long as the investigation takes, is always easily possible, correct? One memory relates to this. A police officer's cruiser was T-boned by a car that had run a stop sign. The collision forced the cruiser across double yellow lines, where it then also struck an oncoming car. The officer was badly injured and later retired because of his injuries. Wanting to do our best for a brother officer, we wanted to do the best job we could with the accident investigation and the documentation of the accident scene. One problem: it was rush hour on our most heavily traveled road. Our Chief of Police drove by and told us we had 45 minutes to clear the scene. Traditional measuring methods went out the window.

- When, for whatever the reasons, the investigator must hurry, or does not have the time to take traditional measurements. We have all been working an important crime scene, and have just begun, when an even more urgent call comes out, and no one else is available. Street cops become experts at shortcuts. Sometimes, the "normal" way just is not a choice.

- There are many "minor" crime and accident scenes that do not seem to call for the full battery of what is routinely done at "major" scenes. We routinely apply the amount of effort at scenes as they deserve. Time management: every scene is not a triple homicide, and manpower and effort are usually appropriate for the occasion. Unfortunately, some of these "minor" scenes later turn out to be "major" scenes. Unexpectedly, several days or weeks later, someone dies, and what was originally just an assault and battery is now a homicide. Whenever necessary, you can go back and take the photographs and measurements and notes that would have been done had everyone known it was a homicide the night of the original call, correct? With a knowledge of photogrammetry, sometimes you can go back, and with the original photographs, you can get more information out of the photographs.

- "Insurance shots." Knowing the possibilities of photogrammetry, it is advisable to take one or two more photographs "just in case." Then, if it becomes necessary later, these photographs can be used to gather more information than you intended to collect initially. We are always being told we cannot go back again and re-do many of the things we had wished we had done. With photogrammetry techniques available to us, this attitude may have to be reevaluated. Sometimes, we can go back again; we can use the original photographs to glean more information than we have been used to having.

This author is aware there are some very sophisticated measurement systems available to some departments. Total Station is one of these crime scene and

---

[1]Baker, J. S., and Fricke, L. B. (1986). "The Traffic-Accident Investigation Manual." pp. 30–33. Northwestern University Traffic Institute, Evanston, IL.

accident scene measurement systems. Laser sighting devices are married to computer crime scene diagramming software, and 3D computer-generated diagrams are the result. This is a fantastic tool. However, many agencies may not have access to this kind of equipment. Between hand-drawn crime scene diagrams and the Total Stations systems available to some agencies, photogrammetry can fill a very big void.

PERSPECTIVE GRID PHOTOGRAMMETRY

We have already been introduced to the concept of a scale being an essential item of equipment when taking close-up photographs. A **perspective grid** can be considered a scale for an overall photograph.

In overall photographs, much of the entire scene can often be seen in one photograph. Rather than just give the viewer an impression of the layout of the entire scene, another photograph can be taken from the same position, but this time a perspective grid can be included in the scene. Then, with the grid in the scene, all items of evidence in the photograph with the grid can ultimately be linked to the grid. How accurately linked to the grid? As accurately as if each item of evidence had been measured traditionally with a tape measure.

There is not much required of a perspective grid; minimally, it is just a rectangle of known dimensions. However, it is often a $1' \times 1'$ floor tile. Better yet, put four $1' \times 1'$ floor tiles together, and you have a $2' \times 2'$ perspective grid.

Figure 7.1 is this author's set of four well-loved floor tiles used as a perspective grid. My first perspective grid was made of cardboard. It was quickly discovered that cardboard warps in the rain and blows away in the wind. Vinyl floor tiles lay flat, laugh at the rain, and remain in place during windy conditions. When selecting floor tiles, tiles that are nonglossy are preferred, because having glare on any floor tile will make it harder to see the lines on the grid. With larger scenes, all four $1' \times 1'$ tiles are used together. With smaller scenes, like evidence on a bed, a single $1' \times 1'$ floor tile can be used. Figure 7.2 is such a $1' \times 1'$ grid. It was a 'thank you' gift from several officers grateful to have learned how to use photogrammetry.

When the perspective grid is photographed at the crime scene, it is composed so that it is low and centered in a photograph, with evidence beyond it and to the left and right of the grid. This can be regarded as a Rule of Thumb.

Rule of Thumb 7.1: Where the grid is placed within the scene is up to us. Position it so that as many items of evidence as possible can be composed in a single photograph with the grid. If all items of evidence cannot be composed in one photograph with the grid because furniture or other obstacles block the view of some of the evidence when viewed from one position, it will be possible to reposition the grid so that the remaining items of evidence can be linked to the grid in a second or third photograph. When composing these photograph(s), try to arrange the grid so its bottom edge runs parallel to the bot-

Perspective grid: This can be considered the scale of an overall photograph. With the grid in the scene, all items of evidence in the photograph with the grid can ultimately be linked to the grid as accurately as if each item of evidence had been measured traditionally with a tape measure.

Figure 7.1
Four separate 1′ × 1′ floor tiles.

Figure 7.2
1′ × 1′ Perspective grid (Courtesy of Metropolitan Washington Airports Authority PD).

tom of the composed photograph and the grid is close to the middle of the bottom of the image.

Figure 7.3 shows a 1′ × 1′ perspective grid on a bed. It is, as suggested, low and centered in the composition. This image was taken to relate all the items on the bed to the body of a homicide victim. This was originally taken as a "back-up" photograph. All the evidence was going to be measured traditionally. As the items of evidence were photographed, measured, and removed from the bed, a supervisor came into the room and asked a question. The crime scene investigator, this author, stopped to answer the question. Returning to the evidence collection, the next item of evidence was picked up. It was then realized that this particular item of evidence had not been measured yet. After an initial panic attack swept over the crime scene investigator, a sigh of relief was uttered when it was remembered that Figure 7.1 had just previously been taken. If this author had not been convinced of the value of photogrammetry before then, he has been ever since.

Rather than explain the concepts of perspective grid photogrammetry as abstract theory, it has been decided it would be best to walk the reader through a real practical exercise.

Figure 7.4 is an image with a perspective grid in it. It is also included on the companion website that accompanies this text. Feel free to print it out onto a full

Figure 7.3
Perspective grid in a homicide scene photograph (Courtesy of the Arlington County Police Department, Arlington, VA).

sheet of paper and work through the exercise as you are walked through the process. This image has three items of "evidence" in view: a set of keys, a pair of sunglasses, and a purse. Each will be linked to the grid.

MARK THE EVIDENCE

When evidence is normally measured at crime scenes, the center of mass is usually measured. Working from a perspective view makes this difficult or impossible at times. When viewing the evidence in a photograph, the center of mass can often not be determined. If the center of mass cannot easily be determined, rather than guess where it is and risk haggling over this issue with a defense attorney, a new method of measuring the evidence must be developed. From the perspective point of view, it is easy to see the center of an edge of the evidence closest to the viewer of the photograph. Select the center of one edge of the evidence where it

Figure 7.4
Practical exercise image.

comes into contact with the surface it is on, and place a dot there. That dot will now represent the evidence. In Figure 7.4, the purse has been marked by this method. When later locating the purse in the scale crime scene diagram, just remember the point in the diagram is not the center of mass, which we are used to measuring, it is one side of the evidence.

If the center of mass can still be determined from a perspective viewpoint, place a dot at this location. This dot will now represent the evidence. In our example, both the keys and sunglasses can be marked this way.

GRID EXTENSION

We now begin the process of taking the perspective grid and extending it over the evidence within the scene. We will put a checkerboard of known dimensions over the scene to make it easy to determine the location of the evidence. To make this easier for the novice to photogrammetry, a sheet of paper was placed over the original Figure 7.4, with cutouts in the area of the perspective grid and the dots representing each item of evidence. This is not necessary in real cases, but when teaching the perspective grid technique for the first time, it simplifies things quite a bit. Many lines will have to be drawn out over the crime scene image, and it will be easier to make sense of the lines if they are on a clean white background. Figure 7.5 shows the crime scene photograph covered with paper that has cutouts for the grid and evidence.

Because many lines will be drawn to the left, right, and above the original photograph, additional paper is taped to the original image. Figure 7.6 shows additional paper taped to the original photograph.

Next, extend the four sides of the original 2′ × 2′ grid. It is important to use care when doing this, because any inaccuracies in extending the grid sides will result in

Figure 7.5

Covered with paper with cutouts for the grid and evidence.

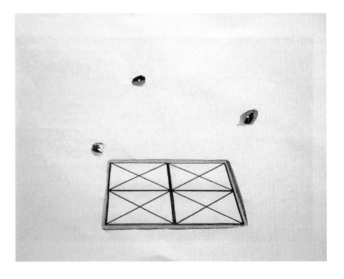

Figure 7.6

Additional paper taped to the original photograph.

Figure 7.7

The four sides of the grid extended.

progressive inaccuracies in the eventual measurements that are determined. The black tape on the edges of the grid is there to help you locate the outside edges of the grid. Do not extend lines from the inside edge of the black tape or the middle of the black tape. Make sure to extend the edge of the grid or the outside edge of the black tape. Where the left and right sides of the grid meet can be labeled the Y-intersecting point, because this will eventually be the top of the Y-line of an (x,y) coordinate system. The Y-intersecting point has also been called the 'Vanishing Point' by some authors writing about photogrammetry. The lines extending the top and bottom of the original grid can now be referred to as horizontal parallel lines. The lines extending the left and right sides of the original grid can now be referred to as vertical parallel lines, although, in perspective, they do not appear parallel. The line that is the extension of the base of the original grid can now be considered the X-line of the (x,y) coordinate system. Figure 7.7 shows the four sides of the grid extended to the left, right, and above the original image.

Measure the distance of the base of the original 2′ × 2′. Be as accurate as possible. If your scale has them, measure the base of the original grid to an accuracy of 1/32nds of an inch. After determining this distance from the lower left and right corners of the grid, measure out this same distance to the left and right of the grid along the baseline. Place a hash mark on the baseline at these two locations. Measure one more increment of the grid's base length to the left and right of the first two hash marks and place two more hash marks on the baseline. You should now have four hash marks on the baseline. Including the original grid's length, there will now be five lengths on the baseline, all the length of the base of the original grid. Figure 7.8 shows this.

Connect each of the four hash marks to the Y-intersecting point, or Vanishing Point, which was determined by extending the left and right sides of the original 2′ × 2′ grid. Notice there are now five 2′ × 2′ squares: the original grid and two to the left and two to the right. The original grid must be extended until all the evidence falls within a 2′ × 2′. If any item of evidence still lies outside the "teepee" drawn so far, it will be necessary to extend the original grid's base length to the left or right on the baseline, make a new hash mark, and connect that hash mark to the Y-intersecting point. Figure 7.9 shows the "teepee" formed by connecting the four hash marks with the Y-intersecting point.

Cross the corners of the original 2′ × 2′ grid and extend these lines until they reach the sides of the "teepee." This will allow the formation of 2′ × 2′ squares above the original row of five 2′ × 2′ squares. Figure 7.10 shows the original grid's corners crossed and extended.

Note where these diagonal lines intersect the vertical parallel lines coming from the first two hash marks on the baseline from the original grid. These two intersecting points can now be connected with a new horizontal parallel line. When this is done, note there are now 10 2′ × 2′ squares. If all the evidence is not

Figure 7.8

The grid base length duplicated four times.

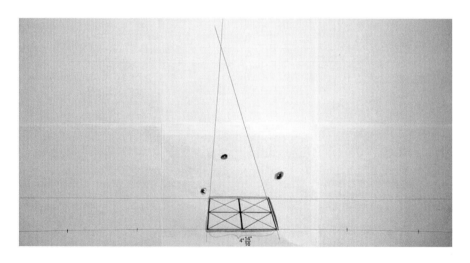

$4''\frac{14}{16}$

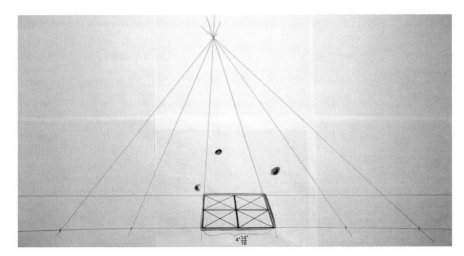

Figure 7.9

Hash marks connected to the Y-point.

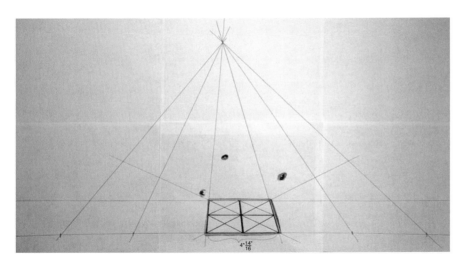

Figure 7.10

Original grid corners crossed and extended.

enclosed inside a 2′ × 2′ square, more horizontal parallel lines will have to be added. Figure 7.11 shows this new horizontal parallel line added.

Note where the diagonals intersect the vertical parallel lines coming from the second two hash marks on the baseline. These two intersecting points can now be connected with a new horizontal parallel line. When this is done, note there are now fifteen 2′ × 2′ squares. If all the evidence is not enclosed inside a 2′ × 2′ square, more horizontal parallel lines will have to be added. Figure 7.12 shows this new horizontal parallel line added.

Two of the items of evidence are now enclosed in a 2′ × 2′ square, but the grid has to be extended to the left, right, and above the original grid until every item of evidence is within a 2′ × 2′ square. To extend the grid system higher, note the 2′ × 2′ square above the original grid, and cross its corners, extending the lines out to

Figure 7.11
Third horizontal parallel line added.

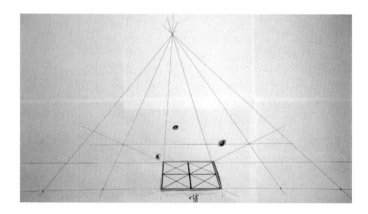

Figure 7.12
Fourth horizontal parallel line added.

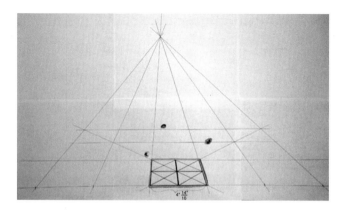

the edge of the "teepee." An accuracy checkpoint is now available. When aligning the opposite corners of the 2′ × 2′ square above the original grid, if you were to extend these diagonal lines downward (you do not have to actually draw the lines downward), you should note that the diagonal lines would intersect the first two hash marks on the baseline. If they would, you can be assured your work up to this point has been accurate. If extending the diagonal line downward would not intersect these first hash marks, this is a sign something has not been done properly, and any measurements derived will not be accurate. If the diagonal line extended downward is only about 1/32nd or 1/16th of an inch off meeting the hash marks, you may continue, but expect your measurements to be up to 1″ to 2″ or more off. If the inaccuracy is farther off, it would be best to begin the grid extension from the beginning. Figure 7.13 shows the corners of the grid above the original grid with its corners crossed and extended to the edges of the "teepee." Note that one diagonal has been extended down to the baseline, where it intersects one of the hash marks. That it intersects the hash mark is a confidence builder.

Where the diagonal lines intersect the sides of the "teepee," you can connect these intersecting points with a fifth horizontal parallel line (Figure 7.14).

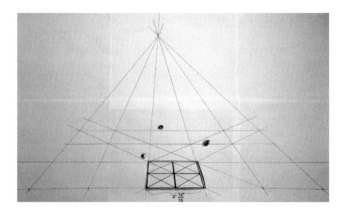

Figure 7.13
Second grid's corners crossed.

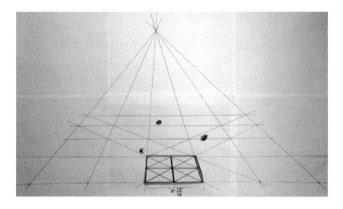

Figure 7.14
Fifth horizontal parallel line added.

If all the items of evidence are not now located within a 2′ × 2′ square, continue crossing the corners of 2′ × 2′'s above the original grid, and use these new diagonals to draw more horizontal parallel lines.

Once all the evidence is located within a 2′ × 2′ square, it is now necessary to label all the lines drawn. Indicate horizontal parallel lines in increments of 2′. Now is a good time to draw a line straight down from the Y-intersecting point/Vanishing Point into the middle of the original grid. This will be called the Y-line. Where the Y-line intersects the X-line is the (0,0) point of an (x,y) grid system. From the (0,0) point to both corners of the original grid are 1′ distances. Where the first two hash marks are located are 3′ marks, and where the second two hash marks are located are 5′ marks, as in Figure 7.15.

GRID REDUCTION

Now that every item of evidence has been enclosed in a 2′ × 2′ square, it is necessary to reduce the size of each grid to more accurately determine the exact location of the dot representing each item of evidence. Cross the corners of every 2′ × 2′ square that has an item of evidence within it. These lines do not have to be

Figure 7.15

All lines marked.

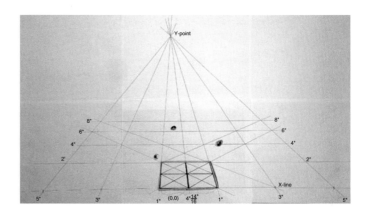

Figure 7.16

Cross the corners of every grid with evidence in it and an adjacent grid.

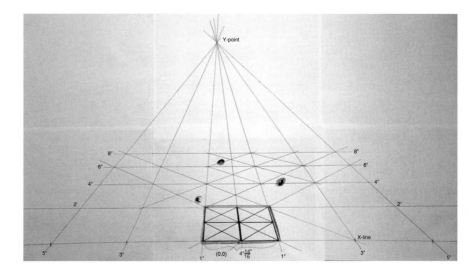

extended outside the $2' \times 2'$s involved. Also, cross the corners of an adjacent $2' \times 2'$. For every $2' \times 2'$ square that has evidence within it, pick another $2' \times 2'$ square, either to the left or right of it, and cross its corners also. Figure 7.16 shows all these $2' \times 2'$ squares crossed.

Each $2' \times 2'$ square with evidence in it can now be bisected, both vertically and horizontally. To bisect a $2' \times 2'$ square vertically, align the X formed by the corner crossing within the $2' \times 2'$ with the Y-intersecting point. Only draw a line that bisects the $2' \times 2'$ square with the evidence in it; it is not necessary to draw the line all the way down from the Y-intersecting point. To bisect a $2' \times 2'$ square horizontally, align the X of the $2' \times 2'$ square with the evidence in it with the X formed when crossing the corners of an adjacent $2' \times 2'$ square. Just bisect the $2' \times 2'$ square with the evidence in it. Figure 7.17 shows the grid with the keys in it bisected both vertically and horizontally. The grids with the sunglasses and purse are also bisected both vertically and horizontally in like manner.

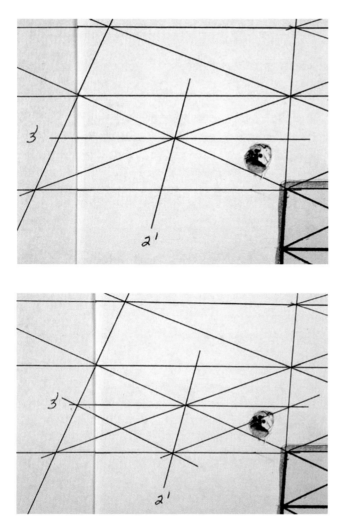

Figure 7.17
2′ × 2′ with keys bisected vertically and horizontally.

Figure 7.18
1′ × 1′ with keys with corners crossed.

From now on, when new vertical and horizontal lines are drawn, immediately label them as to their distance from the X-line and Y-line. If each new line is immediately labeled, it will be easier to later determine the location of the evidence. The 2′ × 2′ grid with the keys in it has now been bisected into four 1′ × 1′ squares. The dots representing the evidence are all now located within a 1′ × 1′ square.

To reduce the size of each 1′ × 1′ grid any evidence is within, repeat the preceding directions. Cross the corners of each 1′ × 1′ grid with evidence with it, and an adjacent 1′ × 1′ grid. Figure 7.18 shows the 1′ × 1′ grid with the keys in it after its corners have been crossed, and an adjacent 1′ × 1′ grid has had its corners crossed also.

Each 1′ × 1′ grid with evidence in it can now be bisected both vertically and horizontally. To bisect a 1′ × 1′ grid vertically, align the X formed by the corner crossing within the 1′ × 1′ grid with the Y-intersecting point. Only draw a line that bisects the 1′ × 1′ grid with the evidence in it; it is not necessary to draw the line all the way down from the Y-intersecting point. To bisect a $1' \times 1'$ grid horizontally, align the X of the $1' \times 1'$ grid with the evidence in it with the X formed when crossing the corners of an adjacent $1' \times 1'$ grid. Just bisect the $1' \times 1'$ grid with the evidence in it. Figure 7.19 has the $1' \times 1'$ grid with the keys in it bisected into four $6'' \times 6''$ squares. When new vertical and horizontal lines are drawn, immediately label them as to their distance from the X-line and Y-line. If each new line is immediately labeled, it will be easier to later determine the location of the evidence.

This entire process is usually repeated one more time. Cross the corners of each $6'' \times 6''$ square with evidence with it and an adjacent $6'' \times 6''$ square. This is shown in Figure 7.20.

Each $6'' \times 6''$ square with evidence in it can now be bisected, both vertically and horizontally. To bisect a $6'' \times 6''$ square vertically, align the X formed by the corner crossing within the $6'' \times 6''$ square with the Y-intersecting point. Only draw a line that bisects the $6'' \times 6''$ square with the evidence in it; it is not necessary to draw the line all the way down from the Y-intersecting point. To bisect a $6'' \times 6''$ square horizontally, align the X of the $6'' \times 6''$ square with the evidence in it with the X formed when crossing the corners of an adjacent $6'' \times 6''$ square. Just bisect the $6'' \times 6''$ square with the evidence in it. Figure 7.21 shows the $6'' \times 6''$ square bisected into four $3'' \times 3''$ squares.

Once all the dots representing the evidence are located within $3'' \times 3''$ squares, no further square reductions are necessary. On one vertical $3''$ line, draw two hash marks, each approximately $1''$ from the other. On one horizontal $3''$ line, draw two hash marks, each approximately $1''$ from the other. There is no need to precisely measure these hash marks. Careful approximations are all that is required. It should now be easy to approximate the location of each item of evidence (a dot) as to their relation to both the X-line and Y-line accurate to the closest inch. Figure 7.22 shows the $3'' \times 3''$ square with the dot in it divided into thirds both vertically and horizontally.

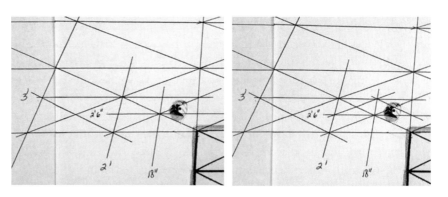

Figure 7.19
$1' \times 1'$ bisected into four $6'' \times 6''$ squares.

Figure 7.20
$6'' \times 6''$ with corners crossed.

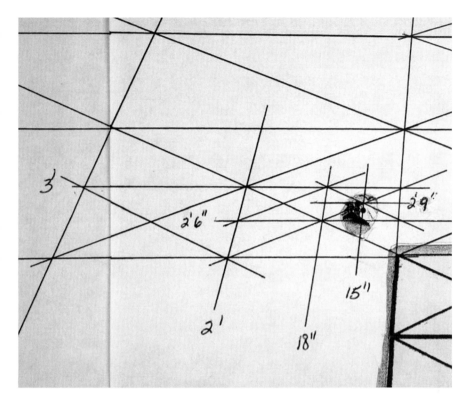

Figure 7.21
6″ × 6″ bisected into four
3″ × 3″ squares.

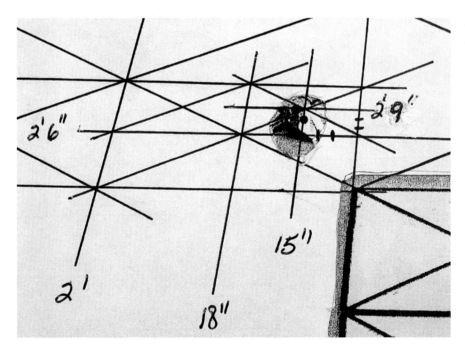

Figure 7.22
3″ × 3″ with evidence divided
into thirds.

The dot representing the keys is now determined to be 2′ 8″ north on the Y-line, and 15″ west on the X-line. This can also be written as (−15″, 2′ 8″), for those more comfortable with traditional X-Y coordinate system notation.

When teaching perspective grid photogrammetry at a local police academy, it was pointed out that quartering 2′ × 2′ squares and 1′ × 1′ squares and 6″ × 6″ squares is relatively easy when the evidence is low enough in the original photograph to end up being located in a large initial 2′ × 2′ square. When the evidence

Figure 7.23

Small original 2′ × 2′ enlarged.

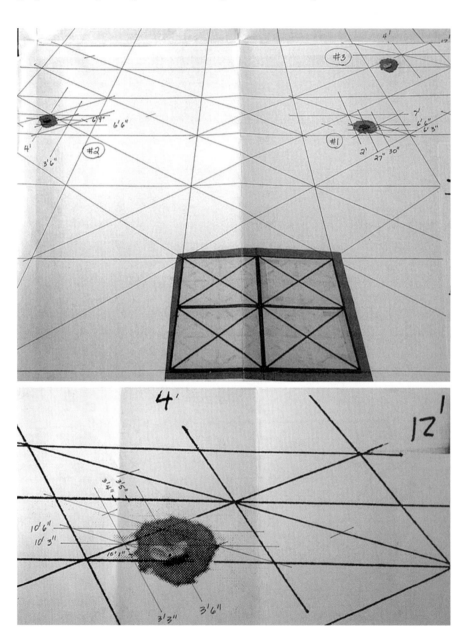

is higher in the photograph, it will eventually end up in a much smaller $2' \times 2'$ square, and all the required quartering becomes difficult, because all the diagonal and bisecting lines get extremely close together. This can be difficult. In these cases, then, the simple answer is to make the original small $2' \times 2'$ square larger. Simply enlarge these $2' \times 2'$ squares on a copy machine as large as you wish. Once they are larger, they will easily allow the required reductions. Figure 7.23 shows such an example taken from a different scene.

Item 3 is a bottle of White Out, and being relatively high in the photograph, it was expected that many small lines could cause difficulties. That portion of the diagram was enlarged and the evidence was easily determined to be 10'1" N and 3'6"W of the grid's (0,0) point.

Many who took the trouble to work along as the directions were being given will want to know the "correct answers" for the sunglasses and purse.

Figure 7.24 shows the final breakdown for the sunglasses, and Figure 7.25 shows the breakdown for the purse.

The "correct answers" for the sunglasses are 4' 2" N, 21" E. The "correct answers" for the purse are 6' 7" N, 5.5"W. It is normally recommended to round all crime scene measurements to the nearest inch. It was indicated that the purse was 5.5"W on the X-line just to show how easy it is for the perspective grid to be that precise.

Skeptics will correctly argue that if the original grid extension over the crime scene is drawn inaccurately, the final determination of the location of an item of evidence may be inaccurately extrapolated. This is true. The initial four lines drawn, the extensions of the sides of the grid, have to be very carefully drawn. If they are extended out over the crime scene haphazardly, the determination of the location of the evidence can be inaccurate. It is recommended the perspective grid technique be tried several times over scenes created for practice and that it not be attempted at actual crime scenes until the grid measurements are routinely accurate with practice scenes.

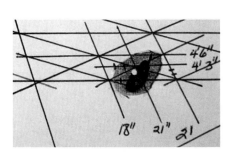

Figure 7.24
Sunglasses: 4'2"N, 21"E.

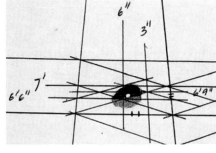

Figure 7.25
Purse: 6'7"N, 5.5"W.

LOCATING THE EVIDENCE IN THE SCALE DIAGRAM

All the evidence is now associated with the (0,0) point of the original $2' \times 2'$ grid. It is known how far up the Y-line the evidence is and how far along the X-line the evidence is. The perspective grid photogrammetry measurement technique is just another version of the baseline coordinate measurement technique. The question before us now is this: how is this information used to locate the evidence in a scale crime scene diagram?

If the evidence is now linked to the perspective grid, it should be obvious that the grid itself has to be linked to the crime scene somehow. How is this to be done? The grid has to be measured and fixed to the crime scene. First, the (0,0) point of the grid has to be measured. Choose the measurement technique of your choice for this. Measure the (0,0) point by the baseline coordinate method, the triangulation method, or any other measurement technique you feel comfortable with. The point is that the grid must have a known location within the crime scene.

Once the (0.0) point of the grid has been measured, one other measurement must be measured. With only the (0,0) point of the grid located in the crime scene, it would be possible for the grid to rotate around the (0.0) point, unless the spatial relationship of the grid is somehow locked in. The second measurement necessary is a point on the Y-line several feet away from the grid. It would technically be possible to measure the (0,2) point on the grid to lock its orientation into the crime scene, but it is recommended to pick some further point on the Y-line for a more accurate determination of the Y-line. When attempting to construct a line, two points are all that are required. The farther apart those two points, the more accurate the line's direction.

Figure 7.26 shows why the line formed by connecting two dots that are close together can be less accurate than the line formed by connecting two dots farther apart. Rather than just measuring the (0,2) point, it is advised to measure some point farther down the Y-line. Its exact distance from the (0,0) point is irrelevant; the farther it is from the (0,0) point, however, the better. Place a coin on the Y-line 5' to 10' from the (0,0) point. Double check it truly is on the Y-line. If you do not trust your ability to eyeball the coin being on the Y-line when viewed from both the (0,0) point toward the coin and when viewed from the coin toward the (0,0) point, you can lay down a tape measure down the Y-line from the (0,0) point to fix the Y-line and lay a coin under the tape measure. Again, it is not necessary to note the coin's distance from the (0,0) point. Its only purpose is to lock the grid's orientation in the crime scene. Measure the coin's position to lock it to the crime scene, by whatever measurement technique you wish. You now have two points measured: the grid's (0,0) point and a point on the Y-line.

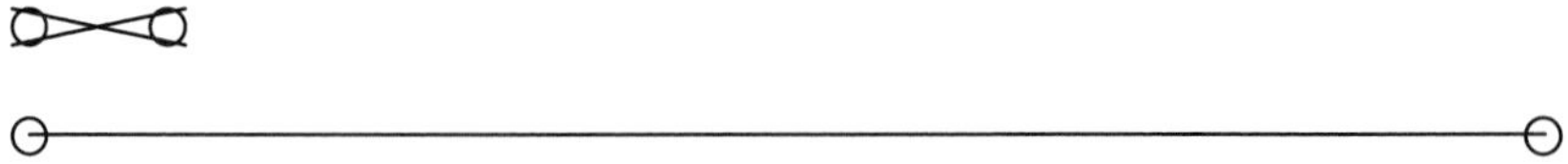

Figure 7.26

Two points determine a line.

These two points are placed into the crime scene diagram in scale. They are only inserted into the crime scene diagram as two lightly drawn dots, because they will eventually be erased. All the evidence associated to the grid can now be located into the crime scene. Each item of evidence has a known distance out the Y-line, measured from the (0,0) dot toward the dot that is a point on the Y-line. From this Y-distance, the evidence is also a known distance perpendicular to the imaginary line connecting the two dots. This distance to the left or right of the imaginary Y-line is the X-distance of each item of evidence.

So, the perspective grid measurement technique requires two measurements, the (0,0) point and a point on the Y-line. If there are only two items of evidence in the crime scene, using a perspective grid saves no time, because both items of evidence will also have to be measured traditionally. However, when there are three items of evidence, or more, within view of a perspective grid laid down in the crime scene, using the perspective grid technique saves time. Just measure the (0,0) point and a point on the Y-line, and you can then pick up all the evidence.

Again, it is critical that the perspective grid measuring technique not be used at actual crime or accident scenes until the practitioner is routinely obtaining accurate measurements when the perspective grid technique is applied to mock crime scenes, when all the evidence has previously been measured by a traditional measuring technique, and all the correct answers are already known.

One restriction regarding perspective grid photogrammetry must be mentioned. This technique can only be applied to evidence on the same plane as the perspective grid. It is only to be used on flat surfaces. That is because evidence lower than the same plane as the grid will appear "lower" in the photograph, and items lower in the photograph will be interpreted as being closer to the grid. Evidence on a raised area will appear higher in a photograph, and evidence higher in a photograph is interpreted as being farther from the grid. Only use the grid for scenes on flat surfaces.

This author has not used the perspective grid at very large crime scenes and would suspect its accuracy in such situations. In practice, one can feel comfortable as to the accuracy of the perspective grid technique at crime scenes no larger than 40'. Beyond that, inaccuracies with the extension of the original grid over the crime scene may seriously affect measurement determinations. As mentioned earlier, photogrammetry is not supposed to be regarded as a replacement for all other measuring techniques. The Total Station types of measuring systems cannot be beat for large scenes and scenes with a rolling topography.

PERSPECTIVE DISC PHOTOGRAMMETRY

Should you ever have the need to testify about the measurement technique used to produce a particular crime scene diagram and happen to mention that you used the perspective grid photogrammetry technique to determine the measurements of some of the evidence, be prepared for the following. An attorney with extensive experience working with professional accident scene reconstruction companies may be very familiar with photogrammetry. Many professional accident reconstructionists use variations of photogrammetry to do their jobs. In court, an attorney may attack your credibility with some variation of this approach:

"So, I understand you used perspective grid photogrammetry to determine the location of "x" item of evidence. Sounds like a very nontraditional measurement technique. Well, I have heard of it before, and there is just one thing I'd like to know. There are many variations of photogrammetry available to anyone interested in determining measurements at accident and/or crime scenes. That being true, I'd just like to know why you chose to use a method known by all to be inferior to another photogrammetry technique? I mean, if you are going to apply photogrammetry techniques at a crime/accident scene, why use the perspective grid technique, when it is well accepted in that community that the perspective disc method is far more accurate? Why gather evidence with a poor substitute for a more highly regarded technique?"

This could go on and on. The purpose is to undermine your credibility and your expertise. What is the foundation for this attack?

True, there are variations of photogrammetry. One of those variations is called the perspective disc method.

The perspective disc method is an alternative method that can be used instead of the perspective grid technique. It is believed to result in more accurate measurement determinations because the horizon line is more precisely located during the process. The improvement in measurement accuracy is sometimes real but frequently too small to make a difference in crime scene diagrams. With the perspective disc method, three circular discs are used instead of a square or rectangular grid. Figure 7.27 shows the same scene we are familiar with, but this time it has three perspective discs in it. Each of these discs is 1′ in diameter. The three discs can each be 2′ in diameter. In any case, the three discs should be the same size.

With this technique, one of the three discs is arranged low and centered in the composition, just as the perspective grid was. One of the other discs is put somewhere on the right side of the composed image, and the other is placed on the left side. Their precise location is irrelevant; one just has to be to the left, and one to the right.

The next step is to draw tangential lines from the lower grid to both the other two discs. Draw lines from the right side of the lower disc past the right side of each other disc. Draw lines from the left side of the lower disc past the left side of each of each other disc.

Figure 7.28 is the result. Both pairs of vertical parallel lines will eventually intersect. A line can connect these two intersecting points, which is equivalent to the horizon at the scene.

Figure 7.27
Perspective discs in a scene.

Figure 7.28
Tangential lines extended.

This is the basis for the claim to more accuracy when using the perspective disc method. With the perspective grid technique, the horizon was determined by extending the left and right sides of the perspective grid. The Y-intersecting point was a point on the horizon line. It is claimed by some that a more precise way to determine the horizon is to use three discs rather than one square or rectangle. That may be true, but let us continue a bit more.

From the horizon line, a perpendicular can be drawn that intersects the center of the lower disc. This is the equivalent of the Y-line. From this point on the horizon line, two more lines can be drawn that are tangential to both sides of the lower disc. Figure 7.29 is the result.

From this position, two lines that are perpendicular to the Y-line can be drawn that are also tangential to the top and bottom of the lower disc. The result is Figure 7.30

Around the original disc with a 1′ diameter, there is now a 1′ × 1′ square. From this point, the perspective grid technique is used to determine the location of all the evidence. The main difference between the perspective grid technique and the perspective disc technique is the accuracy by which the horizon is determined. What real difference does this make to the accuracy of any crime scene diagram resulting from both techniques? Because the perspective grid technique can result in measurements within 1″ of the known measurements at

Figure 7.29

Lines from the horizon line to the lower disc.

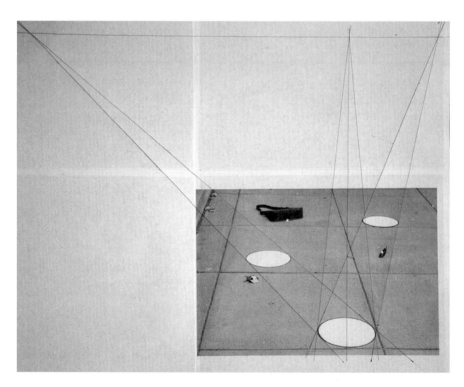

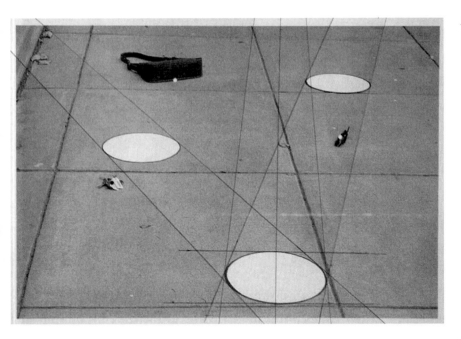

Figure 7.30
Horizontal tangential lines on lower disc.

mock crime scenes set up to check the accuracy of the technique, it is suggested the perspective disc technique can result in more accurate measurements.

Is "more accurate" necessary for crime scene measurements? No. Most law enforcement agencies recommend rounding off crime scene measurements to the nearest inch. Why? Mainly because a more accurate measurement standard, $\frac{1}{2}''$ for example, will not make a difference in the final crime scene diagram. The width of your lines drawn with pencil lead or pen ink will often "absorb" crime scene measurements of less than one inch.

There is another reason that measuring more precisely than whole inches is often not recommended. Trying to be more precise can often be considered absurd. Many readers may recall an article on the accuracy of crime scene measurements where a huge distance was measured and recorded as something like 199′ 11.75″. Why not just indicate the distance in question was 200′? Over that distance is another quarter inch even relevant? Most would guess if that same distance is measured five times by different people that five different distances would be recorded. For a measurement to be a "fair and accurate representation of the scene," how precise does it have to be? Most believe rounding off to the nearest inch is sufficient. Of course, many attorneys may try to argue to the jury that measuring to the nearest $\frac{1}{2}''$ is more accurate than measuring to the nearest inch. Of course, that is true. If you fall for this trap in court and state you measure to the nearest $\frac{1}{2}''$, then that same attorney will ask, with a grin, aren't we aware that measuring to the nearest $\frac{1}{4}''$ would be more accurate? Of course it is. And this circle can continue forever, because, mathematically speaking any distance/mathematical increment can be divided in half. The real point is, at what distance increment do crime and accident scenes need to be measured to be "fair and accurate representations of the scene?" To the nearest inch!

Also, it should be mentioned that if care is exercised in extending the original four sides of the perspective grid, then measurements as accurate as can be determined by the perspective disc method *can* be obtained.

Therefore, in the preceding scenario, when the attorney asks why we did not use the perspective disc measurement technique instead of the perspective grid measurement, because the disc method is considered to be more accurate, our answer should be something like this:

"Of course I know the perspective disc method is thought by some to be a more accurate method for determining the position of evidence. When I have applied both techniques to the same staged crime scene, the measurements produced by the perspective disc method have sometimes been slightly more accurate. The difference in accuracy, however, is so minimal that it makes no meaningful difference in the actual crime scene diagrams produced by both techniques. It sometimes produces an increase in accuracy equivalent to being more precise by ½″ or ¼″, which is not going to show up in the finished crime scene diagram. And, if done carefully, the perspective grid technique can produce measurements just as accurate as the perspective disc method."

Of course, to be able to testify to this, you do have to work several mock crime scenes with both techniques. If you find you actually prefer the disc method, by all means continue to use it. It is a very good technique.

NATURAL GRID PHOTOGRAMMETRY

There will be times when there are natural features within the crime scene that will make the addition of perspective grids or perspective discs unnecessary to apply photogrammetry techniques to derive the measurements of evidence within the scene. The natural features themselves can be used to construct a grid over the evidence. Figure 7.31 is an example of such a scene.

Natural grid: There will be times when there are natural features within the crime scene that will make the addition of perspective grids or perspective discs unnecessary to apply photogrammetry techniques to derive the measurements of evidence within the scene. The natural features themselves can be used to construct a grid over the evidence.

What is necessary to use the natural grid photogrammetry technique?

Clearly visible in the scene there needs to be three lines, two that are vertical parallel lines, and one that is perpendicular to the vertical parallel lines. While at the scene, you will have to measure one of the vertical parallel lines and the horizontal parallel line. In Figure 7.31, the two requisite vertical parallel lines are the AC line, and the line going down from the B point that is cut off by the right side of the photograph. That the image was intentionally composed to cut off this line to emphasize its length does not need to be known. Line AB is the required horizontal parallel line. The two necessary measurements are provided at the bottom of the image.

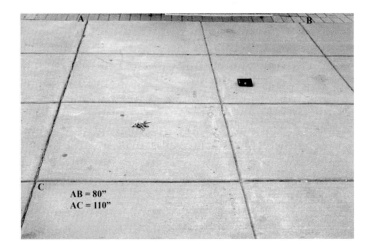

Figure 7.31
The natural grid technique would work here.

If such an image is not composed and photographed by you at the original crime scene, the natural grid technique can still be applied to the photograph if you can return to the scene to collect the two required measurements. In other words, we are now moving into an area where you may be able to extrapolate measurements from photographs you did not take. If anyone's photographs have these necessary elements in the photographs, you will be able to ultimately derive the measurements of the evidence within the photographs.

In Figure 7.31, the wallet and the keys are enclosed in a huge rectangle measuring 80″ × 110″. Rectangles expand and reduce just as 2′ × 2′ squares do. All the same principles apply.

In Figure 7.32, both the left and right vertical parallel lines have been extended until they intersect. Although the letter "D" has been drawn in at the lower right of the image, resist the temptation to assume the obvious line in the cement running along the bottom of the image will eventually be the CD line. The workmen who poured the cement pavement may not have been as precise as crime scene diagrammers may wish, as we will soon see. We will use photogrammetry to construct the 80″ × 110″ rectangle, and not rely on concrete pourers.

Figure 7.33 shows how the rectangle will eventually be completed. With the perspective grid technique, the original 2′ × 2′ grid already had a midpoint, formed by crossing the corners of the grid. In this example, the fourth corner is currently missing. To locate it, we can begin by measuring the length of line AB on the image we are working on and put a hash mark on the AB line at its halfway point. From the Y-intersecting point, a line can be drawn through this hash mark, continuing to the bottom of the image. Notice how this new vertical parallel line does not align with the vertical line in the cement "midway" between point A and

Figure 7.32

Vertical parallel lines extended until they intersect.

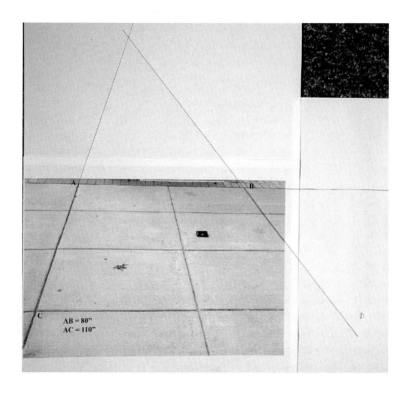

Figure 7.33

Determining the midpoint of the AB line and the larger rectangle.

point B. Somehow the cement pourers did not locate that line properly. Now, point B and point C can be linked by a line. Finally, by connecting point A with the intersection of the BC line and the line coming from the Y point through the hash mark, the D point can now be determined. What we thought might be the CD line is not.

Figure 7.34 shows the position of the D point of the rectangle. It is easier to trust the principles of geometry before relying on concrete pourers to help with our crime scene diagramming. Once the rectangle has been determined, it is smart to begin labeling the distance of each line. If one waits until many lines are drawn, it may be difficult or impossible to determine which line is which. From now on, as each new line is drawn, it will be labeled.

The $80'' \times 110''$ rectangle has been bisected vertically. It is now necessary to bisect it horizontally. The $80'' \times 110''$ rectangle has had its corners crossed; to bisect it horizontally, another rectangle will have to have its corners crossed.

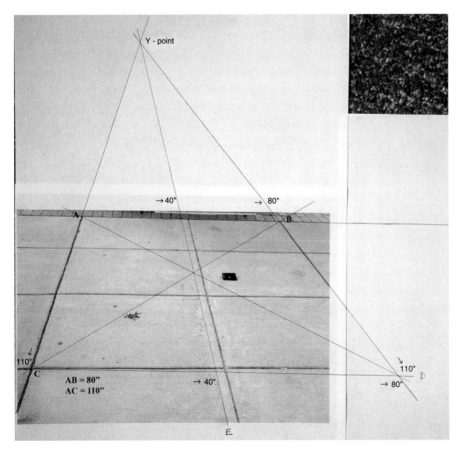

Figure 7.34
The D point located.

Rather than constructing another $80'' \times 110''$ rectangle adjacent to the original rectangle, notice there are two $40'' \times 110''$ rectangles already drawn, separated by the line coming from the Y point through the middle of the original rectangle. In Figure 7.35, the corners of the left $40'' \times 110''$ rectangle have been crossed, but rather than making a large "X" through the entire rectangle, only a small "x" has been made in the middle of the rectangle. With that "x" aligned with the "X" of the larger rectangle, the original rectangle can now be bisected horizontally. Both of these new lines can be labeled. The wallet and keys are now in $40'' \times 55''$ rectangles, and each is below the midpoint of the original rectangle.

Figure 7.35

Bisecting the original rectangle horizontally, resulting in $40'' \times 55''$ rectangles.

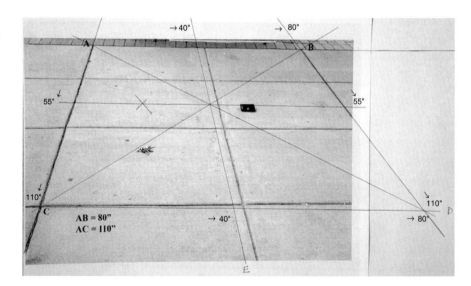

Figure 7.36

Bisecting rectangles again, resulting in $20'' \times 27.2''$ rectangles.

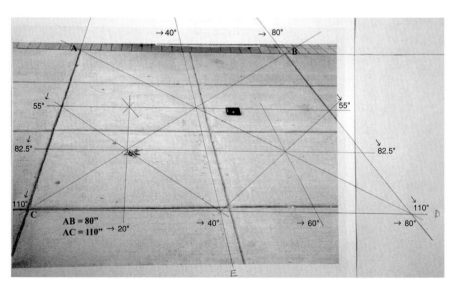

Figure 7.36 is the result of crossing the corners of each $40' \times 55'$ rectangle with evidence in it, and bisecting each both vertically and horizontally. Each item of evidence is located in a $20'' \times 27.5''$ rectangle. Do not round the 27.5″ measurement up or down to an even 27″ or 28″. A rounding to the nearest inch is only permissible at the end, when the exact measurement of the evidence within the scene is determined.

Figure 7.37 is the result of crossing the corners of each $20' \times 27.5''$ rectangle with evidence in it and bisecting each both vertically and horizontally. Each item of evidence is now located in a $10'' \times 13.75''$ rectangle. Again, retain all the decimal

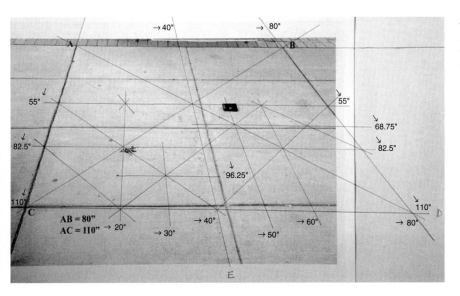

Figure 7.37

Bisecting rectangles again, resulting in 10″ × 13.75″ rectangles.

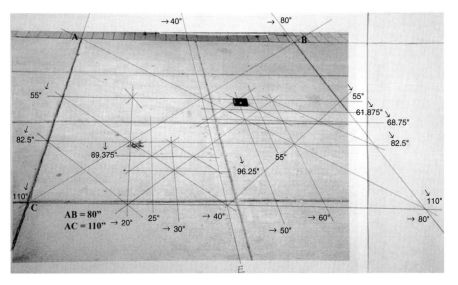

Figure 7.38

Bisecting rectangles again, resulting in 5″ × 6.875″ rectangles.

places of these measurements until the final determination of the evidence is made.

Figure 7.38 is the result of crossing the corners of each 10″ × 13.75″ rectangle with evidence in it and bisecting each both vertically and horizontally. Each item of evidence is now located in a 5″ × 6.875″ rectangle. Retain all the decimal places.

Figure 7.39

Bisecting rectangles again, resulting in 2.5″ × 3.4375″ rectangles.

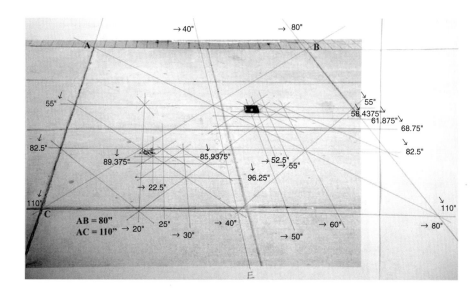

Figure 7.40

Rectangle with keys divided into thirds horizontally and vertically.

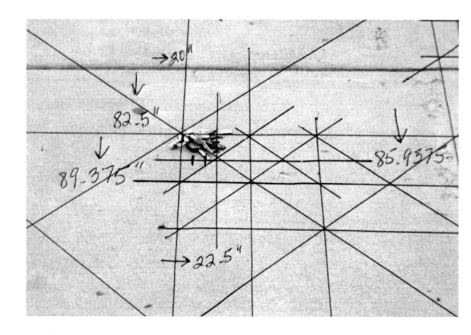

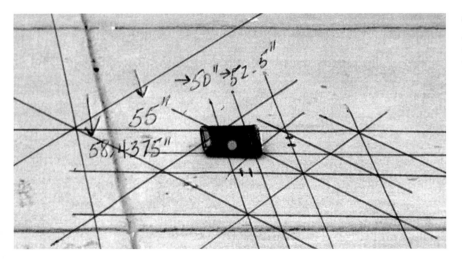

Figure 7.41
Rectangle with wallet divided into thirds horizontally and vertically.

Figure 7.39 is the result of crossing the corners of each $5'' \times 6.875''$ rectangle with evidence in it and bisecting each both vertically and horizontally. Each item of evidence is now located in a $2.5'' \times 3.4375''$ rectangle. Retain all decimals.

Figure 7.40 is the result of placing three hash marks on both vertical and horizontal sides of the $2.5'' \times 3.4375''$ rectangle the dot representing the keys are in. These hash marks can be approximated. Once this is done, it will be easy to arrive at the whole numbers for the vertical and horizontal position of the keys. If we are seeking the distance from the A point in the original image, each horizontal hash mark is $20.833''$ and $21.666''$, respectively. Because the dot of the keys lies between them, $21''$E is the closest answer. Vertically, each hash mark stands for $83.6458''$ and $84.7916''$, respectively. As the dot of the keys aligns with the first hash mark nicely, that rounds up to $84''$ as the closest answer. The keys are then, $21''$E and $84''$S, or $(21'', -84'')$ from the A point.

Figure 7.41 is the result of placing three hash marks on both vertical and horizontal sides of the $2.5'' \times 3.4375''$ rectangle the dot representing the wallet is in. These hash marks can be approximated. Once this is done, it will be easy to arrive at the whole numbers for the vertical and horizontal position of the wallet. If we are seeking the distance from the A point in the original image, each horizontal hash mark is $50.833''$ and $51.666''$, respectively. Because the dot of the wallet would intersect the first hash mark, that rounds up to $51''$. Vertically, each hash mark stands for $56.1458''$ and $57.2916''$, respectively. Because the second hash mark would intersect the dot, that rounds to $57''$ as the answer. The wallet is then $51''$E and $57''$S, or $(51'', -57'')$ from the A point.

REVERSE PROJECTION PHOTOGRAMMETRY

Reverse projection:
A large transparency of an original crime scene image can be looked through after returning to the original crime scene, and the analyst can move around the scene until the fixed features of the scene align with those same points in the transparency. This takes considerable trial and error. The only way for the fixed features of the scene to line up perfectly is if the original position of the original photographer was eventually duplicated. However, once the fixed features do line up, an assistant can place markers into the crime scene that correspond to the evidence in the transparency. These markers can then be measured to determine their original location within the crime scene.

At times your knowledge of photogrammetry will have others seek you out to help them extrapolate measurements from images relating to their crime scenes. You will first hope to find a perspective grid or a set of three perspective discs positioned in their images. If these are lacking, you will next look for natural grid elements in the crime scene image. If you come up with three strikes, are you out? No. There is still hope. The reverse projection photogrammetry technique may be applied to the crime scene photograph.

Some necessary elements have to be located within the crime scene photograph for this technique to be successful. In the crime scene image you are examining, there have to be sufficient fixed features present. Fixed features are scene elements that would still be present and easily visible if you were to return to the crime scene. The crime scene photograph cannot just include a number of items of evidence all positioned on an endless expanse of green grass or in a large parking lot with nothing but indistinguishable concrete surrounding the evidence. Clearly identifiable markings must be around the evidence. Figure 7.42 is an example. The four walkways are fixed at this scene and will be there until the entire area is bulldozed for a new building. One could return to the scene and easily find this alignment of walkways. If so, my badge case could easily be reinserted into the crime scene by reverse projection photogrammetry, and then it could be measured just as accurately as it could have been on the day of the original incident.

The term "reverse projection" needs some preliminary explanation. Isn't it true that a "reverse projection" implies an "original projection?" Certainly. What is this "original projection?" The original projection was light rays reflecting from the original crime scene into a camera lens, where they formed an image on film or on a digital sensor. Projected light rays from the crime scene formed an image.

We can now take an $8'' \times 10''$ enlargement of the original image and copy it onto a sheet of acetate with a copy machine. The result will be a transparency of the original $8'' \times 10''$ image. If we were to return to the original crime scene, we

Figure 7.42
Outdoor scene perfect for reverse projection photogrammetry.

Figure 7.43

Viewing through transparency at a "crime scene" (Courtesy of GWU MFS Student Kelly Brockhohn).

could walk around the scene with the transparency held up in front of our eyes, as in Figure 7.43, until all the fixed features aligned perfectly. The only way for the fixed features of the scene to line up perfectly is if the original position of the original photographer was eventually duplicated.

This is much easier said than done. To reapproximate the position the original photographer stood in when the original photograph was taken can take quite a bit of trial and error. Moving to the left and right; moving closer to the scene or farther away; and changing the distance the transparency is held in front of you can all be very frustrating. In addition to those gross positioning changes, there is one other consideration that must be duplicated. The only way the scene will align perfectly is to view through the transparency from the height of the original photographer's eyes. If you are taller or shorter than the original photographer, the scene cannot align perfectly. Part of this process is to find out the eye height of the original photographer and to duplicate that height when viewing the crime scene through the transparency. Of course, view the transparency with only one eye open. If you were to take turns closing one eye at a time, you will notice the elements in the transparency "jump' to the left and right.

When teaching the various photogrammetry techniques at GWU, reverse projection photogrammetry seems to be the student's least favorite technique. Mostly, it is presumed, because many have to stand on a pile of books or bend over for extended periods of time, making the process a bit cumbersome and uncomfortable. This technique is, however, just as accurate as the other photogrammetry techniques. It also has the benefit of being accurate on multiple planes rather than just the one plane a perspective grid is placed on.

Once the fixed features have been aligned, while viewing through the hand-held transparency, the process will progress more smoothly if the transparency can be locked into place. This is done by placing the transparency in a clear plastic holder and clipping the plastic holder onto a tripod. Figure 7.44 shows two clamps clipping

Figure 7.44

Plastic holder with transparency clipped to a tripod (Courtesy of GWU MFS Student Kelly Brockhohn).

the plastic holder onto a tripod head. The tripod quick-release, which screws onto the bottom of the camera, has been removed from the tripod to facilitate this.

Notice the alignment of the transparency. In effect, it must duplicate the angle of the film or digital sensor when the original image was captured. Because the camera was aimed down into the scene when the original image was captured, the transparency must also be placed in this position. It is important that you avoid "forcing" the alignment of the transparency and the crime scene by twisting the plastic holder so either side of the plastic holder is closer to the crime scene than the other side of the plastic holder. Although the transparency is tilted down, it must remain in the same orientation as the film or digital sensor when the original image was captured.

Once the fixed features of the crime scene are aligned with the fixed features of the transparency, the evidence in the original image can now be repositioned back into the crime scene. Figure 7.45 shows the original scene with three items of "evidence" in it.

The transparency has those same items of evidence visible. At this point, you can have an assistant move markers around the crime scene as you view the scene through the transparency. Direct them with "forward," "backward," "left," and "right" until all the markers are aligned under the evidence in the transparency. Double check that the fixed features of the transparency remain aligned over the fixed features of the crime scene. Once this has been completed, you have recreated the location of the evidence within the crime scene. Now, the evidence can be measured as you would have measured it at the original crime scene.

The FBI uses a more sophisticated variation of reverse projection photogrammetry to determine the height of bank robbers and to determine the length of sawed-off shotguns photographed by surveillance cameras in banks. The version explained in this chapter is very simplified.

Rhino photogrammetry: By use of Rhinoceros 3D modeling software, a crime scene image can have a grid system superimposed over it. Once the software grid system is locked to the fixed features of the crime scene, every point in the combined image can be measured to any other point.

RHINO PHOTOGRAMMETRY

As intriguing as these photogrammetry techniques are, for years this author has been searching for a way to simplify the entire process. At the beginning of this

Figure 7.45
Aligning evidence in
transparency with the crime
scene (Courtesy of GWU MFS
Student Kelly Brockhohn).

chapter, this question was asked: "would it not be nice if we could superimpose a uniform checkerboard grid system, like $1' \times 1'$ squares, over the entire crime scene?" Many software programs have grid systems as part of their packages. Perhaps one of these grid systems could be adapted so they could be superimposed on any crime scene photograph.

The first task was to determine the minimum number of reference points needed in a crime scene photograph to "lock" a grid system to the photograph. Of course, a perspective grid has four corners. Requiring four rectangular points of reference in a crime scene photograph returns us to the natural grid technique. Could this be simplified? Perhaps three points of a right triangle would enable a grid system to be "locked" to a crime scene diagram. It was then realized that three points of a right triangle are really just a variation of a rectangle, with one corner missing. I wanted to be less restrictive. How about any three points of any triangle? It would be enormously convenient to be able to develop a method of extrapolating crime scene measurements from any photograph with only three fixed features anywhere in the image.

A search was begun for robust grid systems in software programs that might fill the bill. After much searching, I found Rhinoceros®.*

Figure 7.46 shows objects in their software viewed in their "top" window, which is equivalent with a birds-eye view of the objects; and in their "perspective" window, which is the same view we would have of the objects standing nearby. These two grid systems will be used with a mock crime scene to determine the accuracy of the software. The evidence will be measured traditionally, so the "correct" answers are already known.

Figure 7.47 shows the top view of the grid system again. It is very much like our crime scene diagrams should be laid out; the crime scene and evidence are all

Figure 7.46

2D and 3D figures in Rhinoceros® (Courtesy of Robert McNeel & Associates, Seattle, WA).

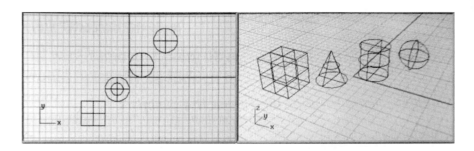

Figure 7.47

Rhinoceros®, top view of grid (Courtesy of Robert McNeel & Associates, Seattle, WA).

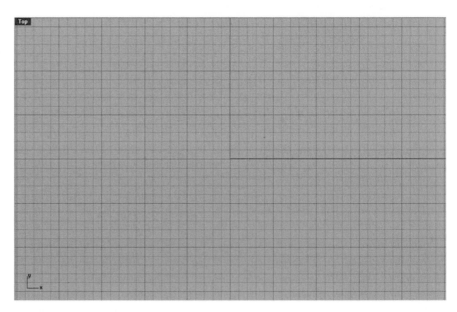

*Rhinoceros is a modeling software program used primarily to draw 3D objects. It was the software program used to model the Titanic for the movie. http://www.rhino3d.com/

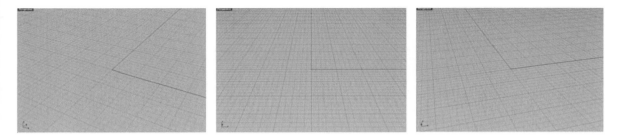

Figure 7.48

Rhinoceros®, perspective view of grid (Courtesy of Robert McNeel &Associates, Seattle, WA).

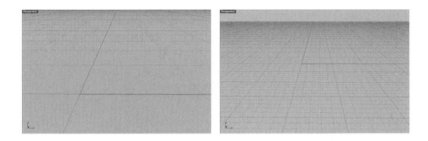

Figure 7.49

Rhinoceros®, perspective view of grid, zoomed in and zoomed out (Courtesy of Robert McNeel &Associates, Seattle, WA).

that are lacking. It is also very much like draft paper with grids on it. On the top view, we will eventually locate the three known reference points in scale. The squares shown can be considered meters, feet, or any "units" of our choice. When the cursor is placed anywhere over the grid system, x and y coordinates are indicated at the bottom of the scene, with an accuracy to three or four decimals. The (0,0) point of the grid is the intersection of the primary vertical and primary horizontal lines as shown.

Figure 7.48 shows the perspective view again. Tools allow the perspective grid to be rotated to the left and right, as in Figure 7.48, or zoomed in and zoomed out as in Figure 7.49.

Figure 7.50 shows three points of a triangle added to the top view. These will be the three known reference points from the crime scene. When anything is done in the top view, it is also done in the perspective view. In the perspective view, we can see the same points.

While in the perspective view, a crime scene photograph can be imported as a background, with the grid system now visible over the photograph. Figure 7.51 shows a mock crime scene photograph before and after being imported into the perspective view. Three labels were placed on the floor at the three reference points to be used to check the accuracy of the software. For more precision, "Xs" were also marked on the floor at the precise location of the three reference

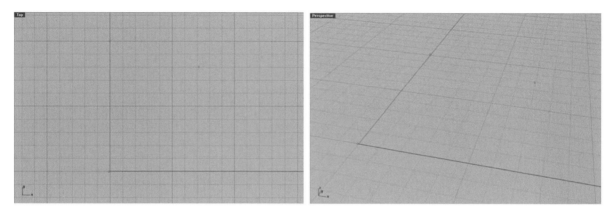

Figure 7.50

Rhinoceros®, triangular reference points added to the grid (Courtesy of Robert McNeel & Associates, Seattle, WA).

Figure 7.51

Rhinoceros®, perspective view of grid superimposed over a mock crime scene (Courtesy of Robert McNeel & Associates, Seattle, WA).

points. In the perspective view, the three small dots are the perspective view known dots related to the known triangle originally created in the top view.

Without much difficulty, the perspective view grid is rotated, zoomed, pushed, and pulled until the three known triangle dots are aligned with the reference points in the photograph. Once that has been done, place the cursor over any item of evidence, and the (x,y) position of each item of evidence is indicated at the bottom of the screen. A "distance" tool also allows the distance between any two points to be determined. Select any of the reference points as the starting point, and each item of evidence can be triangulated. Selecting a second reference point as another starting point, then putting the cursor over all the evidence completes the triangulation.

How accurate were the measurements? All the measurements were less than 1″ off the known measurements. The question can be asked, "How accurate were the measurements obtained with the tape measure?" The original measurements were rounded to the nearest inch, as has been my practice for decades. Of course, we could debate the accuracy of the original measurements, but what technique could be used to check the accuracy of the tape measurements? Bring in another tape measure? Total Station? Regardless, I am satisfied.

It must be added that a camera with a full-frame digital sensor was used for this exercise. A Canon EOS 5D, a 12.8 mp camera, with a zoom lens set to 50 mm was used. Similar exercises with film cameras have also been as accurate. It has been noticed that digital cameras with focal length multipliers have posed some problems, which cannot be explained yet. A Canon D60 has a smaller digital sensor, with a 1.6 focal length multiplier. That means a 50-mm lens gets cropped similar to the view from an 80-mm lens. When 80 mm is indicated in the Rhinoceros® software as the focal length of the camera used, the measurements have not been correct. At times, it has been necessary to "tweak" the software program and indicate either a 79-mm or an 81-mm lens has been used to obtain measurements close to the known measurements. One possible obvious explanation may suffice. When camera manufacturers attribute a focal length designation to a lens, just how precise are they? It may be that the precision of the Rhino software is able to detect 'nominal' focal length variations.

Figure 7.52

Perspective grid used at a triple homicide scene (Courtesy of the Arlington County Police Department, Arlington, VA).

Figure 7.53

Perspective grid used at a homicide scene (Courtesy of the Arlington County Police Department, Arlington, VA).

This author first learned about photogrammetry in 1988, while studying at the Virginia Forensic Science Academy. Realizing its potential to help get the job as a crime scene investigator done, it was used at all important crime scenes. Figures 7.52 and 7.53 are just two more examples of its use. Figure 7.52 was at a triple homicide, and the perspective grid was used to locate the bloody shoe print near one of the victims. Figure 7.53 was a different homicide scene, with evidence around the victim. Neither defendant will be able to view these images in this book, because they are no longer with us.

Photography can record life, as well as crime and accident scenes. With photogrammetry techniques, photography can also be used to extrapolate the precise measurements of the items within a photograph.

We just need to figure out how to get DNA from photographs.

SUMMARY

Most crime scenes, and the evidence within them, are measured with tape measures so that crime scene diagrams can be constructed. As the crime scenes or accident scenes become larger and more elaborate, many departments use sophisticated laser sighting systems combined with computer software, like Total Station, to create very detailed maps of the scenes. Highly trained FBI image analysts can also extrapolate bank robbery subject heights and shotgun lengths with very sophisticated photogrammetric techniques.

On many occasions, however, very elementary photogrammetric techniques can effectively be used by the personnel of local law enforcement agencies. Basic photogrammetry techniques can help determine the location of evidence at smaller crime scenes. This chapter demonstrated various simple photogrammetry techniques that can help the local crime scene technician "get the job done." Photogrammetry is another tool on the Bat-Belt of the crime scene photographer.

DISCUSSION QUESTIONS

1. Briefly describe the general concept of photogrammetry.
2. All the basic photogrammetry techniques rely on the ability to superimpose a grid system over a crime scene and the evidence within it. Discuss two of the variations by which this can be done.
3. Some of the photogrammetry techniques mentioned apply to a crime scene photographer originally present at a crime scene, when a grid or discs can be inserted into the crime scene photographs. Some apply to attempting to obtain measurements from a photograph after the scene has been cleared, and no grids or discs are in the images. Which techniques apply to the latter?

EXERCISES

1. In an area approximately $10' \times 10'$ square, place three items of "evidence." Place a $2' \times 2'$ perspective grid in the same area. This can be four $1' \times 1'$ floor tiles laid adjacent to each other. Ensure that the full perspective grid and all three items of evidence can be composed in the viewfinder by use of an SLR camera with a 50-mm lens. Compose the scene so the grid is low and centered in the field of view. Take that photograph. Lay a tape measure from the (0, 0) point of the perspective grid (the middle of the lower edge) up through the middle vertical line of the grid into the scene. The tape measure is both the Y-line of the perspective grid, and it is the baseline by which all three items of evidence will be measured by baseline coordination. With the tape measure positioned like the Y-line of the grid, determine the measurements of each item of evidence. Rather than measuring "center of mass," place a coin at the middle of the side of each item of evidence closest to the grid. These coins will represent each item of evidence. These measurements will be the "known answers" that the grid measurements will be compared with.

   Have the photograph printed and then enlarge it to approximately $8'' \times 10''$ on a copy machine. Follow the directions in the chapter for extending the $2' \times 2'$ grid squares over the scene until all three items of evidence are within a $2' \times 2'$ grid extension. Reduce the $2' \times 2'$ squares containing each item of evidence to $1' \times 1'$ squares, then $6'' \times 6''$ squares, then to $3'' \times 3''$ squares. Determine the location of each coin representing the evidence to the nearest inch. Compare with your known measurements. The first time this is done, your answers should be within $2''$ of the known distances.

   Repeat this exercise, having the mock crime scene larger each time, until areas of $25' \times 25'$ routinely are used and the accuracy of each item of evidence is within $1''$ of the known.

2. Find an area where there are two parallel vertical lines and at least one horizontal line that is perpendicular to the parallel vertical lines. Place three items of evidence within this scene. Compose this scene with an SLR camera with a 50-mm lens so that the horizontal line runs horizontally through your field of view. Take the photograph. Measure one of the vertical lines and the horizontal line. Consider this intersection of these two lines as the (0,0) point of an (x,y) coordinate system and measure all three items of evidence to the (0,0) point.

   Have the photograph printed and then enlarge it to approximately $8'' \times 10''$ on a copy machine. By use of the natural grid technique demonstrated in this chapter, construct a large rectangle around the evidence, and then reduce the size of this rectangle until the distance of all three items of evidence to the (0,0) can be determined.

Repeat this exercise, having the mock crime scene larger each time, until areas of $25' \times 25'$ routinely are used and the accuracy of each item of evidence is within $1''$ of the known.

FURTHER READING

Baker, J. S. (1983). "Perspective Grid for Photographic Mapping of Evidence." 2nd Ed. The Traffic Institute, Northwestern University, Evanston, IL.

Baker, J. S., Fricke, L. B. (1986). "The Traffic Accident Investigation Manual." Northwestern University Traffic Institute, Evanston, IL.

Crone, D. R. (1968). "Elementary Photogrammetry." Fred Ungar Publishing Company, New York, NY.

Ghosh, S. K. (1988). "Analytical Photogrammetry." Pergamon Press, New York, NY.

Hyzer, W. G. (2000). Forensic photogrammetry. *In* "Forensic Engineering." 2nd Ed. (Carper, K., L., Ed.). pp. 327–360. Elsevier, New York.

Hyzer, W. G. (1981). Using a circular scale of reference. *Photomethods* **24,** p. 2.

Hyzer, W. G. (1982). Uses of circular scales. *Photomethods* **25,** p. 14.

Hyzer, W. G. (1982). Perspective grid photogrammetry. *Photomethods* **25,** 6–7.

Moffett, F. H., and Mikhail, E. M. (1980). "Photogrammetry." 3rd Ed. Harper and Row, New York, NY.

Slama, C. C., Editor. (1980). "Manual of Photogrammetry." 4th Ed. American Society of Photogrammetry, Falls Church, VA.

Whitnall, J. (1984). Unimpeachable witness: The grid. *Photomethods* **27,** 44–61.

Whitnall, J., and Millen-Playter, K. (1988). Repeating history by 'reverse projection' reconstruction. *Law Enforcement Technol.* 24–28, May/June.

Whitnall, J., and Millen-Playter, K. (1985). The nitty 'griddy' of accident reconstruction. *Law Enforcement Technol.* 20–29, January.

Wolf, P. R. (1983). "Elements of Photogrammetry." 2nd Ed., McGraw-Hill, New York, NY.

CHAPTER 8

DIGITAL IMAGING

David "Ski" Witzke

LEARNING OBJECTIVES

On completion of this chapter, you will be able to . . .

1. Explain the components of a digital image.
2. Explain what image quality means as it relates to a digital image.
3. Explain the difference between dots per inch and pixels per inch.
4. Explain the difference among image resolution, monitor resolution, and printer resolution.
5. Explain how pixels per inch (PPI) are converted to dots per inch (DPI).
6. Identify and explain the differences between the image file formats used in digital cameras.
7. Explain the difference between lossy compression and lossless compression.
8. Explain the best practices for photographing evidence to maximize image quality.

KEY TERMS

Airbrushing
Artifacts
Bayer pattern
Bit
Byte
Cluster of dots (or device dot cluster)
Digital evidence
Digital image

Dynamic range
Film recorder
Interpolated
JPG
Pixels
Lossless compression
Lossy compression
Metadata
Monitor resolution

Noise
Process
RAW format
TIF

IN THE BEGINNING

Five or six years ago, a television show appeared on the scene that took everyone off guard in terms of its success. The premise was a bit of a leap of faith on the part of the network and producers to present weekly tales of murder and intrigue by following criminal investigations by . . . whom? Street-smart, world-weary detectives with incredible deductive skills and chips on their shoulders? Swaggering District Attorneys presenting the hard truth before bored, yet attentive, juries? Not this time. Instead, the network in question introduced their sleuths as a group of scientists, fighting crime and searching out truth from a laboratory rather than chasing deviants through the streets.

What? A prime-time television show focusing on *crime scene investigation?*

By now you certainly know the success of that show. Millions tune in every week to see computer-aided science in action, indicting bad people and freeing the falsely accused, offering postmortem justice to the victims but little if any kudos to the beleaguered team of forensic gurus. Part of the appeal is that these hi-tech sleuths do not seek recognition, but rather are content that they did their job and did it well, honoring the integrity of the process more than their fame. As a result of this successful prime-time formula, Hollywood has remained true to form in developing numerous clones—some successful, some not, but all sharing the same item of distinction. Most of the science on these shows, especially in the realm of digital imaging, still resides in the realm of science fiction.

This chapter intends to dispel those myths, giving you a foundation of the science of image correction in digital forensic imaging that applies in real-world cases and can be used for real-world courts of law. This chapter is not intended to provide an insight about what you can do in Photoshop, but rather provide ideas about what you can do, what is acceptable, and how you can present digital evidence so that it is accepted in a court of law.

A BRIEF HISTORY OF PHOTOGRAPHY AS EVIDENCE

In the beginning was the photograph, and the photograph was good. Right? Well, not necessarily. For decades, photographs have been accepted in the courtroom as true and accurate representations. Until some of the more recent well-known cases were tried, many defense attorneys would not even consider challenging the use of conventional film, because they believed that it was difficult to commit fraud using silver halide–based film for two reasons: (1) it was very difficult and expensive, and (2) it was believed that image manipulation of conventional film was easily detectable.

As a result, many people in law enforcement have had the mistaken idea that traditional film processing provides a higher level of image integrity and provides a more secure (both physical and long-term) method for storing images. The truth of the matter is that this absolutely is not true.

Photographs have been retouched by use of a number of image capture (camera), development, and printing techniques since the 1800s. For example, **airbrushing** was developed in 1879 by Abner Peeler, and a patent for this technique was issued to Mr. Charles Burdick in 1893.

In its early implementation in photography, airbrushing involved a small, precise tool for spraying paint. In addition to blending colors, such as adding shadows, this technique provided the ability to hide signs that an image had been manipulated.

Until the 1920s, airbrushing techniques were used primarily for photographic (color) retouching. As new photographic tools and technologies were invented, new airbrushing techniques were used to create modified pictures, such as the propaganda posters created and used by the US Army during World Wars I and II. As early as 1929, shortly after the first commercially available cameras were introduced, Lenin had his enemies air-brushed out of photographs, as shown in Figure 8.1.

Airbrushing: A technique for blending colors such as adding shadows by applying ink or paint with a sprayer. This technique provides the ability to hide signs that an image has been manipulated.

Figure 8.1
Stalin frequently had his enemies airbrushed out of pictures.

For a period of time, Nikolai Yezhov, chief of the Soviet secret police, was close to Stalin. In the first picture, he is seen with Stalin. But he does not appear in the manipulated photo. (Pictures taken from The Commissar Vanishes.)

In addition to airbrushing, there have been numerous techniques practiced over the years to enhance and even manipulate photographs. In 1969, John Cohen exhibited more than 40 special effect techniques that he devised on the basis of photographing projected images years before computer manipulation ever existed. You are certainly familiar with the true and accurate images appearing in publications such as *National Enquirer, Celebrity News, Gossip,* and other magazines. Airbrushing and other image manipulation techniques have been used by the print media for years.

The question then is why do attorneys and law enforcement agencies put so much faith in traditional film photography when there has been such a long history of image manipulation? I am endlessly amazed that traditional photographic images are not challenged more often in the courtroom.

For example, the rapid growth of digital technologies for the past 10 to 20 years has made manipulation of traditional silver halide film images even easier. Photographers and graphic artists can digitize traditional photographs with film scanners and flatbed scanners very easily. Once the images are digitized, they can add images into one photograph from another photograph, change colors of individual items in the photograph, and so forth, and then output their new creations to film for printing.

With the proliferation of film scanners and flat bed scanners, images can be digitized by anyone. By using even the simplest imaging tools, they can perform a myriad of tasks, including merging images, thus changing the contents of images, changing colors, restoring old/damaged photographs, and the like. Then they can output their new image to film once again with a laser projection device called a **film recorder**. So the myth that film is more secure than digital is just that; it is a myth.

Film recorder:
A hardware device, similar to laser projector, that writes computer graphic files onto analog film.

This process was used extensively in the early 1990s when presentation slides were created on computers by use of any number of image processing tools available at that time, such as Adobe Photoshop, Paintshop, PhotoImpact, and so forth.

By use of a film recorder, digital images were projected directly onto film from the computer rather than printing a hard copy image that had to be photographed and then processed. The result was a higher quality image instead of a second-generation image created from a low-resolution output device. (Remember, high-resolution photo printers and high-quality, glossy photo papers were not readily available.)

Today, film recorders like the one shown in Figure 8.2 can record up to 16,000 lines of resolution, which represents an addressable resolution of 16,384 pixels by 13,448 pixels, which is approximately 220.3 megapixels.

Furthermore, high-quality LCD projectors have now replaced the use of slide projectors because of their increasing image quality (both lumens and resolution). As a result, it is no longer necessary to create images that must be projected onto film with film recorders, then have the slide film developed, have the slides

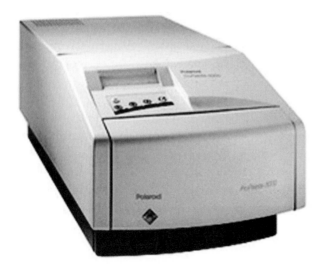

Figure 8.2

High-resolution LCD projectors are replacing the need to create slides for presentations, although film recorders like this one are still readily available.

Today, high-resolution LCD projects coupled with applications like Microsoft PowerPoint are replacing the need to create slides for presentations. However, film recorders like this Polaroid Propalette 8000 digital film recorder are still readily available even on eBay so that you can make your own negatives from digital images. (Picture taken from actual eBay listing.)

mounted into frames, and have the frames inserted into slide trays. Today these images are created, managed, and displayed directly with the computer, giving us higher-quality images much faster with significantly less hassle.

However, film recorders are still readily available . . . and affordable. John Cohen would have been able to produce far more than just 40 special effect techniques with today's technologies. The possibilities today would be practically unlimited.

Although these new technologies have helped improve speed and image quality, the acceptance and use of these technologies have created numerous problems in law enforcement. Not only are we facing questions about the integrity of the original digital image, we are facing challenges about our ability to identify if and where computer enhancements are made. Even our personal integrity is being called into question.

Several film manufacturers, including Kodak, have spent thousands of dollars comparing original film photographs to modified film photographs to determine whether or not they could identify the original image from the modified film image. The results of all these tests were the same: they could not identify which was the original image and which was the manipulated image. In fact, when Kodak released the findings of their study, they stated that they could not discern the existence of pixel clouds within the manipulated image. In other words, they could not identify which image had been captured originally with a camera or captured after the image had been modified and recaptured with a film recorder.

Pixels: An abbreviation for picture element, which is the smallest element of a digital image

Metadata: The data contained within an image file that describe how and when the image was captured, as well as information including date, time, file size, enhancement history, and other data.

The good news for law enforcement is that unlike traditional film cameras, digital cameras do provide us with the ability to identify whether an original image has been altered. For example, most digital cameras manufactured within the past few years capture information about the image **pixels** (also known as picture elements) together with information such as date and time when the image was taken and camera settings such as aperture setting, shutter speed, ISO, flash settings, exposure compensation, lens used, and focus distance.

These data elements (also known as **metadata**) are then stored as part of the image file in a collection of data fields called the file header.

The most common header format used today is the EXIF (exchangeable image file format) header. In fact, EXIF has become the standard for storing camera information, thus providing interoperability between digital cameras and image processing programs.

EXIF data has also proven to be very useful because photographers with digital cameras do not have to worry about recording the settings they used when taking the image on a photo log. The photograph is taken, the image together with the information about the photograph, such as the camera make, model, and serial number, and the camera settings are stored with the image. Later the photographer can analyze these data and determine which camera settings provided the best results, thereby learning from their photographic experiences.

In the past, the FBI had always required two photographers at each scene: one photographer operated the camera, while the second photographer manually recorded the camera settings on a photo log. As you can imagine, this is a very time-consuming and labor-intensive process. Yet it does nothing to preserve and protect the integrity of the image captured on film. As professional as a person or persons may be, people still make mistakes.

Today, many image editing and viewing programs can display, and even edit, the EXIF data. With some image editing programs, the EXIF data are lost when the image file is resaved after editing. Therefore, it is imperative to always preserve the integrity of your original digital image using an application such as Foray Technologies' ADAMS products.

By use of digital image management applications such as Digital Workplace, you not only preserve and protect the integrity of your original digital images, but you also maintain the EXIF data in a format that cannot be edited. As a result, you not only have a more secure imaging solution, but you can also prove what camera settings were used, which is something that simply cannot be done with a film camera.

The bottom line is that today digital imaging provides a faster, more secure imaging solution than film was ever capable of providing. You can now also prove the integrity of any original image, which is something that is physically impossible to do with film.

 Rule of Thumb 8.1: Be sure to maintain the EXIF data in your original file so that you can refer to that information for all of your camera settings.

With a digital imaging process, you will no longer have to ask the question of whether or not your film contains the original image or if it was altered, retouched, and re-recorded. The truth of the matter is that digital imaging actually provides the best line of defense for actually proving image integrity.

THE NEW DIGITAL ARSENAL FOR LAW ENFORCEMENT

Technology changes faster than we can keep up with it. New digital cameras are constantly coming out that are more affordable and provide an even wider range of capabilities.

Digital video recorders are being used for surveillance, crime scene analysis, and investigation, as well as in patrol cars and more. Digital technologies have also provided a new arsenal of weapons for the courtroom, enabling law enforcement personnel to clearly articulate and illustrate their analysis and investigation through the use of digital photographs (PowerPoint presentations) and computer simulations.

The bottom line is that digital imaging has become the weapon of choice for arson investigations, crime scene investigation, fingerprint identification, sexual assault examinations, domestic violence cases, and so forth. Digital technologies have enhanced the way we do crime scene reconstruction, calculate blood spatter trajectories, create photo lineups, create wanted and missing persons fliers, and manage evidence and property. Medical examiners are using digital imaging technologies during autopsies; nurses are using digital technologies during sexual assault investigations.

Just a couple of years ago, it would take days or even weeks for domestic violence or sexual assault cases to be heard, and photographs were used that were often blurry or failed to show the extent of the actual injuries.

With digital technologies, images of injuries such as black eyes, bruised cheeks, handprints around the neck, scratches, and abrasions can be sent by means of computer from the detective to the prosecutors and judges before arraignment of the suspect. In this case, the judge can decide whether or not to grant bail or issue a restraining order to safeguard the victim before arraignment of the suspect.

Although these digital technologies enable criminals to be prosecuted faster and, in many cases, save lives, new challenges are also being made in the courtroom. Allegations and insinuations have been made against law enforcement that challenge the integrity, quality, and authenticity of digital images. Defense attorneys challenge the use of these technologies saying that digital images can be altered easily or that the technology is not scientific or not proven. Unfortunately, the burden of proof rests not on the defense who made those allegations, but on the prosecutor who must defend the use of the technology and demonstrate that the technology is not only sound but that it is independent, scientific, non-biased, and, most importantly, that it has been used properly.

Figure 8.3

Digital technologies assist in catching and convicting the bad guys.

Digital photographs are frequently used to help identify possible suspects.

Lossy compression:
A technique for reducing the file size of a digital image. When the compressed image is decompressed, the missing information can be reproduced exactly as it was captured.

Lossless compression:
A type of compression that ensures that the original image information can be recreated exactly as it was captured.

For example, one of the most frustrating challenges that law enforcement agencies face internally is that many supervisors do not appreciate or understand the difference between a **lossy compression** (JPG) capture format and a **lossless compression** (RAW or TIF) capture format.

Cameras that use lossy compression, such as the compression inherent in a JPG format, discard actual picture information to create smaller file sizes. When these images are subsequently decompressed, displayed, and enhanced, artifacts actually are added into the image. This opens a huge door for attorneys to challenge both the quality of the image and the analysis of the image.

When presenting these images in court, the forensic expert must be knowledgeable of these facts and must be able to articulate the effects of a lossy compression scheme compared with a lossless compression on the image. Certainly, in some instances, such as traffic accidents and general crime scene photographs, the effects of a lossy compression will not have a significant effect on the content of the image. Any image, however, that requires analysis should be captured by use of a lossless compression scheme.

ARRESTING NEW DEVELOPMENTS

During the past decade, we have witnessed a phenomenal surge in the development, use, and acceptance of digital technologies within national security, antiterrorism efforts, law enforcement, and criminal investigation. Not only has the use of digital cameras become widespread, but the use of digital photographs and digital video in the courtroom has flourished as well.

In addition to increasing productivity, digital technologies have afforded law enforcement agencies the ability to reduce and even eliminate film, processing, and printing costs while minimizing development/processing time. More importantly, digital technologies provide immediate access to vital information.

Some tradeoffs also exist. A growing mandate exists for better (higher-resolution) images, standardized image processing techniques, better image management processes, and training.

Although these mandates have helped create leading edge hardware and software products, they have also let us adapt technologies to meet specific program requirements and provide innovative and effective information technology solutions for the courts and law enforcement agencies worldwide.

The greater acceptance of digital technologies has primarily been the result of new, high-resolution digital cameras. These cameras have changed drastically the way we do image analysis, image enhancement, and image processing, making the images more reliable and more dependable.

Since the digital imaging revolution began in the early 1990s with the release of software applications such as Dr. Halo, Adobe Photoshop version 1.0, imaging technologies have changed, and image-processing capabilities have improved significantly. Digital cameras have a greater dynamic range and include a vast array of new capabilities; new digital video recorders have improved the way we analyze surveillance videos or document and analyze crime scenes.

As shown in Figure 8.4, capturing the image is only one piece of the puzzle. To complete the analysis, we must often process the image for better visualization. Adobe Photoshop is one of the oldest, most widely used, and most widely accepted image-processing tools available today. This application has been used by law enforcement personnel to integrate latent prints with automated fingerprint identification systems, perform crime scene analysis, create mug shot lineups, perform ballistics analysis and comparisons, create facial composite drawings and perform facial reconstruction of unidentified victims, and prepare court exhibits.

As is the case with all technologies, baseline standards, recommended techniques, and generally accepted guidelines exist. Although mandated procedure or technique is required, the question that must be answered is: Would an individual with comparable training, expertise, and ability be able to create a reasonable likeness of the same image?

Figure 8.4

Image capture is only one piece of a multifaceted puzzle.

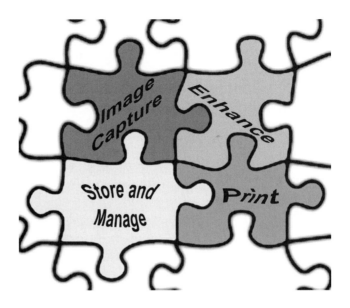

Our experience with conventional photography has taught us that not everyone has the same skill when it comes to photographing a difficult object. The angle of the light, the density and intensity of light, the filter(s) being used, and so forth will drastically affect image quality. It is only through experience that we develop the ability to capture those types of images.

Similarly, it is only through training and experience that we develop the ability to **process** (enhance) digital images as well.

Even the printer (ink jet, laser, or dye sublimation) that we use can affect the quality of the image because of the different types of inks (dye versus pigment) and the different types of paper (plain, matte, and glossy).

Today, most ink jet printers use multi-size dot technology together with multi-pass technology, where the color tonal value for each output pixel is created by multiple passes of the print head that spray varying size ink droplets directly onto the paper to create an image.

Ink jet printers can also use from 4 to 12 ink colors together with varying sizes of ink droplets, typically ranging from 1.0 picoliter droplets to 20.0 picoliter droplets. Today, a number of ink colors are available for ink jet printers compared with the old days when cyan, magenta, yellow, and black were the rule. For example, depending on the printer that you are using, you could have any combination of red, blue, green, cyan, photo cyan, magenta, photo magenta, yellow, photo yellow, regular black, matte black, light black, and light-light black. (Some manufacturers refer to light black as gray, and light-light black as light gray.)

Ink jet printers provide a range of options regarding the number of nozzles per print cartridge, ranging from 180 nozzles per print cartridge to 2560 nozzles

Process: The steps involved in improving the perceived quality of the image or preparing the image for printing.

per print cartridge. Although we are still limited by the printer's rated DPI, image quality can be improved by choosing a printer with more ink colors, smaller picoliter droplets, and a higher nozzle count (per print cartridge).

Many people are not aware that ink jet printers automatically adjust the volume of ink that is sprayed onto the paper based on the type of paper that is used. As shown in Figure 8.5, ink jet printers typically use an output resolution of approximately 720 dpi when you use plain paper because the ink bleeds (is absorbed) into the surface of the paper. When glossy photo paper is used, the ink actually beads up on the surface of the paper and retains its dot (round) shape and provides a clearer image. In addition, more ink droplets can be sprayed closer together, thus providing a significantly higher dpi.

Laser printers, on the other hand, use a highly focused beam of light to place dot arrays onto the photoreceptor drum, which in turn transfers the image to the paper. The image itself is actually created with static electricity, dry ink (toner), and heat to bond (melt) the toner onto the paper. Unlike ink jet printers with varying ink droplet sizes, there is only one size of toner powder (particles) and the laser has only a single, unvarying diameter. Moreover, most laser printers are limited to only four colors of toner: cyan, magenta, yellow, and black.

Dye sublimation printers (also referred to as dye sub printers) use a roll of film that resembles sheets of red-, blue-, and yellow-colored cellophane stuck together end to end. Embedded in this donor film are dyes that correspond to the colors used in printing: cyan, magenta, and yellow. Most dye sub printers use equal parts of cyan, magenta, and yellow to create black, thus delivering all four elements of the CMYK color mode. In contrast, some dye sub printers actually include a black-colored segment to complete the CMYK elements of color. Similar to a laser printer, the print head on a dye sub printer contains a heating element.

Figure 8.5

Ink droplet on plain paper versus ink droplet on glossy paper.

Ink droplet on photo glossy paper keeps its shape, while the ink droplet becomes irregular in its shape as it is absorbed into plain paper.

Figure 8.6

The ultimate goal is a reliable and accurate representation of the facts.

 The original image

 An accurate representation

 A not-so-accurate representation

Digital image: The process of creating an electronic recording in a two-dimensional form of a physical object, such as a person, place, or thing.

Digital evidence: Evidence, such as latent prints and impressions, that can only be recorded or collected by photography when the image itself is not recoverable. Also, electronic files that give rise to the cause of action, such as child pornography; audit trails in the case of computer hackers; and images of computer hard drives, cell phones, and PDAs that are seized during the investigation of a crime.

However, the heating elements in a dye sub printer provide a wide range of temperatures to bond the appropriate amount of each color of dye onto the surface of the paper. Typically, dye sub printers make a complete pass over the paper for each of the colors contained on the ribbon, thus little by little building the image on the page.

As a result, the same image printed from the same computer can appear different simply because of different output devices, as shown in Figure 8.6.

Our goal throughout the entire process is to produce the most reliable, accurate representation possible.

DIGITAL IMAGE OR DIGITAL EVIDENCE? THE DEBATE CONTINUES

Today more and more people are using computers, watching digital television, wearing digital watches, using digital cell phones, and so forth. The terms and words associated with these technologies are infiltrating our day-to-day language. As a result, the use (or should I say abuse) of the word digital to describe these technologies has created a great deal of confusion and consternation.

For the purposes of this book, we will use the term digital image to refer to the process of creating an electronic photograph (one existing inside a computer) from a physical object (a document, photograph, or other three-dimensional object).

We also need to make a distinction in this book between a **digital image** versus **digital evidence**.

According to the American Society of Crime Laboratory Directors, Laboratory Accreditation Board (ASCLD/LAB), when evidence, such as latent prints and impressions can only be recorded or collected by photography and the image itself is not recoverable, the photograph or negative of the image must be treated as evidence.

As was stated earlier, a digital image is a representation of a physical object. From a purely legal argument, the physical object is the actual evidence; the photograph is merely documenting its existence, such as recording the existence of a fingerprint found on a piece of evidence for historical purposes, or recording an object that may change over time, such as a latent print on paper developed using a chemical process (ninhydrin) in which the actual fingerprint will fade over time.

In the latter case, the line between digital image and digital evidence becomes extremely thin.

To clarify the distinction between digital evidence and documentation a little better, think of a digital image as a recording of a location or event as documentation and digital evidence as physical proof of the crime. To make this distinction a bit more clear, a digital image could be a photographic recording of where the blood appeared on a wall at the crime scene, where the drywall with the fingerprint in the blood are the actual evidence.

To use another example, the existence of a sexually explicit digital photograph involving young children created with the primary aim of eliciting significant sexual arousal can give rise to the actual cause of action (i.e., child pornography). In this case, the presence of the digital image on the computer would be the actual evidence. (Some people also refer to the contents of a computer hard drive as a disc image, which is considered the evidence.)

Digital evidence is not always a digital image. A breach of network security, a list of victim names on a suspect's computer, or a list of credit card numbers and/or other personal information obtained illicitly and stored on a computer are all considered to be incriminating and, therefore, digital evidence.

As illustrated in Figure 8.7, the way digital images are handled, how they are processed, how they are stored, and how they are described can make a difference in the courtroom.

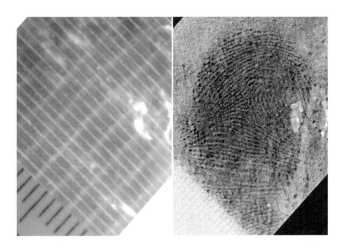

Figure 8.7

Images are evidence: properly tracking the techniques used during the enhancement process and explaining which tools were used can make all the difference in the courtroom, especially when the original image appears to be unidentifiable and the processed image can be identified easily.

The bottom line is that numerous, complex issues are associated with the use of digital technologies within the law enforcement community. Law enforcement agencies have been incorporating new computer forensic disciplines into their infrastructures in an effort to fight criminal activity by use of digital image processing or developing policies and procedures for handling incriminating electronic evidence.

The key is remembering that the same procedures, definitions, and guidelines may not always be applicable in all circumstances. For example, the image capture tips and techniques described in this book are neither case-setting mandates, policy directives, nor are they the only correct image processing techniques.

Through the use of the information contained in this book, we hope that more law enforcement personnel will be trained to work effectively and efficiently with digital images and that they will maximize the reliability of those images, resulting in more cases solved.

Regardless of whether you are dealing with digital images or digital evidence, nothing is more easily damaged, corrupted, or erased than electronic data. Therefore, you must be able to prove beyond a reasonable doubt that the digital image or digital evidence is true, accurate, and reliable and that it has not been significantly modified in any way since you acquired it.

Thus, we set the scene: In the beginning was the photograph, and the photograph was good. Right? Well, not necessarily. It all depends on you.

EXPOSE YOURSELF TO DIGITAL IMAGING CONCEPTS: BITS, BYTES, PIXELS, AND DOTS

As mentioned earlier, we have proven to be very willing participants in the development, use, and acceptance of digital technologies: digital watches, digital cell phones, digital televisions, digital video recorders, digital music recorders/ players, and on and on.

Although the use and acceptance of these technologies have advanced rapidly in recent years, the terms and understanding of these technologies has not made the same quantum leap.

For example, walk into a computer store such as CompUSA, Best Buy, Circuit City, or Staples and ask two people in each store to explain to you the difference between the resolution of a digital camera and the resolution of a printer. One is measured in pixels and the other is measured in dots. So what is confusing about that?

To eliminate confusion and frustration in a meaningful discussion about digital imaging technologies, you should never use the terms dots per inch (dpi) and pixels per inch (ppi) interchangeably. These terms have totally separate and distinct meanings. Generally speaking, pixels are square, pixels are continuous tone, and each pixel represents a unique color value. Dots, on the other hand, are irregular in shape (the degree to which the dot bleeds or is absorbed into the paper depends on the quality and thickness of the paper on which the dots are placed and the absorption rate of the paper), use white space to simulate a specific color value, and use multiple dots to represent a single pixel color value.

Rule of Thumb 8.2: The terms dots per inch (dpi) and pixels per inch (ppi) may not be used interchangeably because these terms have totally separate and distinct meanings.

Let us see whether we can clear up some of the confusion. The most commonly used and misunderstood terms in the digital world are:

- Bits
- Bytes
- Pixels
- Dots

IT'S ALL BITS AND BYTES TO ME

A **bit** is the smallest element used by a computer or a digital camera. It can have only one of two values: one or zero. No values can exist between these two values. In its most simple, basic form, all digital information is stored as a series of ones and zeroes.

A **byte** is made up of 8 bits, hence the term 8-bit processing. As shown in Figure 8.9, you can use 8 bits to a byte to represent a total of 256 different tonal range values, such as 256 different shades of gray, ranging from pure black to pure white.

Bit: The smallest element used by a computer or a digital camera. It can have only one of two values: one or zero.

Byte: A measurement of computer data that consists of 8 bits.

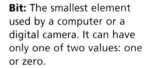

Figure 8.8

Inside a computer, a picture appears as nothing but ones and zeroes.

Figure 8.9

This table illustrates how each additional bit doubles the possible number of combinations of 1 or 0, which is expressed in powers of 2. Therefore, 2^8 would provide 256 possible combinations... or tonal values.

Bit	Supporting equation	Possible combinations of 1 or 0
2^1	2 x 1	2
2^2	2 x 2	4
2^3	2 x 2 x 2	8
2^4	2 x 2 x 2 x 2	16
2^5	2 x 2 x 2 x 2 x 2	32
2^6	2 x 2 x 2 x 2 x 2 x 2	64
2^7	2 x 2 x 2 x 2 x 2 x 2 x 2	128
2^8	2 x 2 x 2 x 2 x 2 x 2 x 2 x 2	256

As shown in Figure 8.10, the density or the contrast between varying shades increases as the number of bits increases, which is important to know, because different digital capture devices capture light values from 10-bit to 16-bit grayscale and then resample those values to 8-bit grayscale so that the grayscale values can be displayed and adjusted. (Photoshop CS and CS2 AKA Photoshop 8 and 9 provide the ability to enhance 16-bit images; however, you may not see a noticeable difference in a 16-bit image and an 8-bit image because of the limited display capabilities of most video cards/monitors.)

Figure 8.10

When talking about bit depth, the dynamic range (contrast) between 4-, 8-, and 16-bit grayscale is significant.

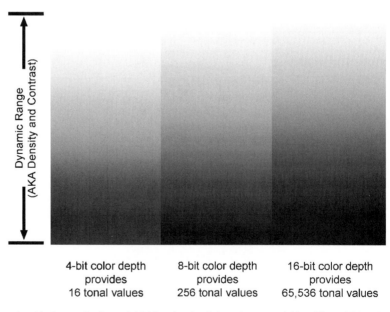

4-bit color depth provides 16 tonal values

8-bit color depth provides 256 tonal values

16-bit color depth provides 65,536 tonal values

The contrast between 4-, 8-, and 16-bit color depth is rather astonishing. The additional contrast provides more detail within the image.

8-Bit Grayscale

Number of bits	Tonal Values
8	256 shades of GRAY
TOTAL BITS 8	

24-Bit Color

Number of bits	Tonal Values
8	256 shades of RED
8	256 shades of GREEN
8	256 shades of BLUE
TOTAL BITS 24	

48-Bit Color

Number of bits	Tonal Values
16	65536 shades of RED
16	65536 shades of GREEN
16	65536 shades of BLUE
TOTAL BITS 48	

Figure 8.11

With 8 bits per color channel, there is a total of 24-bit color. Similarly, with 16 bits per channel, there is total of 48-bit color.

Now that we know 8 bits allow us to display 256 shades of gray, let us see whether we can add to the complexity of terms without confusion.

Many people refer to 8-bit color and 24-bit color as the same thing. Although 8-bit color is typically used to describe the gradient tonal range from black to white, it is also used to describe 24-bit color, because there are 8 bits or 256 tonal values for each color channel—red, green, and blue.

Earlier we said that 8 bits equal 1 byte. Therefore, each picture element (also known as a pixel) has 1 byte for each color: red, green, and blue. Each byte allows us to represent up to 256 shades. Similarly, as shown in Figure 8.11, 16-bit and 48-bit color are often used interchangably to describe tonal value in digital images.

Regardless of whether your digital camera uses a CCD (charge-coupled device) or a CMOS (complimentary metal-oxide semiconductor) imaging sensor, the sensor starts at the same point. It has to convert light into an electronic signal that is broken down into a digital value for each sensor, which is also commonly referred to as a pixel. During this process, the computer assigns a red, green, and blue color value for each pixel. This means that there are 3 bytes for each pixel.

This means that if you have a 6 million pixel (aka, megapixel) camera, your noncompressed image size would be 18 million bytes.

When converted into megabytes, this would be approximately 17.16 megabytes.

Let us see whether that can be explained without going too deep into computer science.

A bit can have only one value: a one or a zero (a bit is on or off, as in a computer instruction); eight bits make up a single byte, thus providing 2^8 or 256 possible values.

In contrast, bytes are used to represent the storage capacity of a string of bits such as the byte required to store the value of a single pixel within a digital image. Because we use a base 10 numbering system, we use a multiple of 10 bytes to describe the storage capacity of 8 bits per byte. Although not to be confused with bits, bytes are also defined in powers of 2 to 2^{10}. If you do the math, you will see that there are 1024 bytes in a kilobyte, 1024 kilobytes in a megabyte, 1024 megabytes in a gigabyte, 1024 gigabytes in a terabyte, 1024 terabytes in a petabyte, and 1024 petabytes in an exabyte.

To convert the millions of pixels in a digital image into kilobytes and then megabytes, you must first divide the total number of bytes by 1024 to determine the total number of kilobytes, then divide that number by 1024 to determine the number of megabytes.

In addition, please remember that this is only an approximate number of megabytes. Each camera stores varying amounts of information about the camera and the image (metadata), which will cause the stored file to be slightly larger.

Let us summarize where we are just to make sure you are not lost. A digital image is defined as a numerical representation of a physical object recorded electronically in a computer as a series of binary digits (bits); each bit is either a one or a zero, and there are no values in between.

Therefore, a digital image is made up of picture elements more commonly known as pixels. Each pixel is stored on a digital camera, a computer, a CD or DVD, etc. as a byte, which consists of a series of 8 bits.

When used to describe storage capacity, bytes are grouped as 2^{10} or 1024 kilobytes; and 1024 kilobytes equal 1 megabyte.

Now that we have taken a picture and we know how it is stored, we have to understand how that image is displayed on a monitor or printed.

Here is where the real confusion begins.

ARE ALL PIXELS EQUAL?

When a digital image is captured, the image is digitized into a series of pixels. Each pixel has a specific color value, such as teal, fuchsia, orange, or brown, based on the composite values of red, green, and blue for each pixel.

On the basis of 256 possible shades of red, 256 possible shades of green, and 256 possible shades of blue, you can have a total of 16,777,216 possible shades for each pixel, which would yield a specific color value—teal, fuchsia, burnt umber, whatever.

Computer screens also have a resolution of their own that has no direct relationship to the pixels in the image that you captured. For example, a standard **monitor resolution** is 1024 pixels by 768 pixels.

Monitor resolution:
A measurement of the number of pixels, both width and height, that a monitor is capable of displaying.

This monitor resolution refers to the number of color pixels on the screen. These display pixels are made up of a combination of red, green, and blue lights. (Although meaning the same thing, these lights are displayed by use of different technologies, depending on whether you are using a cathode ray tube [CRT], liquid crystal display [LCD], light-emitting diode [LED], or plasma display.)

The bottom line is that a combination of red, green, and blue lights coupled with the intensity of each individual light creates an optical illusion that causes your eye to perceive the specific color value of an image pixel contained within your digital image.

Let us refer again to the 6-megapixel digital camera that was discussed earlier. We said that the width of the sensor was 3000 pixels and the height of the sensor was 2000 pixels for a total of 6 million pixels. So how do we display 6 million pixels on a display screen that only contains 786,432 pixels (1024 pixels × 768 pixels)?

Simple. We zoom out on the image (reduce image size). Instead of looking at every pixel across the image, we sample the image and display one pixel value for every 2.93 pixels contained in the original image.

In contrast, if we want to see more of the detail contained within the actual digital image, we zoom in on the image (enlarge the image size). If we zoom in too much, we can cause the image to lose image quality and appear pixilated. For example, we can use those same display pixels (1024 × 768) and display only 640 pixels × 480 pixels of our digital image. Because we are displaying fewer image pixels, they will appear large and boxy.

Rule of Thumb 8.3: Resolution by itself does not imply image quality or size because the same number of pixels can be displayed as a small area (zoomed out) or as a large area (zoomed in) on a monitor or printout. Therefore, you must be careful about enlarging low-resolution images too much on a monitor or in a printout.

The moral of the story is to not get all hung up on trying to figure a ratio of digital image pixels to display pixels because as soon as you zoom in or zoom out on the image, the ratio will no longer be the same. Just bear in mind that even a low-resolution image can appear clear when it is displayed on a monitor or it can appear very ugly if it is enlarged too much on the screen. (Starting to feel like Goldilocks? This one is too big. This one is too small. This one is just right.)

PIXELS AND DOTS ARE NOT THE SAME

In the previous section, we mentioned that on a computer monitor a single image pixel is composed of a series of display pixels that create an optical illusion to make your eye see a specific color value.

Cluster of dots (or device dot cluster): The number of dots in a cluster that varies based on the number of inks, varying sizes of droplets, and number of nozzles per print head in your printer.

Similarly, when an image is printed, each pixel is made up of a **cluster of dots**.

As we described earlier, the number of dots in a cluster can vary based on the number of inks used by your printer, especially if you are using an ink jet printer. For example, perhaps your printer uses four inks—cyan, magenta, yellow, and black—or a combination of 6, 7, 8, or 12 ink colors. The bottom line is the same whether you are using an ink jet printer or a color laser printer: the printer combines a series of dots to create the illusion of a specific color value, such as wheat, egg shell, sky blue, or candy apple red.

So how do we convert pixels to dots? When an image is printed, each pixel is converted into a series of halftone dots. Each halftone dot is then converted into a series of device dot clusters based on the available ink colors used by the printer, the number of nozzles per print head, the size of droplets, and the type of paper used.

Figure 8.12 helps to show how this process works.

Just remember that the number of halftone dots used for each pixel and the number of device dot clusters used for each halftone dot will vary on the basis of the type of paper used—quick-drying, premium photo glossy paper provides a better-quality image than either matte finish or plain paper. With glossy paper, the ink actually floats on the surface of the paper, and because of its quick drying capabilities, the dots can actually be overlapped. Using matte finish or plain paper, the ink cannot be placed as accurately, because the ink actually sinks into the surface of the paper and the ink colors bleed into one another.

Today's digital cameras match or slightly exceed the performance (image quality) of silver halide film, and computer graphics have achieved the goal of photorealism. So what is the problem? It is all about output.

To date, I have not found a laser printer that provides the same quality of output provided by ink jet printers. After conducting image quality tests using an approved FBI test target at 500 ppi on a wide variety of printers for more than 6 months, the Epson R1800 and the Canon i9900 provided the best output quality for forensic digital imaging today.

However, some newer printers have not yet been tested, such as the new Canon image PROGRAF iPF5000 ink jet printer, with its 12 inks and 30,720 print nozzles.

Regardless of the type or model of printer you use, the image quality is still not as good as the traditional silver halide–based photographs used by many latent print examiners for comparison. With traditional photographic prints, you are looking at all of the information contained within the negative; with digital imaging, output devices use only a portion of the available information when printing calibrated, 1:1 life-size images. For example, a latent print scanned on a flatbed scanner has a calibrated resolution of 1200 ppi, and the output device prints 400 ppi, then the latent print image must be resampled to eliminate a total of 800 ppi. As a result, approximately 60% of the image data is lost during resampling.

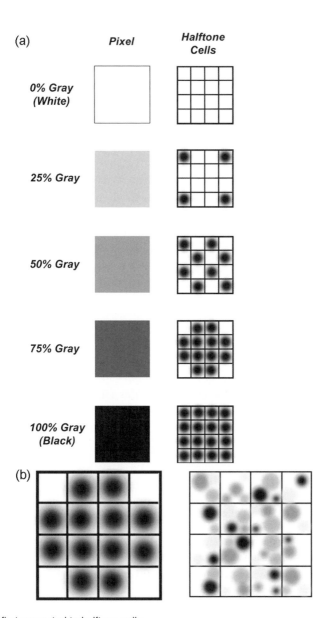

Figure 8.12

Pixels are first converted to halftone dots, which are ultimately converted to device dot clusters.

(a) Pixels are first converted to halftone cells.
(b) These halftone values are then converted to device dot clusters, which are based upon the number of inks used in the printer as well as the type of paper used.

As a result, latent print examiners must retrain their eyes for doing comparisons using lower-quality images, doing comparisons on screen, or using larger than life-size images. For example, let us say that you are using a 6-megapixel camera, which has a sensor with a resolution of approximately 3000 pixels wide by 2000 pixels high. If you capture an area that is 3 inches wide by 2 inches high, you would have a resolution of approximately 1000 pixels per inch.

If you are using a dye sublimation printer that has an output of only 300 pixels per inch, the life-size output (3 inch by 2 inch printed area) would only consist of 900 pixels by 600 pixels. In other words, you would only be viewing approximately 30% of the total pixel values contained in the original image. If you were to view the image at three times its life-size, 3 inches would become 9 inches, and 2 inches would become 6 inches. On the basis of an output of 300 pixels per inch, 9 inches would contain 2700 pixels and 6 inches would contain 1800 pixels, thus providing the ability to see almost every pixel in the originally captured image.

As described earlier, not all types of printers work the same way, and they do not provide the same image quality. However, every printer has a purpose:

- Some printers provide very good image quality, but at the cost of slow output.
- Some printers provide high-speed output, but at the cost of lower image quality.
- Some printers cost very little to buy, but the cost per print is exceedingly high.
- Some printers have a hefty price tag, but the cost per print is very low.

The moral of the story is that you must decide which printer will work the best for your specific needs. You must determine:

- Average number of prints required (output capacity)
- Use of prints (image quality desired for 1:1 prints, crime scene photos, evidence, documentation, and so forth)
- Size of prints (what print sizes are most commonly used, such as 4×6, 8×10, and 12×18)

Then it is just a matter of determining what tradeoffs and/or sacrifices you are willing to make. In most law enforcement agencies, the issue boils down to speed and cost versus image quality.

Comparing prints per minute (speed) and costs per print is very easy; comparing image quality can be both subjective and confusing. Although all ink jet printers are rated at a specific dpi, the printer's rated resolution is only one of the elements concerning image quality.

Therefore, although no given equation will always give the exact corresponding number of dots to pixel ratio, at least you can determine the approximate ratio of dots to pixels.

Dots are arranged into dot grid cell blocks to reproduce more than two tonal (halftone) values. In many instances, printers use a 4×4 dot grid to produce 17 monotones (grayscale values) as shown in Figure 8.13. With seven ink cartridges, the 4×4 dot grid can produce a total of 119 color tones.

In contrast, most of the newer ink jet printers use an 8×8 grid, which produces 65 monotones. With multiple ink colors, varying size droplets, and multipass technology, the 8×8 dot grid yields a significantly greater number of

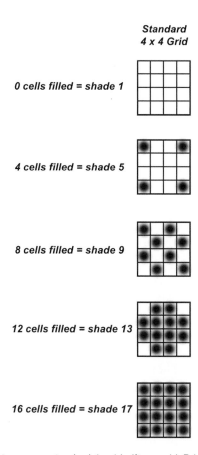

Standard
4 x 4 Grid

0 cells filled = shade 1

4 cells filled = shade 5

8 cells filled = shade 9

12 cells filled = shade 13

16 cells filled = shade 17

Figure 8.13

By use of a simple 4 × 4 grid, the printer can produce a total of 17 shades of gray.

Most inkjet and laser printers use a standard 4 x 4 halftone grid. Printing a dot in each grid produces a different shade of gray; leaving all grid spaces empty represents white. With 16 grids available, plus all grids empty, the printer can produce a total of 17 shades of gray.

color tones. For example, an 8×8 grid using 3 varying size droplets with 7 inks would yield a total of 455 color tones.

$8 \times 8 = 64$ plus 1 for all white = 65 monotones

65 monotones $\times$ 3 varying size ink droplets = 195 monotones

195 monotones $\times$ 7 inks = 1365 possible color tones

Studies have shown that the healthy human eye, under optimal lighting conditions, can distinguish somewhere between 16 and 32 shades of gray in a range from black to white.

By use of 8-bit grayscale, computers can distinguish 256 shades of gray between black and white, which is approximately eight times greater than the ability of the human eye.

Using 8 bits per color channel, there are 256 shades of red, 256 shades of blue, and 256 shades of green; the computer can render a total of 16,777,216 different color combinations for each pixel. Although 1365 possible color tones is nowhere close to the more than 16 million possible shades available, it does produce a nice, picture-quality print.

Ultimately, the tradeoff is that a larger dot grid cell block produces more tonal values but has a lower resolution, whereas a smaller dot grid cell block produces a higher image resolution at the expense of tonal quality.

Now that we have the color tonal values defined, we can continue our digital imaging math class and move on to determining approximate dot to pixel ratios.

Keep in mind that it is impossible to determine the exact ratio of dots per inch (dpi) to pixels per inch (ppi) because each printer manufacturer uses its own proprietary algorithm for different paper types, ink droplet sizes, nozzles per print head, different number of inks, and so forth. You can, however, identify an approximate ratio of dots to pixels using your printer's specifications.

To begin, you must first identify the number of halftone values (dpi) specified for width and height. (This is the easy part. They almost always tell you this information right on the box that the printer arrives in.) Next, you must perform the following calculations: (The examples contained in the following steps are based on the Epson StylusPhoto R1800 Ink Jet Printer.)

1. Divide the width dpi by the number of dots in the width of the dot grid (i.e., 5760 divided by 8 = 720)
2. Divide the height dpi by the number of dots in the height of the dot grid (i.e., 1440 divided by 8 = 180)
3. Multiply the dividend of the width times the dividend of the height (i.e., 720 × 180 = 129,600)
4. Determine the square root of the sum from step 3 (i.e., square root of 129,600 = 360)

Although no firm, fixed, fast rule exists as to when to use a dot grid of 4 × 4 or 8 × 8, the specifications for your printer will often give you a hint. For ink jet printers that have a resolution greater than or equal to 2400 dpi in the width or height dpi measurement, you can reasonably be assured that the dot grid is a standard 8 × 8 dot grid. Anything less than 2400 dpi typically uses a 4 × 4 dot grid.

Once again, remember that larger dot grid cell blocks produce more tonal values but have a lower resolution, whereas a smaller dot grid cell block produces a higher image resolution at the expense of tonal quality.

So what does all this mean in the real world? Let us take a latent print that is to be submitted to IAFIS, which requires a minimum resolution of 1000 pixels per inch (in accordance to Appendix F of the FBI's Integrated Automated Fingerprint Identification System (IAFIS) Image Quality Specifications [IQS]). To visualize all 1000 × 1000 pixels contained within the image, we would have to print as 2.28 inches:

360 pixels for the first inch
360 pixels for the second inch
280 pixels for the remaining .28 of an inch

Dye sublimation printers are a printer of a different color. Dye sublimation printers actually print pixels, not dots. The color value for each pixel represented is the buildup of the colors contained on the dye sub donor ribbon. Therefore, it represents an actual color value; it is not made up of dot clusters used in conjunction with white space on the paper. It is for this reason that dye sub printers are commonly referred to as continuous tone printers.

Similarly, dye sub printers only print a limited amount of image data when the image is printed as a life-size 1:1 image. By use of the previous example, the 1000 pixel per inch image would have to be printed out as 3.333 inches instead of 1 inch to visualize all 1000 pixels. For example:

300 pixels for the first output inch
300 pixels for the second output inch
300 pixels for the third output inch
100 pixels for the last one-third output inch

In addition to getting very high-quality images from dye sub printers, you do not even have to do the math to figure how many pixels are represented in 1 inch of output; the printer's specifications actually tell you the number of pixels output for each inch. The bad news is that the newer ink jet printers are producing high-resolution results that are as good as, if not better than, most dye sub printers.

This should help shed a little light on confusion about dpi, ppi, droplet size, dot pitch grids, colors of inks, etc.

MAKE A STRATEGY FOR YOUR ENTIRE IMAGING PROCESS

Some people still wonder why latent print examiners complain about the image quality of 10 print cards from live scan devices. First, take a high-quality image, then compress it (losing image detail intentionally), then print approximately 300 pixels of the total number of image pixels in that image. The result is a digital output from a known print that is often of lower resolution and lower image quality than the corresponding latent print that was photographed at a crime scene.

A vast number of automated fingerprint identification systems (AFIS) administrators want to cut costs, and I am recommending higher-resolution images that require more storage space and better output devices that produce somewhat more expensive prints at a slower rate of speed, which means it will require either more printers or a more expensive printer with a faster speed.

As described throughout this chapter, our goal in forensic imaging is to ensure maximum image quality when possible. The thing to remember is that this is not simply accomplished in one single area within all of the related imaging processes in law enforcement.

As illustrated in Figure 8.14, the world of digital imaging creates a big picture. Not only do you have to take into consideration what camera you are using, but you have to account for the following:

- Image size based on camera resolution and capture format, which affects the amount of images that can be stored on your storage medium in the camera and on the computer itself.
- Image quality based on type of lenses used, such as macro lens or wide-angle lens.
- Processing speed and storage requirements of your PC, including impacts on network performance and throughput if you are storing images on a network server.
- Backup and archiving routines to ensure data security and prevent catastrophic loss of data.
- Output requirements that include life-size, 1:1 images and enlargements for court displays.

Even though we can define what a digital image is and we can articulate what image resolution means, we are still confronted by the confusion about what image quality means for capture, display, and output.

It is up to every one of us to clear up the confusion and uncertainty surrounding the definitions of digital imaging and help everyone understand that image resolution equals image quality. Once we understand and can explain the effects of these terms on the entire digital imaging process, we will be able to reduce the fear and

Figure 8.14

Digital imaging includes input, processing, and storage, as well as output. Miss one element in your planning and you miss the big picture.

The digital imaging process includes input, storage and management, processing, and output. Ignore one of these elements in your planning process and you create a hole in the big picture!

frustration of the technology and remove the dark clouds of suspicion hanging over the world of digital imaging.

TAKE A PICTURE...IT LASTS LONGER. OR, DOES IT?

In addition to many people having the mistaken idea that traditional film processing provides a higher level of image integrity, they also think it provides pictures with a longer life span. The truth of the matter again is that this absolutely is not true.

TRADITIONAL NEGATIVES ARE LOSING QUALITY

Photographs can still be produced from the glass or tin plates used in the photographic processes of the 1800s. Unlike the photographs printed from these negatives originally, the new prints will have a degraded image quality.

Traditional silver halide-based negatives of the past 50 to 75 years have also begun to fade. The effects of heat and humidity and level of care that these negatives have received affect the quality of the images produced from these negatives.

One type of photograph that has a significantly limited life span is the Polaroid photograph of the past 50 plus years. Many of those images from 35 to 40 years ago will not survive solely because of the limited technology that was available when those films were made. To make a long story short, for some of us, the photographic images that we took when we were younger will not be available within our own lifetime.

DIGITAL BITS LAST LONGER

The good news is that this is not true for those people who use digital cameras. The only way that the image will not be available within their lifetimes is in the event that they actually lose the image file. For example, when an image is printed from an electronic file, the actual data bits used for the print process do not change. Each copy produced from an electronic copy is created using a series of bits (ones and zeroes) that will not change from the first copy to the 500th copy.

Negatives, on the other hand, will fade because of the light striking the negative when it is printed over and over again. In addition, the handling that the negative receives will also affect the life span of the negative itself and will jeopardize image quality for the future.

FILM, DIGITAL CAMERA, OR FLATBED SCANNER?

Which device should you use? Film or digital? Which digital camera (resolution) should you use? Or should you consider using a scanner?

The first step in determining which device you should use is to determine the size of the object (also commonly referred to as the area of interest) to be captured. This will help you determine the final resolution of the captured image and help identify the best technique for capturing your images.

For example, if you are using a 6 mega pixel digital camera (a camera that has an imaging sensor with a resolution of 2000 pixels high by 3000 pixels wide), you can capture an area that is 2 inches high by 3 inches wide to produce an image that has a resolution of 1000 pixels per inch, the NIST standard.

Similarly, if the area that you are capturing is 4 inches high by 6 inches wide, each inch of the area of capture will then only use 500 pixels of the sensor—2000 pixels divided by 4 inches of capture area, and 3000 pixels divided by 6 inches of capture area—thus yielding an effective resolution of 500 pixels per inch, no longer meeting NIST standards.

Therefore, on the basis of the size of the object being captured, a digital camera may not produce the desired image resolution.

As stated in the preceding section, scanners produce a digital image with a consistent resolution over the entire area of the image. For example, if the scanner has a CCD with a resolution of 1200 dots per inch (in this instance, it is permissible to use the term dots per inch and pixels per inch interchangeably, although the appropriate term for use with a scanner is SPI or samples per inch), the scanner would produce a higher resolution image than a digital camera.

As will be discussed later in this chapter, the term dots per inch is used to define output. In that discussion, you will learn that dots can have only one color value; the color value of each dot is dependent on the type of printer that you use. In the case of scanners, the CCD has a series of light-sensitive diodes in horizontal rows across the width of the CCD array. Each diode acts like a miniature digital camera in that it digitally captures a specific area of the item laying on the face of the scanner. Unlike a digital camera, however, each diode captures three samples: one color sample for red, one color sample for green, and one color sample for blue. Once the samples are processed for each sensor, the sensor values are passed to the computer and represent a single pixel within the entire image.

More specifically, each row on the CCD array contains 1200 diodes for each inch of the total scan width, which is commonly 8.5 inches wide. So regardless of whether you are capturing an object that is 1 inch wide, 2 inches wide, or 4 inches wide, you are still using 1200 diodes for each inch (3600 diodes per inch or 3 rows of 1200 diodes per inch for a color image). In other words, a 1-inch area will use 1200 diodes, a 2-inch area will use 2400 diodes, and a 4-inch area will use 4800 diodes, thus yielding a resolution of 1200 samples (pixels) per inch.

In addition, digital cameras and flatbed scanners use two very different techniques for assigning color values to each pixel. Digital cameras use a variation of the **Bayer pattern** in which each photo sensor collects only one grayscale value, which is based on the density of light that strikes the photo receptor, which is filtered by a red, green, or blue sensor.

Bayer pattern: A pattern of color filters that overlays the imaging sensor's photoreceptors. Also commonly referred to as a mosaic pattern. Twice as many green-filtered pixels than red or blue exist, because the human eye is more sensitive to green.

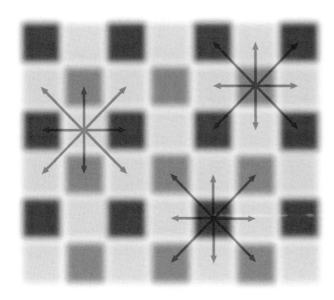

Figure 8.15

The photoreceptors in the Fuji-S-3 Super CCD capture a grayscale value on the basis of the filtered light density. Therefore, each photoreceptor needs to store only one color value, which is a grayscale value. Color is added when the image is processed or when it is stored on the camera as a JPG image. RAW images are simply the unprocessed grayscale values captured by the photoreceptors.

The photo receptors on a camera sensor capture brightness only, not color. In other words, they "see" only grayscale values based on the red, green or blue filter. The color value for each pixel is calculated based on the nearest neighbors illustrated by the arrows.

The resulting pixel color value is based on its receptor value together with the values of the eight surrounding neighbor sensors. As illustrated in Figure 8.15, 50% more green sensors (4 of 8 surrounding neighbor sensors) are present than red and blue sensors (2 of 8 are red and 2 of 8 are blue), which means 50% of the green color values are **interpolated**, and 75% of both blue and the red color values are interpolated.

Typically, interpolation is used to describe a mathematical process that creates missing pixels or to improve perceived image quality by increasing the number of pixels within a given image. However, digital cameras use interpolation to create missing color values on the basis of the values of the neighboring sensors.

Some imaging sensor manufacturers, such as Foveon, use CCDs that capture three separate color samples (red, green, and blue) for each sensor, similar to the multi-sample process used in scanners. According to Foveon, the resulting color images are more accurate in terms of color balance and more accurately represent the actual item being photographed. They cite tests in which photographs from cameras by other leading digital camera manufacturers contain moiré patterns, false color values, and so on.

The bottom line is that two major issues should be considered when deciding what device to use: the desired resolution and the most accurate method for capturing color values. Of course, limitations to this exist; for example, three-dimensional objects such as light bulbs cannot be scanned; they must be photographed. If you are using an alternate light source, scanning is not an option. So you have to decide which device will work the best in a particular scenario.

Interpolated: Interpolation is a mathematical process that creates missing pixels (when decompressing a JPG file) or to improve perceived image quality by increasing the number of pixels within a given image. Digital cameras also use interpolation to create filtered color values on the basis of the Bayer pattern, which uses the filtered light density values of the neighboring sensors.

In addition, capturing superglue (cyanoacrylate process) prints on plastic baggies is a snap with a flatbed scanner with a transparency adapter. These images are captured as positive images—black ridges on a white background versus white ridges on a black background—which eliminates the need to invert the images, thus providing faster, easier, painless image capture. That is what digital imaging is all about: lower costs, faster processing, fewer backlogs, and faster transmission of images to the court.

BEST PRACTICES FOR CAPTURING FINGERPRINTS, SHOE PRINTS, TIRE MARKS, BITE MARKS

We digitally capture numerous types of images every day. The challenge is: How do we capture those images at the highest possible resolution with the greatest degree of accuracy?

Regardless of the image that you capture, the goal with a digital camera (or a film camera for that matter) is to fill the frame with the object. When sensors on a CCD or CMOS sensor capture excess white space, you are actually losing information about the image. The frame must be dedicated to the object and the scale only. As shown in Figure 8.16, image quality is affected both by the size and number of pixels within the image.

Failure to optimize image resolution creates problems removing background patterns, such as backgrounds on checks or money orders where gradient color values are used as a means of security. By increasing the area of capture, a single photoreceptor must capture a larger area, using all of the color values in the area covered by that single sensor and the light intensity to determine the resulting pixel value recorded by that photoreceptor. By capturing a smaller area, you minimize the area covered by each photoreceptor, which also means that you get better color management and have a greater opportunity to eliminate backgrounds.

In addition to blending multiple color values to make a single pixel color value, capturing too large of an area with an imaging sensor often creates moiré patterns within an image.

 Rule of Thumb 8.4: In digital imaging, resolution and color go hand in hand. The higher your resolution, the better your image quality, because resolution helps to maintain color values and keep them separated, and color values provide better image detail.

CAPTURING FINGERPRINTS AND BITE MARKS

When capturing a single fingerprint, palm print, or bite mark, you should set the lens at its closest focusing distance. Then, move the camera as close to the object as possible, using the distance of the lens to the object as your primary method for focusing. This allows you to get as close to the image as possible, filling the entire frame with the object and ensuring the highest possible resolution for image capture. Then you can use the focus ring on the lens to fine-tune the focus.

Once you have captured the image, you can use the scale in the digital image to calibrate the image for accurate one-to-one, life-size output. For example, using Adobe® Photoshop®, you can accurately calibrate any image with a scale in it following these simple steps:

1. Crop the image based on a known distance using crop tool, type the letter C or choose the crop tool from the toolbar. (NOTE: The larger the selection, the more accurate the results.)
2. From the Image menu, choose Image Size. When the image size dialog box appears:
   a. Ensure that the Resample function has been disabled—a check mark should not appear in the box to the left of the word Resample.
   b. Enter the known distance of the cropped area. (NOTE: Please be sure to verify that you enter the appropriate measurement—width if the scale is going lengthwise across the page and height if the scale is going vertically from top to bottom on the screen.)
   c. Highlight the value contained in the resolution field by placing your cursor over the contents of that field and double clicking the left mouse button.
   d. Copy the resolution value by pressing Ctrl and typing the letter C.
   e. Click Cancel to close the Image Size dialog box.
3. Undo the cropping by pressing the Ctrl and Alt keys and typing the letter Z (also known as Step Backwards).
4. From the Image menu, choose Image Size. When the image size dialog box appears:
   a. Ensure that the Resample function has been disabled—a check mark should not appear in the box to the left of the word Resample.
   b. Highlight the value contained in the resolution field by placing your cursor over the contents of that field and double clicking the left mouse button.
   c. Paste the resolution value obtained in Step 2 by pressing Ctrl and typing the letter V.
   d. Click OK to close the Image Size dialog box.

CAPTURING SHOE PRINTS AND TIRE MARKS

You do not have to have a medium- or large-format film camera to capture shoe prints. Admittedly, these large-camera formats are very nice and they provide excellent comparison images. Unfortunately, those cameras and their corresponding films are becoming more and more difficult to obtain. Along with the digital revolution came a change in business practices for film and camera manufacturers, such as Kodak, Nikon, and FujiFilm.

- In January 2004, Kodak announced it would stop selling traditional film cameras in the United States, Canada, and Western Europe.
- In January 2006, Nikon Corporation (Japan) announced it would stop manufacturing most 35-mm film SLR bodies, as well as all large-format lenses and enlarging lenses, most manual focus 35-mm lenses, and related accessories.
- Also in January 2006, KONICA Minolta, the third largest maker of photographic film, pulled out of the camera and photo businesses to stem growing losses . . . selling out to Sony.

However, many experts in the field today still say that you cannot use digital imaging for the capture of shoe prints and tire marks. If the proper procedures are followed, you can most definitely use digital technologies for the capture, analysis, and output of these objects.

A common misunderstanding about the use of digital technologies to capture shoe prints, for example, is that you must capture the entire shoe print as a single image. Although this is partially true, we cannot rely on this image for our analysis. For example, from our earlier discussion in this chapter you will recall that the larger the area you must capture, the lower the resolution of the image. The more that you can fill the frame with the object, the higher your resolution will be. In other words, extra space around the object in your viewfinder is not a good thing.

Even with all of the tools in Adobe Photoshop, you still cannot improve the image quality of a low-resolution image. That only happens on "CSI". . . and that is television.

To capture a shoe print or tire tread, you should begin by taking a full-length photograph of the object being captured. This will provide you with a reference point should you have any question about the position, size, or location of the image.

Once you have successfully captured your full-length photograph, the next step is to capture a closeup of the object, which in turn increases the resolution of the image. For example, instead of using 124 pixels—camera sensors—to capture a 1-inch area, you can increase the resolution—using 250 or more camera sensors—just by getting closer to the object.

In other words, to maximize image quality, you should capture the image in segments, allowing for at least a 1- to 2-inch overlap from one segment to the next. Once the images are calibrated, you can stitch the images together to create a single image using the full-length image as proof that the print existed as an entire (whole) object. Law enforcement agencies in Nebraska, Iowa, and Alaska have successfully used high-resolution, stitched images for more accurate and reliable analysis. Alternately, other agencies have used the low-resolution images together with high-resolution images of a specific area of interest within the footprint or tire tread.

Typically, when capturing shoe prints, three photographs are required: one photograph of the entire shoe print, one photograph of the front portion of the shoe print, and another photograph of the heel or back portion of the shoe print. The last two images are then calibrated and stitched together to make a single, high-resolution image.

When capturing tire tread impressions, you should not capture an area wider than 11 or 12 inches with an 8 or 10 megapixel digital camera for optimum image quality. Attempting to capture a wider area will degrade image quality below the desired threshold (again, capturing a larger area lowers image resolution). Furthermore, you should not attempt to capture a footwear impression or tire tread impression with anything less than an 8 megapixel digital camera.

In addition, if you capture a standard footwear impression or tire tread impression in 11- or 12-inch segments, it typically will allow you to capture the full width of the tire tread and allow for 1- to 2-inch overlap. By use of this procedure with a high-end, high-resolution digital camera, you can capture footwear and tire tread impressions with resolutions approximately 500 ppi, which is the same resolution used on many AFIS systems for capturing latent print images. Imagine having the same high-quality image detail for a shoe print that measures 8 inches by 12 inches as you do for a fingerprint that measures 1 inch by 1 inch.

When capturing a shoe print or tire tread, it is imperative that you use a high-resolution digital camera. The use of a four or five megapixel digital camera with JPG compression will not provide the desired results, and is not acceptable.

Rule of Thumb 8.5: To photograph footwear impressions or tire tread impressions properly, you must use a high-resolution digital camera.

THE SCIENCE OF DIGITAL IMAGING IS GETTING A BIT DEEPER

As discussed earlier in this chapter, the number of bits used per color channel (also commonly referred to as bit depth or color depth) has increased from 8 bits to 16 bits per color channel. The result is that we can now acquire images with significantly more accurate color representations. These 16-bit color images also assist in our image enhancement processes, as well as producing higher-quality outputs.

Today, most digital cameras have a dynamic range that enables them to capture between 8 and 12 bits per channel. All these images must be reprocessed when the image is downloaded from the camera to convert them to 8-bit color information (except RAW images, which are typically stored with their original 12-bit color depth).

In contrast, images acquired with a flatbed scanner are captured at up to 16 bits per color channel. Once the acquisition device captures these images, the image is then reprocessed and saved in a format that can be used on the computer.

NOTE: Adobe Photoshop CS and CS2 provide the ability to import, read, and process 16-bit images, as well as 8-bit images. However, only limited functions can be performed with 16-bit color images.

The ability to capture images with a greater bit depth provides both optimum image quality and maximum opportunity to eliminate backgrounds and so forth during image processing. For example, having an image with a higher bit depth often can help you identify separate color values, such as a ninhydrin print on a check that uses a similar color value for the check background, overlapping fingerprints, or notice very light detail contained in a poor-quality latent fingerprint.

The bottom line is that before you can take the first step in image processing, you must ensure that you have followed the guidelines for ensuring that you have a good image to work with. Contrary to popular belief, Adobe Photoshop cannot turn a proverbial pig's ear into a silk purse unless you begin with good images. Put another way, the possibility of enhancing, analyzing, and identifying a fingerprint, shoe print, tire tread, ballistic comparison, etc. begins with good image capture.

IMAGE S&M...STORAGE AND MANAGEMENT OF YOUR DIGITAL IMAGES

Have you ever heard a conversation like this one before?

"Digital imaging is better than film imaging."
"No, it's not. Film is better than digital."
"No. You're wrong."
"No. You're wrong."

Would you believe that they are both right and wrong? It really depends on what you are trying to accomplish, how you are trying to achieve the end results, and what camera you are using (that goes for both film and digital cameras).

To genuinely appreciate digital image quality, you must also appreciate:

- The size of the imaging sensor (CCD or CMOS chip)
- The size and number of photo cells on the sensor
- The size of the area (object) being acquired

Today, most digital camera manufacturers share some basic goals. They want to provide a higher resolution (more pixels), a wider frame coverage (larger CCD/CMOS sensor), and additional features all at a lower cost.

Unfortunately, pixel count and sensor size are not tied together. Most digital camera manufacturers try to place more photo cells onto imaging sensors that are smaller than the fingernail on your pinky finger; several pro-consumer digital SLR cameras have imaging sensors that are now approximately the size of a standard 35-mm frame.

To put more receptors on a smaller imaging sensor, most digital camera manufacturers make the individual photo cells smaller (approximately 3 to 4 microns in size versus the typical 6- to 10-micron size of the individual photo cells in a digital SLR camera). Unfortunately, these smaller photo cells have reduced sensitivity to light, lower **dynamic range,** and are more susceptible to **noise**.

Smaller photo cells are also subject to diffraction (a type of distortion that makes the image look smeared because of light spreading out from a small sensor) and other undesirable secondary effects.

As shown in Figure 8.16, both the number and size of pixels can affect image quality significantly.

Many digital camera users further degrade image quality by use of compressed file formats such as JPG to store larger volumes of pictures with limited storage space and use compressed file formats so the camera will operate faster. Unfortunately, high speed and image quality do not go hand in hand. Like digital cameras, printers cannot do both high speed and provide high image quality—one suffers at the expense of the other.

IMAGE FORMATS: JPG VERSUS RAW VERSUS TIF

After all these years of teaching digital imaging and providing consulting services to law enforcement agencies, I am still surprised when I hear the question: "which image format should we use?" You would think that by now we could reach agreement on something as simple as an image format. We have not done that.

Dynamic range: The sensitivity to the difference between the values of gray from pure black to pure white. Also referred to as density, there are 256 shades from black to white in an 8-bit grayscale image; there are 65,536 shades of gray between black and white in a 16-bit image.

Noise: Unwanted artifacts within a digital image that affect image quality, such as dust or scratches.

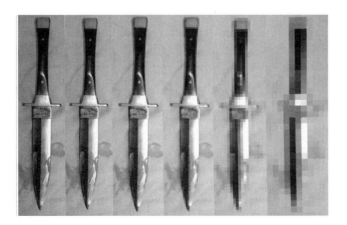

Figure 8.16

In digital cameras, the color value for each pixel is interpolated on the basis of the filtered light density of the neighboring sensors.

It really should not be a surprise, because after 10 years people still do not know the difference between a dot and a pixel either. I have found that it all comes down to opinions, and everyone has one.

Today, most digital cameras can save images in one or more of three primary image formats:

- JPG (joint photographic expert group)
- RAW (camera raw [read unprocessed] sensor data)
- TIF (tagged image file)

Some camera manufacturers also provide the ability to save images simultaneously in both RAW and JPG formats.

JPG

JPG: Short for joint photographic experts group. An image format that reduces the image file size—sacrificing image quality—so that the image does not require as much storage space.

JPG (also referred to as JPEG) is the most commonly used (or misused) digital camera image format.

When a picture is taken with a digital camera, the image sensor uses the camera's settings, including, but not limited to, white balance, sharpness, and hue and saturation to process the actual image data. Once the actual raw sensor data has been processed, it is discarded, and the processed image data are then compressed and stored. Size does matter when it comes to the storage space required to store your images and the speed of downloading and opening images.

Of the three file formats, JPG provides the smallest file size. Because of its smaller size, it is also the fastest to download both from the sensor in the camera to the camera's storage device and the fastest to download when you are downloading images from the camera (or the camera's memory device) to your computer. In fact, most digital cameras provide multiple image quality (compression) settings, such as fine, normal, and basic. Depending on the processor speed, memory, and software within the camera, some digital cameras actually compress images more than others. For example, images generated from the highest JPG setting in some digital cameras are difficult to distinguish from their uncompressed counterparts.

Therefore, some people would have you believe that a JPG image gives you outstanding image quality while delivering the fastest speed for all camera functions. There is no such thing as a fast, high-resolution totally lossless image quality.

As demonstrated in the pictures in Figure 8.17, JPG compression removes features; relocates boundaries; alters the size, shape, and color of features; and reduces resolution differently for different locations in the same image. This process also alters pixel values differently, depending on whether they lie at the center or boundaries of the pixel grid into which the image is divided. This produces **artifacts** and loss of details that are not consistent and cannot be predicted.

Artifacts: Disturbances within a digital image that affect image quality, such as pixilation, dust, scratches, moiré patterns, printing distortions, or other imperfections that interfere with image quality.

Although high-speed capture and rapid file saving on the camera might be useful if you are photographing a sporting event such as WWF Knockdown or a NASCAR event, it is not that important for photographing shoe prints, tire treads, latent prints, or blood spatter at a crime scene.

Zooming in too much on a low-resolution compressed image shows the boxy artifacts caused by resampling.

In addition, almost all digital cameras offer three different JPG image quality options. As was mentioned earlier, most cameras provide fine, normal, or basic image quality options for saving JPG images. The fine image quality option keeps more of the actual image data (less lossy compression), and basic keeps the least amount of image data (very lossy compression).

Many people do not realize that the JPG format does not actually remove the entire pixel during compression; it simply removes the pixel values. (In other words, the pixel is a mere shell of itself.) For example, a 6-megapixel (MP) camera (3000×2000) stores approximately one third of the actual pixel values, resulting in a file size slightly less than 2 megabytes (2 MB). When the image is decompressed, the file is returned to its former 6-MP size, which yields an RGB image that is approximately 17.2 MB in size.

Because of all of the compression and decompression, you should not use a JPG format to save original images that you expect to enhance later. For example, saving pictures of shoe prints in a JPG format can cause artifacts that would preclude the visualization of unique details such as wear patterns.

Similarly, pictures that are captured and stored initially as JPG images should be saved as TIF images after they have been processed so they will not be recompressed when they are resaved. Typically, this conversion is done immediately (using the "Save As" command) on opening the file in Photoshop before any enhancement processes are initiated. (Once again, be aware that once you have lost the original image data from the camera sensor when the camera saved the image as a JPG file, you will never get the original pixel values back.)

It is never a good thing to throw away evidence. High-resolution, noncompressed image files do show more detail and they provide better image quality; unfortunately, they also require significantly more storage space.

Another major problem with the JPG format is that every time you open a JPG file and change even so much as one single pixel within the image and then resave the file, the entire image is recompressed. In other words, if you were to go through a series of enhancements with an image and resave the image after each step, the image would become more and more degraded with each successive resave. Therefore, you should always save enhanced files in a lossless format such as TIF at maximum (color) bit depth.

Another problem with the JPG format is that the image on the screen does not reflect the new compression of the saved image. To view the actual saved image, you would have to close the saved image and then open the newly saved image.

With JPG files, you actually are discarding the true image (pixel) data. Some crime scene photographers will even argue that you do not lose enough detail to make a difference when JPG formats are used for crime scene photographs. That may be true if you are just taking overall crime scene photographs. It is not an acceptable practice to use a JPG format to photograph objects at the crime scene, especially where those photographs may be needed for analysis, such as footwear comparisons, bloodstain pattern analysis, or latent print or palm print analysis.

From a totally technical perspective, once you discard the original information, you cannot get the same information back. You merely get an interpretation of that information. That is why you would not be able to use JPG to compress anything that must be restored precisely. For instance, lossy compression formats cannot be used for database files, spreadsheets, and so forth, because these files must be restored back to their original, 100% accurate value when they are decompressed. If you compressed a spreadsheet and could not restore it accurately, you could have all sorts of problems. This same issue applies if you are trying to zoom in on detail in an aerial photograph that has been compressed or enlarge a footprint for comparison.

So what are the advantages of using a JPG format?

- The images require less space not only on the camera media, but they require less space on the computer.
- The JPG format provides faster processing not only on the camera but when downloading and viewing the image files on the computer.
- JPG files are a compatible format and can be shared easily. Because of their smaller file size, they are easily transmitted by e-mail, and they can be imported into word processing files (for reports) and PowerPoint presentations (for courtroom presentations) much easier. (Document files are much smaller as are PowerPoint files, but the use of JPG files in PowerPoint presentations also enables slides to be displayed much more rapidly.)

In many instances, the advantages of the JPG format outweigh the disadvantages. Disadvantages include the following:

- Images are compressed by discarding actual sensor data (exact image cannot be recreated... simply an interpretation of the original image data).
- Each time file is opened, processed, and then resaved, the image is recompressed, which causes even more data loss.
- Certain types of images, such as blood spatter and footprints, compressed using a JPG format are not good for detailed, in-depth analysis. When pixel values are compressed, their true color values are lost. When images are decompressed, interpolated pixel values cause backgrounds to become commingled with the foreground. When zoomed and enlarged on the screen, the degradation of image quality and visibility of squares (artifacts caused by a combination of a nearest neighbor/bicubic resampling algorithm) can become quite noticeable.
- Resampling techniques are used to restore missing pixel values when digital images are decompressed for display. Depending on the level of compression used when the image was captured, these techniques often cause image degradation with squares, as shown in Figure 8.17.

RAW (Which Literally Means Unprocessed "RAW" Image Data)

A number of digital camera manufacturers also provide a camera **RAW format** as one of the methods for storing images on your camera.

These RAW camera files are simply the unprocessed sensor data. The good news is that the RAW files are smaller than a noncompressed TIF file, but they are larger than a JPG, because none of the actual sensor data are discarded during compression.

The bad news is that RAW image files have to be processed before they can even be viewed. For example, you would have to use a RAW image file converter (one is typically included with the software when you buy a digital camera) to view images captured by that camera in a RAW format. Some applications, like Adobe Photoshop CS and CS2, provide a RAW image converter for many of the more commonly used digital cameras.

When you open a RAW image file, you can manually adjust white balance, temperature, exposure compensation, brightness, contrast, sharpness, lens aberration, and hue and saturation. With all this flexibility comes a higher level of complexity. Because some people do not like to fiddle with the RAW options, they argue that JPGs can provide a better image easier, because the JPG file uses the white balance and sharpening settings of your camera when you actually take the picture. No postprocessing of the JPG image is needed just to open the image.

The funny thing is that many forensic photographers prefer to use the RAW format because they have the power to adjust the various parameters to produce the image quality and tone that they desire.

RAW format: One of the methods for storing images on a digital camera. RAW files are simply the unprocessed sensor data from a camera.

In addition, each camera manufacturer states that their proprietary RAW format delivers an image that is superior in quality to that of their competitors. Because these proprietary formats store information in a different manner, they must use different file extensions to identify which format they are in, such as NEF, DCR, CRW, and RAW.

The advantages of using RAW files over JPG or TIF files include the following:

- RAW files are smaller (in size) than TIF, yet they do not discard any of the true, captured sensor values.
- RAW files are stored in their native 12-bit grayscale mode (4096 tonal range values) compared with the 8-bit grayscale values (256 tonal range values) used in the processing of JPG and TIF files.
- RAW files are not processed within the camera; color processing is deferred until the file is opened on the computer. Therefore, files are approximately one third the file size of a noncompressed TIF image and can be written to the camera's storage medium faster than a TIF file. In addition, they can be downloaded much faster than TIF files from the camera storage media.

Disadvantages of using RAW files include the following:

- RAW files cannot be shared easily because RAW file formats are not standardized, which means that the files may not be able to be opened by the recipient. Today, most camera manufacturers use a proprietary RAW format; therefore, the compatibility between individual camera models and imaging software is also limited. Adobe Photoshop CS and CS2 have seamless, integrated RAW file converters for most digital camera manufacturers. The good news is that Photoshop automatically logs any adjustments made to the RAW file when it is opened.
- The computer does all the processing rather than the camera. This means that you must store not only the original RAW file but also the converted file and in many instances the processed (enhanced) image file. Maintaining the relationship between all of these can be troublesome, not to mention the challenges related to storing and retrieving the images.
- Accuracy of tonal range can be argued, especially in the case of bruising. For instance, in domestic violence or sexual assault cases, the person processing the image files may not have been the person taking the pictures. Because the person opening the file can adjust the hue and saturation subjectively, the true nature of the picture has the potential of being compromised.
- Software developers such as Adobe and Microsoft have developed digital imaging applications that include their proprietary RAW file conversion utility. Because they handle adjustments such as temperature, hue, saturation, lens aberration, and so forth differently, the color values, contrast, and sharpness potentially can be different each time the file is opened.

TIF

Over the years, **TIF** files (also referred to as TIFF, which is short for tagged image file format) have been identified as the truest interpretation of an image. None of the data captured by the camera sensors is lost, plus all processing is done within the camera. All camera settings, including white balance and sharpening, are saved within the processed image file.

Several years ago, some digital camera manufacturers, such as Kodak, used a proprietary file format with the TIF extension as part of their naming convention. Unfortunately, these TIF files could not be viewed or processed without the use of the DCS Twain interface software. This caused a great deal of confusion, because TIF files are supposedly compatible with most imaging applications and operating systems. The Kodak TIF files were neither.

Advantages include the following:

- Most accurate interpretation of the image—not dependent on subjective interpretation of color values and not subject to compression.
- Most reliable format when forensic analysis of the images is required.
- The only reliable method for storing processed images that were captured as either RAW or JPG.

Disadvantages include the following:

- Large files require vast amounts of storage space and significantly more time to download image files from the camera.
- Large files cannot be easily shared across the Internet.
- Larger files require more time to print.

Although you would think that image quality would win out over smaller, less accurate formats, it has not. Most law enforcement agencies today make a distinction between evidence and documentation photographs. Evidentiary photos are captured by using either TIF or RAW file formats. Photographs used solely for documentation purposes are captured and stored as JPG files.

Arguments for the sake of smaller files have often won out over the argument for accuracy. In some agencies, IT departments have mandated that digital photographs only need to be captured and stored in a JPG format. It is clear that those individuals do not have to put their career on the line and make life or death decisions. (Think about how you might feel if you could not solve a case just because you did not have good enough images.) The need for the best possible image quality should always outweigh the argument of budget deficits, reductions in manpower, and the like. Photoshop is a very powerful tool for image enhancement, but it is no substitute for good photography and good image quality.

TIF: Short for tagged image file format. A noncompressed format used for storing digital images. Commonly referred to as the truest interpretation of an image, because the image can be reproduced exactly as it was captured and stored bit by bit.

Rule of Thumb 8.7: After enhancing an image, you should use a lossless compression format such as TIF, because compression formats, such as JPG, will degrade the image.

IMAGE PROCESSING (THE PROCESS FORMERLY KNOWN AS "ENHANCEMENT")

As described at the beginning of this chapter, photography lost its innocence many years ago. Photographs have been retouched by use of a number of image capture, development, and printing techniques since the 1800s.

With the availability of high-resolution digital cameras; fast, powerful personal computers; and sophisticated photoediting software, the manipulation of digital images has become more common. We have all seen the marvels (commonly referred to by defense attorneys as manipulations) created by computers for movies and the hoaxes that have arisen through image manipulation. One of the most recent hoaxes was the photograph (shown in Figure 8.18) that was supposedly taken by an individual on top of the World Trade Center on 9/11.

As a result, the law enforcement community has found itself under fire by concerns (both inside and outside of the law enforcement community) about the admissibility of digital images in the courtroom. In fact, many law enforcement agencies and prosecuting attorneys are concerned about converting to digital imaging technologies today because of misleading opinions expressed in articles such as "Reasons to Challenge Digital Evidence and Electronic Photography" by Michael Cherry and "Fingerprint Evidence in the 21st Century" by Edward J. Imwinkelried and Michael Cherry published by the National Association of

Figure 8.18

This hoax photograph showing an airliner approaching the WTC on 9/11 was seen around the world.

Criminal Defense Lawyers. Other informal opinions, such as the editorial entitled "Beware the Pitfalls of Digital Evidence" by Scott Christianson and articles such as "Digitized Prints Can Point Finger at Innocent" published in the Chicago Tribune still feed the controversy today.

So how can this technology withstand the charges of impermissible tampering with the original image to produce images that are not "clear and accurate representations of the original scene?" Just because we have the technology at our fingertips to do some pretty remarkable manipulations of images does not mean that all our images will be altered or manipulated in some way. Admittedly, much of the skepticism revolves around the lack of knowledge about and the lack of training in the technology.

Film negatives have always been subject to many of the same manipulations that are now the focus of digital imaging critics. A skilled darkroom technician can do miracles with a negative. Yet, it has been extremely rare for traditional photographs to be challenged and excluded from the courtroom. Why? Because, if necessary, the darkroom technician can come to court to testify that no illicit tampering was done during the development and printing process.

In the past, the admissibility of traditional photography simply rested on the integrity of the photographer and the darkroom technician. The same criterion is now being applied to the admissibility of digital imaging; however, we have not achieved this lofty goal. A general fear factor is that technology allows for the possibility of manipulation; therefore, many defense attorneys want the jury to assume that it has been tampered with even though the individual presenting the digital images in court testifies under oath that no such manipulations were done to any images being offered in court as evidence.

So, what is permissible, and what is not permissible? Or, what are the permissible techniques that can be used with digital images, and what are the impermissible manipulations that will not survive court challenges?

The answer is quite simple: permissible digital image processing involves those techniques that enable us to visualize the image more clearly. The core of permissible processing is that (1) nothing has been added to the original image; and (2) nothing essential has been removed from the original image.

Manipulations, however, do those things: they add data that was not originally there, and they take away data that was originally present.

Ultimately, the judge and jury must be convinced that only permissible digital image processing techniques were used in the images presented in court. The bottom line is that it still comes down to the integrity of the witness offering the image in court as evidence.

Unfortunately, many people are skeptical and tend to believe the worst. Until digital image processing is as easily accepted in court as are traditional photographs, it is our job to make it easier to accept such processing.

Rule of Thumb 8.8: To ensure admissibility of enhanced images in court, you may use only those processes that do not add anything to the original image and that do not remove anything essential from the original image.

Strategies include some of the following ideas:

- Written standard operating procedures (SOPs) must be in place, which spell out what is and is not permissible. (Following standard guidelines, such as those provided by SWGIT, can also help to ensure that the procedures used are consistent from one agency to another.) Then, all operators of digital image processing hardware and software must adhere to the SOPs.
- Those with access to this technology must be adequately trained, and training records must be maintained.
- Limited access to this capability is recommended. This limits the chain of custody problems that may occur. Passwords must be issued to selected personnel, and restricted access is recommended.
- Some recommend that digital authentication software should be used to ensure images are authentic and uncompromised.
- Original, tamperproof images should be archived to a secure storage medium such as CDs or DVDs. Only copies can be processed and then offered in court. The original image is considered the equivalent of the film negative, which should always be secure and as originally captured.
- The enhancement history should be acquired and maintained. As each processing is made, a detailed list of (1) the type of processing and (2) the degree of that processing is maintained. (Photoshop CS and CS2 can record the enhancement history for you automatically.) If a challenge in court is directed at a subsequent image that was digitally processed, a new copy of the original, authenticated image can be produced, and the imaging process can be performed again following the enhancement history to prove that the subsequently processed image was, indeed, derived from the original, not just a new image substituted somewhere is the process.

When all is said and done, the reliability of the imaging process; the digital image; and the integrity, training, and experience of the technician doing the imaging process are crucial elements regarding the admissibility of enhanced images when the enhancements are done with a computer. Under existing rules of evidence, the main requirements for admissibility of a photograph (whether digital or film) into evidence are relevance and authentication.

The integrity of these images is assured to some degree by the chain of custody, but the most important requirement is the ability to verify that the image is genuine and valid. In most cases, existing rules of evidence are applied to photographs that are admitted as evidence. In some cases, the technician must be prepared to testify that the images are accurate, reliable, and correct.

DIGITAL S&M: THE STORAGE AND MANAGEMENT OF YOUR DIGITAL IMAGES

Even though we have all become more photo aware because of the plethora of digital cameras and digital camera phones, many law enforcement agencies still refuse to use digital cameras. They argue that digital imaging involves expensive, rapidly changing technology. They also cite concerns about system upgrades/technology replacement and concerns about reliable, long-term data storage.

On the other hand, more and more law enforcement agencies are using digital technology. Some of the reasons for this switch are that digital imaging:

- Is more affordable.
- Provides instant gratification.
- Guarantees better images (This is not totally true. It has been my experience that detectives and crime scene technicians are simply taking more pictures—approximately 70 to 80% more—because of the idea that no cost of film or processing is associated with digital images. Because there are more images, a higher probability exists that there ARE better images, which means more cases solved.)
- Allows easier information sharing.
- Offers the possibility of a paperless photographic solution.

Regardless of whether they are making the switch to digital imaging or not, both sides agree that image storage, retention, and retrieval are major concerns. The foundation for this concern is that image storage and management are the least standardized components in the world of digital imaging.

For the IT department, dealing with vast numbers of image files (of varying file sizes and formats from cameras and scanners) can be a very painful process, whereas most crime scene technicians and detectives are enjoying (and benefiting from) taking more pictures at the crime scene.

Questions that must be addressed for a successful digital imaging program include the following:

- Where are the images to be stored?
- How (in what format) are the images to be captured?
- Who downloads the images?
- How do you account for multiple cameras being used at the same crime scene? (More specifically, if two or more detectives are using a digital camera at the crime scene and both detectives reset their digital cameras to start with the

number 1, how do you ensure that you can account for multiple image files with the same file name and how do you track which images were taken by which detective or photographer?)

- Who has access to the images (chain of custody also commonly referred to as the image integrity), both before and after the images are downloaded from the camera to the computer?
- How do you authenticate your original images?
- How do you track and maintain the relationship between the original image and subsequent enhancements of that image?
- How do you account for all images (originals and enhanced copies) for a single case? (Moreover, digital images can come from multiple departments within a single agency. They can come from evidence, crime scene unit, toxicology, serology, fire arms, etc. When preparing for court, the agency must know what images exist in each of these departments to comply with discovery requirements. Finding and retrieving images from a vast number of departments can be a real problem for these agencies.)

In addition, you should always maintain at least two independent copies of your image files regardless of what image format you use. One of these copies should be stored offsite in a different location to protect against adverse incidents such as fire, tornados, and floods.

If you are storing images on CDs or DVDs, you should also verify that the images were written to the CD or DVD properly before deleting the files from your computer. You can verify the content of your CDs or DVDs either by opening one or more images on the disk after the images have been burned to the disk, or you can use the Windows Explore function and display thumbnails of the contents of the disk.

Although you can develop and implement a digital imaging solution yourself, there are also turnkey solutions available (such as Foray Technologies' Authenticated Digital Asset Management System (ADAMS), Linear Systems' Digital Image Management System (DIMS), and Data Works Plus' Digital Crime Scene products) that can eliminate many, if not all, of the frustrations shared by the IT department and the digital imaging end users. Whatever solution you choose, whether it is your own homegrown solution or a commercially available turnkey solution, flexibility, reliability, and maintainability are crucial elements that will help ensure success with your digital imaging program.

The bottom line is that the goal of implementing a successful digital imaging program can be achieved through appropriate planning, training, and management. We have at our disposal the latest and greatest digital imaging technologies available. Not only can we reach for the stars, but now we can even see the stars more clearly than ever thanks to the wonderful world digital imaging technologies.

SUMMARY

The widespread acceptance of digital imaging has advanced significantly over the past decade. The technology has gone from costly low-resolution cameras to inexpensive high-resolution cameras. In fact, today many people even have digital cameras on their cell phones.

The adoption of digital imaging technologies by law enforcement has also grown by leaps and bounds. Digital cameras are being used at the crime scene investigation to photograph evidence such as latent prints, footwear impressions, blood spatter, and more.

The resulting digital images are used in courtrooms across the country. However, ground rules must be followed regarding the ability to authenticate the images and preserve and protect the integrity of the images primarily to safeguard the right of defendants to challenge their reliability.

Following established case law precedence, we can submit a picture that was photographed by an unknown photographer as evidence. The photograph does not even have to be the original. All that is required is that it must be a fair and reasonable representation of the scene.

Over the past 10 years, legal challenges have also been raised because of concerns about the reliability of science and technology in areas such as DNA and fingerprint identification when the new technologies were first used with evidence. Over time, the questions have become few and far between. With the level of acceptance of digital imaging technologies today, I truly believe that the courts will take judicial notice of the more technical aspects of digital photography, and digital imaging will be as readily accepted as traditional photography.

In the meantime, it is crucial to understand the elements of this technology to ensure its successful use both in and out of the courtroom. Unfortunately, far too many people do not understand the impact of resolution on image quality, because resolution varies from camera to monitor to printer. So you might say, "it's about the resolution," but then again, it is not just understanding resolution. Digital imaging is more than just a bunch of pixels.

FURTHER READING

Baxes, G. A. (1994). "Digital Image Processing: Principles and Applications." John Wiley & Sons, Inc., New York, NY.

Galer, M. and Horvat, L. (2003). "Digital Imaging." Elsevier Science, Burlington, MA.

Johnson, H. (2004). "Digital Printing Start-Up Guide (Digital Process and Print)." Course Technology PTR, Boston, MA.

Johnson, H. (2004). "Mastering Digital Printing." Muska & Lipman Publishing, Cincinnati, OH.

Resciker, D. R. (2001). "The Practical Methodology of Forensic Photography." CRC Press LLC, New York, NY.

Russ, J. C. (2001). "Forensic Uses of Digital Imaging." CRC Press LLC, Boca Raton, FL.

Sawyer, B., and Pronk, R. (1997). "Digital Camera Companion." The Coriolis Group, Inc., Scottsdale, AZ.

Sheppard, R. (2005). "Complete Guide to Digital Printing." Lark Books, Asheville, NC.

CHAPTER 9

SPECIAL PHOTOGRAPHY SITUATIONS

LEARNING OBJECTIVES

On completion of this chapter, you will be able to . . .

1. Explain the two most critical elements involved in working an accident scene.
2. Explain why prioritizing perishable evidence is important at accident scenes.
3. Explain noncritical priorities at accident scenes that usually require early attention.
4. Explain how to photograph the approach to the actual accident scene.
5. Explain the necessary photographs required to adequately photograph the roadway or the surface of the accident scene.
6. Explain the required photographs to photographically document the vehicles involved in the accident.
7. Explain some additional concerns when photographing injuries to drivers and occupants of the vehicles involved.
8. Explain how to determine the focal length of the telephoto lens used for a particular surveillance job.
9. Explain how to select the appropriate shutter speed to eliminate the potential blur from hand holding the camera during surveillance situations. If this is not acceptable, then using a tripod is recommended.
10. Explain how the appropriate shutter speed is selected to freeze the movement of the suspect under surveillance.
11. Explain how it is predetermined that the surveillance suspect will be in focus, even if they will be moving within different distances from the camera.
12. Explain how to select the appropriate ISO film speed for the dim lighting conditions that may be present at nighttime surveillance scenes.
13. Explain the optimal conditions for doing aerial photography.

14. Explain the optimal camera variables that should be used with aerial photography.
15. Understand the safety concerns that apply whenever taking aerial photographs.
16. Understand the safety concerns that apply whenever taking underwater photographs.
17. Explain the difference between reflection and refraction and their effects on underwater photography.
18. Explain the difference between scatter and backscatter and their effects on underwater photography.
19. Explain how color is absorbed by water at different depths and the solution to this problem.

KEY TERMS

Backscatter	Prioritizing perishables	Refraction
Perishable evidence	Push processing	Scatter

ACCIDENT PHOTOGRAPHY

INITIAL PRIORITIES AT ACCIDENT SCENES

Two Critical Elements

If the layman was asked the first job requirement of a law enforcement officer at a major accident scene involving injuries, the most frequently offered answer would probably be to render aid to the injured. Although certainly important, because the survival of critically injured parties is at stake, one responsibility is even more critical: that is, to protect the accident scene. It does not benefit the injured accident victim, the officer, or the general public if conditions are left so that another accident can occur. Even worse would be allowing a situation to exist where another vehicle was able to drive into the original accident scene while the officer was rendering first aid to the first accident victim. Then the first accident victim and the officers present at the scene are imperiled. This can be regarded as this chapter's first Rule of Thumb.

 Rule of Thumb 9.1: The first responsibility of a law enforcement officer has to be to protect the accident scene itself, so that no further accidents can occur.

Then, the responsibility to render aid to the injured becomes paramount.

Prioritizing Perishables

Once the scene has been stabilized and the injured have been dealt with, the next most important responsibility is to protect perishable evidence from being lost or destroyed.

Some evidence may remain at the scene for long time periods and still retain its full value as evidence. Other types of evidence are fleeting or highly perishable and need to either be collected as soon as possible, protected to avoid loss or change, or documented one way or the other. Three primary types of **perishable evidence** exist.

Inclement Weather

Bad weather can destroy evidence. Evidence can be washed away in the rain, blown away by the wind, covered up by snow, and moist tire tracks can evaporate. The initial law enforcement officers at the scene have a responsibility to protect these types of transitory evidence as best as they can. That may mean covering them up with something, or it may mean that a traffic cone must be placed over them. Some affirmative action must be taken to preserve this fragile evidence somehow. If it is fleeting, it will need to be documented as soon as possible by photography and measurements.

Evidence in the Path of Travel

Some evidence may be lying in the roadway in a position where other traffic may destroy it. This may be ordinary traffic trying to make its way around the accident scene, or it may be other vehicles operated by rescue personnel or other law enforcement officers responding to assist at the accident scene. In either case, the initial officers at the scene have the responsibility to protect this evidence also. Traffic may have to be directed around this evidence until it can be documented by photography, measured, and collected.

Evidence Subject to Biological or Physiological Processes

If a driver is under the influence of alcohol or drugs, both of these forms of evidence will diminish over time by being metabolized. The presence of alcohol or drugs in the driver must be documented as soon as possible with appropriate field tests and then confirmatory tests. Deceased subjects are subject to a variety of early physiological changes that need to be documented, because in a short time, they will exhibit different characteristics.

NONCRITICAL EARLY PRIORITIES

The next most important issue is to attempt to reestablish the traffic flow. The accident may be tragic to those involved, but many others are affected by it, and it is said, "Life goes on." Alternate traffic lanes around the actual accident have to be established, or detours have to be determined.

Prioritizing perishables: Some evidence may remain at the scene for long periods and still retain its full value as evidence. Other types of evidence are fleeting or highly perishable and need to either be collected as soon as possible or protected to avoid loss or change.

Besides reestablishing the traffic flow, one other matter has to be considered before the actual business of working the accident is begun. People may be gathering to look at the effects of the accident. "Rubber-neckers" may slow traffic, but pedestrian onlookers may actually make the investigation of the accident even more difficult. Well-meaning but naïve onlookers have actually collected evidence scattered around the scene, trying to be helpful, and have then turned in their evidence to law enforcement personnel at a later time. Some evidence can become someone's souvenir and never will be seen again.

Therefore, it is imperative to safeguard the accident scene from the crowd sure to gather around any serious accident. The use of police banner-guard is the normal means by which off-limit areas are marked. Police cars can block streets, charleyhorse blockades can be placed as needed, and officers physically positioned to direct pedestrian and vehicular traffic can all be used to meet this need.

THE ACCIDENT SCENE

The Approach to the Scene

As with crime scenes, it is easy to get too focused on the actual impact site, with the damaged cars and the injuries to the occupants of the vehicles. Although certainly important, overall aspects of the scene are usually just as important in determining what happened and why. For each vehicle involved in the accident, it is necessary to document the approach to the scene. To do this effectively, it may be necessary to backtrack each vehicle's approach to the actual accident scene by several blocks. It is possible for a driver to lose control of a vehicle several blocks before an actual impact occurs. Sometimes the "cause" of the accident is not near the actual accident scene.

Traffic Control Devices

The photographer may be required to include photographs several blocks away from the actual site. It will be necessary to begin the documentation of the approach to the accident site from a position that includes all the traffic control devices that should have controlled the driving behavior of each driver in the accident. These concerns include the following:

- Speed limit signs
- Lane control lights, signs, or painted lane markers
- Stop, yield, or other caution signs
- Bridge height indicators
- School warning signs
- Pedestrian crosswalk signs
- Red, yellow, and green traffic lights
- Railroad crossing warning signs, etc.

Figure 9.1

Traffic control types (Courtesy of the Arlington County Police Department, VA).

This distance may be as little as 100′ from the point of impact, or it may involve moving beyond sight of the actual impact area. This will vary with the accident.

Weather Conditions

Sometimes the immediate site of the accident does not adequately represent the actual weather conditions that influenced traffic as the vehicles approached the accident scene. Snow, for instance, may have melted from some areas of the highway but still be present in other areas. Wet pavement, in combination with oil residues on the highway, may have made a particular section of the roadway slick. A vehicle may have begun to lose control and begun swerving several hundred feet away from the ultimate impact site, but it is not until the approach to the scene is examined that this is discovered. Portions of the highway may be icy, although the actual accident scene may be free of ice. If no one survives the accident and can report what happened to investigators, it is up to the investigator to examine the entire "scene," and this may include areas out of sight of the actual impact location. Again, it is important to resist the urge to get too focused just on the immediate accident site.

View Obstructions and Sight Distances

Assuming the individual drivers were driving diligently:

- What could they have been reasonably expected to see?
- When could they have seen it?

These are two critical questions often asked of professional accident reconstructionists. The elements of time, distance, and speed calculations are used in conjunction with view obstructions. These concerns may necessitate a return to the scene at a later point for additional photographs and measurements, but these same concerns should be present in the mind of the photographer taking the original photographs. As mentioned previously, the scene photographer may not have the necessary training to figure out the precise positions of each driver at the . . .

- Point of possible perception
- Point of actual perception
- Point of no escape

However, being aware that view obstructions are critical factors in many accidents, the scene photographer can attempt to take photographs at several points along the approach to show the visibility of the overall scene each driver might have had.

Figure 9.2 documents a driver's possible view of a vehicle coming from the left. Even if this driver was required to stop for a stop sign at the intersection, could an approaching driver from the left have been seen? Another set of similar photographs from a stopped position at the intersection can also be taken if the issue is whether the driver should have ventured into the intersection after stopping. In this case, however, the driver ran the stop sign, colliding with the vehicle coming from the left.

With these types of photographs, it is important to also remember the drivers were probably positioned differently than a standing photographer. The driver in the seat of a tractor trailer has quite a different sight line than the driver in a small sports car.

 Rule of Thumb 9.2: Efforts should be made to take photographs from the eye level of the drivers involved.

Witnesses' Viewpoints

Not only are the sight lines of the drivers involved in the accident of importance but also the positions of any witnesses that may be available. An important aspect of many trials is whether the witness could have seen what they said they saw from the position they were in at the time of the accident. Therefore, it is just as critical to determine the position of the witness and their viewpoint. Were they also in a vehicle, a pedestrian on the sidewalk, or did they view the accident from

Figure 9.2

View of approaching traffic from the left (Courtesy of Ofc. Nash Peart, Arlington County Police Department, VA).

a second floor window? Photographs should be taken from these positions as well. It may be wise to also take a photograph from the accident scene looking toward where the witness was supposed to have been. This may help clarify whether the witness had a clear view of the accident scene.

The Actual Accident Scene

Temporarily ignoring the vehicles and occupants of the vehicles, what are some of the most frequently found types of evidence at accident scenes?

Tire Marks

Without being an accident reconstructionist and being able to interpret the tire marks at an accident scene, it is still obvious that they are important, and it is important to photographically document them. Whether they are skid marks, scuff marks, impact scrubs, or any other kind of tire marks, they may be crucial to determining exactly what happened at the accident scene (Figure 9.3).

Recall that when the sun is out, the use of a polarizer filter may be useful for imaging the full length of tire marks. Sun glare may wash out some of the tire marks, including the "shadow" marks at the beginning of skid marks.

Normally, four views of each tire mark will be needed:

- From one of the ends, down its length, to show the directionality, which is often best done with the photographer viewing the roadway parallel to the side of the

Figure 9.3

Skids, scuffs, and impact scrubs (Courtesy of the Arlington County Police Department, VA).

road. Remember to use a normal 50-mm lens to avoid distorting the length of the tire mark or the accident scene.

- Another view from the other end. Again, parallel to the side of the road.
- A third photograph of the length of the mark should be made viewed from perpendicular to the side of the road. This shows the side view of the tire mark in relation to the side of the roadway, but this may not necessarily be the best view of the tire mark to show its true length. The tire mark may be diagonal within a lane.
- A final view of the length of the tire mark from the true side of the tire mark, not from the side of the roadway, to show its length from the best position, which is a view of the tire mark with the film plane parallel to it.

Material on the Roadway

Often a critical element of the accident may be the preexisting condition of the roadway itself. It may have . . .

- Defects: potholes or bumps in the road
- Contaminants: gravel, sand, dirt, leaves, oil
- Weather-related problems: snow, ice, sleet, rain

These all need to be carefully photographed as possible contributing causes of the accident.

Marks in the Roadway; Debris on the Roadway

During the accident, the vehicles will frequently leave marks on the roadway. There may be gouges, scrapes, or tire tracks in the soft shoulder of the road. Marks like this are frequently used to determine the path of travel of each vehicle, or the actual point of impact, when the energy from the impacting vehicles forced them down into contact with the pavement.

The impact between the vehicles may also have knocked off dirt from the undercarriage, effectively positioning the vehicles at impact. During the accident, glass may have broken, paint may have chipped off, vehicle fluids may stain the roadway, and vehicle body parts may have broken off. The positions of each of these types of evidence can help determine the position of the vehicles at the point of impact and afterwards as they come to final rest. This type of evidence must be both photographed and measured for the accident scene diagram (Figure 9.4).

Other Damage

Part of the documentation of the accident scene may be to show the damage to guardrails, sign posts, utility poles, trees, bushes, fences, etc. (Figure 9.5). Not all accidents remain neatly on the roadway. Vehicles have even struck buildings after leaving the roadway.

Figure 9.4

Marks and debris in the roadway (Courtesy of the Arlington County Police Department, VA).

Figure 9.5

Damage to other property (Courtesy of the Arlington County Police Department, VA).

Vehicular Damage

Midrange photographs of each vehicle must be taken to show them in relation to the fixed features of the accident scene itself. They should also be photographed viewed parallel to the roadway edge and perpendicular to it. Then the vehicles have to be photographed to show their relationship to each other, whether they are still in contact with each other, or have separated from the force of the accident.

To adequately document the damage to the vehicles involved in the accident, these kinds of photographs are often required.

Exterior and Interior Views*

Sometimes, "damage" to another side of a vehicle may later become an issue in a trial when one driver claims to have been struck by another "phantom" motorist, lost control, and then caused the accident at issue. To prove that this was not the case, complete exterior photographic documentation is required. Then it can be visually proven there is no other damage to the vehicle in question to corroborate, or refute such alibi attempts (Figure 9.6).

 Rule of Thumb 9.3: It is, therefore, just as important to document the obvious damage, as well as areas of the vehicles that are not damaged.

To the extent possible, these photographs should be taken before the vehicles are moved. Often, this may not be practical because of rescue operations attempting to remove the injured who may be trapped inside the vehicles. Accident damage should be distinguished, whenever possible, into one of these categories:

- Damage from the accident
- Damage from rescue activities
- Damage from moving/towing operations

Photographers and accident reconstructionists not present at the original accident scene have to try to distinguish between these various types of damage. Only adequate photography and proper note taking can provide this.

In addition to the preceding recommended photographs, particular attention should be paid to the following:

- Tires: damaged, worn, signs of skidding, and pattern type (Figure 9.7).
- License plates and vehicle identification numbers (VINs). Documenting the VINs is important because at times a vehicle has no license plates at all, and sometimes the plates that are on the vehicle do not belong to it. Often, tracking the true owner of the vehicle can then only be done with the VIN (Figure 9.8).

*DuBois, R. (1993). "Photography in Traffic Accident Investigation Manual." Institute of Police Technology and Management (IPTM), University of North Florida. Jacksonville, FLA.

Vehicle Exteriors

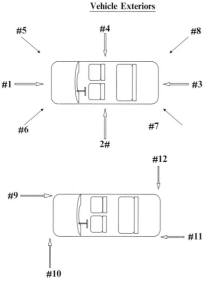

Vehicle Interiors

Figure 9.6

Exterior and interior photographs.

Vehicle Exteriors:

1-4:   All four sides of the vehicle, even if there is no apparent damage to any side.

5-8:   Damage to any of the corners. These are to be done only if that particular corner is damaged.

9-12:   Photos down the side of the vehicle. These are to be done only if that particular side shows crush damage. These are to assist the accident reconstructionist determine speed from crush.

13.   The exterior of the vehicle photographed from straight overhead, if possible.

14.   The undercarriage, as necessary.

15.   Any damage to the wheels and tires, including indications of skidding, wear, or puncture.

16.   License plates.

17.   Paint transfers.

18.   Vehicle lamp filaments. Caution, these may be very fragile, so consider photographing them in place before removing them.

19.   Detached vehicle parts.

Vehicle Interiors:

1.   Driver's seat.

2.   Driver's floor.

3.   Left side of instrument panel; windshield, including the steering wheel.

4.   Left front door and interior panels.

5.   Left rear seat.

6.   Left rear floor.

7.   Right front windshield header; right front A-post.

8.   Left rear door and interior panels.

9.   Right rear seat.

10.   Right rear floor.

11.   Left front windshield header; left front A-post.

12.   Right rear door and interior panels.

13.   Right front seat.

14.   Right front floor.

15.   Right side of instrument panel; windshield.

16.   Right front door and interior panels.

17.   Obvious damaged areas, which may have contributed to the injuries of occupants.

18.   Seatbelts, to indicate if they had been worn by occupants or not.

19.   Any air bags which had deployed. It may also be valuable to show that a particular airbag did <u>not</u> deploy.

20.   Vehicle Identification Number.

21.   Pedals: brake, clutch, accelerator. A shoe tread pattern may be visible on them; or, the pedal pattern may be visible on the shoes or bare feet of an occupant, indicating they were the driver.

22.   Speedometer dial, for possible needle impact point visualization. This may require fluorescent photography.

Figure 9.7

Tire damage (Courtesy of the Arlington County Police Department, VA).

Figure 9.8

Vehicle identification numbers (VINs) (Courtesy of the Arlington County Police Department, VA).

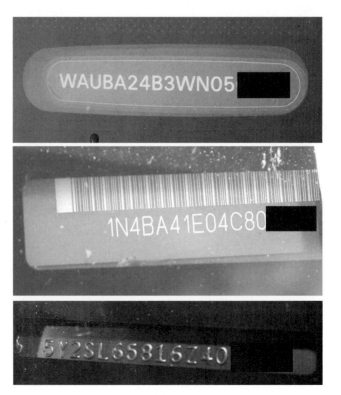

- Pedals: brakes, clutch, accelerator (to match with impressions on the shoes of the driver). With a heavy front-end accident, the driver's shoes, or bare feet, may take on the pattern of the pedal they had been on, and/or the shoe tread pattern may be transferred to the pedal the shoe was on, which is one way to establish the driver of the vehicle, when three occupants are standing outside the car, each pointing to the other when asked who the driver was (Figure 9.9).
- Safety equipment: airbags, seat belts. Photographically document their use or non-use (Figure 9.10).
- Paint transfers, particularly in hit-and-run accidents.
- Vehicle lamp filaments to ascertain whether the lights were on or off at the time of the accident (Figure 9.11 and 9.12).

Figure 9.11 shows several variations likely to be encountered at accident scenes. The top left image shows a straight filament not involved in an accident. The top

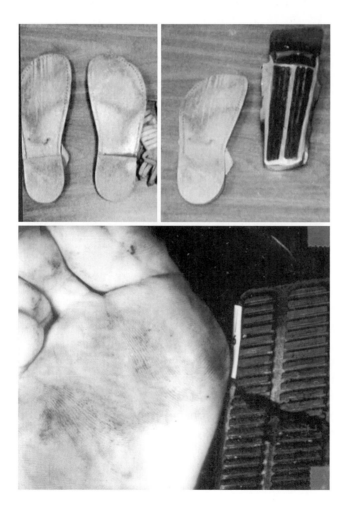

Figure 9.9

Pedal marks on sandal and foot (Courtesy of the Arlington County Police Department, VA).

Figure 9.10

Air bags deployed (Courtesy of the Arlington County Police Department, VA).

Figure 9.11

Single light bulb filament (Courtesy of Cpl. Marc Hackett, Arlington County Police Department, VA).

right image shows some flexing or bending of the filament, normally an indication of the light being "on," and, therefore, the filament was hot and flexible at the time of the impact. A hot filament will bend toward the impact direction. The bottom image shows bending and black oxidation from the bulb breaking and the incandescent filament being exposed to the air. Normally, an inert gas inside the light bulb prolongs the filament's life span, but when the bulb breaks, oxidation covers the filament. Figure 9.12 shows a broken bulb that had two filaments in it. The left image shows an undamaged bulb. The right image shows the far right filament is bent, with oxidation covering it; if you look carefully, you can see broken glass melted onto the filament, which is a sure sign this filament was incandescent at the

Figure 9.12
Double light bulb filaments
(Courtesy of Cpl. Marc Hackett,
Arlington County Police
Department, VA).

time of the accident when the bulb broke. The left filament is exhibiting "rainbow" coloration, without heavy oxidation, indicating it was off but near a filament that was incandescent at the time of the accident. Being near an incandescent filament, the left filament was hot but not incandescent itself. Sometimes high and low beams are made with two different bulbs. Sometimes a single bulb can serve as both a high and low beam with two different filaments.

- Speedometer dial: is the needle frozen in place or did it slap the dial at impact? Before newer digital dials took over the dashboard, dials had needles to indicate speed, rpm, and other aspects of the engine. Frequently, the tips of these needles were coated with fluorescent paint to make them easier to see at night. During heavy front-end accidents, the needle would frequently slap the face of the dial, and with an ultraviolet light the speed at impact could be seen by a faint fluorescence of the dial's surface where the needle slapped it. At times, the needle would also jam at its impact position. Figure 9.13 shows one needle jammed at 6000 rpm. The top two images demonstrate the need for fill-flash on bright sunny days. When the damage is on the shade side of the vehicle or any aspect of the vehicle in the shade needs to be documented, fill-flash comes to the rescue.
- Detached/missing vehicle parts: have they been located?
- Overhead view: sometimes the best way to see which direction the force(s) came from.
- Undercarriage view: sometimes this relates to the damage in the roadway or to objects/people that have been driven over (Figure 9.14).

Human Bodily Injury

At times, the only way to position the occupants of a vehicle within the vehicle before the accident is to match the patterns of the personal injuries suffered during the accident with the damage to the vehicles or to other aspects of the shape of the vehicle, because sometimes the occupants are unwilling or are unable to just tell investigators where they were inside the vehicle at the time of the

Figure 9.13

Speed and RPM dials (Courtesy of the Arlington County Police Department, VA).

Figure 9.14

Four undercarriage combo (Courtesy of Cpl. Lisa Haring, Arlington County Police Department, VA).

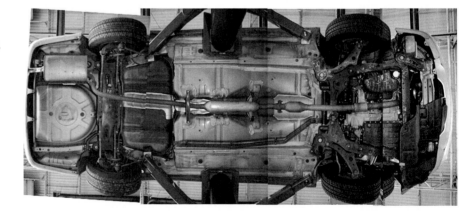

accident. The driver may fear he will be charged with being at fault in the accident. The driver may not have a driver's license, or it may have been suspended. The driver may be intoxicated. The vehicle may have been stolen, and the driver prefers to be thought of as "just a hitchhiker." The occupants of the vehicle may be injured and unable to talk, unconscious, or dead.

In these circumstances, often the parties can be positioned inside the vehicle on the basis of their injuries. In Figure 9.15, notice the patterned injury made by a seat belt on the shoulder of the left subject. This individual will have a difficult time denying he was the driver, just as a person with bruises to the chest in the shape of an oval similar to the steering wheel would. The grill of a truck struck the subject on the top right. The subject on the bottom right was the driver of a car struck on the driver's door.

Damage to the windshield and corresponding bruising/scratches to the forehead of the occupants may be the only way to determine how people were positioned within a vehicle. Pedestrians may also impact a vehicle. Figure 9.16 shows two pedestrian head impacts to the windshields of the top two vehicles. The bottom image indicates the danger of not wearing seat belts and having the vehicle stop abruptly. Three subjects in the front of the vehicle left their marks on the inside of the windshield.

Finally, if occupants have been ejected from the vehicle during the accident, it may be possible to closely inspect the windows or doors through which it is suspected they exited the vehicle and find hair, tissue, or clothing fibers there, confirming that was their exit route. Detailed photographic documentation of all these types of evidence may prevent the need for difficult and lengthy investigations later.

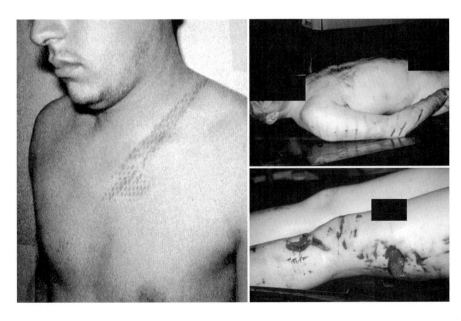

Figure 9.15

Accident injuries (Courtesy of the Arlington County Police Department, VA).

NIGHTTIME ACCIDENT PHOTOGRAPHY

The painting with light technique may serve to light large dimly lit accident and crime scenes. Two other techniques exist that you may wish to keep on your Bat-Belt to use at your convenience. Although the painting with light technique will always ensure the colors of the scene are properly documented, if no bright ambient lights are too close to your accident scene or crime scene, these other techniques may be considered. When bright ambient lights are present, the resulting photographs may contain unwanted color tints.

Time Exposures

When film cameras were still king and seeing an instant display of your photographic talent was not available, it was sometimes possible to do time exposures at large dimly lit outdoors scenes. A precondition to obtaining a well-exposed image was the absence of bright ambient lights too close to the scene of interest. If the predominant light was just from the stars and the moon, this technique frequently produced satisfactory results.

400 ISO color film was the film of choice. The aperture would be set to f/8, and the camera would be hyperfocal focused. That meant that focus was set for 30′, and this resulted in a depth of field range of 15′ to infinity. The shutter speed necessary for a good exposure was pure guesswork, but frequently great results could be anticipated from locking the shutter open for 2 minutes, 4 minutes, and 8 minutes on successive shots.

Figure 9.17 is the result of these three timed exposures. Because there is some tungsten lighting coming from the building on the right side of the images, a yellowish tint becomes more and more obvious. Many, however, would not find this objectionable. If the yellowish tint is considered too offensive, the image can be printed in black and white. Figure 9.18 compares painting with light to the time exposure to show painting with light's ability to capture true colors. The time exposure is again influenced by tungsten lighting coming from the left side of the image.

Figure 9.17

Time exposures: 2, 4, and 8 minutes.

Figure 9.18

Painting with light compared with a time exposure.

Time exposures with a digital camera would be easier to achieve, because after the first image was captured, a pretty accurate "guesstimate" of the second exposure could be obtained from seeing the exposure of the first image.

Aperture Priority Overalls

With digital cameras becoming more pervasive in crime scene and accident scene photography, the entire process is simplified somewhat. Again, 400 ISO film is recommended, as is f/8 and hyperfocal focusing. Now, however, the camera is set to aperture priority exposure mode. Theoretically, when sufficient light comes into the camera through the lens, the shutter will close, completing the image capture. Again, caution should be used if bright ambient lights are in the immediate area. The camera's light meter may preferentially meter the light coming from these bright light sources, ensuring they are properly exposed, but the rest of the scene may be underexposed. To minimize this risk, it is suggested to use a spot meter, if the camera has one. Then, the shutter will close only when the area the spot meter is on has received sufficient light. This provides more control over the exposure where it is needed: in the dimly lit areas. If the initial exposure is not properly exposed, the exposure compensation dial can be set to +1 or −1 as needed.

Figure 9.19 shows a fatal motorcycle accident scene. The top image shows the scene lit only by ambient lighting; the bottom image shows the scene with an aperture priority exposure. The airplane streak in the sky is the telltale sign of an exposure taking several seconds. Figure 9.20 shows another accident scene on a

Figure 9.19

Aperture priority overall (Courtesy of Cpl. Marc Hackett, Arlington County Police Department, VA).

Figure 9.20

Aperture priority overall (Courtesy of Cpl. Lisa Haring, Arlington County Police Department, VA).

Figure 9.21

Aperture priority overall (Courtesy of Photographic Specialist Keith Dobuler, Fairfax County Police Department, VA).

major highway. Again, the top image shows the dimly lit scene photographed normally. The bottom image is an aperture priority exposure. This image also caught a plane streaking across the sky.

Figure 9.21 is an image of the Home Depot where Washington, DC, area snipers, John Allen Muhammad and John Lee Malvo, shot and killed Linda Franklin in Falls Church, Virginia. Linda Franklin was one of 10 killed and 4 wounded by the snipers in 2002. The aperture priority exposure mode was used in this image the night of the shooting.

SURVEILLANCE PHOTOGRAPHY

LENS SELECTION

A Rule of Thumb greatly simplifies the selection of the focal length of the telephoto lens to be used in long-distance surveillance situations.

Rule of Thumb 9.4: Use 2 mm of lens for every foot of distance between the camera and the suspect.

For instance, if it is known that it will be safe to set up in a position that puts the photographer at a distance of 600′ from where the suspect will be, then $600 \times 2 = 1200$. A 1200-mm lens will be required to be sure of recognizing the suspect's face or to be sure that the suspect's license plate number can be read at that distance. This makes it necessary that the intelligence information available for the particular job be fairly accurate. It also means that important surveillance photographs have to be planned well, and they should not be ad hoc enterprises. In other words, this type of photography should be well planned and not thrown together just before the actual photo shoot.

Figure 9.22 shows three images of a relatively close George Washington, who is just 50′ from the camera. George's cropped and enlarged head clearly shows image degradation with the 24-mm lens (0.5 mm of focal length per foot of distance) and the 50-mm lens (1 mm of focal length per foot of distance). With the 100-mm lens (2 mm of focal length per foot of distance), the enlarged head is sharp and clear, which is the basis of the Rule of Thumb 9.4.

SHUTTER SPEED SELECTION TO HAND HOLD TELEPHOTO LENSES

The previous shutter speed recommendation for hand holding any focal length lens applies to surveillance photographs also. The focal length of the lens made into a fraction is the safe shutter speed to use to be sure that there will be no blur from your heartbeat while hand holding the lens. For example, with a 1200-mm

Figure 9.22

The proper lens for the distance.

| 24 mm Lens | 50 mm Lens | 100 mm Lens |

lens that made into a fraction is 1/1200th, the closest shutter speed to 1/1200th on some cameras is 1/1500th of a second if it is necessary to hand hold the camera. In practice, however, that is such a fast shutter speed that it is only usable in very bright lighting conditions. Even in bright lighting conditions, however, it is advisable to put the camera on a tripod to enable slower shutter speeds. As will be explained shortly, slower shutter speeds allow slower ISO films to be used, which is important, because as the ISO film speed goes up, the photo quality (graininess/pixilation) of the photograph goes down, whereas the contrast (a loss of middle gray tones) goes up. Neither is desirable.

In dim lighting conditions, however, the alternative of putting the camera on a tripod to avoid blur from the heartbeat is almost always required. This would enable using much slower shutter speeds, which impacts the ISO film selection. Under the dim lighting conditions of nighttime surveillance photography, using the slowest ISO film rating possible is also a consideration, because this affects the quality of the print because of the grain and contrast of the film.

For example: presume an f/4, 1/1500th, ISO 3200 film, and hand holding a 1200-mm lens under dim lighting conditions results in a good exposure. If the camera were mounted onto a tripod, the shutter speed could be slowed to 1/125th second, which would freeze the suspect expected to walk out of the back door of a residence. Changing from 1/1500 to 1/125 is 3½ stops more light allowed to reach the film. That savings can be applied to the ISO film to reduce the possible graininess of the photograph. That means we would be able to select a film that is 3½ stops less sensitive to light; 3200 to 1600 is 1 stop; 1600 to 800 is another stop; 800 to 400 is the third stop. The other ½ stop we can apply to our f/stop selection, so instead of an f/4, we can use an f/4.5, which is ½ stop dimmer, which improves the depth of field. Therefore, just by switching to a tripod, we can take the photograph at f/4.5, 1/125, ISO 400 in the same lighting conditions, netting a double benefit: more depth of field and a slower ISO film for more sharpness and clarity.

SHUTTER SPEED SELECTION TO FREEZE THE MOVEMENT OF THE SUSPECT

Previous shutter speed recommendations to freeze movement apply to surveillance photography as well:

- 1/125: will freeze a person walking or talking and turning their head to speak to different people.
- 1/250: will freeze a person jogging, jumping, or riding a bicycle.
- 1/500: will freeze a person in a car traveling at 30 mph.
- 1/1000: will freeze a person in a car traveling at 60 mph.

The anticipated movement of the suspect has to be built into the exposure calculation of surveillance photography, which is so important that it is a Rule of Thumb.

Rule of Thumb 9.5: The shutter speed for all surveillance photography must be selected to freeze the anticipated movement of the subject to be photographed. Although certainly an exposure variable, when doing surveillance photography, freezing subject motion is the shutter speed's primary function.

CALCULATING FOCUS OR DEPTH OF FIELD

If you recall, one of the factors affecting depth of field is the camera-to-subject distance. Therefore, even when using wide apertures, a decent depth of field will be available when photographing at normal surveillance distances if you just focus at infinity. Usually, it will be enough to ensure that someone within a 10′ range is in focus. Think of that as the width of a sidewalk outside many businesses. Even with a depth of field scale on the camera lens, this is difficult to determine more precisely.

When more precise calculations of the area of the scene that will come back in focus are required, the recommendation is to take a practice roll of film before the actual surveillance shoot. Or with your digital camera, take some practice shots before the actual photos will be taken. Try to duplicate all the variables that will exist in the required situation:

- Distance from subject
- Shutter speed to stop anticipated movement, tripping the shutter with a shutter-release cable
- Focal length of the telephoto lens
- ISO film
- F/stop selection
- Lighting conditions
- Focused on infinity
- Camera mounted on a tripod

Take a practice roll of film or several digital shots and examine the results.

ISO FILM SELECTION

How can the ISO film speed required for the surveillance conditions that are anticipated be determined? Rather than just loading the camera with the fastest ISO film available or setting your digital camera to its highest ISO setting, it is recommended to use the slowest ISO film you can for the lighting conditions, because the slower the ISO film speed, the better the quality (grain) and contrast of the final photograph. This is so important that it becomes a new Rule of Thumb.

Rule of Thumb 9.6: Use the slowest ISO film possible for the lighting conditions at the scene when taking surveillance photos.

To determine the slowest ISO film speed that can be used, just set all the other exposure variables, mentioned previously, and begin taking meter readings of a surrogate scene lit the same way the real scene is expected to be lit. Even better is to take a meter reading of the actual scene several days before the real shoot. Take a meter reading with the camera set at ISO 400. If the meter indicates that will result in an underexposure, try the next faster ISO film speed: ISO 800. Continue till the meter indicates a proper exposure has been obtained. Using a faster ISO film speed than necessary diminishes the quality of the photographs unnecessarily.

One final observation about these meter reading trials. If the camera has spot-metering capability, use it. Otherwise, the possible bright ambient light around the subject of interest may cause the subject to be underexposed, because an averaging or center-weighted meter is evaluating all the light being reflected from what is viewed in the center of the viewfinder. You are not concerned about properly exposing the inside of the store the subject may be walking in front of. All you are interested in is properly exposing your subject. Average or center-weighted meters react to all the light coming from the central area of the viewfinder. If you do not have a spot meter, several days before the real shoot, position a helper in the spot the suspect is expected to be in and come in close enough to fill the frame with just the light reflected off of their body, excluding all the other light around them, and use that exposure reading for the real shoot.

DAYTIME SURVEILLANCE PHOTOGRAPHY

When taking daytime surveillance photographs, it is obviously necessary to stay out of sight. When the lighting is adequate to see for long distances, it becomes important to hide the camera and the photographer. The question is then how can you hide a camera, yet still get the images required? At first we might consider hiding behind a wall or large tree trunk or even behind a bush. If we have the camera lens peeking around a wall, tree trunk, or bush, won't the lens be obvious enough to be recognized as a camera lens?

It must be remembered that long telephoto lenses are quite large.

Figure 9.23 shows the two lenses used for all the following images. They are not petite lenses. It will be necessary to keep them from being recognized by the suspect as camera lenses.

This task is made easier if we recall the three factors that affect depth of field:

- Aperture selection
- Lens selection
- Camera-to-subject distance

Figure 9.23

Top, Canon 400-mm lens with 2× tele-converter (Courtesy of Tom Beecher, Photografix, Richmond, VA); bottom, Sigma 800-mm lens (Courtesy of Air Force Office of Special Operations-AFOSI).

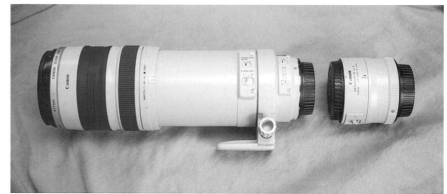

Rephrase this a bit to name the three factors that minimize depth of field:

- Use the widest aperture of the lens
- Use a telephoto lens
- Having great distances between camera and subject ensures small areas that are in focus and large areas that are out of focus.

By a careful use of these three factors that minimize depth of field, you will now learn how to hide your camera and giant telephoto lens behind solid obstructions while still being able to capture an image of your suspect.

Figure 9.24 is a combination of images showing a typical "surveillance" photograph of a subject sitting on a bench at the opposite end of the George Washington University Quad, some 300′ from camera to subject. The top left image shows the wall the camera will be hidden behind in the circle. The bench the "suspect" will be sitting on is in the foreground. The other top images show how much of the lens is behind the wall and how much is required to capture a recognizable image of the subject. Even if the subject were looking directly at the

Figure 9.24

*Photos across George
Washington University's quad.*

camera's position, the small slice of camera projecting from behind the wall would not be recognizable. The bottom left image shows the camera's position behind the wall. The person with the blue shirt is standing where the "suspect" will eventually be seated. This distance between camera and suspect position appears vastly compressed, because another telephoto lens was used to capture this image. The bottom middle image is the actual "surveillance" shot. The bottom right image is an enlargement of the "suspect's" face.

One may wonder why the subject's face is not as sharp as George Washington's face in Figure 9.22? After all, using an 800-mm lens for a 300' distance is overkill, is it not? If there were no obstruction between lens and subject, this would be true. However, by putting so much of the lens behind the wall, some of the possible resolution of the lens is compromised. Although not as sharp as George Washington's face, the sun-glassed smoker would easily be recognized.

Figure 9.25 shows how a series of nighttime photos were set up. During the daytime, student Tonia Busse stood 400' away from a camera with a 50-mm lens, with the camera mounted on a tripod. She is almost impossible to see in the left image. Enlarging the circled area is not much of an improvement. Obviously, the 50-mm lens does not capture sufficient detail to recognize her. The right image depicts Tonia standing in the same position at night but photographed with an 800-mm lens. She and the license plate are easy to recognize. If you have good eyes, you may be able to see that the license frame indicates "Jerry's," the dealership that sold the car. This position, with Tonia 400' from the camera will be seen again soon.

Figure 9.25

400' Away with 50-mm lens (day) and 800-mm lens (night) (GWU MFS Student Tonia Busse).

Figure 9.26

A wild bunny and the disappearing fence.

Using camera equipment properly can make solid object "disappear." First, a simple trick.

Figure 9.26 shows a "wild bunny" in a yard. The scene was intentionally composed so part of a chain link fence crosses the bunny's body. In this case, a 100-mm to 300-mm zoom lens is being used with a 2× tele-converter on a digital camera with a 1.6 focal length multiplier. The short end of the lens would, therefore, be $100 \times 2 \times 1.6$ or 320 mm. The long end of the lens would be $300 \times 3 \times 1.6$ or 960 mm. The top left image is a view from the short end of the telephoto lens, 320 mm. The top right image is the beginning of the lens being zoomed out. The bottom left image has the lens zoomed out even more, and the bottom right image has the zoom lens zoomed about ¾ of the full travel distance of the lens. Notice that the chain link fence is beginning to become "soft" because it is clearly out of focus. The middle image shows the result when the lens is zoomed out all the way to the 960-mm focal length. The chain link fence has "disappeared" because it is so out of focus.

This technique is used by many taking photographs at the zoo. If you cannot find a position to photograph animals without the cage in the way, make the cage disappear.

- Use the longest focal length lens you have.
- Use the widest aperture of your lens.
- Get close to the cage and focus on the animal. Being close to the cage will help ensure the cage is outside the animal's depth of field range.

Let us make the intermediate obstruction a bit thicker.

Figure 9.27 shows the setup in which a neighbor's car across the street is being composed through the branches of a tree. In this case, the front license plate has intentionally been obstructed by two tree branches, one vertical and one horizontal; same 960-mm lens as with the "wild bunny." The top left image is at the 320-mm end of the zoom range. The top right and bottom left show progressing zooms. The bottom right is the 960-mm setting: both branches are now so out of focus they appear to be transparent. The middle image is a crop and enlargement of the tag.

Figure 9.28 shows the application of this concept to a situation frequently used by surveillance photographers. Rather than hide behind a leafless tree, we can

Figure 9.27

Thicker tree branches becoming transparent.

Figure 9.28

Seeing through Venetian blinds.

position the camera behind Venetian blinds that are open only very slightly. To make things interesting, one of the Venetian blinds has intentionally been positioned to be over the license plate of interest. It would be too simple to position the camera so it "saw" between the cracks. The same procedure is used: at the short end of the zoom lens' range, the Venetian blind still obstructs the view of the tag. At the far end of the lenses zoom range, the Venetian blind becomes "transparent," and the tag can be read easily. The bottom right image shows what the suspect would see if he looked directly at the camera's position. The camera is positioned behind the left window with blinds "closed."

A bush is just a variation of Venetian blinds. In Figure 9.29, the photographer has given away his position by holding up his hand over the bush he has hidden behind in the center of the image.

If you cannot find a bush or Venetian blinds to hide behind, any wall or thick tree trunk will also suffice. Just ensure that 10 to 15% of the lens still can "see" the subject of interest, while the rest of the lens is hidden behind the obstruction. With just a sliver of the lens peeking around the obstruction, the suspect will not recognize it as a camera lens.

DEALING WITH UNDEREXPOSED IMAGES

With nighttime surveillance photography, hiding behind obstructions is not as important. Because of limited visibility in the dark, many times it would be almost impossible to see the camera and photographer, even if both were totally unobstructed. Of course, this should still be avoided, in case a changing lighting situation,

Figure 9.29
Hiding behind a bush.

like a car coming around a corner, temporarily threw light directly on the photographer. For nighttime surveillance photography, the main problem to be overcome is underexposure.

When the light becomes dimmer, the fast-speed films come out to play. Sure, 1000 ISO speed film and even faster 1600 ISO and 3200 ISO films are also available. Some digital cameras also have similar fast ISO settings. Are these sufficient to properly expose all nighttime scenes? No.

Unfortunately, even these fast ISO films and digital film equivalents cannot provide proper exposures under extremely dim lighting conditions. That is when infrared lighting or light-intensifying nite scopes become invaluable. These lighting solutions are clearly beyond the regular crime scene photography scope of this book, so we will confine this text to more "normal" solutions.

"Push Processing" Film

When it is known that the film will be underexposed, one can ask that the film be "**push-processed**" in the darkroom. This can sometimes bring out underexposed detail on the film.

This means that the film will be developed differently than normal film. Usually, it means the film will be allowed to remain in the chemical developing baths longer. For this to work, it is essential that the photographer ascertain the exact level of pushing.

For example, presume the widest aperture of your 800-mm lens is f/5.6. Because your target will be a subject sitting at an outdoor café table, talking to coconspirators, you set the shutter speed to 1/125 seconds. You have loaded the

Push processing: This can sometimes bring out under-exposed detail on the film. Usually, it means the film will be allowed to remain in the chemical developing baths longer.

camera with 3200 ISO film, and several days before the actual shoot, you had a friend sit at the table known to be favored by the suspect. Your meter readings on your friend have all indicated an underexposure. How is the precise level of underexposure determined? Change the camera's shutter speed setting to 1/60th shutter speed and take a new meter reading. If the new meter reading still indicates an underexposure, change the shutter speed to 1/30th, and take another meter reading. Still underexposed? Change the shutter speed to 1/15th. Now the meter indicates a proper exposure. Of course, you cannot actually use the 1/15th shutter speed, but now you know precisely that the exposure with 1/125 is exactly 3 stops underexposed. You can take photos at 1/125th shutter speed, and then tell the darkroom operator the film needs a "3-stop push." Each "push" requires a longer time the film will remain in the developing chemicals.

Two issues need to be understood: "Pushing" film results in both an increase in graininess and an increase in the contrast of the images. Mid-tones will be lost. Therefore, "pushing" film has its limits. At some point, the resolution of the desired image will be totally lost. Research indicates that "pushing" film about 3 stops is usually the practical limit, and this presumes you are dealing with a highly skilled professional laboratory, not your local drugstore.

Figure 9.30 shows this sequence of progressive contrast and increased graininess with more and more "pushing," which is the continuation of the previous shots with Tonia Busse 400′ from the camera with an 800-mm lens being used. During the nighttime, the proper exposure was determined, and then a series of progressively dimmer exposures was set on the camera. The 2-stop "push" shows that her face has become extremely contrasty, with mid-toned grays completely gone. The 3-stop "push" begins to lose the face to graininess, while her blouse has gone contrasty. A license plate was added, anticipating the ability of the "pushed" film to record fine details of the tag longer than facial details. The 3-stop "push" shows the tag clearly. With 4 stops of "pushing," both the face and the tag are no

Figure 9.30

Push processing limit (GWU MFS Student Tonia Busse).

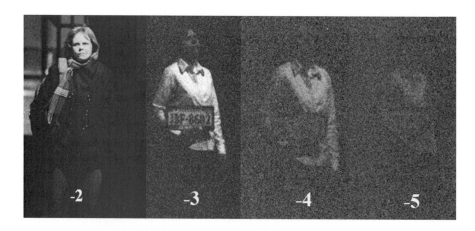

longer recognizable, but the gross blouse shape can still be made out. At 5 stops of "pushing" not much can be identified.

Beyond 3 stops of "pushing" it would be wise to implement infrared lighting with infrared-sensitive cameras or to begin using nite scopes, which are variations of military light intensifiers.

Digital Processing of Underexposed Images

As the march toward converting crime scene cameras from film to digital cameras continues, the natural question is whether underexposed digital images can be successfully salvaged? If so, is the digital processing of underexposed images more or less successful? The answers are "Yes" and "More".

Figure 9.31 shows a series of underexposed digital images salvaged in Photoshop®. It was the intent of this exercise to keep underexposing digital images to the point of failure, just as was demonstrated with the film "pushing." The night of this exercise, the images shown on the camera's LED screen were totally black, so it felt that goal had been achieved. However, when the images were eventually put into Photoshop®, detail was retrieved from all the underexposures. We had failed to find the "failure" limit of Photoshop® processing.

Between the −6 and the −9 exposures, you might even notice a slight improvement in image quality. Confident that the −6 exposure would fail to have any detail recovered, a new experiment was embarked on. This experiment could be done with a digital camera, but not with a film camera. With the digital camera, after the −6 exposure with the camera set to ISO 3200, further underexposures were made by changing the ISO setting to slower film equivalents, rather than faster shutter speeds; −7 was taken with the ISO set to 1600, −8 was taken with the ISO set to 800, and −9 was taken with the ISO taken at 400. The experiment was to see whether the continued loss of contrast and the increase in digital "graininess" would be offset by the reduction in digital "noise" with the use of slower ISO settings. This seems to be the case, but more experimentation needs to be done. Suffice it to say that the recovery of underexposed detail is much easier in a digital environment than in a film environment.

Figure 9.31

Digital processing of underexposed images (GWU MFS Student Tonia Busse).

AERIAL PHOTOGRAPHY

Many times, having an image from a higher point of view than is normal can be extremely helpful. These images can supplement our normal natural perspective images in several situations. Aerial photos can be a great addition to our normal crime scene photography, because higher views of the general area can show interrelationships between different locations much more effectively. Having a jury view the entire scene can make complex explanations of building relationships and traveled paths much easier to understand.

Figure 9.32 is such an example. Having completed one burglary in a shorter time than expected, the suspect decided to try another burglary in the same apartment complex. A confrontation with the second resident ended with that resident being stabbed to death. A very bloody suspect ran from the victim's apartment to his own apartment nearby. A police canine tracked the suspect's trail over the route indicated in the photograph. Such an aerial photograph can make it easier to explain all these relationships to the court.

The Marine Corps Marathon is run through the streets of Arlington County, Virginia, and Washington, DC. As part of the planning for this event, aerial photographs were taken of the race route. With these aerial views of the entire race route, it was much easier to plan which streets had to be closed, where water stations were to be located, and where first aid stations would be set up.

If an arrest warrant or search warrant will be served at a location with a high potential for armed resistance, aerial photographs can assist with the planning of

Figure 9.32

Homicide aerial photograph: canine track from victim's apartment to suspect's apartment (Courtesy of the Arlington County Police Department, VA).

the police action. Possible entry and exit routes can be ascertained and blocked, and the best approaches for law enforcement personnel can be determined.

Fixed-wing aircraft and helicopters are frequently used to take these kinds of photographs, but they are not the only way to take aerial photographs.

Aerial photographs can be taken from balconies, rooftops, cherry pickers, fire department ladder trucks, overpasses, and bridges. Look around the scene in question. What is the highest structure in the immediate vicinity? Consider using that structure as a platform for photographing the area of interest. When these will be insufficient for your purposes, aircraft can be considered. Many larger law enforcement agencies have access to aircraft used for a variety of situations: search and rescue, emergency movement of the critically ill or injured, transportation of dignitaries, and surveillance and aerial photography for law enforcement.

Figure 9.33 shows a balcony and rooftop view of a crime scene and accident scene. Figure 9.34 shows two aerial views of the area around the Falls Church, Virginia, Home Depot where DC area snipers, John Allen Muhammad and John Lee Malvo, shot and killed Linda Franklin. Linda Franklin was one of 10 killed and 4 wounded by the snipers in 2002.

Figure 9.33

Balcony and roof top views (Courtesy of the Arlington County Police Department, VA).

Figure 9.34

One of the DC area's sniper shooting sites (Courtesy of Photographic Specialist Keith Dobuler, Fairfax County Police Department, VA).

Figure 9.35

Accident scene: from the
ground and an aerial view
(Courtesy of Arlington County
Police Department, VA and
Pictometry International Corp,
Rochester, NY).

Another possibility for obtaining aerial photographs is that many jurisdictions survey their land areas with traditional maps and sometimes with aerial views purchased from professional aerial photographers. These aerial views help with locating utility lines and their access, zoning issues, and construction planning. Local law enforcement also has access to these views.

Figure 9.35 shows two views of an accident scene, one from the ground and one from an aerial view of the same area. The aerial view was provided by Pictometry International Corporation, which provides aerial views of all of Arlington County, Virginia, to all county agencies. Figure 9.36 shows three views of the Pentagon. The first is just before the September 11, 2001, attack, provided by Pictometry International Corporation. Superimposed on this image is a red arrow depicting the flight path of American Airlines Flight #77. The middle image was taken on 9/11. The right image was taken 3 days after the terrorist attack.

Figure 9.36

Pentagon, just before the attack
on September 11, 2001 and
3 days later (Courtesy of the
Arlington County Police
Department, VA and Pictometry
International Corp, Rochester,
NY).

Many readers may be familiar with Google Earth (earth.google.com). It provides satellite views of most populated areas of the world. It is fun to find your own residence photographed from space. But whose car is that parked in my driveway?

Although the military has made the use of Unmanned Aerial Vehicles (UAVs) commonplace in Afghanistan and in Iraq, the use of UAVs has begun to filter down to local law enforcement agencies. Several agencies, including the Los Angeles County Sheriff's Department, are testing UAVs for their practicality in the field, which can include highly populated areas. With the cost of UAVs in the neighborhood of $30,000, they make very economical alternatives to the cost and maintenance of regular aircraft. It is easy to predict the use of UAVs by law enforcement agencies will grow tremendously in the near future.

Figure 9.37 shows two UAVs and examples of the photos they are capable of taking. The arrow in the top right image points to the strapped-in 35-mm camera. In addition to still cameras, video cameras can be used, so that ground personnel can get a "real time" view of areas on the ground.

LIGHTING AN AERIAL PHOTOGRAPH

The sun will obviously be considered the sole light source. If it is our intent to have the sun light a scene viewed from overhead, it is best to have the sun as high as possible in the sky. With the sun high in the sky, long oblique shadows will be minimized, if not totally eliminated. We do not want areas of concern obliterated by dark shadows. This means that we will usually want to do most of our aerial photography between 10 AM and 2 PM. Of course, bright sun produces shadows. Being responsible for your overall image means preventing the shadow of the plane or helicopter being in view when you snap the shutter.

Figure 9.37

Unmanned aircraft vehicles and their photographs (Courtesy of Senior Criminalist Thomas W. Adair, Westminster Police Department, CO).

An alternative to having the sun high on a bright day is doing the aerial photographs on an overcast day. In this situation, a uniform cloud cover produces bright light without any shadows at all.

When an immediate need exists, however, the weatherman rarely cooperates. Sometimes aerial photos will have to be taken in less than ideal conditions. If the aerial photography can be planned ahead of time, and optimal conditions are the goal, bright sunny days at midday are the time to fly.

CAMERA CONTROLS FOR AERIAL PHOTOGRAPHY

Motion Control

The primary concern with most aerial photography is to eliminate the blur associated with motion. Various types of motion are our concerns:

- The motion of objects on the ground: Vehicular traffic may be moving in the same direction as the aircraft, or in other directions, at various speeds. Moving one way, such vehicles may be blurred, while moving the other way, they may be successfully frozen by the camera's shutter speed. Vehicular motion, however, may be irrelevant to your documentation of a crime scene.
- The motion of the aircraft: Even a helicopter is moving at all times. It is virtually impossible to have the aircraft completely motionless. Besides moving forward and backwards, the aircraft is constantly moving upward and downward, either intentionally, or because of updrafts or downdrafts.
- The motion of the camera: Our primary concern is with the vibration of the airframe itself being transmitted to the photographer and camera. To minimize this vibration, avoid contact with hard surfaces within the aircraft. Do not brace yourself on any part of the aircraft. Your only contact with the aircraft should be where you sit on the padded seat (avoid the seat frame and arm rail) and where you come into contact with the seatbelt or harness system used in some situations.
- The solution to all these motion concerns is to use a minimum of 1/500th second shutter speed; 1/1000th is better, if it can be obtained while maintaining proper exposures. This deserves to be a Rule of Thumb.

 Rule of Thumb 9.7: The shutter speed for aerial photography should be at least 1/500th second, if not 1/1000th sec.

Early in this author's law enforcement career, the best view of an accident scene was from a nearby bridge. Feeling steady on the bridge, the camera was mounted on a tripod, and a 1/60th shutter speed was used. All the photographs from the bridge came back blurred. That was when it was first learned that bridges are designed to be flexible, and the bridge was moving during all the photographs.

Learn by my mistake: consider every bridge to be moving, and use a faster shutter speed to freeze this motion.

Focus

Focus on infinity. The depth of field requirement is small at aerial heights, usually a minimum of 500'. The depth of field range will be adequate with the widest aperture of the lens. The use of a wide aperture will help maintain faster shutter speeds. Better yet, use an aperture 2 stops in from the widest aperture: the critical sharpness range of f/stops. Two stops in from the widest aperture puts the camera into the "sweet spot" of the lens. Recall the previous discussion on diffraction. This will ensure both the ground level and the tops of buildings are in focus.

Lenses

It is best to use a zoom lens to allow quick variation in focal lengths. Try to avoid lens changes during the shoot, unless you have an assistant handing you equipment, which is also strapped down. Wide-angle and super-telephoto lenses are not appropriate tools from most aircraft heights.

You will find, however, that short telephoto lenses are particularly useful when taking aerial photographs. But, just as hand holding a 100-mm lens on the ground requires a faster shutter speed, when taking aerial photographs with a 100-mm lens, a faster shutter speed is necessary. If 1/500th to 1/1000th shutter speed is recommended with a 50-mm lens in an aerial photography situation, consider 1/1000th and 1/2000th as the optimal shutter speeds with a 100-mm lens in an aerial situation.

Filters

Citizens of Los Angeles, California, and Phoenix, Arizona, know about particulates in the air, but most of us have not had much experience with such dense air. As the distance between the camera and the subject being photographed lengthens, the contaminants in the air can begin to degrade the resultant image. For most of our photography, this is insignificant. When doing aerial photography, it begins to matter.

Low-level aerial photography, below 5000', usually does not require the use of filters. With long oblique views of the scene, or with higher vertical distances between the camera and the scene, consider the following: light scatters as it travels through the atmosphere. Increase the atmospheric distance the light has to travel, and the image may begin to show a lack of contrast because light will scatter when it strikes particulate matter and water droplets suspended in the air, which is often called "haze" and is worsened by auto exhaust, industrial pollution, airborne dust, and smog (smoke and fog). All these result in a scattering of the light, lowering the contrast, and blurring scene details. In these situations, consider using a true haze filter. Simple ultraviolet filters are not effective for aerial photography.

Yellow, orange, and red filters are known to be able to effectively cut through haze when black-and-white film is used. They also filter some of the light coming through them. If you plan to use them to cut through the haze, consider using ISO 200 or ISO 400 films.

Composition

As previously mentioned, avoid having your own shadow in the photo (plane or helicopter). Also, avoid aircraft wings or other aircraft components being in the field of view. This may require the pilot to bank or roll the aircraft over a bit, so you can view straight down.

Do not take photographs through windows if it can be avoided. Reflections and glare will usually produce unacceptable results. Windows may not be entirely clean and may have accumulated fine scratches. It is best to have the window, or the door next to where you will be seated completely removed. (Hence, the harness system.)

You may think you know the area to be photographed very well. How many times have you viewed this area from above? It is easily possible to get "lost" when viewing the scene from a plane or helicopter. Make sure you have good maps and have some obvious fixed features that can easily be seen from above to ensure a proper orientation.

To be sure you have captured the images you need, it is essential that the photographer have communication with the pilot. A headset is the ideal, but lacking that, you should work out some simple hand signals. The pilot may think the last pass over your target gave you sufficient time to get the images you wish, but if you did not get what you wanted, you will need some way to tell the pilot to make another pass or to direct the pilot to fly on one side or the other. Communication with the pilot cannot be overvalued.

Film Speed

On a normal sunny day, ISO 100 speed film is recommended. However, it is also normally recommended to use a shutter speed of 1/60th of a second. Try setting the camera to 100 ISO and 1/1000th shutter speed and take a meter reading of a sunny area. If the resulting aperture is close to 2 stops from the widest aperture of the lens, then 100 ISO film can be loaded into the camera. If not, try the same exposure readings with ISO 200 film and then ISO 400 film. Remember, aerial photographs are frequently enlarged, and slower ISO films will enable bigger enlargements without apparent graininess.

One huge benefit to digital imaging is the possibility to change ISO film speeds at any point. Changing film while in mid-air is extremely difficult. However, the recommendation to use the slowest ISO setting still applies to digital cameras. The best enlargements will be made from the slowest ISO films or digital equivalents.

Exposure

The f/16 sunny day rule works during aerial photographs also. An f/16 and 1/125th (a bright sunny day with crisp shadows) can be converted to f/5.6 and 1/1000 because of reciprocity.

Also recall the old Rules of Thumb: It is better to overexpose film and it is better to underexpose digital images, if exposure errors are unavoidable. It may be difficult to do brackets while flying in circles around the scene of interest, but it sure beats having to make plans for a second aerial excursion because your first images were improperly exposed; with digital imaging comes the ability to instantly see your results.

Safety Concerns

For those who have not attempted aerial photography before, and for those who have, safety is a concern more important than getting the images justifying the trip. It is certainly a rush to be in a helicopter, sitting next to a large gaping hole where the door had been before it was removed. Now, when the pilot tilts the "copter" a bit to give you a good view of the ground, the adrenaline begins pumping. Highway signs saying, "Click it, or ticket" are not needed. You will want to be securely strapped into your seat. The amount of air swooshing around you is difficult to imagine on the ground. When it is advised that you strap everything down, it is because not only will gravity pull loose things down, but the air flow will pull things out of the aircraft laterally.

Initially, one may just think about the loss of expensive equipment. After just a second, though, you should also think about when this expensive equipment hits the ground. Actually, hitting the ground would be a good thing. Imagine if it hits a moving vehicle or person. Michael J. Brooks, FBI Senior Instructor of Scientific and Technical Photography, recently lectured on this same topic and mentioned yet another concern. A jacket, backpack, or camera equipment can get swept out of the aircraft and get sucked up into the rear rotors. Do not dwell on this thought. Just ensure everything within the aircraft is strapped down.

UNDERWATER PHOTOGRAPHY

Mention underwater photography to most, and images like those in Figure 9.38 come to mind.

Figure 9.38 shows views of crystal clear water, bold colorful images and artistic compositions. Why should taking photographs of underwater crime scenes or evidence thrown into lakes, rivers, and ponds be difficult? Because most underwater crime scenes are not surrounded by crystal clear water.

In the section on aerial photography, haze and smog were mentioned as occasional problems to be dealt with. If "thick" air can be a problem for aerial

Figure 9.38

Classic underwater photographs (Courtesy of Nancy Olds, Forensic Photographer, United States Secret Service).

photographers, "thick" water is even more a problem for underwater photographers. Figure 9.39 shows "thick" water and what most underwater crime scene photographers have to deal with most of the time. Suspended in the water are many particulates, which obstruct a clear view of the object being photographed. To compound the matter, the swimming motions of the diver often causes material

Figure 9.39

Thick water: Particulates in the water and bottom debris stirred up (Courtesy of Nancy Olds, Forensic Photographer, United States Secret Service).

from the bottom surface to get stirred up and added to the particulates already in the water.

These problems, and others frequently associated with underwater photography, are discussed in the following.

LIGHTING AN UNDERWATER PHOTOGRAPH

Water is approximately 600 times denser than air, so light travels through it very differently. There appears to be less light underwater from . . .

Reflection

Underwater photography also requires light, either ambient light or the light we bring with us in the form of electronic strobes. The best ambient light penetration into the water occurs during the hours around noon. At different times of the day, more light reflects off the surface, and less light makes it beneath the water. Surface conditions also affect reflection. Calm water reflects little light. Choppy water reflects more. The least ambient light reflection occurs on bright calm days between 10 AM and 2 PM.

Because the underwater lighting is much dimmer than when on dry land, electronic strobes will almost always have to be used to prevent underexposures.

Figure 9.40 shows a close-up photograph of a pistol taken to document that the hammer was down when the weapon was recovered. A strobe was required for proper lighting.

Figure 9.40

Close-up photograph of a gun (Courtesy of Nancy Olds, Forensic Photographer, United States Secret Service).

Refraction

Refraction: When light changes from one medium to another, it will bend. Viewing underwater objects will be affected by refraction. The result is that objects underwater appear approximately 25% closer to the viewer than they really are.

Refraction is the bending of light rays as they pass from one medium to another.

A pencil beneath the surface in a glass of water looks both bent and magnified because of refraction (Figure 9.41).

The least amount of bending occurs when the sun is directly overhead. As light bends, it has to travel farther to reach a given depth, and the increased distance the light must travel reduces its intensity. Recall that the inverse square law stated this (Figure 9.42).

Light will refract when it changes from moving through the air when it strikes water. It will also refract when it is traveling through water and strikes a diver's face mask, similar to when light travels through water and strikes the lens cover of an underwater camera in a protective underwater housing. Refraction is the phenomenon that makes underwater objects, viewed through either normal flat lens covers and flat diving masks, appear approximately 25% closer than they really are.

This creates two underwater problems. If an object appears to be at x distance from the photographer, but we know it is really at y distance from the camera, how do you focus, and where do you aim the flash?

Because the camera "sees" like our eyes do, focus on the apparent distance of an object underwater, which will be a closer distance than the object really is; focusing on the true distance will result in out-of-focus photos. However, it is necessary to aim the strobe at the actual object's distance. Figure 9.43 graphically demonstrates both of these concepts.

Figure 9.41

Refraction bending and magnifying a pencil.

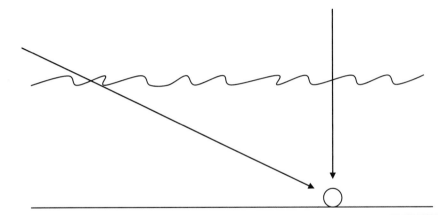

Direct light from above has a shorter distance to travel to light an object. Light coming from a low early morning or late afternoon sun has to travel farther through the water to light the same object, making the light dimmer once it finally reaches the object.

Figure 9.42
Refracted light, traveling farther, is dimmer when it arrives.

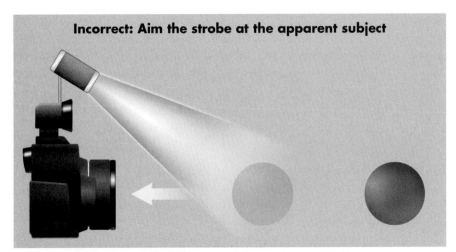

Incorrect: Aim the strobe at the apparent subject

Figure 9.43
Focus on the apparent distance; aim the flash at the actual distance (Courtesy of Jeff Robinson, Graphic Artist, scamper@scamper.com).

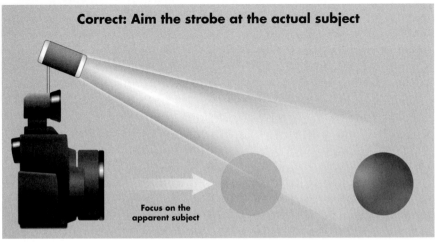

Correct: Aim the strobe at the actual subject

Focus on the apparent subject

One solution to the apparent magnification of underwater objects is to use an underwater camera housing with a domed port that covers the camera lens. The use of a domed port over the camera lens "corrects" the distortion, and objects appear at their true distance. These are much more expensive that the normal flat ports of an underwater camera housing.

The flat port, Figure 9.44, is most commonly used for underwater photography, but the "purists" may prefer the domed port variations (Figure 9.45).

Scatter

Depending on the underwater location, the time of the year, the weather, and water currents, the amount of suspended particles, silt, and minute underwater organisms can vary quite a bit. Light that is reflected from objects in the water travels through this thick medium filled with floating debris; this causes the light to be scattered and diffused. Because of **scatter**, we do not get a clear view of the subject.

Scatter: Light that is reflected from objects in the water travels through a medium filled with floating debris; this causes the light to be scattered and diffused.

Figure 9.44

Underwater camera housing, flat port (Courtesy of Nancy Olds, Forensic Photographer, United States Secret Service).

Figure 9.45

Underwater camera housing, domed port (Courtesy of "Dori" Charmichael, Splash Dive Center, Inc, Alexandria, VA).

Figure 9.46 shows just how unclear an image can be because of scatter, which can be thought of as underwater "smog." All the "classic" underwater photographs most of us are used to seeing are taken in regions known for crystal clear water. Unfortunately, Murphy's law dictates that evidence ends up in the murkiest water possible. Murky water is not only an intermediate obstruction, making it difficult to see the evidence we are looking for; once some of it settles over the evidence, the evidence can become almost impossible to see.

Figure 9.47 shows a pistol almost completely covered by silt. Fortunately, the searchlights caught part of the surface and reflected so the diver could see it.

Figure 9.46

Scatter: Underwater "smog" (Courtesy of Nancy Olds, Forensic Photographer, United States Secret Service).

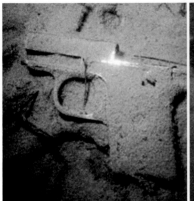

Figure 9.47

Silt partially obscuring a pistol (Courtesy of Officer Julius Wiggins, Underwater Photography Instructor, City of Miami Police Department, Retired).

Backscatter

Backscatter results from minute underwater particles that also affect underwater flash photography, where reflections off of these particles will result in bright spots between the flash and the prime subject matter (like snow).

Figure 9.48 shows various degrees of **backscatter** obstructing the view of a sunken boat. The lower image shows that if the flash does not strike the particulates, the appearance of backscatter is less evident.

Figure 9.49 shows a victim struck by a boat. The top view shows backscatter. The closer view shows paint transfer from where the boat struck the victim. With less water between the camera and the subject, and the flash held in the proper position, backscatter is eliminated.

Backscatter: Flash reflections off of underwater particles will result in bright spots between the flash and the prime subject matter (like snow).

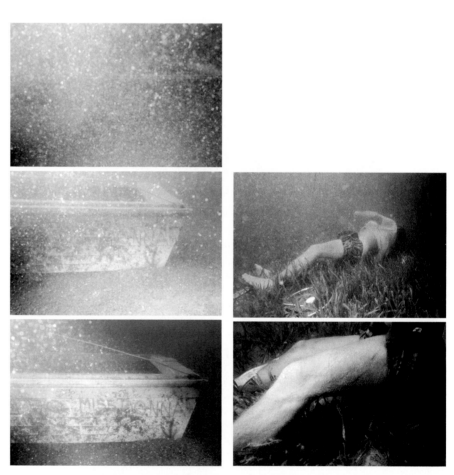

Figure 9.48

Various degrees of backscatter (Courtesy of Nancy Olds, Forensic Photographer, United States Secret Service).

Figure 9.49

Struck by a boat. Trace paint on victim's thigh. (Courtesy of Officer Julius Wiggins, Underwater Photography Instructor, City of Miami Police Department, Retired).

Because neither scatter nor backscatter are desirable features of our photographs, is there a way to avoid both of them? To avoid them, keep your light source as far away from the camera as possible and light your subject from an angle of approximately 45°.

Figures 9.50 and 9.51 show the theory and the practice of backscatter causation and avoidance. Bounce light off of the front of the suspended particulates in the water and the result is underwater "snow." Remove the flash from the immediate area of the camera, and the side lighting of these same particulates minimizes their appearance.

Because Nick Oliver, one of our graduate students at The George Washington University, took the images in Figure 9.51, it is a perfect opportunity for a bit of unabashed institutional self-promotion. Several years ago, a group of our enterprising students asked if there was any opportunity to obtain credit toward our

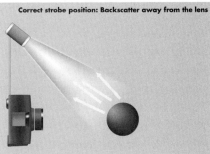

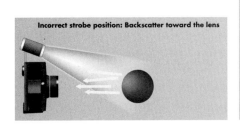

Figure 9.50

Direct flash and the flash removed from close proximity to the camera (Courtesy of Jeff Robinson, Graphic Artist, scamper@scamper.com).

Figure 9.51

Direct flash and the flash removed from close proximity to the camera (Courtesy of Nick Oliver, GWU MFS Student).

Master of Forensic Science degree for work done relating to underwater crime scene searches and underwater photography. The idea did seem to align with our Independent Research Course, where students could take any aspect of our core curriculum courses and delve deeper into that subject matter. Their idea seemed a perfect marriage of our Crime Scene Investigation and Forensic Photography courses. Two stipulations were mandated. They would have to become SCUBA certified, if they were not already so certified. We contacted the Splash Dive Center in Alexandria, Virginia, to facilitate this. Then, they would have to pass the Underwater Criminal Investigators (UCI) course taught by First Sergeant Mike Berry, a Virginia State Trooper and highly regarded local diving instructor. Having passed those two hurdles, the students would then write a lengthy term paper relevant to their training.

Figure 9.52 shows George Washington University's first UCI graduating class, with Mike Berry. Their final dive was in the James River, alongside Richmond, Virginia, where they recovered a rifle that had not been planted for the occasion.

Absorption of Colors

Water acts as a filter, absorbing the different colors of "white light" at different depths. Depending on the clarity of the water, the color red will be absorbed in the first 5 to 10'. Blue is the last color to be absorbed, which will occur at about 100'. It is not necessary to memorize different distances for different colors. The bottom line is that the only way to ensure the capture of proper colors at any but the shallowest depths is to use electronic strobes. Because the light from an electronic flash travels much shorter distances underwater, you can be sure that the backgrounds in most of your shots will likely be green or blue if you are deeper than 25' or so.

Figure 9.53 shows the predominant green casts that will be the result of light absorption at most medium depths. At deeper depths, the only color remaining will be blue, as in Figure 9.54. Please visit the book's companion website (http://books.elsevier.com/companions/9780123693839) for a color version of this figure.

Figure 9.52

George Washington University's first graduating class of underwater criminal investigators (Courtesy of Sergeant Mike Berry, Virginia State Police; Amie Balle and Elizabeth Toomer, GWU MFS Students).

Figure 9.53

Most of the world is green underwater (Courtesy of Nancy Olds, Forensic Photographer, United States Secret Service).

Figure 9.54

Unless you are deeper, where it is blue (Courtesy of Nancy Olds, Forensic Photographer, United States Secret Service).

Solutions to Problems of Underwater Photography

Underwater photography poses a number of problems, mentioned previously. Here is a short list of recommendations that will correct most of these problems.

- Stay near the surface, if possible. When the evidence is in deeper water, this advice cannot be followed.
- Add auxiliary lighting. A Rule of Thumb is frequently offered here.

Rule of Thumb 9.8: When possible, stay within 4′ of the subject to be photographed. Electronic flash will not be effective at the same distances as it is on dry land.

- Shoot when the sky is clear and the sun is out. If possible, dive between 10 AM and 2 PM.
- Shoot when the water is calm.
- Use a higher ISO film. Water absorbs some of the light as well as certain colors. Faster films like ISO 400 will help prevent underexposures.
- An underexposure error will occur when taking exposure readings on a light-colored sandy bottom, just as it will when metering on light-colored scenes on dry land. Knowing this will occur, set the camera's exposure controls to at least a +1, and then bracket with a +2.
- Do the same with backlit scenes. Be careful when aiming the camera toward the surface. The bright surface will tend to make your subject a dark silhouette. To avoid this, an electronic flash, set to the manual exposure mode, must be used. The use of either the automatic flash exposure mode or the dedicated (TTL) flash exposure modes will result in underexposed silhouettes, because their respective sensors will react to the backlighting and shut the flash off early.

LENSES

Wide-angle lenses are regarded as the lens of choice for most underwater photography. Because a wide-angle lens tends to make an object seem farther from the camera compared with a 50-mm lens, to fill the frame with the object, you will have to get closer to it than when using a 50-mm lens. Getting closer to the object means less water is between the camera and the object, therefore there are less particulates to diffuse the lighting. Additional benefits:

- Wide-angle lenses have better depth of field than other lens types.
- Wide-angle lenses force you to get closer to your subject when you fill the frame; therefore, with less light loss because of distance, the result is better colors.
- When you are closer to your subject matter underwater, there is less diffusion of light and more contrast in your image.
- With fewer particulates in the water between the camera and the object being photographed, the less backscatter will be a problem.
- A wide-angle lens is more tolerant of the motion of the photographer, because underwater currents sway their body. Of course, using a flash also helps "freeze" the bodily motion of the photographer.

SAFETY CONCERNS

The requirement to document evidence underwater cannot supersede the diver's awareness of his location, his depth, his total time in the water, and his air supply.

- The underwater environment is inherently a dangerous location.
- The additional weight and drag of the equipment are fatigue factors important to remember.
- Dive only with a partner.
- Do not use a camera neck strap. Instead, consider using a wrist strap that you can easily slip free from if it becomes entangled in underwater debris.

SUMMARY

Working at an accident scene is potentially very dangerous to all concerned. Before photographs can be safely taken at major accident scenes, several priorities have to be dealt with. Law enforcement officers have many responsibilities: to protect the accident scene, manage the traffic flow, and document the scene as part of their responsibility to work the accident and complete an accident report. The same aspects of the scene need to be documented whether the photographer has been trained in accident reconstruction or not. This chapter presented an overview of the concerns of the photographer at a major accident scene.

This chapter covered the normal surveillance photography techniques required to capture recognizable images of a suspect from a distance in both daytime and nighttime situations. The main two differences are that much longer focal length (telephoto) lenses are used to recognize a suspect at a greater distance, and faster ISO films are used to properly expose the suspect under dim lighting conditions. This chapter explained how all the photographic variables are selected for surveillance photography from a distance.

Aerial photography presents many challenges to successful photography. It is important to be aware of the potential problems that may arise and the methods that have been developed to ensure a satisfactory result. This chapter indicated the factors that must be dealt with and the problems that must be overcome when doing aerial photography.

Underwater photography is also potentially dangerous and presents unique problems that have to be solved. The issues encountered with underwater photography and the solutions to these problems were covered in this chapter.

DISCUSSION QUESTIONS

1. Before a major accident scene can be photographed, safety issues have to be addressed. What are they?
2. The immediate accident scene cannot be isolated from its surroundings. How does this apply to the photographic documentation of a major accident?
3. How should the vehicles involved in a major accident be photographed?
4. How can photography document the wounds and injuries to the occupants of the vehicles in major accidents?

5. Large accident scenes can be lit by the painting with light technique during the nighttime. Discuss two other photography techniques sometimes used for large dimly lit accident scenes.

6. Discuss the surveillance photography issues related to lens choice and shutter speed choice.

7. How is the ISO film speed optimally determined in nighttime surveillance photography situations?

8. Explain the camera variables that allow a photographer to blur intervening obstructions to the point they begin to "disappear."

9. Compare film and digital sensors regarding the ability to recover details from underexposed images.

10. What are the main lighting concerns related to aerial photography?

11. How is all the motion related to aerial photography controlled?

12. Indicate some safety concerns related to aerial photography.

13. Indicate some safety concerns related to underwater photography.

14. Effectively lighting underwater objects with electronic flash differs from using a flash on dry land. What are the main issues and solutions?

15. How can you be sure the colors of underwater object are accurately captured?

EXERCISES

1. Choose any car, and do a series of photographs to completely document its exterior. You may omit the undercarriage.

2. Go to a large dimly lit outdoor area at night and photograph the same area, at least 60′ to 80′ long, with (a) painting with light, (b) aperture priority, and (c) time exposures.

3. During the day, photograph a vehicle's license plates while the camera is in a room with Venetian blinds down to conceal your actions. The Venetian blinds may be slightly cracked to facilitate this photography. Then, from the vehicle's position, photograph the room with the Venetian blinds in the same position they were in when you took the original photos.

4. At night, set up at least one block away from an outdoor café setting where people are sitting outside. Photograph the face of one individual facing the camera. Use the appropriate ISO film speed, shutter speed, and focal length to ensure the face is recognizable. It may be very wise to bring this exercise with you in case you are reported as a suspicious person, and local law enforcement taps on your shoulder demanding an explanation of your actions.

Almost every time I have practiced surveillance techniques outside, I have had the pleasure of discussing my activities with our guys and gals in blue.

If the appropriate equipment is available, aerial and underwater exercises will be provided by photography instructors.

Compare your images with those in this chapter and with other images on the supplemental website of images, referring particularly to the following image folders: Accidents and Surveillance.

FURTHER READING

Accidents

Dubois, R. A. (1991). "Photography in Traffic Accident Investigation." Institute of Police Technology and Management: University of North Florida.

Fricke, L. B. (1985). "Photography for Traffic Accident Investigation." Northwestern University Traffic Institute, Evanston, IL.

Surveillance

Eastman Kodak Company. (1972). "Photographic Surveillance Techniques for Law Enforcement." Kodak Publication No. M-8, Eastman Kodak Company, Rochester, NY.

Eastman Kodak Company. (1984). "Using Photography for Surveillance." Kodak Publication No. M-10, Eastman Kodak Company, Rochester, NY.

Siljander, R. P. (1975). "Applied Surveillance Photography." Charles C Thomas, Publisher, Springfield, IL.

Aerial

Colins, M. R. (1998). "The Aerial Photo Sourcebook." The Scarecrow Press, Inc., Lanham, MD.

Eastman Kodak Company. (1985). "Photography from Light Planes and Helicopters." Kodak Publication No. M-5, Eastman Kodak Co., Rochester, NY.

Eller, R. (2000). "Secrets of Successful Aerial Photography." Amherst Media, Inc., Buffalo, NY.

Graham, R., and Read, R. E. (1986). "Manual of Aerial Photography," Focal Press, Boston, MA.

Lloyd, H. (1990). "Aerial Photography: Professional Techniques and Commercial Applications." Amphoto, New York, NY.

Underwater

Becker, R. F. (1995). "The Underwater Crime Scene." Charles C Thomas, Publisher, Springfield, IL.

Edge, M. (2001). "The Underwater Photographer." 2nd Ed. Focal Press, Oxford.

Fine, J. C. (1986). "Exploring Underwater Photography." Plexus Publishing, Inc., Medford, NJ.

Jackson, R. M. (2000). "Essentials of Underwater Photography." Best Publishing Company, Flagstaff, AZ.

Kohler, A., and Kohler, D. (1998). "The Underwater Photography Handbook." New Holland Publishers Ltd, London.

Rebikoff, C., and Cherney, P. (1975). "Underwater Photography." 2nd Ed. Amphoto, Garden City, NJ.

Schulke, F. (1978). "Underwater Photography for Everyone." Prentice-Hall, Inc., Englewood Cliffs, NJ.

Skerry, B., and Hall, H. (2002). "Successful Underwater Photography." Amphoto Books, New York, NY.

Taylor, H. (1977). "Underwater with the Nikonos & Nikon Systems." Amphoto, Garden City, NY.

Webster, M. (1998). "The Art & Technique of Underwater Photography." Fountain Press Ltd., Surrey.

CHAPTER 10

LEGAL ISSUES RELATED TO PHOTOGRAPHS AND DIGITAL IMAGES

LEARNING OBJECTIVES

On completion of this chapter, you will be able to . . .

1. Explain the elements for a photograph to be considered "a fair and accurate representation of the scene."
2. Explain what is meant by requiring the photograph to be "authentic."
3. Explain why all photographs taken at a crime/accident scene may not be admissible in court: to be admissible they have to be: "relevant" and "material."
4. Explain the requirement that a photograph has to be more probative than prejudicial.

KEY TERMS

Authentic	Manipulations	Relevant
Fair and accurate repre- sentation of the scene	Material Prejudicial	Repeatable SWGIT
Interpolation	Processing	
Lossless compression	Reasonable expectation	
Lossy compression	of privacy	

CRITERIA OF PHOTOGRAPHS AND DIGITAL IMAGES AS EVIDENCE

A review of the literature and the case law provides the crime scene photographer with the distinction between when photographs and digital images are clearly admissible in court as evidence and when it would be reversible error to admit them. Between these extremes, it is well known that the trial judge has the ultimate discretion to either admit images as evidence in a court proceeding or to refuse to admit them as evidence.

To be admissible as evidence, an image must be relevant to an issue being contested. The image must also tend to either prove or disprove a disputed or material issue. The image must be established to be "a fair and accurate representation of its subject matter."

Conversely, it would be improper to admit an image as evidence if it were not relevant to issues before the court or if it did not assist the trier of fact in understanding an issue essential to the outcome of the trial. If the image is an unfair representation of the subject of the image, it should also not be admitted as evidence. How can an image be an unfair representation of its subject matter? If it grossly distorts pertinent aspects of the photograph, if it falsely represents the subject matter, if the image is deceptive, and if the image tends to mislead the viewer, then the image should not be admitted in evidence during the trial.

At times these distinctions are clear. At times judges must use their discretion, and their ruling generally will not be overruled unless they clearly abused their discretion. A third possibility exists. The image may be admissible, but the jury may then decide what weight they choose to give the image. The jury may consider the image as an extremely important aspect of a case, or the jury may totally disregard the image, even though it may technically be admissible. "It is well to remember, however, that there are good photographs and poor photographs just as there are good witnesses and bad witnesses."[*]Once admitted in evidence, the jury may assign an image the weight they believe is appropriate.

"A FAIR AND ACCURATE REPRESENTATION OF THE SCENE"

Fair and accurate representation of the scene:
To be admissible in court as evidence, an image or photograph has to be "a fair and accurate representation of the scene." If the image is inaccurate, it may be inadmissible.

All photographs offered in court as evidence are required to have the proper foundation laid before they can be accepted. Someone will have to testify that the photograph is a "fair and accurate representation of the scene" as it was at the time the photograph was taken. This does not necessarily have to be the photographer who took the original photograph, but it does have to be someone who was present around the time the photograph was taken. Several aspects have to be considered for a photograph to be considered "fair and accurate."

Exposure

Normally, a photograph has to be properly exposed for it to survive challenges in court. A grossly overexposed or underexposed photograph may be held to be inadmissible simply because it does not accurately represent the scene as it was.

[*]Scott, C. C. (1969). "Photographic Evidence: Preparation and Presentation." 2nd Ed. Vol. 1. p. 29. West Publishing Company, St. Paul, MN.

There are two well-known exceptions to this requirement that a photograph be "properly" exposed.

1. At times a witness will testify as to what they saw in dimly lit conditions. If a photographer is attempting to photograph the scene from the viewpoint of this witness and attempting to duplicate the lighting conditions that originally existed, a well-lit photograph is not the "proper" exposure. In this situation, a dimly lit photograph may be the "proper" exposure.
2. When critical photographs are being taken, with the ultimate view of having the photographs serve as a standard for a comparison with a known item of evidence, the photographer often "brackets" the exposures. This means that additional photographs are taken that intentionally underexpose and overexpose the original image. When this series of photographs is turned over to a crime laboratory examiner to attempt to compare the photographs to an item of real evidence, the examiner may decide an otherwise "improperly" exposed image may actually show the identifying feature they are interested in better than the "well-exposed" version of the photograph.

For instance, examine these three photographs of a bite mark to a hand. One was taken with the camera's proper exposure, and the other was intentionally underexposed, a −1 bracket. A third exposure, a −2 bracket, is also included. The three images were taken so that the forensic odontologist could make up their own mind as to which image best shows the outlines of the teeth necessary for a comparison with the known teeth of a suspect. Camera systems try to provide proper exposures. In this case, the subject matter most prevalent in the viewfinder was Caucasian skin. It is not the Caucasian skin, however, which is the most important aspect of this image; it is the bite mark in the skin. Underexposing the skin actually shows the definition of the edges of the bite mark better. The examiner may, therefore, go to court with one of the underexposed photographs offered as evidence, and it will be admissible even though one aspect of the photograph is underexposed.

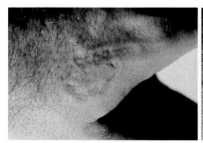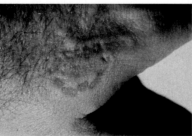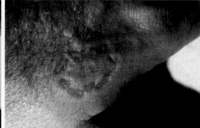

Figure 10.1
Bracketed bite marks: Normal, −1 and −2.

Color Accuracy

It is normally critical for a photograph being offered in court as evidence to have accurate colors. Imagine a case of a fatal hit-and-run accident, where a paint transfer from a green suspect vehicle to the blue victim vehicle is part of the probable cause by which the officers decide to arrest a suspect in a car with fresh damage and a similar-colored paint as is found at the accident scene. Ensuring the colors of the paint in the photograph are accurate can become a pivotal aspect of the trial.

Three methods ensure color accuracy with color film.

1. If color film is exposed by midday sunlight, color film will record the proper colors. Early morning and late afternoon sunlight will transmit a colored tint to the photograph that does not accurately portray the colors present at the scene. This light has a tendency to be tinted orange or red.
2. If midday sunlight is not available, the colors can be accurately captured if an electronic flash is used as the light source. The light emitted by an electronic flash is color balanced so that it will accurately render colors on color film. It is also necessary to use electronic flash whenever indoor photographs are taken, because tungsten and fluorescent lighting will also tint daylight color films. These tints will normally be yellow and green, respectively.
3. Probably the best way to ensure that photographs show an accurate range of colors is to photograph a color scale somewhere on the roll of film containing the critical photographs. Better yet, have the color scale appear in the specific image where color will be an issue. If the digital image or film photograph depicts the color scale accurately in the print being offered in court, it can be assumed the other colors in the image are also colored properly. Of course, at times a witness may have seen an event under poor lighting conditions or in lighting conditions that would produce tints onto objects. In these situations, the photographer would be attempting to replicate the poor lighting conditions.

In Figure 10.2, a white vehicle has been tinted to appear yellow because of the ambient lighting. If the issue in court is why the police stopped and detained a suspect driving a white vehicle when the witness reported seeing the suspect driving away from the crime scene in a yellow vehicle, these two images will explain the matter clearly. Of course, it will be necessary for the photographer to duplicate the yellow tint in one of the photographs. That is not the accurate color of the vehicle, but it is the accurate color of what the witness saw.

Distance Relationships

Photographs should accurately portray the correct distance relationships present at the original scene to be admissible in court as evidence. When the portrayal of an accurate distance is an issue in the court, normally two methods ensure that the distances depicted in the image are accurate.

Figure 10.2

A white vehicle tinted yellow because of ambient lighting. Please visit the book's companion website (http://books.elsevier.com./comp anions/9780123693839) for a color version of this figure.

1. It is normally critical to have taken the original photograph with a 50-mm focal length lens, which will not distort the relative distances. Either wide-angle lenses or telephoto lenses will create a perspective distortion that falsely represents reality. As mentioned previously, wide-angle lenses will elongate the relative distance between foreground and background, and telephoto lenses will compress this same area. The result is that the perceived distances between vertical features of the scene will be inaccurate with any lens other than the 50-mm lens. If the lens being used on the camera at the time is a zoom lens, it is critical that the 50-mm focal length setting be used when the image is captured.

2. Even when a 50-mm lens is used, it is then essential that the distance in question be viewed with the photographer positioned so that they are equal distances from the two items in question. This has been expressed as the photographer being perpendicular to the imaginary line drawn between the two items in question; or, the photographer forming an isosceles triangle with the two items of interest, with the two items of interest equal distances from the photographer. If the photographer uses a 50-mm lens, but then orients himself or herself so the evidence and the photographer forms a straight line,

which is called a linear perspective, the two items of evidence will appear closer together than they really are.

Accurate distance relationships require both the proper lens and the proper perspective view of the evidence.

Focus/Depth of Field

It is crucial that all important aspects of the photograph are in focus. If an important aspect of the photograph is out of focus, the photograph may be declared inadmissible as evidence.[†] Therefore, it is essential that the range of what appears to be in focus, the depth of field, covers all the areas in the photograph that are important. In many cases that means everything in the picture.

A brief return to the first chapter on composition may be required here. If you know an area will be out of focus, you should recompose the scene so that those areas that will be out of focus will no longer appear in the composition. The photographer at a crime/accident scene must constantly be adjusting his camera controls to maximize the depth of field. When it is known that the depth of field will be inadequate to ensure the entire expanse will be in focus, the photographer should consider recomposing the scene to include only the areas that will be in focus.

Creative photographers frequently intentionally throw unimportant aspects of their composition out of focus, usually to force the viewer to look at the main subject who is in focus. Crime scene photographers usually do not have this option. Intentional blurring of parts of the scene, although perfectly acceptable for creative photographers, should not be a photographic technique used by crime scene photographers. The crime scene photographer should first try to maximize their depth of field so that everything in the composition is in focus. After that, if background or foreground areas are known to fall outside the depth of field range, recomposing the images is called for.

[†]"A photograph that is not clear and sharp may be likened to a witness who does not speak distinctly and is hard to understand. Both the blurred photographs and the inarticulate witness may contribute something of value despite their imperfections...a photograph ordinarily is admissible for what it is worth even though it is somewhat out of focus and therefore of little value. However, it is not an abuse of discretion for the trial court to exclude out of focus photographs even though it would not be an error to admit them." (Miss. Butler *v* State, 179 So.2d 184, 253 Miss. 760 (1965); W.Va. Elswick *v* Charleston Transit Co., 36 S.E.2d 419, 128 W. Va. 241 (1945) (court did not err in refusing to admit photograph that was obscure and not sufficiently clear to aid jury.) Indiana: Taylor *v* State, 295 N.E.2d 600, 260 Ind. 264 (1973), cert denied 94 S. Ct. 377, 414 U. S. 1012, 38 L. Ed.2d 250 (fact that pictures are far from good examples of photography goes merely to weight of evidence rather than to admissibility). S. C.–Brooks *v* Brooks, 339 S.E. 2d 531, 288 S. C. S. C. 71 (App. 1986) (whether photographs were of sufficient quality to be admissible was a matter up to the discretion of the trial judge) (Tex. Jimenez *v* State, 787 S. W. 2d 516, 524 (App. 1990) (appellate court suggested that event of retrial the trial court should reconsider its ruling admitting blurred photograph in evidence).Scott, C. C. (1969). "Photographic Evidence: Preparation and Presentation." 2nd Ed., Vol. 2. p. 395. West Publishing Company, St. Paul, MN.

An obvious exception to the usual prohibition of crime scene photographers to intentionally blur aspects of photographs is if a fingerprint is on a window, and the photographer cannot place a sheet of paper on the opposite side of the window to eliminate background views behind the fingerprint, the crime scene photographer can then photograph the fingerprint with the widest aperture of the lens to intentionally blur the background elements so they do not compete with the fingerprint ridge details.

Size

At times it is critical that a photograph be compared with another item of evidence; for instance, if a shoe print pattern is located at a crime scene, it would ultimately need to be compared with a suspect's shoe when that is recovered. For this comparison to be made, it is necessary for the film negative containing the image of the shoe print from the crime scene to be enlarged to a life-size print. If the original image was captured with a digital camera, the original digital image, originally the size of the digital sensor, will have to be enlarged to life size. This can only be done if the original photographs included a scale and if that scale is placed on the same plane as the shoe print. That means if the evidence is a partial shoe print on a raised surface and the photographer is focusing on the top of the raised surface, the scale needs to be raised to the same level of the top of the raised surface by placing objects under the scale to raise it to the proper height. If the evidence is a shoe print in dirt, and the photographer is focusing on the depth of the shoe print, for instance, $1/2''$ below the dirt's surface, the scale needs to be lowered to the depth of the shoe print in the dirt. The photographer must dig a trough next to the shoe print and place the scale in the trough at the same distance from the film plane as the evidence.

These are images of two scales, placed at different distances from the film plane. When positioned equidistant from the camera, the scales are the same size; or $1''$ looks like $1''$ on both scales. However, even when one of the scales is just $1/4''$ from the other, a size difference already exists. This difference in relative sizes becomes more apparent as the scales are moved $1/2''$ and $3/4''$ from the focal plane. A scale that is placed on a different plane than the evidence will inaccurately represent the actual size of the evidence. The questions is this: How can a scale any distance from the actual plane of the evidence enable the negative to be enlarged to the true size of the evidence? It cannot. Therefore, it is critical, when taking images of evidence that will eventually be compared with a known item of evidence that we ensure that the scale is placed at the same plane, high or low, that the evidence is in.

One other interesting aspect should be pointed out when scales are not on the same plane as the evidence: Look at Figure 10.3 again carefully, and you should be able to notice that the further scale is just a little bit out of focus. It becomes progressively more apparent as the scales become further apart, but even at $1/4''$

Figure 10.3

Scales at different levels.

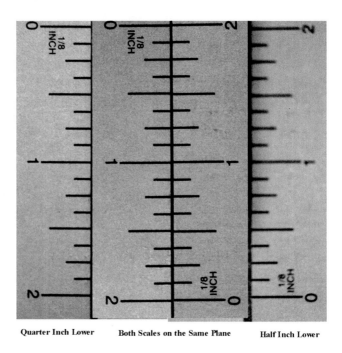

Quarter Inch Lower   Both Scales on the Same Plane   Half Inch Lower

difference, the further scale is just a little bit "soft." The scale is still readable, but not quite as sharp as the scale that was focused on, which is a second reason to place so much emphasis on ensuring the scale is on the same plane as the evidence. Not only will a misplaced scale misrepresent the size of the evidence, the scale will also be slightly out of focus. This may make a photographer seem less than competent at trial. Camera-to-subject distance is one of the factors that affects the depth of field range, and as these distances become closer and closer together, the depth of field range shrinks until virtually no depth of field is present in extreme closeups other than what is actually focused on. In other words, if you focus on a bloody fingerprint on a knife blade and place the scale on the floor next to the knife, there will be two consequences. The scale will be both slightly out of focus and it will not enable a life-sized photo of the fingerprint to be made, because the scale lies on a different plane as the fingerprint.

AUTHENTIC

Authentic: Photographs taken at crime/accident scenes should portray the scene as it was originally found before any alterations.

An **authentic** photograph consists of a fusion of two concepts that will be discussed in the following paragraphs.

"As Found"

For the most part, photographs taken at crime/accident scenes should portray the scene as it was originally found before any alterations, which is sometimes referred to as being "as-is," and sometimes this has been referred to as photographing the scene "in situ" or in the situation as it was originally found. Nothing should have been moved or removed before the photographs are taken. Nothing should have been added to the scene before the photographs were taken. As with everything in life, a few well-known exceptions to this mandate exist.

1. Police banner guard, yellow crime scene tape, is frequently added to secure the perimeter of a scene to protect it and to keep unnecessary people out of the scene. However, just before the exterior overall photographs are taken, this banner guard can sometimes be temporarily removed so that artificial elements are not introduced into the scene. When this is impossible or impractical, the courts have normally "understood" this obvious alteration of the perimeter of the scene.

2. Traffic cones are frequently placed around a scene for the same reasons. Because they are easily accessible to law enforcement agencies, they are frequently used to quickly mark the positions of evidence throughout the scene. Again, they should be removed before the photographs are taken, because the original scene did not have them present.

3. At times, evidence may have been picked up and taken into custody before the photographs of the scene "as-found" have been able to be taken. One situation that may mandate the early collection of some of the evidence is the presence of a hostile crowd when weapons or drugs are obviously present within the crime scene. Some in the crowd may have vested interests in removing evidence from the crime scene. As valuable commodities, many may be tempted to "acquire" the evidence for possible resale later; the "souvenir" collector is always interested in gathering a memento of the occasion. For these reasons, and others, it may be prudent to collect some of the evidence before the crime scene photographs can be taken. These types of evidence should not be replaced in the scene just before the photographs are taken. This may falsely give the viewer of the photograph, usually the judge or jury, the impression that the photograph is of the crime scene before any alteration. It is better to train all first responders that if early evidence collection is justified, a surrogate item, like a crime scene numbered or lettered marker, can be placed in the same location of the evidence just picked up. Then, the scene can be photographed with the surrogate evidence in view, and the testimony of the officers removing the evidence "early" can justify their actions and the photograph with surrogates in the crime scene. Everyone understands and accepts the need to occasionally pick up a golf ball and temporarily replace it with a ball marker left in its place. In the crime scene scenario, however, no effort will be made to replace the evidence where the marker has been placed.

In situations in which evidence has been picked up and collected and nothing has been left in its place, the scene will then be photographed without the evidence within the scene, and it will be up to the evidence-collecting officer to justify the collection and try to verbally explain to the court the location of the evidence within the scene. No attempt should be made to replace the evidence within the scene in the location the officer believes most likely is the position the evidence had been in.

In one jurisdiction at the scene of a police shooting, a rookie officer located the suspect's gun hidden behind a garage. So pleased with this discovery, the rookie proudly brought the gun to his supervisor standing at the curb in front of the incident. When the supervisor informed the rookie that crime scene investigators were almost at the scene, the rookie disappeared. When crime scene technicians arrived at the scene, they were directed to the rear of the garage, where the suspect's gun was photographed "as found." During the subsequent trial, this entire scenario was brought to light. Police disciplinary proceedings quickly followed the trial of the suspect: not just for the rookie, but for the supervisor as well.

The approach of bad weather can also justify the early collection of evidence. We do not have to idly stand by as high winds blow away the evidence, heavy rains wash it down the street, or warm weather evaporates the evidence. Early collection is mandated in such circumstances. If evidence has been removed before the photographs have been taken, the scene is photographed "as found" by the photographer, with some of the evidence already missing. It will then be up to the individual removing the evidence to justify doing so and to have it admitted in court solely on the basis of their testimony, because photographs will not be available showing it in the scene. Perhaps a well-planned photograph with a perspective grid in view might be considered in such situations. If time does not permit photographing all the evidence in place, perhaps one photograph can be composed with many items of evidence in view with the perspective grid.

Recreations

Reenactments or recreations of crime scenes or accident scenes are permissible only as long as no effort is made to suggest those photographs are of the original scene. Recreating scenes has many legitimate purposes; for instance, to allow a jury to determine whether a witness may have been able to actually see an event from a particular vantage point under certain circumstances. Such restagings, however, must always be well documented as such to avoid them ever being mistaken for the original event.

More and more, the technology exists to create mini-movies and animations of one side's version of the sequences of a crime scene that are proposed recreations of the original event. These are not misperceived as being original scene reproductions. However, at times, still photographs are created to supplement the recreations. Efforts must be in place to avoid having the juror misunderstand these images as original "as-is" images.

RELEVANT AND MATERIAL

All the photographs taken at a crime/accident scene may not later be admissible in court as evidence. How can this be? How can all the homicide crime scene photographs, for example, not be admissible in court? For photographs, or any type of evidence for that matter, to be admissible in court as evidence they must also meet another two-pronged test: they have to be both **relevant** and **material**.

"Relevant evidence is any evidence that tends to prove or disprove any disputed fact in the case. No evidence can be admitted to prove facts that are not at issue."[‡] By the time of the actual trial, there may be many issues that are not being contested or are being stipulated to. In these situations, it is not necessary to "prove" a nondisputed issue, and it would be a waste of the court's time to permit such evidence being admitted. Many photographs are taken at a major crime scene "just in case" they will later be needed at trial. Many of the images taken at major crime scenes are no longer relevant to the trial that is being presented. For instance, many images of the wounds to the victim will be taken, both at the crime scene and at the autopsy, in case the manner and cause of death are disputed in the subsequent trial. If the defense offers an insanity plea and does not deny their client caused the death or the death mechanism, then all these photographs may be declared irrelevant to the trial. The only issue will be whether the defendant is sane or insane. The entire crime scene scenario offered by the prosecution may be stipulated to. However, if images suggest rational planning, those may still be admitted as relevant to the issue of sanity.

The Federal Rules of Evidence, Rule 401 defines relevance somewhat differently: "Relevant evidence" means evidence having any tendency to make the existence of any fact that is of consequence to the determination of the action more probable or less probable than it would be without the evidence. Perhaps the same meaning with different legalese. The evidence has to have some probative value to the court proceedings.

Material evidence, to the layman, is frustratingly similar to relevant evidence. "Material evidence is evidence that is important to the case; it cannot be too remotely connected to the facts in issue."[*] Material evidence is offered to prove or disprove a fact in issue. If the evidence does not relate to any issues pertaining to the case, it is not material.

At one homicide scene, the back door was kicked in. Much effort was used to photograph the damage to the back door, anticipating it to be proof of a forced entry to the crime scene. Many midranges and closeups were taken of the back door, both from the inside and outside. Several hours later, a neighbor who was interviewed stated the deceased had previously locked themselves out of the

Relevant: Relevant evidence is any evidence that tends to prove or disprove any disputed fact in the case.

Material: Material evidence is evidence that is important to the case; it cannot be too remotely connected to the facts in issue.

[‡]Kaci, J. (1995). "Criminal Evidence." 3rd Ed. p. 51. Copperhouse Publishing Company, Incline Village, NV.
[*]Kaci, p. 51.

residence and had damaged the door trying to regain entry to their own house several days before the homicide. All those photographs of the back door immediately became irrelevant and immaterial to the homicide investigation.

"In practice, relevant evidence is material only if it meets both of these tests:

1. It must be relevant to some fact that is at issue in the case.
2. It must have more than just a remote connection to the fact."*

As stated previously, many of the original photographs taken at a particular crime scene "just in case" may or may not be needed later for the trial. Their use will be based on what issues are being disputed during the trial. For example, perhaps 200 photographs were originally taken at a particular crime scene. The only issue currently in dispute in the trial is whether the defendant shot the victim at close range during a struggle for the gun or whether the defendant shot the victim from a greater distance. It is not disputed that the defendant shot the victim. In this situation, photographs used to establish the range of fire, the gunpowder stippling pattern around the entry wound on the victim, are still admissible, but photographs of the trigger impression on the forefinger of the defendant may no longer be relevant to the case.

It was not wrong to have taken the original 200 photographs. However, by the time the case goes to trial, only a dozen or so images may still be relevant and material to the case.

MORE PROBATIVE THAN PREJUDICIAL

Prejudicial: Evidence that is otherwise relevant and material to the case may not be admissible in court if the evidence is judged to tend to adversely prejudice the jury against the defendant. If the judge believes such photographs would tend to upset the jurors to the point they may lose their objectivity and forget the defendant has not yet been found guilty, the images may be excluded.

Evidence that is otherwise relevant and material to the case may not be admissible in court if the evidence is judged to tend to adversely prejudice the jury against the defendant or inflame the jury against the defendant.

Examples of such photographs might be those that show mutilations or wounds on small children who are universally regarded as innocent victims. The judge has the right to exclude such images that may prejudice the jury against the defendant in court more than they tend to prove an issue being contested in that trial. If the judge believes such photographs would tend to upset the jurors to the point that they may lose their objectivity and forget the defendant has not yet been found guilty, the images may be excluded. These same images, however, may later be deemed to be admissible after the defendant has been found guilty, during the sentencing deliberations, to set the penalty, if a range of sentencing options exists that the jury can choose from.

Sometimes, gory or gruesome images also fall into this same category. The judge may reason that, although the images certainly tend to prove that the defendant did what he or she is charged with doing, the images still have an over-

*Kaci, pp. 51/52.

whelming tendency to upset the normal sensitivities of the jury members. In these cases, the judge may exclude the images on the basis of their tendency to prejudice the jury against the defendant. Different judges may view the same images differently. Sometimes it is difficult to anticipate how the judge may react to the images offered in the case. It is often prudent to have a "Plan B" if you draw a judge willing to exclude such images. Many crime scene photographers come prepared for such possibilities. The first image offered to the court as evidence will be a full-color photograph. Just in case the judge objects to the color photograph, a black-and-white version of the color photograph is ready to be offered to the court.

Figure 10.4 shows a close-range shotgun wound to the neck, both in its original color version and in a black-and-white version. If a judge decides the color image is too upsetting for the jury, it may be possible for the same judge to accept the black-and-white version. Better the black-and-white version than nothing. Always have a "Plan B."

"The court is cautious about admitting evidence if it would be likely to arouse either hostility or sympathy for either side."* This quote suggests two additional ideas. The image cannot overly prejudice the jury in favor of the victim, and the image should not prejudice the juror against the victim. Usually, the latter issue does not occur often. Why would the prosecution offer an image in court as evidence that did not show the victim in a favorable light? As the previous image of a rape victim demonstrated, it is possible to do this unknowingly. The tattoos on the victim's hands may not mean much to most people, but it only takes one juror to hang up a jury, and if one juror saw that image and acquired the wrong impression about the victim, the case may be in jeopardy.

Is it not the job of the prosecution to show the victim in a favorable light? Certainly. But if an image goes over the top while doing this, the image may be

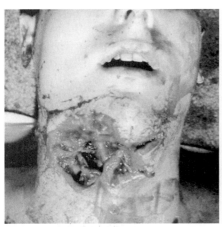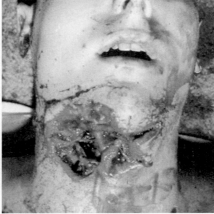

Figure 10.4

Color and black-and-white version of the same image. Please visit the book's companion website (http://books.elsevier.com./companions/9780123693839) for a color version of this figure.

*Kaci, p. 52.

held to be inadmissible. Photographs are routinely shown to a variety of people while trying to determine whom the victim really is. If the recognition of the victim from a photograph becomes a disputed point in the trial, having a young girl in her white Confirmation or First Communion dress does more than help identify the child. It produces undue sympathy toward the victim. That photograph may be held to be inadmissible.

The judge is the final arbitrator when challenges to the relevance, materiality, or the prejudicial nature of photographs are raised. Prepare the images intended to be presented in court as evidence with this in mind.

THE PURPOSE OF CRIME SCENE PHOTOGRAPHS

DOCUMENTATION OF CRIME/ACCIDENT SCENES AND EVIDENCE

Photographs are one of the basic techniques to document crime/accident scenes, the other techniques being sketches/diagrams and note taking. As a documentation technique, we remember there are three basic categories of crime/accident scene photographs: overalls, midranges, and close-ups. A previous chapter on these techniques will suffice for their explanation.

TO PROVIDE INVESTIGATIVE LEADS

Photographs are also used throughout the investigation. Other investigators, not present at the original crime/accident scene, can review the original scene photographs to gain an understanding of what happened and then assist with the investigation. Photographs can be shown to a variety of people to identify both the people and objects critical to the particular crime under investigation. Crime scene photographs can be shown to crime scene reconstructionists, who were not able to be present during the original crime scene processing; with the proper photographs, they can aid in the reconstruction of the events of the scene.

TO REFRESH MEMORY, SUBSTANTIATE TESTIMONY, AND CLARIFY UNDERSTANDING

During the trial, photographs can serve a variety of purposes. When a witness is trying to explain a complicated idea, often a photograph or two may assist in making the point. When the trial is being conducted a long time after the original crime occurred, photographs of the scene can easily and quickly refresh the memory of personnel present at the original scene. If contradictory explanations are being presented about a particular material fact about the scene or the evidence within the scene, photographs can help decide the issue or make it easier to determine who may be telling the truth.

PHOTOGRAPHS OF SUSPECTS AND EVIDENCE

If a crime scene photographer approaches a person on the street with a camera in hand, the person may indicate they do not want to be photographed. They may even say that if they are photographed without their permission, they will sue the photographer for infringing on their rights. Is the consent of the person being photographed required? What other "rights" might such a person have?

IS CONSENT REQUIRED?

If the person or object being photographed is in a public place, available to be viewed by anyone else present, then photographs can be taken by law enforcement personnel, even if someone says they do not want you to photograph them or the object, which is not true of commercial photographers attempting to take photographs to sell to a variety of publishers. In many jurisdictions there are "model" contracts that would have to be signed by the person being photographed, giving the photographer permission to use the photograph for gain. If the photograph is being taken for law enforcement purposes, and not for profit, the rules are different. If the evidence is in a public place, photographs can be taken even if the subject objects to them. If the subject chooses to expose their face or property in a public venue, anyone around can look at them or their property. If the general public can look at them or their property, then it is also legal to photograph them for law enforcement purposes.

DO MIRANDA RIGHTS APPLY?

Miranda *v* Arizona, 384 U.S. 436 (1966), applies to testimonial evidence only. It protects a person against involuntary, compelled self-incrimination, and statements made without the benefit of legal counsel being present. It does not apply to photographs or other aspects of physical evidence. A subject about to be photographed may assert their Miranda rights will be violated if the photographer takes their photograph without their permission. They are incorrect.

TRESPASSING

It is illegal to trespass on the property of another without their permission. If the photographer trespasses on restricted property to get the photograph in question, he may be charged with this crime, and the photographs taken from an illegal position may eventually be excluded from court. A person's reasonable expectation of privacy can be violated with the proper legal instrument: a search warrant. Without a search warrant, or the consent of the property owner, entering on restricted property puts the photographer in legal jeopardy and will make the resultant photographs inadmissible in court.

Legal flying restrictions exist, such as how close you can fly to the ground in various situations. There have been cases in which law enforcement personnel, when attempting to photograph marijuana fields, have flown over the property at illegal heights to recognize and photograph the plant material. If the flying height has been too low, the courts have held the resultant photographs to be a product of an illegal search and have held the photographs to be inadmissible. Fly at the proper height, and get a longer telephoto lens.

SURVEILLANCE PHOTOGRAPHY

Surveillance photographs are frequently taken without the knowledge of the person being photographed. The same rule applies as stated previously: if a person or object is in the public view at a place open to the public, they can be photographed whether or not they know they are being photographed. Even if they later learn of the photographs, they cannot object to them being used as evidence at trial.

Reasonable expectation of privacy: Items seized after unreasonable searches and seizures from places where one has a reasonable expectation of privacy will be held to be inadmissible in court.

If the subject, however, is in a place where they normally have a **reasonable expectation of privacy**, the rules are different. Private places do not fall within the meaning of public places. The public cannot stroll into a private home, even if the front door is unlocked. The courts have protected these and other private places. Items seized after unreasonable searches and seizures from places where one has a reasonable expectation of privacy will be held to be inadmissible in court.

This expectation of privacy also applies to photographs in/into such places. Photographs of the interior of a home, even if it is a crime scene, cannot be used in court as evidence unless the reasonable, and legal, expectation of privacy has been legally overcome. This can be done with a search warrant. The search warrant requirement applies to both standing in a subject's living room and taking photographs of a homicide victim there, and it applies to setting up a camera two blocks away with a super-telephoto lens to photograph the people sitting around a dining room table while planning a bank robbery. In both situations, a search warrant would be required. A search warrant is the legal instrument by which an otherwise legal expectation of privacy can be violated. Therefore, photography that may be deemed to be invasive would require a search warrant.

Even though a person's home is usually regarded as a restricted area for the collection of evidence, limits to this exist. The reasonable expectation of privacy has to be manifested by positive actions. The doors have to be closed and the blinds or drapes have to be drawn. If a nude person walks in front of an unblocked window where anyone on the front sidewalk can easily see them, even though they are in their "castle," they are still exposing themselves to the public. If the general public can see them from the sidewalk, law enforcement personnel can also take enforcement actions and arrest them for indecent exposure.

If someone chooses to grow their prize marijuana plant so the light from the front window will nourish it, and it can be seen by any of the public on the front sidewalk, law enforcement eyes can also legally see it and use it for probable cause for a search warrant.

LEGAL IMPLICATIONS OF DIGITAL IMAGING

Despite the fact that digital imaging is quickly becoming the norm for most law enforcement purposes, many agencies are still fearful of making the jump into digital imaging. Just because digital imaging can show dinosaurs roaming isolated islands and can bring to life our old friend, Superman, it may not be appropriate for digital images to be offered as evidence in court. No one can deny the marvelous possibilities of the technology. Just because the possibility exists for making the otherwise impossible seem possible, it is not a logical consequence that such "magical" manipulations must be applied to images intended for use as evidence in court. Nuclear energy can be used to construct bombs to kill people. Some nuclear energy provides essential power to cities. It cannot be denied that nuclear energy can be used for both purposes. It is not, however, necessary that all nuclear energy is used for destructive purposes. There can be other, socially acceptable and responsible uses for that technology.

If one use of the technology is deemed to be destructive and "evil," how can anyone be assured every use of the technology is not also destructive and "evil." The answer is obvious. Set up rules and abide by them.

The same goes for digital imaging. Do not deny that digital images can be manipulated. Of course the technology makes that a possibility. Assert that despite this possibility, in the context of law enforcement imaging, there will be a clear set of digital procedures that will be acceptable and a clear set of digital procedures that will be unacceptable. Standard operating procedures (SOPs) will be written that clearly point out the manipulations that will be prohibited and the processes that will be permitted. Then, follow your SOPs.

Writing SOPs may feel daunting for some. Others know that assistance is available for generating your agency's SOPs. In fact, the more your agency takes on the role of a "Lone Ranger," working independently from others doing the same thing, the more you set yourself and your agency up for a fall. The more your agency aligns itself with the digital imaging community as a whole, the more likelihood your agency will succeed when digital images are offered to the courts as evidence.

The Scientific Working Group on Imaging Technologies (SWGIT) is a group of digital imaging practitioners organized under the umbrella of the FBI to determine the best practices for law enforcement agencies and individuals who use digital imaging as part of their work flow.

SWGIT: The Scientific Working Group on Imaging Technologies has been "created to provide leadership to the law enforcement community by developing guidelines for good practices for the use of imaging technologies within the criminal justice system."[*]

---

[*]SWGIT.

Just as there are SWGs for other forensic science groups, DNA analysts (SWG-DAM), latent print examiners (SWGFAST), firearms and tool mark examiners (SWGGUN), trace examiners (SWGMAT), questioned document examiners (SWGDOC), just to name a few, crime scene photographers and forensic photographers working in a laboratory environment have SWGIT to represent them. Perhaps SWGIT, itself, explains their reason for being:

> The Scientific Working Group on Imaging Technologies (SWGIT), was created to provide leadership to the law enforcement community by developing guidelines for good practices for the use of imaging technologies within the criminal justice system.[7]

All these guidelines represent the work of more than 30 photographers, scientists, instructors, and managers from more than two dozen federal, state, and local law enforcement agencies, as well as from the academic and research communities.

BACKGROUND

Although digital imaging technologies have been used in a variety of scientific fields for decades, their application in the criminal justice system has been relatively recent. Consequently, there has been a need to gather and disseminate accurate information regarding the proper application of this and other imaging technologies (including silver-based film and video) in the criminal justice system.

MISSION STATEMENT

The mission of the SWGIT is to facilitate the integration of imaging technologies and systems within the criminal justice system (CJS) by providing definitions and recommendations for the capture, storage, processing, analysis, transmission, and output of images.

DOCUMENTED PROCEDURES

Personnel engaged in the capture, storage, processing, analysis, transmission, or output of imagery in the criminal justice system should ensure that their use of images and imaging technologies is governed by documented policies and procedures.[*]

Reinventing the wheel is not necessary. As an individual supervisor tasked with developing your agency's SOPs for implementing digital imaging technology,

---

[*]SWGIT documents.

where do you begin? SWGIT has already done the hard part. The more you comply with the guidelines developed by SWGIT, the more success you will have when challenged in court when you offer digital imaging evidence the defense will be sure to characterize as "demonic, and sure to lead to the downfall of civilization." However, should you choose to reinvent the wheel and develop your own SOPs in a vacuum, the more likely the defense will be successful when attempting to throw suspicion on your work and work products. SWGIT is one bandwagon worth jumping on.

A quick synopsis of some of SWGIT's guidelines follow. To be as up-to-date as possible, it is recommended the reader review SWGITs current Guidelines as published on the Internet. When this text was first organized, it was thought that appending all of SWGITs Guidelines at the end of the textbook would be a valuable resource for the readers. As with all things digital, however, there are and will be constant changes and updates to the SWGIT Guidelines as they currently exist. Any published list of all the SWGIT Guidelines will soon be out of date. But the general flavor will remain.

The following are some of the general recommendations for creating SOPs for your agency.

MANIPULATIONS PROHIBITED

What are digital manipulations? By definition, a **manipulation** involves either the addition of details that were not in the original image, or the deletion of essential details that were in the original image. Hollywood's creative community thrives on these activities. They do not belong in a courtroom manifesting themselves in digital images being offered to the court as evidence. Every agency's SOP should contain a clear definition of what a manipulation is, quickly followed by a prohibition of the use of manipulations on digital images destined for the courtroom.

Manipulations: A manipulation involves either the addition of details that were not in the original image, or the deletion of essential details that were in the original image.

Adding and Deleting Digital Data

Figure 10.5 shows the range of manipulations that are possible. The original coin was copied and multiplied to appear in various locations around the original coin. Digital details were added. The image of the pistol, magazine, and cartridge was **manipulated** by cloning areas of the carpet, and then superimposing carpet over the magazine and cartridge. Digital details were eliminated. The license plate image was manipulated by copying the number "0" and then superimposing the "0" over the number "2," which is a combination of adding to and deleting details from the same image. None of these kinds of activities are proper with digital images going to court. They would be neither "accurate" renditions of the scene nor "authentic" depictions of the original scene "as found."

Figure 10.5

Manipulations: Additions, changes, and deletions.

Additions   Changes   Deletions

Interpolation: Increasing the original resolution of an image by having the computer software expand the original pixels of the image. The software will then "create" new pixels to insert between the original real pixels. The result is a new larger image that looks indistinguishable from the original smaller image. The difference is that the new image contains digital data that did not come from the original crime scene; those new pixels were created by computer software.

Interpolation

When a digital file currently has sufficient resolution to enable a $4'' \times 6''$ image to be printed, but an $8'' \times 10''$ image is requested, the original image can be resized upward or interpolated. Enhancement software will separate the original pixels to cover the desired size of the new image. To fill in the gaps between original pixels, the computer software will "create" new pixels and insert them in between the original pixels. When done in moderation, most digital images can be enlarged beyond their original resolutions, and the results are virtually indistinguishable to the viewer of the image. However, none of the computer-generated pixels are digital details from the crime scene. They are all computer constructs. It is suggested that no image going to court as a critical part of the prosecution should be a product of this kind of interpolation.

Part of that image is merely a computer construct and is not detail from the crime scene or digital data originating from the evidence originally photographed.

Some may believe that just because the interpolated image cannot be distinguished from the smaller original image, the use of interpolated images is justified. Sometimes yes, and sometimes no.

If the image in question is an overall view of the crime scene or a midrange view of an item of evidence and a fixed feature of the scene, the use of interpolation may be perfectly justified. In this case, resizing the image to make a larger print makes it easier for the jury to view the image without seriously affecting the detail of the image. Many closeup photographs of evidence not destined to be the subject of critical comparisons also benefit from interpolation.

If the image is destined to be used for critical comparisons or is considered an examination quality photograph, interpolation does not, in fact, increase any of the details needed for the examination or comparison and may, in fact, make an examination more difficult. In this situation, interpolation should be avoided. This author is aware that some have charged that if any digital image is altered in any way, even to the changing of individual pixels, the images should not be admissible as evidence. This author agrees with most, who believe that various "processes" designed to improve the visibility of essential details in a digital image, which do in fact change individual pixels, are perfectly justified, as long as the essential details of the digital image are not compromised. "Proving" all 6, 8, or 10 megapixels are exactly the same before and after digital "processing" is not necessary. Interpolation certainly does not add critical details to an image destined for comparisons and should not be used in those situations.

If one were to offer an interpolated image to the court as evidence, it would be necessary to qualify the image as "partial data from the crime scene, and partial data created by a computer program." The defense attorney will not remain in his or her seat for long if such a statement is made.

Figure 10.6 graphically demonstrates the effects of interpolation. At the top left, several individual pixels were changed to specific colors so the effects of interpolation could easily be seen. The other two images show varying effects of different levels of interpolation. The essential idea to be understood is that interpolation does not just enlarge the image by an exact multiplication of the original data. The software attempts to fill the gaps by "guessing" the color and tonal values that would best complement the original "real" pixels. The effect is that colors and tonal values that were not in the original image are being created and added to the original image, just so that it can be printed larger than the original. On the pixel level, the changes are dramatic. Unfortunately, or fortunately if you are a proponent of this digital manipulation technique, the gross image rarely suffers to the extent most people would notice. If the original image is des-

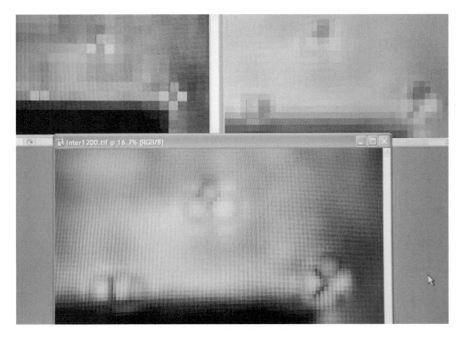

Figure 10.6
Interpolated changes to pixels.

tined for "critical comparisons" or will eventually be subject to analysis, interpolation should never be applied to the image.

If the image were just a documentation of the front façade of the crime scene building, there would be no such prohibition to the interpolation of that image. That is not a crucial image by which the guilt or innocence of the defendant will be judged. If the image were a shoe print within the crime scene or a bloody palm print, these images should have higher standards applied to them, and interpolation should not be done to them.

Rotations

Anyone having any digital enhancement software is aware that the irritating occasional image in which the horizontal alignment of the image is a bit off can easily be fixed by simply rotating the image until it appears horizontal again. What a wonderful and essential bit of digital imaging correction this is. For your images of the family dog, birthday parties, and most images created that are not going to court as "critical comparison" evidence, this is true. If the image is destined to be used for a "critical comparison" between the crime scene image and another item of evidence, rotations that are not either 90°, 180°, or 270° rotations should be avoided. Why is this? Most digital pixels are square. If other rotation amounts are considered, the content of each square original pixel is now partially superimposed over its neighboring pixels. Figure 10.7 shows a single pixel rotated just 1°.

Part of the original content of the blue pixel is now falling outside the perimeter of the red pixel, which represents the orientation of all the pixels, which will not rotate. When rotated in other than 90° increments, each pixel will now acquire color and tonal values from all four of its original neighbors. To visualize its new color or tonal value, each pixel must now average its original digital data with these four new additions. Every pixel will change, and the changed product is not a result of crime scene data, it is a result of computer algorithms.

Again, the result will not be visible when viewing the gross image. Why the big deal, then? It is important to remember because you should not be offering an image to the court as an examination quality image if it is not solely a result of data

Figure 10.7

Pixel alignment after a 1° rotation.

from the crime scene. If it is a product of a computer's averaging of data from pixel neighbors, because the image has been rotated in increments not 90°, the image is no longer original digital data. Figure 10.8 demonstrates this concept well.

Individual pixels were altered to make the changes more obvious. Even with just a 1° rotation, the result is colors and tonal values that are not related to the original images, but rather are constructs of the software's programming. The image of the front façade of the crime scene building can be rotated with no problems. Images used for "critical comparisons" should not be rotated unless they are in 90° increments.

This author is aware of many who will maintain that if the essential details of the image used by examiners and analysts in their examinations are not compromised even if some slight changes are made by the computer software, my protestations should be moot. This author is certainly to be counted in the camp of those who believe that "minor" changes to individual pixels that do not adversely affect an examiner or analyst's ability to affect their comparisons should be permissible. If a "process" does not substantially add to the likelihood of successful

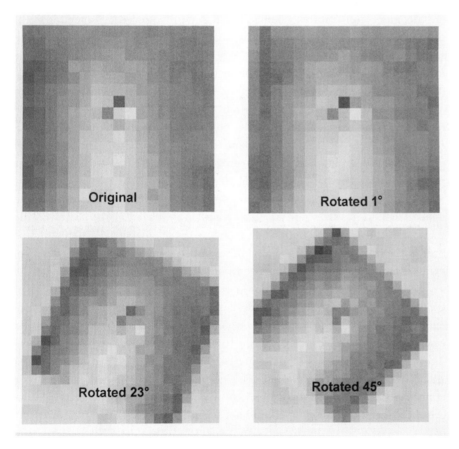

Figure 10.8
Changes to rotated pixels.

examinations and comparisons, it should used with reservation. It is good to know just what the effects are that a "process" is having on an image.

COMPRESSION CHOICES

When the digital camera is set up to capture images, usually the crime scene photographer must choose between a variety of **"lossy"** compression formats or "lossless" compression formats. Lossy compression formats, like JPEGs (Joint Photographic Experts Group), come in a variety of increments, such as Max, Hi, Med, and Low.

These choices allow the photographer to store more images on their media card. To store images on the same card, some of the data of each image is discarded to make a smaller file size, which is done primarily by two techniques. When many adjacent pixels of the same color or tonal value are present, the JPEG software will discard all but one of the pixels, which will stand as a surrogate for all the discarded pixels. When the compressed file is used to reconstitute the image, that single surrogate pixel will be duplicated everywhere discarded pixels had been. This procedure can drastically reduce file sizes. A second approach to JPEG compression is to take similar, but differently colored or toned, adjacent pixels and average them to arrive at homogenous adjacent pixels. As before, one surrogate pixel can substitute for all the other similar pixels that can now be discarded. Once these digital data are eliminated by a JPEG compression format, the data can never be retrieved again; they are permanently lost.

An alternative to the JPEG compression format is the TIFF (Tagged Image File Format) compression format. TIFFs are considered a **"lossless"** file format. TIFF compressions reorder the digital data into a smaller file size without deleting any pixels at all. When many similar adjacent pixels are present, such as in the blue sky, similar blues can be reordered like this.

A string of 10 exact blues, for example BBBBBBBBBB, takes up more space than $10 \times B$, which takes up only four spaces on the digital flash card or on the computer hard drive. When the file is used to form a digital image, all the original blue pixels will assume their original locations. Nothing will be lost during compression and the creation of an image from that lossless file.

JPEG compressions are an option only because they do a magnificent job of compressing digital files, usually without any noticeable degradation to the original image. Just because you cannot see a change in the image does not mean no change to the image exists. The distinction between what is proper for the documentary image of the façade of a crime scene and what should be allowable for a "critical comparison" photograph needs to be made. The higher standard only applies to the "critical comparisons," because decisions about the guilt or innocence of a defendant in court should not be based on images that have had part of their original detail discarded, even if "you cannot see the difference."

Lossy compression: JPEGs (Joint Photographic Experts Group) To store more images in a given space, some of the data of each image is discarded to make a smaller file size. Once these digital data are eliminated by a JPEG format, the data can never be retrieved again; they are permanently lost.

Lossless compression: TIFF (Tagged Image File Format) compressions reorder the digital data into a smaller file size without deleting any pixels at all.

Figure 10.9 shows five images of a bruise to an arm. They all look alike, do they not? If they look alike, why all the fuss? Should any of the five images not be able to be used in court as evidence? In this case, it is not what we can see that is important, it is what we know has happened to the content of the image that is important. When writing your agency's SOPs for digital imaging, if it is known any process results in the loss of significant digital data, that process should not be permitted when applied to examination quality images or images considered "critical comparisons." How would you like to have to admit to the judge that the image being offered in court to help convict the defendant has knowingly had much of its digital data discarded by a process you used? Especially when a process is available that does not discard digital data or change it.

Let us replace one pixel with a red one and see what happens on the pixel level. Figure 10.10 demonstrates these changes. Compressing the data with a TIFF format results in the original red pixel being duplicated when the full image is recreated from the compressed file. The Max JPEG reconstitution of the image also looks very good, but an examination of the new file size compared with a TIFF file size of the same image would convince anyone that somewhere data has been lost by JPEGing the image. JPEGs Hi, Med, and Low show obvious changes to the original red pixel. Your agency's SOP should clearly indicate when JPEG compressions are allowable and when only TIFF formats should be used to store image files.

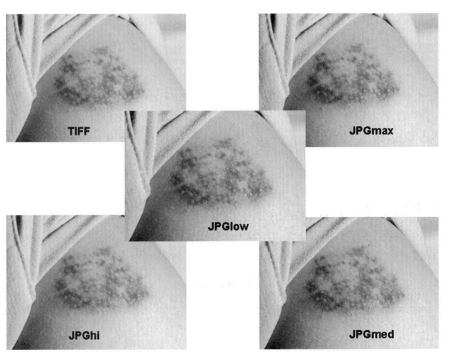

Figure 10.9

A bruise with TIFF and JPEG compressions (full size).

Figure 10.10

Bruise shown in Figure 10.9 enlarged, a single pixel changed to red, then saved with TIFF and JPG compressions.

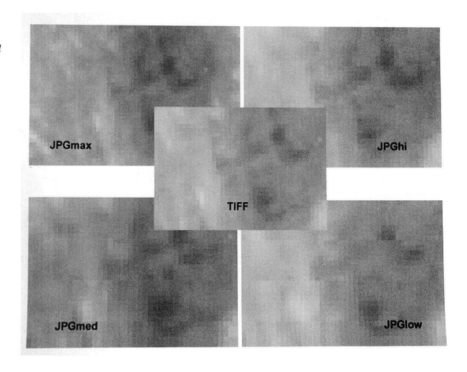

A HISTORY OF PROCESSED IMAGES SHOULD BE MAINTAINED

If manipulations are to be avoided with "critical comparisons," what digital "changes" are permissible? The term for acceptable "changes" to a digital image is **processing**.

Processing: The essence of processing is to take the essential detail that is in the original image and make it easier to see.

Before the advent of digital cameras, film cameras captured images on the film in the camera. This film would have to be processed in a wet-chemistry darkroom to produce a print or what everyone knows as a traditional photograph. Film was processed; digital images are also now processed.

Processing differs from manipulation in this way. Whereas manipulations either adds details that were not originally present or deletes details that were originally present, the essence of processing is to take the essential detail that is in the original image and make it easier to see. Many of the processing techniques in a digital darkroom have analogous techniques in a wet-chemistry darkroom. SWGIT has conveniently arranged these digital processes, which are similar to traditional wet-chemistry processing, into category 1 processes. Much that can now be done with digital images has also been able to be done to film. Many of the film processes are mirrored now with digital processes. In fact, that is exactly why the older term "enhancements" has been changed to processes. Because the two types of processing are so similar, the digital imaging community wanted to assure the public of this heritage and reassert that much digital processing is merely updated digital versions of what has been commonly accepted with film processing.

Digital imaging technology has advanced the art of improving an otherwise marginal image. Category 2 processes are on an additional list. Their purpose is the same: to improve the visibility of the essential details in the image. However, these techniques do not have a clear-cut heritage in the wet-chemistry darkroom. Category 2 processing is also considered "advanced" processing technique. At the end of a category 2 processing, the resulting image may not look similar to the original image. Anticipating the wailing to come from the defense side of the court, strategies have been developed to ensure even these processes will ultimately be acceptable to the courts. The goal? It is to ensure the court that the resultant images are in fact improvements of the essential details already contained in the original image. The tactic? Make sure the entire process is **repeatable** by any other specialist in the field of digital imaging, be that another law enforcement digital imaging expert or even a prosecutor's digital imaging expert.

If it is ever asserted by any expert that he or she is the only one who is gifted enough to be able to affect the necessary identification, red flags should immediately come up. An identification that can be made only by one person is worthless. Unless it can be verified by other equally competent peers in the relevant scientific community, any identification should be suspect. If only one imaging analyst can produce the processed image necessary for comparison purposes, it, too, should be viewed with skepticism. Only when digital imaging processes can be achieved by a variety of competent digital imaging experts should the courts feel comfortable the processing is legitimate and a commonly acceptable practice in the relevant community.

To ensure the courts that no "magic" is being done to the original image or that no substitution of one image for another is taking place, a precise history of all "changes" done to the image must be recorded and be able to be presented to the court should the court wish to have the repeatability of the process demonstrated.

A clarification is warranted here. Under circumstances in which a particular digitally processed image has been challenged, it may be necessary that another expert follow the precise history of all the steps taken by the initial analyst who produced the first processed image. It can then be said that the initial processing steps are repeatable by any other competent expert to assure the court the processed image did, in fact, come from the original image from the crime scene.

Another meaning of "repeatable" has sometimes been confused with this first meaning. When an identification has been made by one examiner, it is then necessary that the identification be verified by a second examiner. The second examiner then begins a new examination from scratch and may frequently use entirely different aspects of the evidence to establish the first identification is warranted. The first identification has been "repeated" by the second examiner, but the second examiner has made an independent examination of the evidence and has not just reconsidered the same aspects of the evidence as the first examiner. The original identification has been verified or repeated. The exact steps used by the first examiner were not necessarily repeated.

Repeatable: An identification able to be made only by one person is worthless. Any identification should be suspect unless it can be verified by other equally competent peers in the relevant scientific community.

Who Did the Processing and When?

As with any other processing of evidence, a record must be kept as to who had access to the digital evidence, which is a standard chain of custody requirement. Critical evidence must not be left in an open place where it is accessible by any without legitimate access to the evidence. If access to the evidence is not carefully managed, the defense can question whether inappropriate personnel gained access and changed or altered the evidence in some way. There should be restricted access to the room in which the computer with the digital files is maintained. The computer housing the evidence should record who accessed the computer. The computer can require an authorized password to activate the software. A log of who accessed the digital evidence and when must be maintained and routinely checked by a supervisor.

Types of Processing

Detailed notes must be kept as to the exact kind of advanced digital imaging processing that was done to the image. Some software automatically records any changes to the image and adds this information to the metadata in the image file. Otherwise, written notes have to be maintained, so that, if challenged, another expert can apply the same sort of processes to the image.

Amount/Degree of Processing

In addition to the type of processing applied to any particular image, the amount of the processing technique must also be maintained and recorded. Most digital imaging techniques have a numerical increment of the process that is also indicated in the software. These numerical values must also be recorded. Only if this is done can any other competent expert come in to duplicate the changes made during the original processing.

TRAINING/COMPETENCY AND PROFICIENCY TESTS

Only personnel properly trained to competency should be performing category 2 processes on evidence images. A complete training record must be maintained on all personnel deemed digital imaging experts. As with all crime laboratory specialists, each digital imaging specialist/examiner should also have to successfully complete a proficiency examination.

AUTHENTICATION/ENCRYPTION REQUIRED?

Each agency must also address whether it will subscribe to a "requirement" to include authentication/encryption software to all images processed to improve the visibility of details in images destined to be used as evidence in court, which some believe is an unnecessary requirement for digital images, because there was never the equivalent requirement for film prints, and film images were acceptable in court as evidence. Others acknowledge the added scrutiny digital images have to undergo until they are as easily accepted by the courts as evidence. The requirement to authenticate images as being unaltered may be only a passing need, which eventually becomes unnecessary. This issue has not been resolved, so the individual agency must address it one way or the other. Ignoring the issue is not the proper response to the issue.

ORIGINALS ARCHIVED; PROCESSING ONLY APPLIED TO COPIES

A film negative was the "original" evidence. Individual prints were processed to have more or less contrast or to filter out improper color tints. Individual prints were burned and dodged when more or less light was required in selected areas of the image. The negative, however, was never changed. The digital equivalent is to archive the original to a CD or DVD and then to make a copy and apply processing to the copies only. Multiple copies of the original can be worked on, but the original digital image must never be altered.

These are the minimal recommendations SWGIT suggests should be included in an agency's SOPs related to processing digital imaging. It is suggested you view the actual SWGIT documents and guidelines available on the Internet. Updates and changes to these documents are inevitable, and one must view the most recent documents to be assured of full compliance with SWGIT recommendations, which is extremely important. In one recent court case, the judge made it clear the processed digital image was acceptable in his court as evidence simply because the processing of the image had closely followed SWGIT guidelines. The more you can cloak yourself and your agency under the SWGIT umbrella, the more likely your processed digital images will be accepted into court as evidence.

INTERNATIONAL ASSOCIATION FOR IDENTIFICATION (IAI) RESOLUTION 97-9

Defense attorneys are famous for overstating or understating the facts if it suits a position they are currently advocating. Many may try to assert that digital imaging is such a new technology that it cannot meet Frye Standards of being generally accepted in the relevant scientific community. Here is one more tidbit you might keep in your back pocket if this is ever mentioned in court. In 1997, the IAI (International Association for Identification) adopted Resolution 97-9. In that resolution, the IAI "recognized digital imaging as a scientifically valid and proven technology." The IAI, by the way, is the body that certifies forensic photographers and crime scene photographers. Besides latent print examiners, crime scene analysts, forensic artists, and blood spatter analysts, the IAI also certifies crime scene photographers and digital imaging specialists. It is to be expected that becoming certified in the field will be another criterion by which a judge can decide that digital imaging has been properly done. Follow SWGIT guidelines and be an IAI-certified forensic photographer.

CASE LAW CITATIONS RELEVANT TO FILM PHOTOS AND DIGITAL IMAGES

Please note: The same case may be listed under multiple categories if relevant to each category.

FAIR AND ACCURATE REPRESENTATION OF THE SCENE

Franklin v State, 69 Ga. 39, 42 (1882)

We cannot conceive of a more impartial and truthful witness than the sun, because its light stamps and seals the similitude of the object on the photograph put before the jury; it would be more accurate than the memory of witnesses, and because the subject of all evidence is to show the truth, why should not this dumb witness show it?

Almond v State

This case relates to the admissibility of digital evidence. Almond raised the issue of digital images as evidence at his trial. The Georgia Supreme Court stated that because the record showed "that the pictures were introduced only after the prosecution properly authenticated them as fair and truthful representations of what they purported to depict," they were properly admissible.

Cunningham v Fair Haven & Westville R. Co., 1899

McCarvel, Roderick T. "You all won't believe your eyes: Digital Photography as legal evidence." Available at: www.seanet.com/~rod/digiphot.html. This case deals with photographic evidence being inadmissible because not only was it inaccurate, it was also misleading in nature.

US v Hobbs, 403 F.2d 977, 978 (1968)

When photographs first began finding their way into judicial trials, they were viewed with suspicion and received with caution. It was not uncommon to place on the offering party the burden of producing the negative as well as the photograph itself and of proving that neither retouching nor other manual or chemical intervention was reflected in the proffered print. That burden has now shifted, and the proponent of a proffered photograph has established a prima facie case for its admissibility when he has shown it to be an accurate representation of the scene in question.

Blair v the Inhabitants of Pelham (1875)

This case determined that a photograph is admissible if it is "true representation" of the location. In this case, the plaintiff was suing the town because he was injured in a carriage accident because of a defect in the road, which the town was responsible for maintaining.

People v Boston, 139 NE 880 (1923)

Boston was convicted of carnal knowledge of a minor. Boston believed the court erred in not admitting a photograph that did not show a hollow in the ground where the victim claimed the attack took place. The court stated that the photograph did not show the hollow because it was taken at an improper angle. The court decided that the photo must be taken at an angle that provides an accurate representation of a person's view to be admissible.

Luco v United States, 64 U.S. 515 (1860)

This case involved a dispute over a 270,000-acre grant of valuable land in California supposedly made to Jose de la Rosa on December 4, 1845, and purporting to have been signed by Pio Pico as acting Governor that in actuality was found to be a forgery. This case was remarkable for this one thing, among others: that in the trial below, Mr. Vance, a photographer, was examined, who attached to his deposition photographs of original documents, of impressions of genuine seals, and of the signatures of Pio Pico. These were exhibited during the argument in this court. Id. 64 U.S. at 541. After oral argument and the examination of the signatures of acting Governor Pio Pico and secretary of state Jose Maria Covarrubias by the Court with known photographic samples of the governor's signature and state seal, the Court stated, "We have ourselves been able to compare these signatures by means of photographic copies, and fully concur that the seal and the signatures of Pico on this instrument are forgeries. . . ." Id. 64 U.S. at 541, which is the first legal case known to have used photographic evidence in the United States.

Potts v People, 158 P.2d 739 (Colo. 1945)

The admissibility of photographs does not depend on whether the scenes depicted or objects portrayed could be described in words, but on whether it would be helpful to supplement his or her testimony by their use.

U.S. v Meyers, 972 F.2d 1566 (Ga. 1992)

In a case involving allegations of violation of an arrestee's civil rights by police, the trial Court admitted into evidence photographs of a bunk bed inside a jail cell where the arrestee was placed after police used a stun gun on him, even though the photographs were not of the actual jail cell. On review, the Court held that the photos fairly and accurately depicted the bunk bed and jail cell and therefore were properly admitted.

U.S. v Garvin, 88 Fed. Appx. 542 (Pa. 2004)

No abuse of discretion was found in allowing photographs of a crime scene into evidence that were taken during daylight hours even though the alleged crime occurred at night. The photos were certified as fairly and accurately depicting the scene by a state witness, and defense counsel was afforded a full and fair opportunity to cross-examine the witness on the differences in the crime scene during the day versus at night. Detailed testimony was also offered regarding lighting conditions during police surveillance, including visibility, whether cars were moving back and forth, how far the officer was from the defendant when he saw him, and the type and strength of binoculars used by police.

Trexler Lumber Co. v Allemannia Fire Insurance Co., 136 A. 856 (1927)

A photograph is not accurate if the scene has been altered before the photograph. In this photograph, the plaintiff cleared the debris from the damage off the property before taking the picture.

EXPOSURE/LIGHTING

State v Heinz, 223 Iowa 1241 (1937)

Proper lighting must be used to prevent distortion In this case, poor lighting created shadows, which the judge ruled could be mistaken for bruising.

Kauffman v Meyberg, 140 P.2d 210 (1943)

In this case, an infrared photograph was admissible to show erased writing. Infrared photography was used to show erased writing on stock certificate stubs.

State v Thorp, 1934

Case may be found at: www.epic-photo.org/become4.htm. For the first time, the NH court approved an ultraviolet photograph as evidence. It was a footprint left in blood on the floor.

Parish v State, 165 S.W. 2d 748 (1942)

Parish was convicted of swindling. The Court ruled that the photograph was admissible when side lighting was used to display indented writing even though the expert was not in court. However, the equipment used was presented in court along with procedure.

Green v City and County of Denver, 268 Ky. 475 (1943)

One photograph was underexposed but accepted. This case dealt with putrid meat. One photo of many was underexposed; however, because the photographer explained it, it was accepted along with the others.

U.S. v Dombrowski, 877 F.2d 520 (III. 1989)

The reviewing Court spoke on the admissibility of several photographs that were ultimately found to have been properly admitted into evidence at the time of trial. These photographs included (1) pictures of the crime scene without snow on the ground even though on the night of the crime there was snow on the ground and (2) pictures taken across the street from the crime scene at a point that was close to a bright street light creating an artificially bright image of an alley where the crime occurred. The photographs portrayed the alley as much darker than the street seen in the foreground.

Scott v New Orleans, 75 F. 373 (1896)

A photograph should be taken under similar conditions to the time of the incident. In this case, a man tripped over a portion of a broken sidewalk on a dark, gloomy day. The photograph of the scene was taken on a bright, sunny day but was admitted by the court with a warning to the jury.

COLOR ACCURACY

People v Moore, 310 P. 2d 969 (1957)

In this case of murder committed during the course of a rape, the state introduced into evidence both black-and-white photographs and color slides showing the victim's body and the area in which it was found. Various state witnesses used these photographs to explain the injuries and the crime scene. The defendant challenged the admission of the colored slides showing the victim's body after testimony by the photographer who conceded that one transparency had a red or orange effect on it from light and the type of film used. However, the photographer also certified that the body and other objects pictured were accurately reproduced and that the wounds on the body were not distorted. In light of the fact that the jury heard the testimony regarding how the red or orange effect was made, the trial judge appropriately used his broad discretion to allow the photographs into evidence.

Commonwealth of KY, Department of Highways v Harvey, 396 S.W. 2d 311 (1965)

This case involved the amount awarded to a property owner to make way for a highway. The defendant claimed that the photograph introduced to show property value should have been excluded, because the color of the barn appeared blue instead of brown. Although the color was off on a trivial part of the photograph, the court ruled it was admissible.

People v Simms, 143 III. 2d 154 (1991)

This case determined that color photographs may be used during the sentencing phase of a trial.

In this case of murder, the judge stated that color photographs might be clearer and more persuasive than oral testimony in some cases. The court agreed with the prosecution that the photographs of the inflicted wounds spoke to intent to commit murder.

US v Gibson, 17 CMR 911 (1954)

A nurse was charged with the premeditated murder of a baby girl. "The color slides of the deceased were taken under the supervision of Captain Bridgens, the pathologist, as a matter of pathological interest, and although stated not to be 100% accurate as to color, they were illustrative of the pathological lesions found."

Brockman v State, 79 N.W. 2d 9 (1956)

In this case of rape, a photograph was introduced of the victim and her injuries, but the colors in the resulting print were not exact. The Court ruled that the colors in the photograph need not be exact, only "faithful and accurate."

DISTANCE RELATIONSHIPS/OPTICAL ILLUSIONS

State v Givens, 28 Wis. 2d 109 (1965)

Proper focal length should be used to provide an accurate representation of the scene.

In this case, a photograph was taken of a lobby area to show that a sit-in was disruptive to the office. The court stated that the photograph was admissible because the normal lens gave an accurate picture of how small the lobby area was with the demonstrators in it.

McLean v Erie RR, 54 A. 238 (1903)

This was a suit for injury caused by a train hitting the defendant's wagon. The photographs of the scene were introduced with a word of caution by the judge about perspective and how a photo can be deceptive when determining distance. The appeals court ruled that the camera's perspective affects the photograph as evidence and, therefore, the judge's remarks were sound.

US v Hansen, 652 F.2d 1374 (1981)

Hansen, Means, and Bryant were charged with possession and intent to distribute cocaine and marijuana. The search was improperly performed, and the one photograph admitted into evidence was suppressed. The court ruled that "the objection to the admissibility of the one photograph offered was sustained at the suppression hearing on the grounds that the picture was not a true and accurate representation of the scene because of the wide-angle lens and speed at which the picture was shot, and the agent's own testimony that the picture showed more than he actually saw. (II R. 65-66). As stated later, we find no error in this ruling."

Archer v Gage, 270 P. 521 (1928)

The wrong lens may make the photo distorted and difficult to authenticate. This case involved a bus that ran over a child. The photograph of the bus admitted into evidence was taken with a wide-angle lens that caused distortion. Because of the distortion, a witness asked to identify the bus was unable to do so.

Puleo v Stanislaw Holding Corp., 213 NYS 601 (1926)

This accident case saw the admittance of a photograph that distorted distance. The courts later found that the photograph should not have been admitted. "It is a matter of almost common knowledge that photographs may be taken from different angles, so as to exaggerate certain distances or, on the other hand, to make space more compact."

Floeck v Hoover, 195 P.2d 86 (1948)

This case involved the wrongful death suit by the family of Floeck, who was involved in an accident with Hoover. The location of the accident was determined by using a photograph taken at the scene. The court stated that to show the distance between two objects, the camera should be perpendicular to a line between the two. Because of error in the taking of the photograph, the court stated that the picture should not have been admitted as evidence.

SIGHT PERSPECTIVE

McLean v Erie R. Co., 54 A. 238 (1903)

In an action for damages stemming from a collision between defendant's train and the plaintiff's wagon as it was attempting to cross the train tracks, the issue was whether photographs depicting the location of the accident were properly admitted. The subject photographs were taken approximately 3 weeks after the accident in January 1903, resulting in the issue of whether the photographs could show proper distance perspective and whether the shrubbery and trees had changed in the interim. During the trial, the judge admitted the photographs, although he stated they were "unavoidably misleading" in showing distance and the condition of the trees and shrubbery. The appellate court concluded that the trial judge's comments were legitimate and appropriate on the basis of the photographic evidence. The Court agreed that photographs can alter perspective, depending on the viewpoint from which the photograph is taken.

People v Boston, 139 N.E. 880 (1923)

Boston was convicted of carnal knowledge of a minor. Boston argued the court erred in not admitting a photograph that failed to show a hollow in the ground where the victim claimed the attack took place. The court stated that the photo did not show the hollow because it was taken at an improper angle. The court held that the photo must be taken at an angle that provides an accurate representation of a person's view to be admissible.

Chicago v Vesey, 105 Ill. App. 191 (1902)

In this case, Vesey sued because he tripped over a rail at a railroad crossing. His daughter took a photograph of the rail 2 years later from a low vantage point. The court ruled that the photograph should not have been admitted, because it was not taken at eye level and, therefore, had distortions and was "dangerously misleading."

Stone v Northern Pacific Railway, 151 NW 36 (1915)

A picture was inadmissible because it was taken from the wrong point of view. In this case, a photograph was taken that was supposed to show what the train engineer saw as he approached an intersection; however, it was taken at the eye level of a person standing on the ground, so it was ruled inadmissible.

Farmer v School District No. 214, King Co., 17 P. 2d 899 (1933)

A traffic accident case hinged on whether the driver could see the bus driver signal with his left hand. The photograph was taken from the same perspective (same height, same lateral position) as the driver claimed to have been in. Therefore, the court ruled the picture admissible and found the driver at fault.

SIZE AND SCALE

Selleck v Janesville, 104 Wis. 570 (1899)

Photographs that create a distorted size of an object are inadmissible. This case was a personal injury lawsuit caused by a defective sidewalk. In this case, a picture of a foot was taken improperly, so its size was greatly enlarged.

Arkansas Power & Light Co v Marsh, 195 Ark. 1135 (1938)

Items used to show size should be chosen so as not to make the object being photographed appear smaller or larger. In this case, a large pole was placed as a marker in a hole. The judge ruled that the pole made the hole appear smaller than it actually was, so the photograph was ruled inadmissible.

Great Atlantic & Pacific Tea Co. v Lyle, 351 S.W. 2d 391 (1961)

This case involved a personal injury suit. Lyle fell because of a depression in a parking lot. The photograph admitted into evidence had a measuring stick placed on the depression to mark the depth.

TIME FRAME

McLean v Erie R. Co., 54 A. 238 (1903)

In an action for damages stemming from a collision between the defendant's train and the plaintiff's wagon as it was attempting to cross the train tracks, the issue of whether photographs depicting the location of the accident were properly admitted. The subject photographs were taken approximately 3 weeks after the accident in January 1903, resulting in the issue of whether the photographs could show proper distance perspective and whether the shrubbery and trees had changed in the interim. During the trial, the judge admitted the photographs, although he stated they were "unavoidably misleading" in showing distance and the condition of the trees and shrubbery. The appellate court concluded that the trial judge's comments were legitimate and appropriate on the basis of the photographic evidence. The Court agreed that photographs can alter perspective depending on the viewpoint from which the photograph is taken.

Hancock v State, 47 So. 2d 833 (1950)

In this case, the father was convicted of killing his daughter, but the defendant challenged the photographs of the victim because of changes. The Court ruled that some changes were inevitable between the time of death and the taking of the photographs, because rarely are photographs taken at the exact time of death. Therefore, oral proof is sufficient in explaining changes between time of death and the photograph.

Fox v City of Kansas City, 343 SW 2d 200 (1960)

This was a suit for injuries caused by tripping on a raised portion of a sidewalk at night. The plaintiff claimed he could not see it because of a shadow cast by foliage and a streetlight. The Court determined that the lighting present at a scene at night cannot be accurately captured with photography and that the defendant's photograph was taken in December without any foliage and was therefore inadmissible.

Bodam v City of New Hampton, 290 SW 621 (1927)

This was a suit against the city because of property damage caused by a change of the street grade. The photograph used as evidence was taken 10 years earlier. The scene had undergone many changes, but the Court ruled that there was no substantial change to the "lay of the ground."

Grimm v East St. Louis & Suburban Ry., 180 Ill. App. 92 (1913)

This case concerned a traffic accident that occurred when there was four feet of snow on the roads. When the picture was taken, the snow had melted, so the court ruled it did not accurately portray circumstances for driver.

Sherlock v Minneapolis, St. P & SSM Ry., 138 N.W. 976 (1912)

The case concerned an accident at a railroad crossing in December and photographs taken in June. The Court ruled that snow made no difference in the case and, therefore, the photo was admissible.

Dallas Ry & Terminal Co. v Durkee, 193 S.W. 2d 222

This case involved a car accident in which the visibility of the driver was at issue. The photograph was taken when there was no foliage on the trees, although the accident occurred when the trees were full. Because of the drastic change in foliage, which affected visibility, the court ruled that the photograph was inadmissible.

Reliance Insurance Co. v Bridges, et al., 311 S.E.2d 193 (Ga. App. 1983)

In a wrongful death action resulting from a multi-vehicle accident, the trial court's verdict was upheld when one of the challenges raised on appeal was whether it was an error to have allowed photographs depicting the scene of the motor vehicle accident into evidence, because they were taken sometime after the actual occurrence. On the date of the collision, it was claimed that the view of the driver was obscured by trees and bushes. In the photographs that were introduced by the plaintiff, taken several months after the subject accident, the trees and bushes had been cut down. The appeal court found that it was not error to have allowed these photographs into evidence. The Court pointed to the testimony of a deputy sheriff, who traveled this road on almost a daily basis in which he stated that he never noticed any change in his ability to see because of the amount of foliage present or absent. When agreeing to admit the photographs into evidence, the trial judge instructed the jury to use the photographs for whatever purpose the jury deemed proper and to remember that certain photographs were taken at a later time than this accident occurred.

U.S. v Dombrowski, 877 F.2d 520 (Ill. 1989)

Here, the reviewing Court spoke on the admissibility of several photographs that were ultimately found to have been properly admitted into evidence at the time of trial. These photographs included (1) pictures of the crime scene without snow on the ground even though on the night of the crime there was snow on the ground and (2) pictures taken across the street from the crime scene at a point that was close to a bright street light, creating an artificially bright image of an alley where the crime occurred. The photographs portrayed the alley as much darker than the street seen in the foreground.

Commonwealth v Best, 740 N.E.2d 1065 (Mass. App. Ct. 2001)

In a case involving charges of distribution of heroin, the trial court's denial of defendant's request to introduce photographs of the area where a heroin exchange took place with a codefendant was not an abuse of discretion. Defendant's photographs were taken at 2:00 PM (daytime), whereas the offense allegedly occurred at 6:00 PM, after it turned dark. Under these circumstances, the trial judge could properly conclude that the defendant's photographs were not a fair and accurate representation of the area at the time of the alleged offense and were, therefore, inadmissible.

Riggs v Metropolitan St., 216 Mo. 304 (1908)

A picture should be taken at same time of day and should clearly show objects to be authenticated. In this case, a photograph was taken exactly 1 year after the incident took place for accuracy; however, the scene was too dark to accurately authenticate the plaintiff in the photo.

Brockman v State, 79 N.W.2d 9 (1956)

The defendant was convicted of rape and among other arguments on appeal, disputed the admission of one of several photographs (photograph No. 14) taken by the police depicting injuries to the victim's face that defendant claimed did not show an "exact reproduction." The Court held that for purposes of admissibility, photographs must be shown to be true and accurate representations of what they purport to exhibit. "By that is not meant that it must be shown that the photograph is a true and correct picture or representation of the object photographed in the minutest details, but it must be made to appear that the photograph is a substantially true and correct picture or representation of the object, and not a distorted or false one." Id. 79 N.W.2d at 14. Here, the Court found that the state properly authenticated the picture through the testimony of the police officer, who photographed the victim and attested that it was a fair and accurate representation of the victim and her injuries.

AUTHENTIC

US v Rembert (1988)

In this case, it was stated that the contents alone of each photograph provide enough circumstantial evidence to authenticate. In this case, the defendant was charged with robbery (along with kidnapping and transporting a stolen vehicle across state lines). The defendant claimed the photographs from the bank surveillance camera were not properly authenticated.

US v Stearns (1977)

The contents of the photographs along with circumstantial/indirect evidence may authenticate it. In this case, the defendants were charged with stealing a boat. The defendants claimed that their boat was disabled. Photographic evidence showed their boat at full sail in the background.

Kleveland v US (1965)

It was decided in the case that the witness authenticating the picture did not need to be the photographer or be present when the picture took place. In this case, Kleveland was suing the government for a faulty ladder rung that broke and caused him injury.

US v Clayton (1981)

A photograph can be authenticated by explaining the method used to capture the image and the chain of evidence. In this case, surveillance photographs inside the bank were used to identify the suspect and obtain accurate measurements using a reenactment.

U.S. v Crockett, 49 F.3d 1357 (Iowa 1995)

Photographs taken of a crime scene during the daytime were properly admitted as evidence in the defendant's trial, even though the crime occurred at night. The photographs were properly authenticated as to the layout of the crime scene by the police officer, who was unable to testify as to the lighting conditions depicted in the photographs and whether certain light fixtures were working on the night of this crime. The Court held that the authentication of the crime scene layout was more important for the jury to understand than the lighting conditions, and the Court held no error was committed by allowing the photographs into evidence.

People v Rodriguez

This case holding indicates that the court seeks to ensure that an individual with first-hand knowledge of the photographed scene attests to the picture's accuracy. This demonstrates that for admissibility, photographs must be relevant and authenticated.

U.S. v Shugart, 117 F.3d 838 (Tex. 1997)

In a case involving prosecution for the illegal manufacturing of drugs, the Court held that photographs depicting laboratory equipment and chemicals were admissible, even though law enforcement agents had rearranged the evidence before the photographs were taken. Here, the photographs were being offered to show what evidence was found at the crime scene and not to demonstrate what the crime scene looked like when agents arrived.

RELEVANT AND MATERIAL

Loftin v Howard, 82 So.2d 125 (Fla. 1955)

The overall purpose of using photographs at a trial is to assist in recreating what took place. If a photograph is misleading in any way, then it should be excluded.

Moeller v Hauser, 237 Minn. 368, 54 N.W.2d 639 (1952); Horne v Vassey, 579 S.E.2d 924 (N.C. App. 2003); State v Phillips, 46 P.3d 1048 (2002), opinion supplemented 67 P.3d 1228, cert denied 124 S. Ct. 469; Baldwin v State, 784 So.2d 148 (Miss. 2001)

Photographs are generally admissible when they accurately portray anything that it is competent for a witness to describe in words or when they help the jury to understand any disputed issue. Photographs act as an aid to verbal descriptions of objects, and conditions are admissible only where they are relevant to some material issue.

Peagler v Atlantic Coast Line R. Co., 234 S.C. 140, 107 S.E.2d 15 (1959)

Photographs are properly excluded from evidence when they are not substantially necessary to show material facts or conditions.

State v Pettigrew, 860 P.2d 777 (N.M.App.1993)

The defendants were convicted of felony aggravated battery and challenged whether photographs depicting the victim after the battery should have been admitted. The defendants argued that the life-size photograph of the victim taken after the battery but before receiving medical treatment was not relevant. The Court disagreed, noting that the photographs were relevant to show the type and extent of injuries and to assist the treating physician's testimony during trial. In further argument, the defendants claimed that the photographs were not clear representations of their crime and actually obscured the true nature of the injuries because of the blood and swelling that was shown before the victim received treatment. Again, the Court rejected such an argument finding that the photographs were aids for the jury to determine whether an aggravated battery occurred and further observed that photographs often give a clearer comprehension of the physical facts than what can be described in words by a witness.

AZ v Paxton, (April 16, 1996, Court of Appeals of Arizona)

In this case, two men suspected of murder were arguing over who had shot the victim. A bloody seat cover was found with the body. A digital photograph taken by the girlfriend was enhanced and showed that it was on the passenger side, proving one of the men wrong. The defense tried to argue that it was irrelevant, but it was allowed in as evidence.

People v Rodriguez

This case holding indicates that the court seeks to ensure that an individual with firsthand knowledge of the photographed scene attests to the picture's accuracy. Again, this demonstrates that for admissibility, photographs must be relevant and authenticated.

State v Rhodes, 627 N.W.2d 74 (Minn. 2001)

The defendant was convicted of the first-degree murder of his wife, who allegedly fell off of their speedboat while the two were taking an evening ride on a lake. The wife was admittedly not a good swimmer and was not wearing a life vest on this particular evening. During the trial, a videotape made by the prosecution of the area of the lake where the defendant claimed he searched for his wife was shown to the jury. The videotape was made during the day while the incident occurred at night, leading the defendant to argue that the video made certain locations on the lake clear to the jury while they had not been as clear to the defendant on the night of the drowning. The Court held that "no event can be perfectly reenacted" and that the daytime video assisted the jury in placing the location of the alleged crime in context to the evidence presented. Id. 627, N.W.2d at 84. The locations and distances on the lake that were shown in the daytime video were relevant separate and apart from the defendant's ability to perceive the same scenes on the night in question that gave the tape evidentiary value.

Henderson v Fields, Id. (Mo. App. W. D. 2001)

The Court restated the well-settled principle that photographs, even gruesome ones, are admissible if they are otherwise relevant. Here, the photographs were relevant to the issues of speed, injuries, level of impact, and point of impact and, therefore, properly admitted.

Cochrane v McGinnis, 160 F.2d 447 (E.D.N.Y. 2001)

The trial court's decision to exclude a fourth photograph that was argued to be exculpatory evidence in favor of the defendant was proper once it was determined that the photograph was untrustworthy. The circumstances in which this photograph emerged were suspicious, having appeared more than 1 year after defendant's murder trial was underway and taken of a "rigged scene" designed to produce a misleading impression on the jury.

MORE PROBATIVE THAN PREJUDICIAL/INFLAMMATORY

Even photographs that depict gruesome images (i.e., autopsy photographs, graphic crime scenes, dead bodies) or show images that would tend to arouse the passions of individuals (i.e., pornographic pictures of children, tattoos that depict symbols of hate, use of drugs) are generally admissible at trial. These types of photographs, however, must still be shown to be more probative than prejudicial to the involved parties to be admissible. In essence, photographic evidence should be excluded only if its sole purpose is to arouse the emotions of the jury or to cause prejudice to some party such as the defendant, plaintiff, or victim. (Remember, photographs that arouse the sympathies of a jury are also properly excluded for the same reasons: for example, when a photograph of a deceased individual is shown with his or her children in a wrongful death case.)

Hayes v State, 85 S.W.3d 809 (Tex. Crim. App. 2002)

In considering whether to admit or exclude photographs on the basis of gruesomeness, the court can look at several factors, including whether the probative value outweighs the prejudicial value, the number of photographs to be offered, whether the photographs are in color or black and white, whether they are close-ups, whether they show a victim's body clothed or naked, and the amount of detail that is depicted.

Hayes v State, Id.

The appeals court found that photographs depicting the coroner pulling back skin around a victim's head to show the gunshot wound did not make the photographs so gruesome as to become inadmissible. The Court reasoned that if the skin had not been pulled back, the jury would not have been able to see the full extent of the injuries inflicted on the victim.

O'Neal v State, 158 S.W.3d 175, 356 Ark. 674 (2004)

Even the most gruesome photographs may be admissible if they assist the trier of fact in any of a number of ways, including by shedding light on some issue, by proving a necessary element of the case, by enabling a witness to testify more effectively, by corroborating testimony, or by enabling jurors to better understand testimony.

State v Hawkins, 58 S.W.3d 12 (Mo. App. E.D. 2001)

A photograph should not be excluded from evidence unless the trial court, in its discretion, finds that the photograph has a greater prejudicial effect than its probative value. Usually, if a photograph is shocking or gruesome, it is because the crime was shocking and gruesome.

State v Duguay, 158 Me. 61 (1962)

Duguay was convicted of murder. The photographs in question were taken by the medical examiner and used to explain the course of the bullet through the victim's brain. The judge stated that whether a photograph is admitted lies in the effect, not in whether it is a color or black-and-white print.

Hrabak v Madison Gas & Electric Co., 240 F. 2d 472 (1957)

The plaintiff was injured while painting the electrical towers (as he was hired to do). Photographs of the injuries included pictures of the plaintiff's face in pain, which was ruled to be prejudicial extraneous material. The court ruled that the photographs were inadmissible.

State v Williams, 565 S.E.2d 609 (N.C. 2002)

The following array of photographs and pictorial evidence was not found to be so gruesome or repetitive that the trial court should have excluded them from the defendant's trial: 11 photographs taken at the crime scene where the victim was murdered, 3 photographs taken at the victim's autopsy, and videotape taken at the crime scene and of the body at the morgue.

Wright v State, 250 So. 2d 333 (1971)

Wright was convicted of kidnapping and murder of a female child. Eight photographs were admitted into evidence. Three of the eight photographs were judged too gruesome and not needed to prove the state's case, so the court ruled that they should not have been admitted into evidence. These pictures show the girl at the gravesite and at the coroner's office.

Albritton v State, 1969

Case may be obtained at www.fastcase.com. In this case, a 16-month-old girl, Stacie, had been repeatedly beaten by her mother's boyfriend. The child died from injuries as a result of the repeated beatings. Photographs were taken in the nude from the front, back, side, three-quarter length, half-lengths, and head view. The reason for this was the gruesome extent of injuries covering the child's entire body. Defense argued they were inflammatory, but the objection was overruled and the photographs were allowed in.

Faught v Washam, 329 S.W.2d 588, 600 (1959) citing Scott on Photographic Evidence (1942) 601, loc. Cit. 475

The vital, mirror-like appearance of a photograph makes it capable of inciting passions and prejudices of a jury and that the danger in this respect increases as photography improves. This was a personal injury action resulting from a car accident that began after plaintiff dozed off while driving on a bridge and bringing his vehicle to a stop on an angle. The defendant's vehicle then struck the plaintiff's unlit and stationary vehicle resulting in various personal injuries. Plaintiff was permitted to introduce six colored photographs of the plaintiff's injured right foot and left thigh (plaintiff had skin grafted from the thigh onto the foot). Two photos were taken six months post accident and the other four photos taken one year post accident showing bloody scabs in high and unrealistic colors. In this, Missouri's first case that called for appellate discussion of colored photographs, the Court held that such photographs should not have been admitted. ". . . as for the six colored photographs in the instant case, we have had sufficient familiarity with male limbs to know that the limbs shown in these photographs are not portrayed in their natural color (and certainly the same is true with respect to the backgrounds). . . ." Id. 329 S.W.2d at 600. As such, the photos were found to have been inflammatory and designed to elicit sympathy for injuries in "high and unrealistic colors." Id. 329 S.W.2d at 600.

Henderson v Fields, 68 S.W.3d 455 (Mo. App. W. D. 2001)

This case involved a wrongful death action against the defendant who killed an entire family while driving drunk. On appeal, the defendant claimed numerous errors, including the admission of certain photographs that (1) depicted the decedents after the accident and (2) of a Mothers Against Drunk Driving ("MADD") calendar ribbon found in the front seat of defendant's vehicle immediately after the collision. The defendant claimed the photographs were unduly prejudicial to him. The Court restated the well-settled principle that photographs, even gruesome ones, are admissible if they are otherwise relevant. Here, the photographs were relevant to the issues of speed, injuries, level of impact, and point of impact and therefore properly admitted. The same result was reached regarding the picture of a MADD calendar photographed in the front seat of the defendant's vehicle after the crash. Such was relevant to the issue of punitive damages and to the defendant's recklessness in driving drunk.

US v Thompson (1984)

Blown-up black-and-white autopsy photograph admitted into evidence to illustrate testimony of the medical examiner. In this case, a 4-month-old boy died of meningitis, which was complicated by malnutrition and starvation.

Duncan v State, 827 So.2d 838, 851, quoting Price v State, 725 So.2d 1003, 1052, quoting Grice v State, 527 So.2d 784,787 (Ala. Ct. App. 1988)

A convicted murder's contention on appeal that pictures of the corpse of his victim might have inflamed the jury is ironic. That risk comes with the territory.

State v Robinson (1970)

"Gruesome" color autopsy photos may be admitted even if not necessary.

In this case, the judge stated that some photographs may end up not being necessary but should be admitted anyway as an anticipatory move to rebut the opposition.

Axelrod v Rosenbaum, 205 A.D.2d 722 (2nd Dept. 1994)

The Appellate Division held that a photograph of the plaintiff taken shortly after a motor vehicle accident should have been admitted on the basis that the photograph could have assisted the jury in determining the amount of damages and in assessing pain and suffering. The photograph was taken while the plaintiff was in the hospital after the accident but was found not to be inflammatory where it fairly and accurately depicted the plaintiff. Because of this error, the case was remitted for a new trial on the issue of damages only.

State v Frazier (1995, Ohio)

It was determined that gruesome photographs are relevant even if cause of death is not an issue.

In this case, Frazier was convicted of killing his step-daughter. He contended that the photographs were irrelevant and inflammatory, because the manner and cause of death were not in dispute.

U.S. v Gladfelter, 168 F.3d 1078 (Neb. 1999)

The Court held there was no prejudice to the defendant when redacted photographs taken of the defendant as he was booked for robbery and carjacking were shown to the jury during trial.

Solomon v Smith, 487 F. Supp. 1134 (1980)

The defendant was convicted of robbery, rape, and sodomy. The victim identified him by requesting a duplicate of his photograph and drawing a hood on him. The court later ruled that this identification was unduly suggestive.

Mann v Oklahoma, 749 P.2d 1151 (1988)

The defendant was convicted of first-degree murder and sentenced to death. Among the appealable issues was whether it was error to allow the jury to view portions of a videotape depicting the recovery of the victim's body from a river and two pictures of the body immediately after it was recovered from the river. The defendant argued the evidence was repetitious and unduly gruesome and allegedly resulted in prejudice to him by arousing the emotions of the jury. The Court held that the portions of the color videotape of the recovery of the victim's body were neither repetitious nor unduly gruesome and found there was no error or prejudice in allowing the jury to view this evidence. Interestingly, although the Court characterized the two photographs as "inordinately grisly" (depicting the victim's body with mud, blood, and slime from the river; a slit in the throat; and entry bullet wounds in the chest and head in sharp focus and at close range) and concluded that they should not have been presented to the jury, having done so was in actuality harmless error by the trial court resulting in no prejudice to this defendant. In its reasoning, the Court found that there was so much evidence presented to the jury implicating the defendant in this crime that the jury would have reached the same verdict even if the two photographs had been excluded.

State v Pettigrew, Id. (N.M.App.1993)

As the Court pointed out "admission has been upheld for far more gruesome and potentially prejudicial and inflammatory photos than the one in the instant case." Id. 860 P.2d at 781.

Simmons v State, 797 So.2d 1134 (Ala. Crim. App. 1999)

The defendant was convicted of the capital murder of a 65-year-old mildly retarded great-grandmother, whom the defendant stabbed and disemboweled. Among the issues raised on appeal, the defendant claimed it was an error for the trial court to have admitted autopsy photographs that were inflammatory and prejudicial. The Court held that it was not an error to admit these photographs, because they were relevant in illustrating the testimony of the coroner as to the extent and cause of the victim's injuries. The Court also upheld the trial court's decision to admit the photographs depicting the crime scene and the extent of the victim's injuries. Such photographs included one depicting the toilet bowl containing pieces of the victim's intestines. Here, the trial court cautioned jurors that they should not allow the photographs to unduly influence their emotions. The photographs were relevant as evidence illustrating how the crime occurred.

Maxwell v United States, 368 F.2d 735 (1966)

The defendant shot and killed the victim, a stranger to the defendant, while the men were patrons at a bar. The trial court accepted into evidence photographs of the victim's body that it found to be neither horrifying nor gruesome and that were relevant to the crime of murder charged against the defendant. The Ninth Circuit agreed with the trial court and reiterated the following standard.

Such photographs should be excluded when their principal effect would be to inflame the jurors against the defendant because of the horror of the crime; on the other hand, if they have a probative value with respect to a fact in issue that outweighs the danger of prejudice to the defendant, they are admissible, and the resolution of this question is primarily for the trial court in the exercise of its discretion. Id. 368 F.2d at 739-740, citing to *People v Chavez*, 329 P.2d 907, 916 (1958).

Harris v State, 843 So.2d 856 (Fla. 2003)

The defendant was convicted of the first-degree murder of his wife and was sentenced to death. The defendant appealed on nine different issues, one of which was whether the trial court erred in admitting what the defendant argued were inflammatory photographs consisting of images of the victim's body (and a short videotape of the area where the body was found). The photographs of the victim's head and skull (referred to as photographs 2 and 6) showed maggots, and photograph 103 depicted the victim's skull with the murder weapon inserted into the hole in the skull. The Court held that the photographs were all admissible. Photographs 2 and 6 were relevant to the medical examiner's testimony as to the cause of death, and photograph 103 was relevant to understand the way the weapon caused the wound in the victim's head.

Commonwealth of Pennsylvania v Duca, 165 A. 825 (1933)

This was a murder case in which the witnesses identified the defendant from a "rogue's gallery" picture that had been altered by removing the prison record numbers. The court ruled that it is proper to cut off prejudicial information from a photograph.

Birmingham Baptist Hospital v Blackwell, 221 Ala. 225

A photograph is inadmissible if includes irrelevant details. In this case, the judge ruled that a photograph of a woman in her hospital bed, which included a view of the entire room, was inadmissible because the series of photographs were intended to show just the injuries and not create sympathy that those injuries had placed the woman in the hospital.

Diel v Ferguson, 138 S.W. 545 (1911)

This was a case of personal injury caused by a defective plank in a sidewalk. The photograph admitted into evidence showed an adjacent stretch of sidewalk that was not part of the suit. The defendant claimed that the photograph should be excluded because of the extraneous material; however, the Court ruled that the photograph needed the extra material to provide orientation and that the material did not prejudice the photo.

Beckman, Inc. v May, 331 P. 2d 923 (1958)

This was a case of two trucks that collided on a narrow bridge, causing injury to the plaintiff. The photograph was of the plaintiff's leg in the hospital; however, the Court chose to exclude the photograph because in the background was a conspicuous picture of him in his Navy uniform.

People v Cruz, 1980

Case may be found at www.online.ceb.com/calcases/c3/26C3d233.htm. Case may also be found at www.fastcase.com. In this case, Cruz was found guilty of murdering his wife, step-grandson, and step-granddaughter by using a piece of pipe. He supposedly had a history of mental illness and was intoxicated at the time. Three black-and-white photographs, each showing one of the three victims, were admitted to show the brutality, force, and nature of the wounds. They were appealed by the defense, because they believed that the pictures were gruesome and lacking probative value. They were allowed in.

DIGITAL MANIPULATIONS V PROCESSING

US v Beeler, 62 F.2d 136 (1999)

The Court compared both the original and enhanced versions of a video surveillance tape before ruling that the enhanced version was accurate, authentic, and a trustworthy representation of the original tape. The Court listened to testimony from a visual enhancement expert and personally viewed all three versions of the videotapes before it was satisfied that the tapes all depicted the same images and had not been manipulated or altered to make the images misleading.

Rodd v Raritan Radiologic Associates, P.A., et al., 860 A.2d 1003 (2004)

In a medical malpractice/wrongful death action brought by the decedent's estate, the trial court was found to have committed reversible error by admitting digitized, computerized, and magnified images of the patient's medical films, and a new trial was ordered. The magnification of the patient's films had a tendency to distort, rather than clarify, the specific condition at issue. Furthermore, the use of such enlargements and magnifications created testimonial evidence rather than an exhibit used to assist in the medical testimony and in the jury's understanding of the issues that led to prejudice to the defendants. The Court further stated that the use of computer-generated exhibits requires a much more detailed foundation than that for ordinary photographs or enlargements. Computer-generated exhibits must be proven to be "faithful representations of the subject at the time in question" and to do so, the process used to create such exhibits must be fully explained using the testimony of a knowledgeable witness who can be examined and cross-examined about the functioning of the computer. (Id. 860 A.2d at 1012)

US v Mosley (August 31, 1994, United States Court of Appeals for the Ninth Circuit)

In this case, an FBI agent used a digital imaging technique on a bank's surveillance tape to match a robber suspect's face to a lineup picture. A unique characteristic on the defendant's face was enhanced. The digitally processed evidence was allowed in after some debate over whether the Court had erred in letting it in.

Knowlin v Benik (2004)

Knowlin was convicted of armed burglary. Knowlin contended that footwear impression evidence found at the scene had been planted, because they were not visible in photographs. However, the technician explained that the prints were "washed out" because of a poorly aimed flash and that the prints can be seen when the photographs are digitally enhanced.

State v Reyes (2003)

In a murder case in 1996, a fingerprint on tape was at first deemed to be unusable. However, in 1999, the print was digitally enhanced by "dodging and burning" and matched to Reyes. The Court believed that the technique met the Frye test, and, therefore, the photograph was admissible.

Dolan v Florida, 743 So. 2d 544 (1999)

Dolan was convicted of sexual battery of a store clerk. He appealed because the surveillance video was enhanced, and, therefore, the stills were not proven to be an accurate representation. However, the Court disagreed and stated that because the videotape was accurate, the enhanced stills were accurate.

"The process, which was the subject of detailed step-by-step testimony by the forensic video analyst at trial, resulted in a computer-enhanced image that was 'bigger, brighter, and better.' The process began with identifying the best video recording machine for playback, playing the tape on that machine, and transferring the image electronically to computer as a digital image. Digital images are composed of millions of tiny dots, referred to as 'pixels.' Next, the most relevant frames of the video images were selected, and computer software was used to add, enlarge, darken, or lighten pixels in the now-digitized picture to clarify and focus the image. Once each image was computer enhanced, still prints were made. These prints were admitted into evidence at trial."

The State properly supported these enhanced images by presenting evidence to establish that the enhanced prints were taken from a videotape that accurately depicted the location of the assault. The State also presented the testimony of the forensic analyst who performed the enhancement who was able to explain about the computer enhancement process and attested that the images were not altered or changed from how they appeared on the original videotape. The State also presented testimony about how the videotape was made, the kind of surveillance system used by the store, and how the system worked.

Commonwealth of PA v Auker (July 31, 1996, Supreme Court of Pennsylvania)

A man accused of kidnapping and murdering his wife was convicted, and some of the evidence was still shots taken by an ATM camera. They were enhanced to show it was, indeed, the husband who she got in the car with the day of her disappearance. The defense tried to argue it should not have been admitted, but it was allowed in.

Commonwealth of PA v Auker (2005)

Burke was arrested for murder, and forensic odontologists used digitally enhanced photographs for comparison to bite marks found on the victim. Burke was eventually determined to be innocent.

Hartman v Bagley, 333 F. Supp. 2d 632 (2004)

Hartman was convicted of kidnapping and murder. He applied and was denied a writ of habeas corpus. In his original case, a digitally enhanced print was identified as his from the victim's bedspread. The court ruled it admissible.

US v Allen, 208 F. Supp. 2d 984 (2002)

This concerns the Daubert hearing where Allen was charged with a bank robbery, and the expert was going to testify that the shoe print found at the scene could belong to Allen. The examiner used software to superimpose a digital image of the print onto an image of Allen's shoe.

Commonwealth of Virginia v Robert Douglas Knight (1991)

In this murder case, a bloody fingerprint on a pillowcase was enhanced digitally. A Frye hearing was called, and the court decided that the application met the requirements.

Fisher v State, 7 Ark. App. 1 (1982)

Fisher was convicted of theft but, on appeal, contended that the court erred in allowing portions of a videotape in. The Court disagreed and stated: "Relevant computer-enhanced still prints made from videotape recordings are admissible in evidence when they are verified as reliable representations of images recorded on master videotapes. . . . The master videotape used in producing the computer enhancement should also be admitted in evidence to help determine the reliability of the still picture."

Nooner v State, 907 S.W.2d 677 (Ark. 1995);
Nooner v State 322 Ark. 87 (1995)

See also, 3 C. Scott, Photographic Evidence §1295 (2nd Ed. 1969 & Supp. 1994). Nooner was convicted of robbery and murder. Surveillance photographs were enhanced and altered during the course of the investigation. Photographs of the victim's face were "mosaicked out" at the request of the family. Images of Nooner were enhanced to make them clearer and brighter. The court ruled that the photographs were admissible. The Arkansas Supreme Court found no error in admitting the still photographs. It highlighted that: (1) both the original videotape and the still photographs were introduced for the jury to view; (2) the jury viewed the videotape before being presented with the enhanced still photographs; and (3) witnesses involved in the enhancement process "meticulously" described their role in the enhancement process and how the process was done (one witness described transferring the still frame from the videotape onto his computer and removed the graininess from the suspect's face by softening the pixels. He attested that he did not add or subtract features from the original photo frame). On the basis of the foregoing, the Court found that the evidence was reliable and, therefore, properly admitted. The Court also highlighted the absence of any evidence of distortion.

Bryant v State of Florida, 810 So.2d 532 (Fla. App. 1 Dist. 2002)

The defendant appealed three convictions of child abuse allegedly occurring at a daycare center, citing improperly admitted evidence of certain enhanced excerpts of videotapes made at the direction of the State from unaltered original videotapes taken from a time-lapse camera installed at the daycare center. The State gave the tapes to a videographer who, in turn, created both partially enhanced excerpts and later fully enhanced excerpts from the original time-lapse tapes. The jury was shown only these enhanced excerpts and was never shown the full time-lapse videotape in its original form. The court was forced to reverse the defendant's conviction and ordered a new trial.

Kennedy v State, 853 So.2d 571 (Fla. App. 4 Dist. 2003)

The defendant was convicted of first-degree murder and robbery. The police, while investigating the crime scene, located a bloody shoe print and lifted several fingerprints found on a comforter and a souvenir. To enhance both pieces of evidence, the crime laboratory used a liquid called Luco Crystal Violet on the bloody shoe print to make it more visible and a computer system called MoreHits Digital Imaging on the fingerprints. The laboratory processed the fingerprints by taking a photograph of the print and putting the photograph into the MoreHits Digital Imaging software, which enhanced the print. The laboratory technician was able to testify at trial that the computer software keeps a record of every change that is made to the photograph and that the original photograph of the evidence is never changed. The defendant challenged the use of both enhancement techniques on appeal, arguing that the Court failed to first assess whether these enhancement techniques met the test under Frye. The Appeals Court upheld the defendant's convictions, citing mainly the fact that neither technique used by the police altered, created, or changed the way the police would compare the evidence. Rather, these techniques were tools the police used to better see and compare the evidence with suspects. Certainly the enhancement of otherwise available evidence is not an innovative scientific theory. Id. 853 So.2d at 573.

WA v Hayden, 950 P. 2d 1024 (1998)*

Hayden was convicted of murder; however, he appealed, stating that digital enhancements of a print should be inadmissible. The court disagreed. The defendant was convicted of felony murder in connection with the rape and resulting murder of a 27-year-old student. Bloodstains were found in parts of the victim's apartment, including bloody hand prints on the fitted bed sheet. The Kirkland Police Department gave the fitted bed sheet to a police department latent print examiner, who performed several techniques to both raise and set the bloody prints left on the sheet*.

The latent print examiner next sent the treated sheet to an expert in enhanced digital imaging at the Tacoma Police Department. Here, digital images of the pieces of bed sheet were taken, and, with the help of computer software, background patterns and colors were filtered out with the goal of enhancing the images. By use of the enhanced photographs of the latent prints, 12 points of comparison on one of the fingerprints were established and more than 40 on one of the palm prints. This led police to connect Hayden to the crime. Defendant Hayden argued that the enhanced digital imaging process used by police to produce and enhance the prints did not satisfy the Frye test and, therefore, the trial court should have excluded this evidence. The Court affirmed the defendant's

---

*This is the seminal case in the issue of enhancement. Before this case, no other Court had thoroughly examined whether digitally enhanced print processes are admissible under the ordinary rules for other photographic evidence, or whether digital images would require a new standard to be admissible. In reaching its decision, the Washington State Court reviewed many sources of information and used the Frye test for scientific evidence.

The enhanced digital imaging process is a totally new process based on research and development done in the late 1960s and early 1970s for the space program. The technology used to enhance photographs of latent prints evolved from jet propulsion laboratories in the NASA space program to isolate galaxies and receive signals from satellites.

*The latent print examiner cut out the bloody areas of the sheet, treated the pieces with a dye stain amido black that reacted with the protein in blood, turning the sheet navy blue. He rinsed the pieces in pure methanol to lighten the background leaving only the protein stains dark blue and finally dipped the pieces in distilled water to set the prints.

conviction and the trial court's decision to admit the enhanced latent print. The Court also settled the issue by stating that the enhancing process for raising latent prints was a generally accepted process in the relevant scientific community when the digital imaging enhancement is performed by a qualified expert using appropriate software. Id. 950 P.2d at 1028. The Court pointed out that nothing appeared in the digitally enhanced photograph that was not already present on the fabric. The digital expert testified that the software used a "subtractive process" in which elements were reduced or removed while nothing was added. The software itself prevented addition to, changing of, or altering the original image. Simply stated, image enhancement makes what is already there more visible. Id. 950 P.2d at 1028.

State v Swinton, 847 A.2d 921 (Conn. 2004)*

The defendant was convicted of murder. During the victim's autopsy, crescent-shaped marks on both breasts, believed to be human bite marks, were discovered. These bite marks were photographed. Once the defendant was identified as a suspect, a mold of his teeth was made for purposes of comparison. At trial, the Court allowed into evidence: (1) computer-enhanced photographs of the bite marks taken during the autopsy and (2) images of the defendant's teeth superimposed on photographs of the bite marks. Both photographs were enhanced using different computer software. On appeal, the defendant challenged these two photographs arguing that the State failed to provide adequate foundation testimony about the reliability of these software programs in making comparisons. In its decision, the Appeals Court held that the enhanced photographs of the bite marks using Lucis software were properly admitted. But the Court also held that the superimposed images onto the mold of the defendant's teeth using Adobe Photoshop should have been excluded.

Photographs of the Bite Marks

The state's introduction of the bite mark photographs came through a very knowledgeable state witness who was able to explain the Lucis program and that it was accepted and relied on by experts in the field of forensic pattern analysis. The witness was well versed in the Lucis program and performed an in-court demonstration of how bite mark evidence is enhanced. Most significantly, the witness was able to attest that nothing was added or removed from the photograph during the enhancement process, because the Lucis software is not equipped with the ability to alter images in these ways.

---

*After the decision in State of Washington v Hayden (cited above), the next Court to take a hard look at digital images and the enhancement processes was in this case. The Court meticulously considered the evidence and noted that even by this time there remained very few cases that actually discussed the admission of digitally enhanced evidence and nearly none that discuss evidence that is computer generated. (Id. 847 A.2d at 938)

Superimposition of Defendant's Teeth

As opposed to the Lucis program, the state failed to supply a knowledgeable witness who could testify to the reliability of the evidence and the processes used to generate it with using Adobe Photoshop software. The overlays were brought into evidence using the testimony of the forensic odontologist who made the comparison. However, the forensic expert lacked the computer expertise necessary to permit a full exploration of whether Adobe Photoshop was a reliable process.

English v State, 422 S.E.2d 924 (1992)

The Court allowed into evidence enhanced still photographs taken from a videotape. Here, the requisite foundational showing was made by the prosecution, when it presented the testimony of the technician who performed the enhancements who described the process used to obtain the still prints. The technician also testified that the enhanced prints were fair and accurate representations of what appeared in the original videotape.

ENLARGEMENTS

State v Dunn, 109 So. 56 (1926)

The photographic evidence in this murder case was projected onto a screen. Then, the projections of the known and unknown palm prints were drawn on to indicate points of comparison.

Bombailey v State (1990, Alabama)

This was a case of child abuse by the parents. The defendant challenged the admissibility of enlarged photographs of the victim's injuries taken at the emergency room as the victim was receiving treatment. "The fact that the photographs were in color and were enlarged is of no particular significance as long as there was no distortion of the depiction of the injuries." Id. 580 So.2d at 46.

Jones v State, 249 Ga. 605 (1982)

Jones was convicted of murder while committing a felony. The medical examiner used enlarged photographs to explain to the jury. The court ruled that enlarged photographs are admissible as long as no distortion or enlargement of individual items in the picture exists.

Epps v State, 216 Ga. 606 (1961)

Epps was convicted of rape. The photographs admitted at trial were enlarged; however, the defendant maintained that because they were larger, the photographs were no longer an accurate portrayal. The court ruled that they were accurate.

People v Jennings, 96 N.E. 1077 (1911)

The defendant was found guilty of murder and sentenced to death. Among the evidence used by the prosecution to convict the defendant were fingerprints found at the crime scene that were subsequently photographed. Print examiners then used the fingerprint photographs to compare to the defendant's prints. (The photographs were also enlarged at some point either for comparison or for purposes of trial.) The four print examiners testified as to the system used to compare fingerprints, which satisfied the Court and upheld the defendant's conviction and sentence.

Duncan v State, 827 So.2d 838 (Ala. Crim. App. 1999)

During the course of the defendant's trial for capital murder, the prosecution introduced an oversized photograph showing the condition of the 37-year-old victim's body after it was discovered, including injuries and mutilations. The witness who testified he discovered the victim's body stated that the picture, although large, accurately depicted the scene he saw on the date of the discovery. Because there was no distortion to the injuries depicted nor did the photographs mislead the jury in any way, the Court held there was no error for the trial court to have allowed the enlargement into evidence.

People v Watson, 629 N.W.2d 411 (Mich. App. 2001)

In a prosecution for criminal sexual conduct, the prosecution introduced an enlarged photograph of the defendant's stepdaughter's buttocks that the defendant carried in his wallet. On review, the Court found no error in the admission of the enlargement. The Court agreed with the prosecution's theory that this photograph was admissible to show the defendant's motive and intent to sexually assault his stepdaughter.

People v Mathison, 287 A.D.2d 384 (N.Y.A.D. 1st Dept. 2001)

Photographs taken that served to replicate the magnification provided by binoculars used by police while watching an apparent drug transaction were relevant to show the jury the power of the binoculars and how close they made objects appear. The fact that the replicated photos were taken under different lighting conditions was something for the jury to consider in giving the evidence weight and did not in any way affect admissibility.

State v Lloyd, 552 S.E.2d 596 (N.C. 2001)

After a conviction for murder, the defendant challenged the admissibility of enlarged photographs showing the front of the victim's bloodstained shirt with a bullet hole, the victim's shirt next to a pool of blood, and of the back of the victim's shirt with a bullet hole on the sleeve. The Court noted that these photographs were used to illustrate the testimony of State witnesses concerning the circumstances of the murder and, therefore, the photographs were relevant and properly admitted.

Rodd v Raritan Radiologic Associates, P.A., et al., 860 A.2d 1003 (2004)

In a medical malpractice/wrongful death action brought by the decedent's estate, the trial court was found to have committed reversible error by admitting digitized, computerized, and magnified images of the patient's medical films, and a new trial was ordered. The magnification of the patient's films had a tendency to distort, rather than clarify, the specific condition at issue. Furthermore, the use of such enlargements and magnifications created testimonial evidence rather than an exhibit used to assist in the medical testimony and in the jury's understanding of the issues that led to prejudice to the defendants. The Court further stated that the use of computer-generated exhibits requires a much more detailed foundation than that for ordinary photographs or enlargements. Computer-generated exhibits must be proven to be "faithful representations of the subject at the time in question" and to do so, the process used to create such exhibits must be fully explained using the testimony of a knowledgeable witness who can be examined and cross-examined about the functioning of the computer. (Id. 860 A.2d at 1012)

People v Cruz, 773 N.Y.S.2d 392, 5 A.D.3d 190 (1st Dept. 2004)

Photographs presented to the jury that were taken during the daytime while the defendant's alleged offense occurred at night were properly admitted into evidence. The police officer's testimony that these photographs were taken under different lighting conditions than when the alleged drug transaction took place and with binoculars was relevant to illustrate the officer's ability to see the drug sale and to the power of his binoculars.

PRINTING ERRORS

Scottish Union & National Ins. Co. v McKone, 227 F. 813 (1915)

This was a property damage claim in which the McKone family sought reimbursement of property lost during a house fire from their insurer, Scottish Union & National. During the trial, counsel for the insurance company introduced photographs of the house that was involved in the subject fire. However, when plaintiff McKone testified, it was learned that the fireplace was depicted on the wrong side. Subsequent to that testimony, the court excluded the photograph as having no relevancy. On appeal, however, the Eighth Circuit reversed and found that the photograph could have assisted the jury in its decision and assessment at the value of the house. The photograph should have, therefore, been admitted into evidence.

EQUIPMENT

Parish v State, 165 S.W. 2d 748 (1942)

Parish was convicted of swindling. The Court ruled a photograph was admissible when side lighting was used to display indented writing, even though an expert was not in court. However, the equipment used was presented in court along with procedure.

MARKING PHOTOGRAPHS

Kenny v Kelly, 254 S.W. 2d 535 (1953)

In this case, Kenny attempted to block the construction of railroad through a portion of land. Photographs admitted into evidence were aerial pictures of the scene with superimposed block lines, numbers, and boundary lines that showed that Kelly had the right to the land. The Court ruled that these markings did not keep the photograph from being admissible, because they were facts and not points in dispute.

Boulden v State, 179 So. 2d 20 (1965)

In this case of murder, the court allowed an overlay to be placed over the aerial photos and for them to be marked on. "Exhibits 26, 27 and 28 are aerial photographs of the scene of the homicide and surrounding territory. They were admitted without error. Aaron v State, 271 Ala. 70, 122 So.2d 360. The same is true of Exhibits 26A, 27A and 28A, which are plastic covers or overlays placed over the aerial photographs on which identification markings were placed showing the location of pertinent points."

State v Kuhl, 175 P. 190 (1918)

Photographs of palm prints were properly admitted into evidence, even though experts who testified in regard to their content had previously placed lines on them. The lines and their significance were fully explained by the witnesses to the jury.

NO PHOTOGRAPH IS AN EXACT REPRESENTATION

Hancock v State, 47 So. 2d 833 (1950)

In this case, the father was convicted of killing his daughter, but the defendant challenged the photographs of the victim because of changes. The Court ruled that some changes were inevitable between the time of death and the taking of the photographs, because rarely are photographs taken at the exact time of death. Therefore, oral proof is sufficient in explaining in changes between the time of death and the photograph.

SUMMARY

This chapter dealt with the legal issues related to photography. It was pointed out how photographs intended to be offered in court as evidence have certain standards by which they will be evaluated. They have to accurately represent the scene as it was originally found. They have to be authentic and not recreations of the original scene, unless they are clearly documented as being recreations. They have to be relevant and material, just like any other evidence introduced in the course of the trial. They cannot unduly bias the juror against the defendant by appealing to the emotions of the juror.

The other normal purposes of photographs as methods of crime/accident scene documentation were pointed out. Photographs are a means to clarify the understanding of the jury during the trial, and they can be used for investigative purposes early in the investigation of the incident.

Finally, concerns about the rights of those being photographed were addressed. When can the subject photographed legally object to the introduction of those photographs in a trial? Is the consent of the person being photographed required? Under what circumstance might the consent of the subject be required? If the photograph violates a person's reasonable and legal expectation of privacy, under what circumstance can it still be offered in court as evidence? These questions were answered.

To supplement the theories being presented in this chapter, numerous court citations related to these issues are presented at the end of the chapter. It is hoped that these will be considered a valuable addition to this text. These court cases have been grouped by chapter sections. In addition to this grouping of cases, a useful list of possible objections to the use of photographs was found in *The Practical Methodology of Forensic Photography* by David Redsicker. In that text, there were 14 possible points of objection to the use of photographs in court.[*] Court cases have also been sorted by those 14 possible points of objection.

DISCUSSION QUESTIONS

1. How would the color accuracy of photographs be established?
2. Which lens best assures the accuracy of distance relationships?
3. Is it permissible to return evidence that has already been collected to the crime scene to photograph it "about" where it had been when originally found?
4. "All photographs taken at the original crime/accident scene will be admissible as evidence later in court." Comment on this statement.
5. If a suspect has been charged with a mutilation homicide, photographs of body parts of the victim in the defendant's residence may not be admissible at trial, even though they are strong evidence of the defendant's guilt. Why? How might it be argued that these photographs should be admissible?
6. In what situations is it permissible to admit an underexposed photograph as evidence in court?
7. Can a person stopped as a possible suspect in a crime legally refuse to have his or her photograph taken (with a digital camera, for instance) to be shown to a witness who cannot respond to the area of the stop?
8. Do Miranda rights include the right to prevent unwanted photographs being taken?
9. Are there times when anyone might be able to legally challenge photographs and have them held to be inadmissible in court?
10. Define "relevant" and "material."

[*]Copyright 1994, From "The Practical Methodology of Forensic Photography" by David R. Redsicker, pp. 246–249, Reproduced by permission of Routledge/Taylor & Francis Group, LLC.

FURTHER READING

Berg, E. (2000). Legal ramifications of digital imaging in law enforcement. *Forensic Sci. Commun.* **2**, No. 4.

Blitzer, H. L., and Jacobia, J. (2002). "Forensic Digital Imaging and Photography." Academic Press, San Diego, CA.

Coppock, C. C., Kruse, S., and Berg, E. (2002). "The Implementation of Digital Photography In Law Enforcement and Government: An Overview Guide." Charles C Thomas, Publisher, Springfield, IL.

Daubert *v* Merrell Dow Pharmaceuticals, Inc., 509 US, 579 (1993).

Duckworth, J. E. (1983). "Forensic Photography." Charles C Thomas, Publisher, Springfield, IL.

Eastman Kodak Company. "Digital Imagery in the Courtroom" available online: http://www.ictlex.net/index.php/1999/10/05/digital-imagery-in-the-courtroom/October 5, 1999.

EPIC (Evidence Photographers International Council). (1999). "Outline for Standard Crime Scene Photography." The Council, Honesdale, PA.

Frye v. United States, 54 App. D.C. 46, 293 F. 1013, 1014 (1923).

Kaci, J. (1995). "Criminal Evidence." 3rd Ed. Copperhouse Publishing Company, Incline Village, NV.

Kumho Tire Co. *v* Carmichael (97-1709) 526 U.S. 137 (1999) 131 F.3d 1433, reversed.

Lyman, M. D. (2005). "Criminal Investigation." 4th Ed. Prentice-Hall, Upper Saddle River, NJ.

Mello, K. (2002). "Photography and Digital Imaging in Law Enforcement." Institute of Police Technology and Management, Jacksonville, FL.

Miller, L. S. (1998). "Police Photography." 4th Ed. Anderson Publishing Company, Cincinnati, OH.

Nickell, J. (1994). "A Handbook for Photographic Investigation." University Press of Kentucky, Lexington, KY.

Redsicker, D. R. (1994). "The Practical Methodology of Forensic Photography." CRC Press LLC, Boca Raton, FL.

Scott, C. C. (1969). "Photographic Evidence: Preparation and Presentation." 2nd Ed. 3 Vols. West Publishing Company, St. Paul, MN.

TIPS, TRICKS, AND MACGYVERS

As mentioned earlier, law enforcement types, and especially crime scene investigators, have a very special "I can do that!" attitude. Not only can they duplicate the efforts of others before them, CSI types often pride themselves on being able to get the job done with make-shift equipment, costing far less than what can be purchased through crime scene product catalogs. Hence, the term "MacGyvers": with a bit of string, a paperclip, and, of course, some duct tape, CSIs can get almost any job done! Just one example will convince the doubtful. When expensive lasers first began to be used at crime scenes to induce fluorescence in many varieties of evidence, a senior CSI in a neighboring jurisdiction found an old slide projector about to be thrown away, glued a blue filter over the lens, and created his own alternate light source. Most items of evidence stimulated by the laser into fluorescence also fluoresced with the modified slide projector.

In addition to these pure MacGyvers, the ingenuity of CSI photographers can manifest in many other ways. While this book attempts to cover most of the photographic needs of the crime scene photographer, this author is convinced there are numerous examples of the resourcefulness of other photographers still undiscovered. Which leads to this challenge: if you have a photographic tip, trick, or MacGyver, share it! Send it to this author, and if this book is fortunate enough to have a second edition, as many as possible will be included in this Appendix. Of course, all submissions will be properly attributed to the sender. Do it not just to get your name published; do it so that others can benefit from your suggestions. Let us use our cameras to put more bad guys away! As has been said before, the more tools one has on their Bat-Belt, the more problems will be able to be solved as they are encountered. Examples follow.

EVIDENCE ON VERTICAL WALLS

In a recent Journal of Forensic Identification (Vol. 56, No. 1, January-February 2006), an author was describing a case where as a victim was going over a roof top, there were bloody finger marks left on the wall, obviously an attempt to grasp at anything to prevent the fall. The article was about attempting to maximize the depth of field so the entire 3′ length of bloody finger marks would be in focus when the photo was taken when standing on the roof and peering down. The JFI author recommended using a Tilt-Shift lens to increase the depth of field with this angled shot. The challenge had been laid out. This author's creative juices began to flow after reading this article, with the need to capture the entire 3′ length of evidence in-focus without using a Tilt-Shift lens being the challenge. The MacGyver deep inside this author surfaced. Using normally available photographic equipment, a paperclip, and some duct tape, the problem was solved!

The solution was to mount a camera onto a tripod, secure the tripod's weight with "ropes" of duct tape, and lower the tripod over the side of a wall with its three feet on the wall and the camera's film plane parallel to the wall. The camera would be pre-focused so a length of 3′ was in focus. The 10-second delayed shutter option would be used, so that once the shutter was pressed, it would be necessary to position the tripod over the evidence in the 10 seconds available. A paperclip was taped to a stick in order to help position each of the three tripod legs over the evidence, so the camera was properly positioned to capture the evidence. If this were successful, then naturally the entire 3′ length of evidence would be in focus. Real rope, of course, was shunned, because MacGyver would never use real rope! Regular duct tape needed reinforcing, so four thicknesses of tape were used

Figure A.1

Collage of images showing the author setting up a tripod on a vertical wall to photograph evidence with a uniform DOF range; close-ups demonstrate ridges are captured in focus.

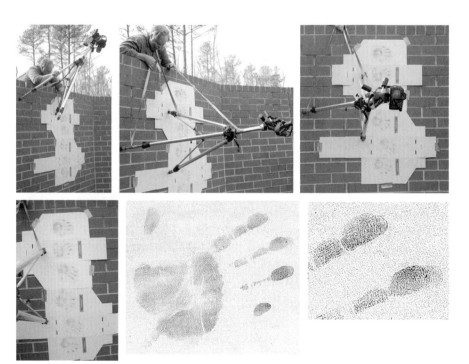

to make sure the tape did not break, taking an expensive camera and lens to a premature death. To simulate 3′ of bloody fingerprints, six full hand prints were used. If successful, it would be good to show close-ups of the fingers with detail in focus. Of course, this entire procedure could have been repeated with the tripod legs shortened to their most compact position for better close-ups, but this wasn't the original challenge. Capturing 3′ of evidence with detail in focus was the challenge.

Figure A.1 shows this author in the process of capturing the shot. Naturally, several tries were needed for success, with only 10 seconds available before the shutter tripped. However, success was obtained.

Have you done anything similar?

COAXIAL LIGHTING

When untreated fingerprints are on glossy surfaces, like credit cards, lighting them with oblique lights is frequently unsuccessful when trying to visualize them for photography. Moving a flashlight around by trial and error usually results in the fingerprint best visualized with the flashlight aimed almost straight down at the print. Of course, when the camera is directly over the fingerprint, with the film plane parallel to the fingerprint, the flash or flashlight cannot also occupy that same position. Or can it? By using a technique called coaxial lighting, both the camera's position and the light used to visualize the fingerprint can occupy the same space. How can the light be positioned so it is shining straight down on the fingerprint, 90° to the surface, when the camera is there also? One must bend the light!

This can be accomplished by using a clean sheet of glass, held at a 45° angle to the fingerprint, with a flashlight positioned parallel to the surface with the fingerprint on it. The light then comes in from the side of the print, strikes the glass, is reflected straight down towards the fingerprint, lights the fingerprint, and is then reflected straight up through the glass into the camera lens.

Figures A.2, A.3, and A.4 show this set-up, as a graphic and with photos of coaxial lighting being used. Figure A.3 demonstrates the technique being used to visualize an untreated fingerprint on a credit card; Figure A.4 show the technique being used to direct light straight down into a film canister to light all the evidence within the canister. Otherwise, with an electronic flash held to the side of the camera, there is partial shadowing within the canister. Of course, it would also be possible to pour the contents of the film canister out onto the table, and photograph them all then. Coaxial lighting is merely another solution to the problem.

Have you done anything similar?

Figure A.2

Graphic showing the coaxial lighting set-up.

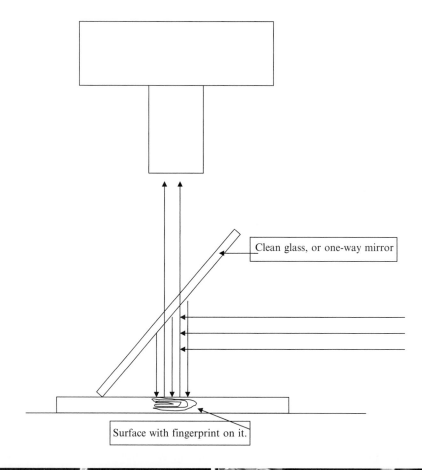

Clean glass, or one-way mirror

Surface with fingerprint on it.

Figure A.3

Coaxial lighting set-up, and results.

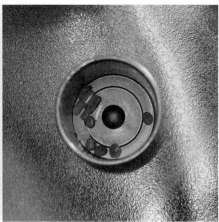

Figure A.4

Coaxial lighting bending light down into a film canister.

MIRRORS

Evidence on mirrors presents other difficulties. To photograph the evidence with the film plane parallel to the surface of the mirror, the photographer will be in his or her own field of view. To avoid this, one option is to photograph the evidence from a slight angle and then use computer software, like Photoshop™, to 'fix' the perspective distortion.

Figure A.5 shows this sequence. But there are times when even a slight photographic angle is to be avoided. If the evidence is a fingerprint or a shoeprint, if a slight angle is used to photograph the evidence, frequently an annoying duplicate image of the evidence will also be photographed. The evidence on the surface of the mirror, and a reflection of the evidence on the mirror's silvering will be captured at the same time!

Figure A.6 shows this possibility. Perhaps the shoe print analyst can still see the detail necessary to make an identification. However, if the photographer were just a bit more resourceful, a better image can be captured. How? By having the film plane parallel to the shoe print when the photo is taken. The trick is to avoid having the photographer and camera equipment in view when this is done. This is managed by cutting a small hole in a sheet of posterboard, and taking the photograph with the posterboard between the photographer and the surface with the shoeprint. The hole in the posterboard only needs to be dime-sized or even smaller. As was mentioned in an earlier chapter, comparison-quality photographs are taken with an f/11 aperture. It is not necessary to have the hole on the posterboard be the entire size of the front of the lens. It is only necessary to have the hole approximately the same size as the aperture used. The posterboard will then block the photographer and the camera from being captured in the photo. What will be captured in the photograph is the hole in the posterboard. That is unavoidable, but not a problem. While viewing the evidence through the viewfinder, you will see both the evidence and the hole. Merely

Figure A.5

Mirror shots: (left) incorrect; (middle) an angled shot to avoid appearing in your own photograph; (right) a Photoshop "correction" of the angled shot.

Figure A.6

Angled mirror shots: (left) angled view of shoe print on mirror; (middle) same shoe print; (right) the angled view captures a "double" view, making IDs difficult.

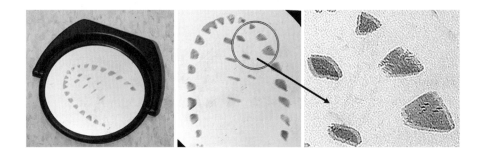

arrange the hole so that it is positioned within the shoeprint in an area without detail. Figure A.7 shows the result.

Figure A.7 shows a side-by-side comparison of an angled view of the shoe print detail and a straight-on shot of the shoe print detail. The straight-on shot is much preferred. Your shoe print examiner will bless you! Since Photoshop™ was used to invert the original light-colored image to a dark-colored image, to make the comparison process easier, some have suggested that Photoshop™ can also be used to 'clone' out the hole altogether. This author disagrees with that suggestion. Leave the hole in the image, proud and prominent. If the defense attorney makes the mistake of asking what that strange shape is, you will then be able to answer the question, explaining your expertise and command of your tools. Since the defense attorney's job is to avoid bolstering your credibility, fewer questions will be coming your way from that defense attorney. Leave the hole there as a trap!

Have you done anything similar?

DISTORTION CORRECTION, REDUX

Despite the fact that this text is not intended to be a Photoshop™ "processing" how-to book, simple "fixes" can be mentioned in this Appendix. It is well known that

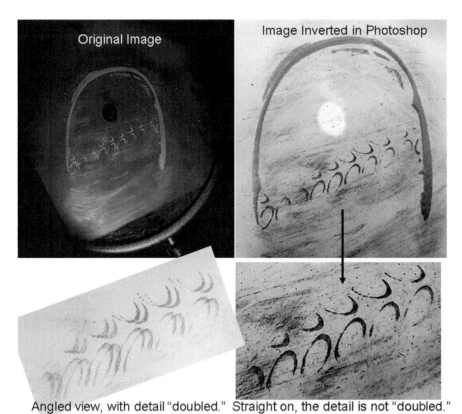

Original Image

Image Inverted in Photoshop

Angled view, with detail "doubled." Straight on, the detail is not "doubled."

Figure A.7

Straight-on mirror shots: (top left) photograph through hole in poster board; (top right) same image inverted; (bottom right) close-up—no "double" image; (bottom left) "double" image from an image like Figure A.6.

photographing the exterior of buildings, particularly with wide-angle lenses, will result in the photographs of the buildings showing some perspective distortion. Look at the two examples in Figure A.8.

This occurs so frequently that we may even fail to notice the distortion, although it is certainly obvious. This is a 'normal' result of viewing the buildings from the ground level. The top of the buildings will always appear to be narrower at the top of the structure. Photoshop CS2™ has very simple "fixes" for this type

Figure A.8

Two "distorted" images of buildings.

of distortion. Using the crop tool, once the cropping rectangle has been placed on the image, it is common for the sides and the top and bottom lines of the cropping rectangle to be moved as desired to crop irrelevant details. With Photoshop CS2™, when the crop tool is selected, a "perspective" box also appears at the top of the screen. When this is checked, the cropping rectangle can be altered by moving the corners of the cropping rectangle. The corners can be aligned with the natural features of the scene, for example, the four sides of the building. See Figure A.9.

Now, when the image is cropped, not only will the material outside the cropping rectangle (which is no longer really a rectangle) be eliminated, but the angled sides of the cropping rectangle will be forced back into a true rectangle, thereby correcting the distortion to the building. See Figure A.10.

Nice straight walls; nice upright trees and light poles!

Have you done anything similar?

Figure A.9

Photoshop™ "crop," showing Perspective box checked to align image edges with building walls (Adobe Photoshop CS2™ screen shot reprinted with permission from Adobe Systems Incoporated).

Figure A.10

Photoshop™ geometric corrections to buildings in Figure A.8.

GLOSSARY

A

Aberrations

- **Chromatic:** Longitudinal and lateral chromatic aberrations. Unless corrected, the different colors of white light will separate and focus at different points. The result is "color fringing," where the edges of objects appear to have strange bands of color.
- **Spherical:** The outer part of a lens will bend/refract light differently than the middle or central part, so that all the light does not focus at the same point. The result is soft focus rather than clear, crisp detail.
- **Coma:** Off-axis detail will not focus sharply on the film or digital sensor.
- **Astigmatism:** Off-axis points in the scene will appear as lines or ovals on the image instead of distinct points.
- **Curvature of field:** A subject that is completely flat will be focused sharply in the center of the image, but toward the periphery of the image, the image is "softer" or slightly out of focus.

Acutance: The rendering of a sharp edge of a subject as a sharp edge in the photograph.

Aliasing: The digital effect of a jagged appearance of diagonal lines, edges of circles, etc. caused by the square nature of most pixels. Also called "jaggies."

Angle of view/field of view: Different focal length lenses allow the camera operator to "see" more or less of the real world. Wide-angle lenses allow a wider field of view; telephoto lenses produce a narrower field of view.

Aperture: The variable size of the lens opening created by the diameter of the diaphragm.

Aperture priority exposure mode: The camera operator selects the aperture/f/stop, and the camera automatically selects the shutter speed to ensure a proper exposure.

ABC (auto backlighting control): Backlighting normally results in an underexposed primary subject. Depressing this button forces the camera to change the exposure setting, usually to a +1½, to have the primary subject properly exposed.

Auto exposure bracketing: With this feature, some cameras will take three images when the shutter button is depressed just once: a 0, +1, and –1. Some cameras will change the exposure settings automatically but require the shutter to be depressed three times in a row.

AE-L (automatic exposure lock): Depressing this button, and keeping it depressed, will allow the camera operator to lock in one exposure setting and then recompose the image. For example, when taking an exterior overall photo of a building façade that is shady, the camera can be aimed down to eliminate the sky from the field of view. With the shutter priority exposure mode, an exposure for the shade is determined and locked in; then, the camera is recomposed with the sky in the background, and the photo is captured.

AF (auto-focus): By depressing the shutter button halfway, the camera will automatically focus on an item in the center of the viewfinder.

AF-L (automatic focus lock): Depressing this button, and keeping it depressed, will allow the camera operator to lock in one auto-focusing position, then recompose, and take the photograph. For instance, one way to hyperfocal focus with an f/8 is to focus at a point 30' from the camera. If the scene of interest does not have anything 30' from the camera to focus on, you could focus on any item 30' away, lock in the focus by depressing the button and keeping it depressed, then recompose the camera on the scene of interest, and take the picture.

ASA (American Standards Association): The previous method for rating film speeds. ISOs are currently used.

AV (aperture value) (aperture priority exposure mode): The camera operator selects the f/stop, and the camera will automatically select the corresponding shutter speed for a proper exposure.

AWB (automatic white balance): The camera will balance colors on the basis of reflected light coming from a white object in the field of view.

B

B (bulb): A shutter speed setting that has the shutter remain open as long as the shutter button is depressed or locked open with a remote shutter release cable.

Barrel distortion: With a wide-angle lens, the sides of buildings or vertical elements can appear to bow out like the sides of a barrel.

Bayer sensor array: Most digital sensors have red, green, and blue filters over different pixels in this particular pattern. The camera's computer software then renders a "colored" image, depending on the different values of light coming from all the digital pixels.

Bit: The smallest unit of digital memory; a contraction from "binary" and "digit."

Blooming: If each pixel is thought of as a bucket collecting light photons, once one "bucket" is full, the electronic charge caused by additional photons will overflow to surrounding pixels, brightening or tending to overexpose them in the process.

Bounce flash: Aiming the electronic flash at a ceiling or side wall to soften the light reaching the primary subject, because the farther light travels to reach a subject, the softer (dimmer) it will be when it gets there.

Bracketing: The intentional overexposure and underexposure of succeeding photographs, taken from the same position, to ensure at least one of the images is properly exposed.

Built-in flash: Some cameras have flash units built into the camera body that can be activated as desired to provide additional lighting when the scene is dim. Usually not very powerful, they have restricted uses at major crime scenes. They also tend to produce red-eye when used to capture images of people.

Burning: Providing additional light to a selected area of an image results in a darker image in that location. Slightly overexposed areas of an image can be improved by burning in those areas. Digital imaging software also allows similar selective darkening to digital images. *See also* Dodging.

Byte: Eight bits of memory in a computer.

- **Kilobyte:** 1024 bytes (roughly 1000 bytes)
- **Megabyte:** 1024^2 (roughly 1 million bytes)
- **Gigabyte:** 1024^3 (roughly 1 billion bytes)
- **Terabyte:** 1024^4 (roughly 1 trillion bytes)

C

Cartridge: Unexposed film is stored in a film cartridge until it is processed.

CCD (charge coupled device): A common digital camera sensor type.

CD-R: Compact disc-recordable.

CD-ROM: Compact disc read-only memory.

CD-RW: CD-rewritable.

Center weighted: A camera metering method by which an exposure is determined primarily from the light coming from a large oval in the center of the field of view.

Circle of confusion: As focused light comes from the lens to the film plane, it travels in a narrowing cone until it meets at a precise point on the film plane, where the image is judged to be in focus. If the light waves focus before reaching the film plane, they begin diverging and form a circle on the film plane. If the light waves have not yet focused to a point by the time they reach the film plane, they will also form a circle on the film plane. The eye perceives some of these smaller circles to still be in focus. Circles larger than a certain size are perceived by the eye to be out of focus, or "confused." The concept of circles of confusion explains why not only a precise object can be seen to be in focus, but also areas in front of that object and areas behind that object can appear to also be in focus at the same time.

Close-up filters: Supplementary filters that can be added to a lens to allow closer focusing and magnification of the subject.

CMOS (complementary metal-oxide semiconductor): A type of digital camera sensor.

CMYK (cyan-magenta-yellow-black): Cyan, magenta, and yellow are the primary subtractive colors. When projected over the same area, they block all light, resulting in black. Many printers also use these dyes to form colors.

Compact Flash™: One type of storage media used in digital cameras.

Compression: To reduce file sizes so that more files can be stored in the same space, files can be compressed by discarding data with similar colors and contrasts. If not compressed too drastically, it is difficult to notice the difference when viewing an image that has been compressed. Images intended to be used for critical comparisons or examination quality images should never be compressed, because this in effect discards part of the evidence.

Contrast: The degree of difference between light areas and dark areas in an image.

Cropping: Resizing an image by discarding parts of the edges of the original.

D

Dedicated flash: An electronic flash unit designed to work with a specific camera, where the camera and the flash share information to ensure a proper exposure.

Depth of field: The variable range, from foreground to background, of what appears to be in sharp focus.

Depth of field preview button: Most lenses default to their widest aperture until the moment the shutter button is depressed, when they close down to the selected f/stop. Wide apertures allow in more light to make it easier to compose and focus the camera. The widest aperture, however, provides the worst depth of field. If it is the camera operator's desire to maximize the depth of field with a small aperture, this depth-of-field range will not normally be seen in the viewfinder. The depth of field preview button forces the lens to close down to the selected f/stop, so that the depth of field can be seen in the viewfinder.

Depth of field scale: This scale is located on many lenses and is frequently shown as pairs of f/stop numbers on either side of the focusing point on the lens. Sometimes they are shown as pairs of colored lines on either side of the focusing point on the lens. They allow the photographer to know the precise depth of field range when either hyperfocal focusing or zone focusing. Unfortunately, many newer lenses do not have this feature.

Diaphragm: This is a system of curved, overlapping metal blades that forms the variable lens opening or aperture.

Diffraction: Light striking an edge will bend and no longer proceed in a straight line. Light coming through a lens with the smallest apertures is more affected by diffraction, because most of the light forming the image is diffracted light. Diffracted light results in soft focusing and the loss of edge detail. The "cure" to diffraction is to open the aperture at least two f/stops from the smallest aperture of the lens. With a wider aperture, more light comes through the center of the aperture, where it is not affected by diffraction.

DIN (Deutche Industrie Norm): A film speed setting used in Germany.

Dioptric adjustment knob: Many camera manufacturers provide a means for individuals to make slight focusing adjustments to the viewfinder. A slide or dial on the right side of the viewfinder allows the individual to adjust the viewfinder to his or her own eyes.

Dodging: At times, only parts of an image are a bit underexposed. Dodging allows the darkroom operator to block some of the light from reaching the light-sensitive photographic paper, which lightens the image in those selected parts of the image. Digital imaging software allows the equivalent for digital images. *See also* Burning.

DPI (dots per inch): Ink jet printers work by spraying minute dots of various dyes onto photographic quality paper. Black dots of various sizes suggest differing shades of gray from white to pure black. Dots of varying colors suggest all the colors of the rainbow to the viewer.

DX (data exchange): At one time it was necessary to "tell" the camera what film was loaded into it by setting a dial on the camera to a specific film speed setting. Most film today has a DX "bar code" on the film cartridge, alternating black and silver squares. Newer cameras have sensors in the camera body that can "read" this "bar code" so the camera automatically knows the ISO of the film loaded into it. The code includes the ISO of the film, the number of exposures on the roll of film, and the exposure latitude of the film.

Dye-sublimation printer: A high-end printer in which the color dyes from three colored ribbons are thermally transferred to the printing media. Dye-sub is considered continuous tone printing, forming any one of 16 million colors, with no discrete dots of dyes used to suggest actual colors.

Dynamic range: The difference between the brightest highlight and darkest value that a sensor or film can detect and record in a single image. Obviously being able to capture detail in brighter and darker areas of an image is a distinct advantage.

E

Exposure compensation: With automatic exposure modes (aperture priority, shutter priority, or program), the exposure compensation dial allows bracketing.

- **Camera:** When using any of the automatic exposure modes, you do not bracket by changing either the shutter speed or the aperture of the lens. The exposure compensation dial is set to +1, +2, −1, or −2 as desired.
- **Flash:** Bracketing may be achieved by altering the flash output as well. The flash intensity can be altered to +1, +2, −1, and −2 as desired.

Exposure latitude: (1) Film has a natural inability to capture details in wide extremes of lighting. With high-contrast scenes, when there are both well-lit areas and areas in deep shadow, only one extreme can record detail. If an exposure for the well-lit area is used, the area in deep shadow will be very underexposed. Fill-flash is used to record details in both extremes. (2) Film that has been either overexposed or underexposed can sometimes be corrected in the wet-chemistry darkroom. Similar corrections can frequently be made to digital images.

Exposure meter: Within the camera body, the reflective light meter measures the amount of light entering the camera and recommends a combination of exposure controls to ensure a properly exposed photograph results. The "standard" used by the light meter is an 18% gray card. If more light is detected by the meter, it will suggest an exposure combination that lets in less light, so that an overall 18% gray image results; if less light is detected by the meter, it will suggest an exposure combination that lets in more light, so that an overall 18% gray image results.

Exposure modes: There are various methods to determine an overall "proper" exposure.

- **Manual:** The photographer sets both the f/stop and shutter speed manually to determine the proper exposure.
- **Aperture priority:** The photographer sets the aperture, and the camera will automatically select a corresponding shutter speed to ensure there is a proper exposure.
- **Shutter priority:** The photographer sets the shutter speed, and the camera will automatically select a corresponding aperture to ensure there is a proper exposure.
- **Program:** The camera selects both the aperture and the shutter speed for a proper exposure.

Extension tubes: A lens-like tube inserted between the camera body and the lens that moves the prime lens elements farther from the film plane, resulting in magnification. Extension tubes of different sizes provide different magnifications.

Eyepiece cover: Many cameras come with an eyepiece cover, which is used to cover the viewfinder when the photographer will not be holding the camera close to his or her eye as is normal; for instance, when the camera is put on a tripod. If light were to come into the camera through the viewfinder, the light meter in the camera body might add it to the light coming into the front of the lens, which would result in an underexposure.

F

FFT (fast Fourier transform): Software that allows repetitive patterns in the background to be blurred or eliminated so they do not interfere with the visualization of the primary subject.

Field of view: The angle of view of any particular lens. A fish eye lens can view 180° of the scene, whereas a super-telephoto 1200-mm lens takes in only approximately 2° of a scene.

Fill-flash: A flash technique used to lighten details in deep shadows while other areas are brightly sunlit. Frequently, the fill-flash is set to be 1 stop less intense than the key/main light at the scene.

Film speeds: Numbers used to relate different films to their sensitivity to light; 100 ISO film is regarded as a "slow" film, requiring more light for a proper exposure; 1000 ISO film is regarded as a "fast" film, requiring less light for a proper exposure.

Filters: Supplemental filters can be used to achieve different photographic effects.

- **UV/Skylight/Haze/A1:** These filters are used to filter out UV light and some of the bright light coming from blue skies. Their main purpose, however, is to protect the outer element of the lens. If the lens accidentally contacts a hard object, the inexpensive filter will get scratched, dented, or broken rather than the more expensive lens.
- **IR/87 IR transmitting:** Blocks visible light while transmitting IR light for IR effects.
- **18A/UV transmitting:** Blocks visible light while transmitting UV light for UV effects.
- **Polarizer:** Blocks polarized light that produces reflections on glass and water. Blocks the glare of the sun, resulting in more accurate colors. Blocks sun glare, so that skid marks can better be seen and photographed. Allows the blue color of the sky to be deepened rather than washed out.
- **ND:** Neutral density filters are like sunglasses for the lens, blocking precise amounts of light without affecting the true colors of the scene.

- **Yellow, orange, and red:** Filters used with fluorescent photography, paired with UV, blue, and green light, respectively.

Flare: If the sun is in front of the photographer, direct sunlight can enter the lens, bounce around the various lens elements, and result in unwanted multicolored geometric artifacts in the photograph.

Flash exposure modes: When the electronic flash is used, there are different means to achieve a proper exposure.

- **Manual mode:** When the flash goes off, exposures are determined by altering the aperture of the lens or the intensity of the flash for the distance of the flash to the subject.
- **Automatic mode:** A sensor on the flash reads the amount of reflected light to decrease or prolong the flash duration for a proper exposure.
- **Dedicated mode:** The light meter in the camera body reads the reflected light coming in through the lens (TTL) to decrease or prolong the flash duration for a proper exposure.

Flash bracket: Usually a mount to affix the flash to, which is also connected to the camera, so that the flash does not have to be held separately. This frees up one hand to make focusing and focal length adjustments. It also positions the flash farther from the lens, reducing the chance of red-eye when taking photographs of people.

Flash diffuser: Direct flash is very harsh, and with many medium-range or close-up subjects, softening the direct flash results in more pleasing images. Diffusers range from multifaceted plastic clip-ons that reflect the flash wider than normal that soften the light, to milk-glass–type covers that enclose the flash head, to bounce surfaces that diffuse the flash beam wider than normal, to handkerchiefs that cover the flash head.

Flip-up flash: Many cameras have built-in flashes that are not visible when not in use. They will flip up when the light is dim and supplemental lighting is needed.

F-number: Like $f/8$, these numbers relate the size of the aperture opening to the focal length of the lens used. The equation, FLL/F-Number = DoD applies, where FLL is the focal length of the lens, and DoD is the diameter of the diaphragm. If the FFL is 50 mm and an $f/8$ is set on the camera, the DoD is: $50/8 = 6.25$ mm.

Focal length: The distance between the optical center of the lens and the film plane, when the lens is focused on infinity.

Focal length multiplier: Many digital SLR cameras use digital sensors that are smaller than a film negative. The result is that only the central portion of the image coming through the lens will be imaged, with the four sides of the original image being "cropped" off. A 1.5-focal length multiplier will make a

50-mm lens act like a 75-mm lens, enlarging the central portion of the image but capturing a narrower field of view of the subject.

Focal-plane shutter: SLR cameras have the shutter curtains immediately in front of the film. Other types of cameras may have leaf shutters located in the lens itself, rather than in the camera body.

Focus: Setting the camera so that light coming in through the lens converges to a precise point on the film plane.

- **Auto-focus:** When the shutter button is depressed half way down, the camera will automatically focus on the object appearing in the middle of the viewfinder or at the position of another sensor selected by the photographer.
- **Manual focus:** The photographer has to manually rotate the focus ring on the lens to focus the camera.

Front-curtain sync: With a dedicated flash, specifically designed to work with a specific camera, the flash can fire either once the shutter first completely opens (front-curtain sync) or just before the rear curtain begins to close (rear-curtain sync). This allows creative effects, such as motion streaks in front of or behind a moving object.

G

GN (guide number): An indication of the relative power of a particular flash unit.

Graininess: When film images are enlarged too much, the individual silver halide crystals making up the image separate and become visible as discrete specks. The degraded image is said to be "grainy."

Gray card: The 18% gray card is the standard used by camera reflective light meters to determine the proper exposure settings on the camera. If a particular scene is not reflecting this "norm," a proper exposure can be determined by aiming the camera at an 18% gray card and taking a meter reading of it.

H

Histogram: A bar graph depicting the relative quantity of dark, medium gray, and light portions of an image to give the photographer a quick assessment of the exposure, contrast, and dynamic range of an image.

History: As digital images are "processed" to improve the visibility of the main subject matter, the photographer should keep notes about both the type of image processing used and the amount of the processing applied to the image. Some software will automatically record these data and link them to the image.

Hot-shoe: Many cameras have a mount on the camera body where the electronic flash will attach so that sensors on the mount and flash will electronically link both.

Hyperfocal distance: When one is hyperfocal focusing to maximize the depth of field in an exterior overall photograph, this is the precise distance the distance scale is aligned with. For example, when hyperfocal focusing with an f/8, resulting in a depth of field range of 15' to infinity, the hyperfocal distance is 30'.

Hyperfocal focusing: This is the method to maximize the depth of field when infinity must also be in-focus. Take a meter reading to determine the f/stop required for the exposure, then align the infinity symbol on the focusing ring with the corresponding f/stop on the depth-of-field scale.

I

Image stabilization: Sometimes located in the lens and sometimes in the camera body, this is technology allowing a camera to be used with a slower shutter speed without incurring blur from camera shake.

Incident light meter: This is a hand-held light meter able to determine the exposure settings required for a proper exposure. This meter is positioned at the location of the subject to be photographed, reading the light falling on the subject.

Infrared: Visible light falls between 400 and 700 nanometers on the electromagnetic spectrum. Infrared light used in photography is between 700 and 1100 nanometers.

Interpolation: A process that enlarges an image beyond its original size by adding extra pixels between the original pixels that are separated to fit a larger area.

Inverse square law: Light diminishes by the inverse square of the distance traveled. As the distance light travels doubles, its intensity is quartered; $2\times$ becomes $(1/2)^2$, which is 1/4th. As the distance light travels is halved, its intensity is quadrupled; $x/2$ becomes $(2)^2$, which is 4.

ISO film speed settings: International Standards Organization. Previously, film speeds had been rated by ASA (American Standards Association) numbers.

J

Jaggies: These are the visible digital "stair-steps" of diagonal lines or edges. Also referred to as "aliasing," this is more pronounced with fewer larger pixels to make up the image or when the digital image is enlarged too much.

JPEG (joint photographic expert group): A "lossy" type of storage of a digital file. Similar color and contrast details can be permanently discarded to reduce the file size, allowing more images to be stored in the same space.

K

Kilobyte: A data unit of 1024 bytes.

L

LCD: Liquid crystal display.

LED panel: Light-emitting diode.

Lenses

- **Close-up lens/macro lens:** Designed to produce magnifications allowing small items the size of a fingerprint to fill the frame.
- **Normal lens:** A 50-mm lens for a 35-mm SLR camera system. Normally, a focal length equivalent to the diagonal of the film.
- **Telephoto lens:** A lens larger than 50 mm.
- **Wide-angle lens:** A lens smaller than 50 mm.
- **Zoom lens:** A lens that can be set to a variety of focal lengths.
- **Achromatic lens:** A lens corrected for chromatic aberration.
- **Anastigmat:** A lens corrected for the aberration "astigmatism."
- **Aplanat:** A lens corrected for spherical aberration.
- **Apochromatic:** A lens that can focus all colors of visible light at the exact same location on the film plane.
- **Aspherical lens:** Reduces spherical aberrations.

Lens shade/hood: An extension of the end of the lens designed to block direct sunlight entering the lens creating flare.

Lossless compression: A compression format that reduces the file size without permanently discarding data, like TIFFs.

Lossy compression: A compression format that permanently discards data to achieve smaller file sizes, like JPEGs.

M

Macro lens: A lens designed to provide magnifications of up to 1:1, where the true size of the object is captured life size on the negative.

Manual exposure mode: The photographer sets both the shutter speed and the aperture manually to achieve a proper exposure.

Manual flash: When the flash goes off, exposures are determined by altering the aperture of the lens or the intensity of the flash for the distance of the flash to the subject.

Manual focus: The focus ring of the lens is manually adjusted to establish focus.

Megabyte: This is usually 1,048,576 bytes; however, it is sometimes interpreted as 1 million bytes.

Metadata: Digital cameras often record more than just the digital image. With a digital image this can include, but is not limited to, exposure settings, information about the camera used, and a record of in-camera processing performed on the image.

Metering: Cameras have various methods to determine the proper exposure of an image.

- **Averaging/center-weighted:** The camera will evaluate the light coming from a large oval in the center of the field of view.
- **Spot:** The camera will only register the light coming from a small circle in the center of the viewfinder.
- **Matrix:** The viewfinder is segmented into various geometric designs, and then the light in each segment is evaluated and weighed on the basis of the computer software in the camera. This usually provides the most precise exposure determination.

Mid-roll change/rewind: Some cameras will allow a roll of film to be rewound when only partially used. To enable the film to be reloaded, the film leader strip is not completely rewound all the way into the film canister. Then, once reloaded, the film can be advanced to the frame last used, and then additional images can be captured on it.

Mirror lock-up: When SLR cameras capture an image, the mirror in the camera body swings up to allow the light entering the camera through the lens to strike the film. This movement generates a small amount of camera shake. When the camera is on a tripod, some cameras allow the mirror to be locked-up before the shutter is depressed. This reduces camera vibrations and potential blur.

Monopod/unipod: A camera support with a single leg as opposed to the three legs of a tripod. Helps steady the camera so that slower shutter speeds can be used, or longer telephoto lenses can be used.

Motor drive: Some cameras will automatically advance the film after each shot. Other cameras will have to be advanced by manually cranking a lever between each shot.

N

Nanometer: One billionth of a meter. Light waves in the visible light range of the electromagnetic spectrum are measured in nanometers. This is the distance between light wave peaks

NEF: Nikon electronic format.

Negative film: Normal print film, where tones are reversed so that highlights appear dark and shadows appear light on the developed negatives.

Neutral density filter: Sunglasses for the camera. These filters uniformly reduce the light coming in to the camera without altering relative colors.

NiCd: Nickel cadmium camera battery.

NiMH: Nickel-metal hydride camera battery.

Noise: Digital image degradation seen as monochromatic grain and/or as colored waves. It can also appear as random groups of red, green, or blue pixels.

Normal lens: For a 35-mm SLR camera, a 50-mm lens. Also, thought of as the diagonal of the film used.

Normal room: Manual flash exposure calculations presume the flash is operated in a "normal room." This is an average-sized living room or bedroom with white ceilings about 8' to 9' high, and light-colored walls. Such ceilings and walls reflect some light toward the subject being photographed. Manual flash exposure calculations presume this additional light is available.

O

Oblique flash: Flash used at different angles, usually to produce side-lighting shadows.

OTF (off-the-film metering): Sometimes called dedicated flash or TTL flash. Light reflected off the subject is reflected back to the camera, comes through the lens, and reflects off the film where the light meter is located.

Out of focus: An image, or a part of an image, is perceived to be out of focus when the light striking the film forms a circle larger than the circle of confusion perceived by the eye to be in-focus.

Overexposure: When the image is lit with more light than normal.

P

Painting with light: A technique to light a large dim scene with multiple flashes while the camera has the shutter locked open so the multiple flashes light a single negative.

Panning: Moving the camera to track with a moving subject to reduce the blurriness of the image.

PC cord: A coiled electronic connection between the camera body and a flash unit used off camera. Also called a sync cord, "PC" is an acronym for Prontor-Compur, the two European shutter makers who, long ago, devised a system that allowed a cable from a studio electronic flash unit to synchronize with a camera shutter.

Pentaprism: A system of internal mirrors designed to allow the photographer to see the same image coming in through the lens while looking into the viewfinder.

Photogrammetry: The method by which accurate measurements may be extrapolated from photographs.

Picture element: Digital camera sensors are made up of individual picture elements, also known as pixels.

Pincushion distortion: The lens distortion that shows linear elements at the periphery of an image that appear to be bending in to the center of the image. Usually most manifest with larger telephoto lenses.

Pixel: A shortened version of "picture element."

Pixelization: When digital images are enlarged too much, the individual pixels making up the image become large enough to see. The image looks boxy, without smooth transitions.

Polarizing filter: Blocks polarized light that produces reflections on glass and water. Blocks the glare of the sun, resulting in more accurate colors. Blocks sun glare, so that skid marks can better be seen and photographed. Allows the blue color of the sky to be deepened rather than washed out.

PPI: Pixels per inch.

Predictive auto-focus: When photographing moving subjects, some cameras enable the camera to predict the location of the moving subject and focus where the object will be when the shutter is fully depressed, rather than where it was then the shutter was depressed only halfway.

Program exposure mode: The camera will select both the shutter speed and the aperture to ensure a proper exposure.

Pulling: Rating a film at a slower ISO rating than it really is to increase color saturation and contrast. For instance, using a 400 ISO film but setting the camera to ISO 320.

Pushing: Rating a film at a faster ISO rating than it really is to extract detail from an underexposed image. For example, using a 3200 ISO film but considering it as ISO 12800, 2 stops faster.

R

RAM: Random access memory.

Raw file format: RAW files store the unprocessed image data. Most digital cameras set to capture images as TIFFs or JPEGs use in-camera adjustments to set various image variables: white balance, exposure value, color values, contrast, brightness, and sharpness. RAW image files must be processed with special software before they can be viewed or printed. The advantage is that you have the ability to alter the image variables as you see fit before you convert these data into the standard JPEG or TIFF format.

Rear-curtain sync: Cameras with dedicated flash units can be adjusted to have the flash fire just before the rear-curtain begins to close. Usually the flash fires immediately after the front curtain has opened all the way. This is used for creative photography where streaking is desired in front of or behind moving objects.

Reciprocity failure: Many different combinations of shutter speeds and apertures will result in the same exposure. Increasing one variable while decreasing the other results in the same exposure. When exposures are either too long or too short, this concept of reciprocity breaks down. Usually exposures longer than 1 second or shorter than 1/1000th of a second will suffer from reciprocity failure. Either extreme results in underexposures.

Reciprocity law: Many different combinations of shutter speeds and apertures will result in the same exposure. Increasing one variable while decreasing the other results in the same exposure.

Red-eye: When the flash is used close to the lens, the light will bounce off the red blood vessels at the back of the eye, and the result will be subjects with red eyes.

Reflective light meter: The light meter in the camera body is a reflective light meter; it reacts to the light reflecting off the subject coming into the camera through the lens.

Refraction: As light moves from one medium to another, it bends. Refraction also is the cause of underwater objects appearing closer to the camera than they really are.

Remote flash cord: An advancement to the original PC cord, this allows the camera body and the flash to interact in additional ways, besides firing the flash when the shutter has opened. The flash will know what ISO film has been loaded into the camera body. The flash will alter its flash throw as different focal lengths are selected on a zoom lens. Changes to the intensity of the flash can be set on the camera body.

Re-sampling: Changing the resolution of an image either by removing pixels (lowering resolution/compressing the image) or adding them by interpolation (increasing the printable size of the image).

Resizing: Changing the size of an image without altering the resolution. Increasing the size will lead to a decrease in image quality.

Resolution: The capability of a photographic reproduction to distinctly record two separate but adjacent elements of an image. In still photography resolution is frequently measured in line pairs for film and pixels per inch for digital files.

Ring flash: A flash unit encircling a lens emitting light onto all sides of a small object. A soft light producing almost no shadows.

ROM: Read-only memory.

Rotation: If a digital image appears slightly tilted, it is possible to turn the image on one of its axes, correcting the composition error.

Rule of thirds: To maximize the depth of field when focusing on medium-sized areas, compose the shot, make a mental note of the distance between the bottom and top of the viewfinder. Focus at a distance point midway between the top and bottom of the viewfinder. Geometrically, this is approximately the same as focusing one third of the way into the length of the scene.

S

Self-timer/delayed shutter times: Many newer cameras allow the camera to set the shutter to trip either 2 seconds or 10 seconds after depressing the shutter button. This allows the camera on a tripod to settle down after the shutter button is depressed, reducing the chance of blur from camera shake.

Shutter: This controls the amount of time light that is allowed to reach the film.

Shutter release cable/remote shutter cable: When the camera is mounted on a tripod, the shutter button should not be depressed manually; doing so will induce movement to the camera and blur the image as a result. These cables dampen this movement.

Shutter priority: The photographer sets the shutter speed desired, and the camera automatically sets the corresponding aperture to provide a proper exposure.

Shutter speed: The time the shutter is allowed to remain open, usually expressed in fractions of a second. Most cameras allow some shutter speeds in full seconds also.

Silver halide crystals: This is the light-sensitive part of common photographic films; the usual light-sensitive chemicals are silver chloride, silver bromide, and silver iodide.

Single-lens-reflex (SLR) camera: This is a camera in which a system of mirrors and prisms shows the user the image in the viewfinder exactly the way the lens captures the image.

Skylight filter: A filter designed to filter out some of the intense light from the sky so that the sky does not appear overexposed. One of several filters commonly used to protect the outer element of the lens.

Slow film: ISO speeds of 100 or less.

Slow sync: Very slow shutter speeds used with a flash, less than 1/60th of a second. If the remote shutter cord broke or was missing while at a nighttime scene, the flash could be fired manually when a 2-second shutter speed was set for the camera on a tripod, allowing the flash to be fired manually while the shutter was open.

SmartMedia™: One of many types of digital memory cards used in some cameras.

Spot metering: The camera will only register the light coming from a small circle in the center of the viewfinder.

Stair stepping: Also called aliasing or jaggies.

Stop down: Normally meaning a change from a wider aperture to a smaller aperture.

Supplementary lenses/close-up filters: A set of three filters that can be used separately or in combination, which produce varying amounts of magnification. Usually designated as a +1, +2, and +4.

Sync cord: A coiled electrical connection between the camera body and the detached flash unit, allowing the flash to fire when the shutter is open.

Synchronize/sync shutter speed: The shutter speed of a particular camera that ensures the flash fires when the shutter is completely open. Sometimes 1/60th, 1/90th, 1/125th, or 1/250th of a second.

- **Front curtain sync:** Dedicated flash units allow the flash to fire immediately after the front curtain has completely opened. This is the default for most flash units.
- **Rear curtain sync:** Dedicated flash units sometimes allow the flash to fire just before the rear curtain begins to close. A creative photography option, allowing streaking in front of or behind moving objects.

T

T ("time" SS setting): The shutter opens when the shutter button is first depressed and closes the second time it is depressed, allowing for various timed exposures.

Tagged image file format: TIFF file format. An uncompressed image file format that is lossless.

Teleconverter: A supplemental lens used with telephoto lens to make them act like longer telephoto lenses. Frequently available in 1.4× and 2× powers, they would make a 100-mm lens act like either a 140-mm or 200-mm lens, respectively. Usually a less expensive alternative to buying a longer focal length lens.

Telephoto lens: For a 35-mm SLR camera, a lens with a longer focal length than 50 mm.

Through-the-lens metering (TTL): TTL metering usually refers to a dedicated flash unit that has its flash duration determined by the reflected light from the scene being metered by the light meter in the camera body.

TIFF: Tagged image file format.

Tripod: A three-legged camera support enabling slower shutter speeds to be used without blur.

TTL flash: TTL flash usually refers to a dedicated flash unit that has its flash duration determined by the reflected light from the scene coming through the lens (TTL) being metered by the light meter in the camera body.

Tv (time value) (shutter priority mode): The photographer selects the shutter speed as desired, and the camera will automatically select the aperture to provide a proper exposure.

U

Ultraviolet: Light energy of the electromagnetic spectrum, invisible to the eye, in the 100-nm to 400-nm frequency.

Underexposure: An image darker than normal, or what is considered properly exposed.

Unipod: A camera support with a single leg as opposed to the three legs of a tripod. Helps steady the camera so that slower shutter speeds or longer telephoto lenses can be used.

Unsharp mask: An image-sharpening filter available with image processing software.

USB: Universal serial bus.

UV filter: Most UV filters block UV light to help prevent overexposures when there are large amounts of sky in the field of view. UV filters are also used to protect the outer element of the lens. The 18A filter, in contrast, transmits only UV light while blocking all visible light.

V

Variable power: Many flash units allow a manual flash unit to be used in varying increments, from 1/2 power to 1/128th power, to enable the flash to be used at closer distances to the subject.

Viewfinder: The opening in the rear of the camera allowing the photographer to compose and focus the photograph. In 35-mm SLR cameras, a series of mirrors allows the photographer to view the subject exactly the same way the film or digital sensor will record it.

W

White balance settings: Digital cameras enable the camera to be adjusted for the ambient lighting to ensure white objects are truly white and improper color tints do not adversely affect the image. When a white subject is captured as white, other colors are also captured faithfully.

- **Auto:** If the scene has a white surface within view, the camera will automatically adjust color levels to ensure it is truly white in the image.
- **Sun:** When sunlight is the primary lighting, whites will be recorded as truly white.
- **Flash:** When electronic flash is the primary lighting, whites will be recorded as truly white.
- **Shade:** When shade is the primary lighting, whites will be recorded as truly white. Otherwise there is a tendency for a bluish tint.

- **Tungsten:** When household light bulbs with tungsten filaments are the primary lighting, whites will be recorded as truly white. Otherwise there is a tendency for a yellowish tint.
- **Fluorescent:** When fluorescent lighting is the primary lighting, whites will be recorded as truly white. Otherwise there is a tendency for a greenish tint.

Wide-angle distortion: Wide-angle lenses tend to elongate the perceived foreground to background distance.

Wide-angle lens: With a 35-mm SLR camera, a wide-angle lens is one with a focal length smaller than 50 mm.

X

X sync: This is an older flash shutter speed that causes an electronic flash to fire in synchronization with the shutter. Some older cameras had an "X" designation on the shutter speed dial to more easily set the shutter speed whenever using flash.

Z

Zone focusing: A focusing technique to ensure a particular area of the scene is included in the camera's depth-of-field range. If the lens has a depth-of-field scale, the rear distance needing to be in focus is aligned with the f/stop on the depth-of-field scale required for a proper exposure. Comparing the depth-of-field scale to the distance scale shows the resulting depth-of-field range.

Zoom lens: A lens with multiple focal lengths.

1:1: Usually, a magnification ratio ensuring the item of evidence is life-sized on the negative. In this case, a single digit fingerprint, or an object the size of a nickel, will fill the frame of the viewfinder. Other times, it is an indication that a print be created so the object is life-sized in the print.

35-mm film: Film that is 35 mm in width, designed to be used with 35-mm cameras.

INDEX

A

Aberrations, 131, 192–194
  chromatic, 194
  coma, 195–196, 196f
  images degraded by, 209–211
  lateral chromatic, 195f
  longitudinal chromatic, 195f
  spherical, 195f
Accident scenes
  aerial view of, 548f
  air bags deployed at, 526f
  approaching traffic viewpoint at, 518f
  bodily injuries at, 527–529
  critical elements in, 514–515
  driver eye level photographs of, 518
  injuries at, 529f
  law enforcement protecting, 514
  nighttime photography at, 530–533,
    531f, 532f, 533f
  noncritical elements of, 515–516
  pedal mark photos at, 525f
  photographs at, 519f, 521f
  polarizer filter used at, 108–109
  tire damage at, 524f
  tire tracks at, 519–520
  traffic control devices at, 516–517
  vehicle approach to, 516
  vehicular damage at, 522
  windshield damage at, 530f
  witness viewpoints at, 518–519
Acutance, 131
ADAMS. *See* Authenticated Digital Asset
    Management System

Aerial photography, 546–553
  of accident scene, 548f
  balcony/roof top, 547f
  camera controls for, 550–553
  homicide, 546f
  shutter speeds for, 550
  sniper shooting sites from, 547f
AFIS. *See* Automated Fingerprint
    Identification System
Air bags, 526f
Airbrushing, 467–468, 467f
Airy disk, 201–202, 201f
ALS. *See* Alternate light source
Alternate light source (ALS), 71
Ambient/existing light, 50–56, 53–56,
    573f
American Society of Crime Laboratory
    Directors, Laboratory Accreditation
    Board (ASCLD/LAB), 476
Amplitude, 384f
Angles of view, lenses, 174f
Angular viewpoint, 26f
Animal bites, 368f
Aperture priority exposure mode, 91
Apertures
  bounce flash and, 299
  circles of confusion and, 143f
  critical comparison photography and,
    204
  digital cameras and, 45
  f/stops and, 37–44, 40f
  lenses designated by, 172–173
  light penetrating, 198f

Apertures (*Continued*)
  narrow, 39f
  small, 140f, 166, 167
  in surveillance photography, 533f
  wide, 18f, 39f, 93, 140f, 143, 143f, 144,
    147, 158, 299, 302
Argon ion laser, 388, 388f
Artifacts, 371f, 500
ASCLD/LAB. *See* American Society of
    Crime Laboratory Directors,
    Laboratory Accreditation Board
Authentic photographs, 576–578
Authenticated Digital Asset Management
    System (ADAMS), 510
Authentication, 596–597, 607–608
Automated Fingerprint Identification
    System (AFIS), 125–126, 488, 489
Automatic exposure mode, bracketing in,
    93–97
Automatic flash exposure mode, 253–259
  automatic flash sensor benefits of,
    253–254
  flash calculator dial and, 255f
  helpful tips for, 259
  non-flash, 256–257
Automatic flash head, 240f
Automatic flash sensor, 253–254
Automatic focus, 134–136
Autopsy, 372

B
Backgrounds, 10f
  composition awareness of, 347–350, 347f
  distracting, 361f
  eliminate irrelevant material in, 8–10
  elongation of, 183–186
  polarizer filters used with, 112f
Back-lit scenes, 83–85, 84f, 265
Backscatter, 560–562, 560f
Ball-bearing, laser lighting with, 206f
Bank robbery, 290f
Barrel distortion, 182–183, 210, 210f
Base length, 430f
Battery consumption, 241
Bayer pattern, 492

Bite marks, 284f, 285f, 286f, 287f, 571f
  capturing images of, 494–495
  photographing, 283–286
  on purses, 287f
Bits, 479–482, 480f, 481f
Black light, 391, 395
Blood
  chemical reagents of, 411–416
  luminol and, 412f, 413f, 414f
  UV light and, 396, 396f
BlueStar, 415
Blur, 61–64
Bodies
  at autopsy, 372
  camera directly above, 364–367, 366f, 367f
  composite shots of, 363f
  crime scene photography with, 357–380
  as fire victim, 370f
  gunshot wound to, 369f, 374f
  side view of, 360f, 361f
Bodily injuries, 527–529, 529f, 593f, 594f
Body area, 377, 379f
Body fluids, 407, 409f
Body panorama, 358, 359f
Bone, 416, 417f
Bounce flash
  dim light and, 298–302
  direct flash hot spot and, 298f
  glass table top and, 298f
  in interior overall photographs, 344f
  mirrors and, 297f, 298f
  optimal, 300f
  when to use, 295–302
  wider aperture for, 299
Bracketing, 81
  in automatic exposure mode, 93–97
  with electronic flash, 93, 97
  in manual exposure mode, 89f, 92–93
  in manual flash exposure mode,
    241–245
Buildings, 23f, 335f–336f, 340f, 634f
Bullet, 164f, 165f, 220f, 353f
Burned writing, 406f
Burning, 60
Bytes, 479–482

C

CALEA. *See* Commission on Accreditation for Law Enforcement Agencies

Cameras. *See also* Digital cameras; Film cameras
  body under, 364–367, 366f, 367f
  controls on, 550–553
  DX coding systems in, 50
  electronic flash/remote cord connecting, 234f
  flash with, 231–236, 236f, 237f
  IR capable, 399f
  ISO settings on, 51f
  with lenses, 16f
  operators of, 310f
  point-and-shoot, 1–4, 3f
  PWL variables of, 305–307
  schematic of, 34f
  shake elimination of, 35
  SLR, 33
  subject distance to, 163–165
  tripod mounting of, 364f
  for underwater photography, 558f
  zoom lens/extension tube for, 191f

Candy bar, bite marks, 287f

Canon equipment, 233f, 235f, 399f, 538f

Cardinal rules
  of composition, 4–6
  of crime scene photography, 25
  fill the frame as, 6–16
  of film frame parallel, 20–25
  maximizing depth of field of, 16–20

Case law citations, 598–624
  authentication in, 607–608
  digital manipulations in, 615–621
  exposure and, 600–601
  fair/accurate representation of scene in, 598–606
  photographic enlargements in, 621–623
  prejudicial /inflammatory photographs in, 609–615
  printing errors in, 623
  relevant/material photography in, 608–609

Castone, 286f

Cathode ray tube (CRT), 483

CCD (charge-coupled device), 481, 492, 493f

Center-weighted light meters, 74f, 75

Chain of custody, 321–322

Chemicals, 407, 411–416

Chromatic aberrations, 194

Circles of confusion, 131, 143f, 144f

Circular aperture, 198f

Circular polarizer filter, 112

Close-up filter set, 191–192

Close-up photography, 344, 352–353, 357f
  altered, 356–357, 357f
  bullet filling frame in, 353f
  film plane parallel and, 25
  filters in, 353f
  with labeled scale, 326–331, 354, 355f, 357f
  partial scale in view in, 355, 355f

Cluster of dots, 484

CMOS (complimentary metal-oxide semiconductor), 481

Coaxial lighting, 629–630, 630f, 631f

Coins, 138f, 188f

Color
  accuracy, 572, 601–602
  channel, 481f
  lighting for, 53
  polarizer filter enriching, 108

Coma aberrations, 195–196, 196f

Commission on Accreditation for Law Enforcement Agencies (CALEA), 127

Composition
  background awareness for, 347–350, 347f
  cardinal rules of, 4–27
  elements of, 26–27
  midrange, 348f, 349f
  proper, 10f
  scissors improper, 10f

Compression, digital data, 592–593, 593f–594f

Consent, 583

Coordinate system (X-Y notation), 429, 438, 464

Corpuscular Theory of Light, 197, 200, 206

Court admissibility, 506–507, 508, 569–570
Creative photography, 574
Crime scene
  accurate representation of, 570–576
  documentation of, 582
  electronic flash/pwl used at, 303f
  exterior overall photographs of,
    332–338, 333f
  interior overall photographs of,
    338–343, 341f
  normal/non-normal lighting at, 76–85
  perspective disc photogrammetry used
    at, 443f
  perspective grid photogrammetry used
    at, 427f
  PWL variable at, 302–318
Crime scene photography
  authentic, 576–578
  with bodies, 357–380
  body area isolated in, 377
  cardinal rules of, 25
  chain of custody through, 321–322
  court admissibility of, 569–570
  creative photography v., 574
  exterior overall photographs for, 338f
  fill the frame for, 6–16
  filters used for, 390
  focusing in, 121
  grid system used in, 457–459
  key concepts of, 6
  maximizing depth of field for, 16–20,
    145, 165–168
  midrange body shot at, 359f
  photogrammetry used in, 421–425
  purpose of, 582
  recreations for, 578
  shutter speed/f/stops used for, 69
  UV light and, 389
  wide aperture options for, 93
Critical comparison photography
  aperture settings for, 192
  film used for, 124–125
  ISO settings for, 50
Cropped images, 175f
CRT. See Cathode ray tube

D
Dark noise, 71, 71f, 72f
Darkroom, 59
Daytime surveillance photography,
  537–542
Decision-making process, 49, 131
Dedicated flash exposure mode, 259–261,
  260f
Defense attorney
  depth of field and, 17–18
  interpolated images and, 589
  objections of, 121
  photograph requests of, 378
  scale objections by, 274
Defense wound, 373f
Depth of field (DOF), 574–575
  defining, 132
  factors influencing, 165–168
  focus calculated for, 536
  focus-f/stops and, 143–144, 145
  f/stops influencing, 30, 69,
    157–160
  impossible range for, 209f
  increasing, 186
  lens choice influencing, 161–165,
    162f
  maximizing, 16–20, 145, 165–168
  preview button for, 169f
  telephoto lenses narrowing, 180–181
  typical scale of, 18–20, 19f
Depth of field scale (DOF scale),
  142, 144f, 150–153
Dial indicators, sync speed, 219f
Diaphragm, 37–39
Diffraction, 196–209
  ball-bearing showing, 206f
  influence of, 204
  linear opaque edge with, 207f
  paper clip showing, 207f
  single discrete point of, 203f
  summarized, 208–211
  two discrete points of, 203f
Digital arsenal, 471–472, 472f,
  509–510
Digital camera, IR light and, 399f

Digital cameras
    apertures changed on, 44
    evidence shot with, 126–127
    EXIF data maintained of, 470
    film cameras v., 123–125
    high resolution photographs required
        from, 472f, 497
    image quality duration of, 491
    information captured by, 470–471
    interpolation in, 499f
    ISO equivalents of, 47–50
    megapixels figured for, 479–485, 492, 497
    reciprocity failures and, 71
    resolution of, 127–131
    storage types for, 49f
    underexposures with, 59
    white balance on, 56
Digital Crime Scene software, 510
Digital equipment, 491–494
Digital evidence, 476–478
Digital Image Management System
        (DIMS), 510
Digital images, 468, 472f
    authentication/encryption of, 596–597
    bit depth of, 497–498
    case law citations concerning, 598–624
    concepts of, 478–491
    digital evidence v., 476–478
    history maintained of, 594–596
    JPEG format of, 500–503
    legal implications of, 585–597
    lossless compression formats preferred
        for, 506
    management of, 470
    manipulation of, 587–593, 588f
    noise in, 499
    planning, 490f
    proficiency testing for, 596
    RAW format of, 503–504
    resolution/image quality in, 494
    rotations of, 590–592, 590f–591f
    storage/management of, 498–510,
        509–510
    strategy for, 489–491
    TIFF format of, 505–506

Digital information, 479f
Digital processing, of underexposures,
        545, 545f
Digital technologies, 471, 473
Digital video recorders, 471
Dim light, 298–302
DIMS. See Digital Image Management System
Diopter dial, 157f
Direct flash
    bounce flash and, 298f
    from electronic flash, 270f
    in interior overall photographs, 344f
    in underwater photography, 561f
Direct light, 269–273
Dirty snow phenomenon, 368
Distance scale, 150–153
Distances, 242–243, 564
Distortion. See also Barrel distortion;
        Perspective distortion; Pincushion
        distortion
    barrel, 182–183, 210
    building showing barrel, 210f
    building showing pincushion, 211f
    correction of, 632–634, 634f
    focal length influencing, 173
    from lenses, 350–352
    pincushion, 210–211
    telephoto lenses causing, 180, 395
    wide-angle lenses causing, 183–184, 350
Dodging process, 60
DOF. See Depth of field
DOF scale. See Depth of field scale
Dots, and pixels, 483–489
Dots per inch (dpi), 478, 479, 488
DPI. See Dots per inch
Drug residues, 409
Dust print lift, 296f
    flashlight movement for, 293f
    left to right reversal, 291f
    photographing, 290–295
    positive/negative, 290, 292f
    recommended sequence for, 294
    from shoe prints, 291f
    from skin, 295f
    from t-shirts, 293f

DX coding system, 50, 52f, 102–103
Dye sublimation printers, 475–476, 486, 489
Dynamic range, 499

E
18% gray card, 88f
   green grass equal to, 87
   lighting with, 72–74, 73f
   metering to, 87
   normal/non-normal crime scenes and, 76–85
Electromagnetic spectrum (EMS), 384–390, 385f
Electronic flash. *See also* Flash head
   battery consumption of, 241
   beyond 30′ nonexistent for, 255
   bounce flash and, 298f
   bracketing with, 93, 97
   built-in, 261
   cameras using, 231–236, 234f, 236f, 237f
   Canon equipment with, 233f, 235f
   crime scene using, 303f
   direct flash from, 270f
   exposure and, 222–223
   fill-in, 58, 86f, 261–269, 264f, 266f, 268f
   film frame parallel and, 22f
   film used with, 224
   filters for, 245f
   flash bracket used with, 235–236, 236f
   flashlight instead of, 283
   ghost and, 221, 221f
   glass table top and, 298f
   guide numbers for, 216–218
   hot shot mounting of, 231f
   manual flash used with, 224–229
   normal room using, 229–231
   proper colors from, 53f
   PWL variables for, 307–311
   recharge/recycle time of, 241
   reducing intensity of, 244f
   reflective ultraviolet photography using, 396
   sensor eye on, 228f, 229f
   shadow control and, 11–14, 13f
   from 6′ away, 233f
   subject matter/ distance and, 232
   subject motion frozen by, 64
   sync speeds of, 218–223
   from 3′ away, 232f
   true colors from, 55f
   underwater photography using, 557f, 559f
   Vivitar 285 HV flash as, 227f, 228f, 229f
Electronic shutter release cable, 36f
EMS. *See* Electromagnetic spectrum
Encryption, 596–597
Equations
   f/stop, 37–39
   ISL, 248
Equipment, photographic, 1–4
Evidence
   digital cameras shooting, 126–127
   digital image as, 476–478
   labeled scales same place as, 327
   on mirrors, 631–632, 632f, 633f
   natural grid photogrammetry with, 447f, 453f
   perishable, 515
   perspective grid photogrammetry and, 427–28, 434f, 437f, 439f, 440–441
   photography and, 466–471, 569–582, 583–585
   prejudicial, 580–582, 581f
   UV/IR light and, 153–156
   vertical walls with, 627–629, 628f
EXIF data (exchangeable image file format), 470
Exposure
   ambient/existing light and, 53–56
   aperture mode of, 91
   aperture priority mode of, 91
   black sweater and, 82f
   case law relating to, 600–601
   changing, 59–60
   darkroom corrections of, 59
   decision-making process for, 49
   electronic flash and, 222–223

film camera errors in, 102–103

film speeds influencing, 44–50

fluorescence evidence and, 407

f/stops influencing, 37–44

latitude, 56–60

lighting and, 50–53, 57–58, 76–77

manual mode, 89–90, 89f, 92–93

modes of, 89–92

program mode, 90–91

proper, 26

shade and, 58f

shutter mode of, 91–92

shutter speed and, 33–37

sunlit grass and, 58f

tricky scenes and, 85–92

Exposure compensation dial, 94–97, 95f,
    102–103

Exposure stops, 30, 31f, 32f

Exposure triangle, 30–60, 34f

Exposure variables, 33–56

    ambient/existing light as,
      50–56

    film speeds as, 44–50

    f/stops as, 37–44

    light as, 50–53

    shutter speed as, 33–37

Extension tubes, 190–191, 191f

Exterior overall photographs, 81

    of building, 335f–336f

    of crime scene, 33f, 332–338

    of foot prints, 338f

    of single family home, 334f

    as street signs, 332f

Eye cup cover, 112–115, 114f

F

F/16 sunny day rule, 97–101, 98f, 99f,
    116, 167

Fair/accurate representation of scene,
    570–576, 598–606

Faxes, resolution of, 126–127

Fibers, 397f, 407–409, 410f

Field of view

    extraneous elements in, 9–10

    primary subject and, 27

of telephoto lenses, 179

wide-angle lens and, 181–183

Fill the frame, 6–16

    close-up photography and, 353f

    irrelevant material eliminated to, 8–16

    with pistol, 7f

    with primary subject, 7–8, 11

Fill-in flash, 58, 86f, 261–269, 264f, 266f, 268f

Film

    critical comparison photography with,
      124–125

    DX codes/canisters for, 52f

    emulsion on, 47, 70, 71

    life span of, 491

    photo identifier on, 322–324

    recorders for, 468–469

    types, 49f

Film cameras

    digital cameras v., 123–125

    exposure errors with, 102–103

    overexposures with, 58–59, 93, 268f

Film frame parallel

    building facade and, 23f

    cardinal rule of, 20–25

    exception to, 21

    flash and, 22f

    isosceles viewpoint and, 24f

    pistol and, 26f, 133f

Film photography

    case law citations concerning, 598–624

    critical comparison photography using,
      124–125

    electronic flash used with, 224

    extreme lighting conditions and,
      57–58

Film plane parallel, 133f, 336f

    close-up photography and, 25

    in full-face photography, 363f

    in interior overall photographs, 343f

    and scissors, 9f

Film speeds

    as exposure variables, 44–50

    f/stop corrections and, 58–59

    knife and, 48f

    light sensitivity and, 44–46

Filters. *See also* Neutral density filter;
      Polarizer filters
   close-up photography with, 191–192, 353f
   commonly used, 103–112
   crime scene photography using, 390
   dent to, 104f
   for electronic flash, 245f
   lenses protected by, 103
   polarizer, 104–112
   screwing on, 105
   for UV light, 103, 104f, 105f, 394f
Fingerprints, 126f. *See also* Bite marks;
      Shoe prints; Tire tracks
   1:7 magnification of, 190f
   capturing images of, 494–495
   image capturing of, 494–495
   on negative, 138f
   1:1 magnification of, 190f
   rhodamine 6G and, 408f
   UV fluorescence of, 397f
Fire victim, 370f
Flash angles, 309f
Flash bracket, 235–236, 236f
Flash calculator dial, 225–227, 226f, 227f,
      242f, 301f
   auto flash exposure ranges and, 255f
   f/stop-distances related by, 242–243
   ISO settings on, 226f
   manual flash and, 230
Flash exposure mode
   automatic, 253–259
   dedicated, 259–261, 260f
   manual, 241–245
Flash head, 231–236
   automatic, 240f
   cameras and, 236, 237f
   filters for, 245f
   focal length and, 238–239
   manual, 239f
   telephoto lens used with, 238–239
Flash operator, 307–311, 314f
Flash sequence, 273–277
Flashlight
   bite marks and, 284f, 285f
   directing light with, 286–288

dust print lift photographing with,
      293f
   indented writing lit by, 288f
   instead of flash, 283
   panning with, 285f
   as PWL light source, 317
Florescent lighting, 54–55
Fluorescence, 387, 411f
   body fluids, 409f
   of bone, 417f
   evidence exposures and, 407
   fibers and, 407–409, 410f
   lighting tint from, 55f
   powders/chemicals and, 407
Fluorescence photography, 397–398
Focal length
   flash head and, 238–239
   full-face photography and, 359
   lens distortion and, 173
   of lenses, 169–170
   for midrange photographs, 351f
   normal lenses in, 173–174
   surveillance photography and, 532–533
   zoom lenses settings and, 170f, 171f
Focal plane shutter, 33, 34f
Focus, 120, 159f, 160f, 165f
   auto, 134–136
   automatic, 134–136
   circles of confusion and, 131, 143f, 144f
   concepts of, 120–122
   depth of field and, 143–144, 145, 536
   entire area in, 139
   hyperfocal, 138–147
   images out of, 156–157
   on infinity, 144f
   infrared/ultraviolet adjustments for,
      153–156, 155f, 392f
   of lenses, 170f
   light rays converging for, 131–132, 132f
   manual, 133
   midway, 153
   pre, 137–138
   ring for, 141f
   rule of thirds for, 150
   visible/infrared light and, 155f

wide/small apertures and, 140f

    zone, 147–150

    zoom lens and, 136–137

Fodis Pro 3D macro lens, 209f

Foot prints, 338f

Foregrounds

    avoid distracting, 9–11

    distracting, 361f

    elongation of, 183–186

F/stops

    aperture and, 37–44, 40f

    crime scene images and, 69

    distance appropriate for, 145–147

    DOF and, 30, 69, 143–144, 157–160

    donuts representing, 43f

    equations for, 37–39

    as exposure variables, 37–44

    f/16 sunny day rule and, 97–101

    film speeds corrected with, 58–59

    flash calculator dial used for, 242–243

    frequent manipulation of, 240

    graphical representations of, 41f

    in hyperfocal focus, 146f

    ISL and, 245–253, 251f

    ISO settings influenced by, 166

    knife and, 45f

    on lens, 46f

    pizza representing, 42f

    shadows and, 99f

    shutter speeds combinations with, 67f, 68f

Full-face photography, 359, 361f, 363f

Full-frontal photography, 375

G

Geometry, 449

Ghost, 221, 221f

Glass table top, 298f

Graininess, 47, 71f

Green grass, 87

Grid corners, 431f

Grid system, 457–459

GSR. *See* Gunshot residue

Guide numbers, 216–218

Gun. *See* Pistol

Gunshot residue (GSR), 403–404, 404f

Gunshot wound, 369f, 373f, 374f

H

Halftone dots, 485f

Hard shadows, 13–14, 14f, 17f, 269, 271f, 272f

High resolution photographs, 497

High-contrast scenes, 83–85

Homemade reflectors, 273f

Homicide scene, 462f, 546f

Horizontal parallel line, 432f, 433f

Hot shot mounting, 231f

Hungarian red fluoresced, 390f

Hyperfocal focus, 19, 138–147

    defining, 141

    DOF maximized by, 145

    f/stops in, 146f

I

IAI. *See* International Association for Identification

Image capture, 494–495

Image processing, 506–509. *See also* Digital images

Image quality, 501f

    digital cameras', 491

    digital imaging and, 494

    photographers responsible for, 134–135, 139

    printers and, 486–489

    resolution and, 482–483

    of silver halide crystals, 491

Image Quality Specifications (IQS), 488

Images. *See also* Cropped images; Digital images

    aberrations degrading, 209–211

    bite marks captured as, 494–495

    burning/dodging of, 60

    capture of, 474f

    court admissibility of, 508

    manipulation techniques tracked of, 477f

    out-of-focus, 156–157

In situ, 352–353

Incident light meters, 74

Indented writing, 286–290, 288f, 289f, 290f

Indoor photography , ISO settings for, 50

Infinity, 144f, 145–150

Information, digital cameras capturing,
    470–471

Infrared focus adjustments, 153–156, 155f,
    392f

Infrared light (IR), 153–156, 386

  burned writing visualized with, 406f

  Canon digital camera and, 399f

  EMS with, 384–390

  GSR and, 404f

  inks and, 402f, 403f

  pistol visualized using, 405f

Ink jet printers, 474–475, 475f, 484

Inks, 401f, 402f, 403f

Insect infestation, 371f

Interior overall photographs, 80, 338–343

  building exterior door for, 340f

  direct/bounce flash in, 344f

  door/hallway for, 340f

  film plane parallel in, 343f

  wide-angle lens used for, 341f

International Association for Identification
    (IAI), 597

Interpolation, 493, 499, 499f, 503,
    588–590, 589f

Inverse square law (ISL), 245–253, 246,
    247f, 251f

  as equation, 248

  physics of light relating to, 39

Inverse square law II, 249f

Inverse square law III, 250f

Inverse square law IV, 251f

Inverse square law V, 252f

Investigative leads, 582

IQS. *See* Image Quality Specifications

IR. *See* Infrared light

Irrelevant material, 8–16

ISL. *See* Inverse square law

ISO settings

  camera dial for, 51f

  determining, 166–168

  of digital cameras, 47–50

  on flash calculator dial, 226f

  f/stop-weather influencing, 166

  lighting conditions and, 536–537

  sunny day/indoor, 50

  surveillance photography and,
    536–537

Isosceles triangle, 23, 24, 346–348, 358

J

JPEG compression, 500–503, 592–593

K

Knife, 31f, 38f, 45f, 48f

L

Labeled scales, 327f, 329f, 330f, 576f

  close-up photography and, 326–331, 354,
    355f, 357f

  partial scale in, 355, 355f

  props and, 329f

  same plane as evidence, 327

Laboratory disciplines, 124–125

Laser printers, 475–476, 484

Lateral chromatic aberrations, 195f

Law enforcement, 471, 509–510, 514

LCD. *See* Liquid crystal display

LCD projector, 468, 469f

LED. *See* Light-emitting diode

Legal implications, digital imaging,
    585–597

Lenses, 168–211, 534f. *See also* Macro
    lenses; Telephoto lenses; Wide-angle
    lenses; Zoom lenses

  angles of view of, 174f

  aperture designations of, 172–173

  camera with, 16f

  distortion from, 350–352

  DOF influenced by, 161–165, 162f

  800 mm view from, 176f

  fast/slow, 170–172

  50 mm, 175f, 180f, 182f, 185f, 193f

  filters protecting, 103

  flare from, 14–16, 15f, 17f

  focal length of, 169–170, 173

focusing, 170f
f/stops on, 46f
macro, 186–192
normal, 173–174
100 mm view from, 180f
optical problems of, 192–211
telephoto/zoom, 161, 171f, 174–181
35 mm view from, 185f
28 mm view from, 182f, 184f, 185f
24 mm distortion from, 182f
2 mm lens per foot distance for,
    176–177, 178f
for underwater photography, 564
wide-angle/telephoto, 62
Leuco fluorescein, 416f
Light, 286–288
    ambient/existing color of, 53–56
    bounce flash and, 298–302
    bulbs, 526f–527f
    circular aperture penetrated by, 198f
    converging, 131–132, 132f
    dim, 298–302
    direct v. oblique, 269–273
    as exposure variables, 50–53
    film/digital sensors sensitivity to,
        44–46
    inks and, 401f
    linear opaque barrier struck by, 197f
    oblique, 269–295
    physics of, 39
    reflective meter for, 74–76, 116
    round opaque barrier struck by, 198f
    speed of, 253
    viewfinder with, 113f
    visible, 385, 386f, 406–416
    wave theory of, 199, 199f
    white, 386f
Light meters, 115
    black sweater using, 82f
    center-weighted, 74f, 75
    18% gray card and, 87
    matrix, 75–76, 75f
    reflective/incident, 74
    sky using, 83f
    spot, 74f, 75

Light-emitting diode (LED), 483
Lighting
    accurate color capture and, 53
    afternoon sun influencing, 54f
    case law relating to, 600–601
    co-axiel, 629–630, 630f, 631f
    18% gray card standard of, 72–74, 73f
    exposures and, 50–53, 57–58, 76–77
    fill-in flash recommendations for, 261–269
    flashlight used for, 286–288
    fluorescent, 54–55, 55f
    ISO settings and, 536–537
    normal/non-normal crime scene, 76–85
    tungsten, 54, 55f
    in underwater photography, 555–564
Linear opaque barrier, 197, 197f, 200, 207f
Linear polarizer filter, 112
Linear viewpoint, 24f
Line-two points determining, 440f
Liquid crystal display (LCD), 483
Longitudinal chromatic aberrations, 195f
Lossless compression formats, 472, 506,
    592
Lossy compression formats, 427, 592
Luminescence, 389
Luminol, 411–416, 412f, 413f, 414f

M
Macro lenses, 186–192
Magnification
    105 mm lens/extension tube for, 191f
    from telephoto lenses, 175–181
Manipulation, 587–593, 588f, 615–621
Manual exposure mode, 89–90, 89f, 92–93
Manual flash, 239f
    calculator dial and, 230
    electronic flash used with, 224–229
Manual flash exposure mode
    bracketing in, 241–245
    non-flash, 256–257
    problems with, 239–241
Manual focus, 133
Material, 579–580, 608–609
Matrix light meters, 75–76, 75f
Megapixels, 479–485, 492, 497

Metadata, 470
Midrange photographs, 7, 8f, 344–352
    of casing/electric box, 346f
    composition of, 348f, 349f
    crime scene, 359f
    focal lengths for, 351f
    isosceles triangle thinking for, 23, 24
    of pistol, 345f
Mikrosil, 285–286, 286f
Miranda rights, 583
Mirrors, 297f, 298f, 631–632, 632f, 633f
Monitor resolution, 482–483
Motion control, 61–67, 63, 63f
Movement, shutter speeds freezing, 535–536
Multiple flashes, PWL, 313f

N
Narrow aperture, 39f
National Institute of Standards and
    Technology (NIST), 126
Natural grid photogrammetry, 446–453, 447f
    with evidence, 447f, 453f
    midpoint/AB line in, 448f
    point located in, 449f
    rectangles bisected in, 450f, 451f, 453f
    vertical parallel lines in, 448f
Natural perspective, 331–332
ND. *See* Neutral density filter
Negative dust print lift, 290–291, 292f
Neutral density filter (ND), 65–66,
    66f, 112
Nighttime accident photography, 530–533,
    531f, 532f, 533f
NIST. *See* National Institute of Standards
    and Technology
Noise, digital images with, 499
Normal room, 229–231

O
Oblique flash, 269, 274–275, 276f
Oblique light
    direct light v., 269–273
    flash sequence for, 273–277
    flash/non-flash, 269–295
Off the film (OTF), 259

1:1 magnification, 126
    of coin, 138f
    of fingerprint, 190f
    of macro lenses, 186–191
    ratio of a nickel, 189f
1:7 magnification, 190f
Optical center, 169–170
Optical illusions, 602–603
Optical problems
    as aberrations, 192–194
    as diffraction, 196–209
    of lenses, 192–211
Original object, accurate representation of,
    476f
OTF. *See* Off the film
Outdoor scene, 454f
Overall photographs, 331–343
Overexposures, 58–59, 93, 268f

P
Painting with light (PWL), 302–318
    camera operators position in, 310f
    camera variables for, 305–307
    crime scene using, 303f
    flash angles in, 309f
    flash variables for, 307–311
    flashlight used in, 317
    multiple flashes in, 313f
    from one side, 315–318, 316f, 317f
    tips for, 311–314
    tripod variables for, 304–305
    from two sides, 303–304
    variables in, 304–311
    when to use, 302–303
    wide angle lenses used in, 308f
Panning, 64, 285f
Paper clip, 207f
Paper cutouts, 429f
PC cord, 233–234
Pedal mark photos, 525f
Pentagon, 548f
Perishable evidence, 515
Perspective disc photogrammetry, 441–446
    crime scene using, 443f
    lines extending in, 443f

perspective grid v., 441–442, 444–445

tangential lines extending in, 443f, 445f

Perspective distortion, 23–25, 183–184, 343–347

Perspective grid photogrammetry, 425–441

1′ × 1′ corners crossed grid in, 435f

2′ × 2′ grid in, 435f, 438f

3″ × 3″ grid in, 437f

6″ × 6″ corners crossed in, 436f

all lines marked in, 434f

evidence located in, 439f, 440–441

evidence marked in, 427–28

exercise image of, 427f, 439f

extending grid in, 428–433

four sides extended in, 429f

grid base length duplicated in, 430f

grid corners crossed/extended in, 431f

grid reduction in, 433–439

homicide scene using, 427f, 462f

horizontal parallel line in, 432f, 433f

paper cutouts in, 429f

perspective disc v., 441–442, 442, 444–445

positioning of, 425–426

tiles used in, 426f

Y-point hash marks in, 431f

Phosphorescence, 389

Photo identifiers, 87, 88f, 324f

changes to, 323

as first frame of film, 322–324

Photo memo sheet, 324–326, 325f

Photogrammetry

crime scene photography using, 421–425

perspective grid, 425–441

Photograph(s)

of accident scene, 518, 521f

of bodily injuries, 593f, 594f

consent required for, 583

court admissibility of, 506–507

digital cameras and, 472f, 497

enlargements of, 621–623

evidence and, 466–471, 569–582, 583–585

high resolution, 497

hoax, 506f

midrange, 7

sensitive, 372–380

vehicle, 521f–523f

viewfinder composing, 4–6

Photographers

decision-making process of, 115

image quality responsibility of, 134–135, 139

keeping steady, 61

portrait, 16–17

primary subject defined by, 5

Photographic documentation, 357–380

Photographic infrared range, 398

Photography. *See also* Full-face photography

aerial, 546–553, 546f

bite marks, 283–286

body area, 379f

dust print lift, 290–295

equipment used for, 1–4

fluorescence/UV, 397–398

full-face, 359, 361f

lens flare in, 14–16, 15f, 17f

nighttime accident, 530–533, 531f, 532f, 533f

rape victim, 376f, 377f

reflected UV in, 393–396

shadow control in, 11–14

surveillance, 533–545, 584–585

tire tracks, 281–283

underwater, 553–565, 554f

Photoreceptors, 493f

Photoshop, Adobe, 201f, 473, 495, 496, 545, 632–634, 634f

Physics of light, 39

Pincushion distortion, 210–211, 211f

Pistol

with background, 9f

filling the frame, 7f

film frame parallel and, 26f, 133f

IR light visualization of, 405f

midrange photo of, 345f

Pistol (*Continued*)
    oblique viewpoint of, 134f
    underwater photo of, 554f, 555f, 559f
Pixels, 470
    dots and, 483–489
    halftone dots converted from, 485f
    monitor resolution from, 482–483
Pixels per inch (ppi), 478, 479, 488
Pizza, f/stops illustrated by, 42f
Point-and-shoot camera, 1–4, 3f
Poisson's spot, 199–200, 200–201, 206f
Polarizer filters, 66–67, 66f, 104–112, 106f,
        107f, 110f
    background cloth darkened by, 112f
    blue sky with, 109
    color saturation enriched by, 108
    glare and, 110f
    linear/circular, 112
    neutral density filter similar to, 112
    reflection elimination by, 105–108
    relative angle using, 111f
    scene using, 111f
    skid marks more obvious using,
        108–109
Portrait photographers, 16–17
Positive dust print lift, 291, 292f
Powders, 407
PPI. *See* Pixels per inch
Pre-focus, 137–138
Prejudicial evidence, 580–582, 581f,
        609–615
Primary subject
    determining, 4–5
    entirely in shadow, 12f
    field of view and, 27
    fill the frame with, 11
    getting closer to, 7–8
Printers, 486–489, 623. *See also* Dye
        sublimation printers; Ink jet
        printers; Laser printers
Prism, 386f
Processing, 594
Proficiency testing, 596
Program exposure mode, 90–91
Props, 329f

Purses, bite marks on, 287f
"Push processing," 542–545, 544f
PWL. *See* Painting with light

R
Rape victim, photographing, 376f, 377f
RAW compression, 503–504, 592–593
Reasonable expectation of privacy, 584
Recharge time, 241
Reciprocal exposures, 67–71
Reciprocity failure, 70–71, 70t
Recreation photographs, 578
Recycle time, 241
Reflected Ultra-Violet Imaging System
        (RUVIS), 391
Reflected Ultraviolet photography, 393–396
Reflections, 555
    bounce flash avoiding, 297f
    eliminating, 105–108
    glass/water, 107f
    window, 109f
Reflective light meter, 72–76, 116
Reflectors, 270–273, 273f
Refraction, 556, 556f, 557f
Relevant, 579–580, 608–609
Remote flash cord, 233–234
Remote Shutter release cable, 35, 36f, 62
Repeatable process, 595
Resolution, 122–132
    chart, 123f, 203f, 204f, 205f
    of digital cameras, 127–131
    digital imaging and, 494
    of faxes, 126–127
    image quality and, 482–483
    standards/guidelines for, 126–127
Resolution chart, cropped, 203f
Reverse projection photogrammetry, 453–456
    as least favorite technique, 455
    outdoor scene using, 454f
    transparencies used in, 455f, 456f
Rhino photogrammetry, 456–462, 461–462
Rhinoceros, 456—462, 458, 458f, 459f,
        460f, 461f
Rhodamine 6G, 407, 408f
Ring-lites, 235

Rotations, 590f, 591f

Round opaque barrier, 198f

Rule of thirds, 150–153

Rule of thumb

  accident scene

    driver eye level photographs of, 518

    vehicular damage at, 522

  aerial photography shutter speeds in, 550

  bounce flash/wider aperture in, 302

  close-up photography in, 356

  color capture lighting in, 53

  crime scene photography

    blue light/orange filter used at, 390

    focusing in, 121

    green light/red filter used for, 390

    shutter speed/f/stops used for, 69

    UV light and, 389

    wide aperture options for, 93

  critical comparison photography

    aperture settings for, 192

    ISO settings for, 50

  digital cameras

    EXIF data maintained of, 470

    high resolution photographs required from, 497

    underexposures with, 59

  digital images compression formats in, 506

  dots per inch (dpi), 479

  electronic flash, 241, 255

  film cameras overexposures in, 58–59, 93, 268f

  film speeds/f/stop corrections and, 58–59

  filters/screwing on, 105

  fluorescence evidence exposure in, 407

  focus

    entire area in, 139

    midway, 153

    rule of thirds for, 150

  full-face photography lens focal length in, 359

  images/court admissibility of, 508

  indoor photography ISO settings in, 50

ISO settings

  f/stop-weather influencing, 166

  lighting conditions and, 536–537

  sunny day, 50

labeled scales plane in, 327

law enforcement accident scene protection in, 514

lenses focal length distortion in, 173

manual flash calculator dial in, 230

perspective disc photogrammetry in, 441–442

perspective grid photogrammetry positioning in, 425–426

photo identifier changes in, 323

photo identifiers and, 323

resolution/image quality in, 482–483

shutter speeds/blur elimination in, 73

surveillance photography

  focal length/distance for, 532–533

  lens per foot distance in, 176–177

  shutter speeds for, 535–536

underwater photography distances in, 564

RUVIS. *See* Reflected Ultra-Violet Imaging System

S

Safety concerns, 564–565

Scatter, 559, 559f

Schematic, camera, 34f

Scientific Working Group on Imaging Technology (SWGIT), 124–125, 127, 128, 585–586

Scissors photos, 9f, 10f, 52f

Semen stains, 390f

Sensitive photographs, 372–380

Sensor eye, 228f, 229f

Shade

  blue tint from, 53f

  cloudy, 53

  deep, 101f

  exposure and, 58f

  open, 100f

Shadow control

  electronic flash and, 11–14, 13f

  sunlight and, 12f, 86f

Shadows, 98f
  bounce flash avoiding, 298f
  f/stops and, 99f
  hard/soft, 13–14, 14f, 15f, 17f, 269,
    271f–272f
  primary subject in, 12f
  reflectors softening, 270–273
  scissors and, 52f
  soft, 15f
Shoe prints
  bloody, 416f
  capturing images of, 495–497
  dust print lift from, 291f
  oblique flash and, 274–275, 276f
  sand/snow with, 277–281, 279f, 280f,
    281f
  tripod setup for, 278f
  wax used for, 277, 279f
Shutter exposure mode, 91–92
Shutter release cable/bulb, 35f
Shutter release, delayed, 62
Shutter speeds
  for aerial photography, 550
  blur eliminated using, 61–64
  crime scene images and, 69
  dial, 35f
  exposure influenced by, 33–37
  as exposure variables, 33–37
  f/16 sunny day rule and, 97–101
  f/stop combinations with, 67f, 68f
  knife at different, 38f
  motion control from, 61–67, 63f
  movement frozen by, 535–536
  off from sync speeds, 222f
  proper exposure and, 37
  rain/snow eliminated by, 64–67, 66f
  for surveillance photography, 535
  telephoto lenses and, 534–535
Sight perspective, 603–604
Silver halide crystals, 46–47, 484, 491
Single family home, 334f
Single lens reflex camera (SLR), 33
Skid marks, 108–109
Skin, 80, 295f, 368
Skull, wound to, 374f

Sky, 83f, 103–104, 109
SLR. *See* Single lens reflex camera
Small apertures, 140f
Sniper shooting sites, 547f
Snow, 64–67, 66f
Soft shadows, 13–14, 15f, 270
Software programs - grid systems from,
  456–457
Sony 717 IR capable camera, 399f
SOP. *See* Standard Operation
  Procedures
Speed of light, 253
Speedometers, 528f
Spherical aberrations, 195f
Spot meters, 74f, 75
Stab wounds, 373f
Stalin, Joseph, 467f
Standard Operation Procedures (SOP),
  127–128
Stati-Lift, 294
Stokes shift, 397
Storage, 49f
Storage/management, 498–510
Street signs, 332f
Subject, 64, 163–165, 232
Sunlight, 12f, 86f
Sunlit grass, 58f
Sunrise/Sunset, 53
Surveillance photography, 533–545, 538f,
  539f, 540f, 543f, 584–585
  aperture settings in, 533f
  daytime, 537–542
  focal length/distance for, 532–533, 541f,
    542f
  ISO settings and, 536–537
  shutter speeds for, 535
  2 mm lens per foot distance for, 176–177
Suspects, 583–585
SWGIT. *See* Scientific Working Group on
  Imaging Technology
Sync speeds
  bullet and, 220f
  dial indicators for, 219f
  of electronic flash, 218–223
  shutter speed off from, 222f

T

Tangential lines, 443f, 445f

Telephoto lenses, 62, 161, 173, 538f
  distortion from, 180, 350
  DOF narrower from, 180–181
  field of view of, 179
  flash head used with, 238–239
  magnification from, 175–181
  shutter speeds hand holding, 534–535
  view compression from, 179–181

Theory of reciprocity, 70

3D modeling software, 425

Through the lens (TTL), 259

TIFF compression, 505–506, 592–593

Tilt-Shift lens, 628

Time frame, 604–606

Tire damage, 524f

Tire tracks, 281–283, 282f
  at accident scenes, 519–520
  capturing images of, 495–497
  photographing, 281–283

Total station, 424–425

Traffic control, 517f

Traffic flow, 515–516, 516–517, 518f

Transparencies, 455f, 456f

Trespassing, 583

Tricky scenes, 85–92

Tripod
  camera mounted on, 364f
  shoe prints photographed using, 278f
  variables, 304–305

T-shirts, dust print lift from, 293f

TTL. *See* Through the lens

Tungsten lighting, 54, 55f

U

UAV. *See* Unmanned Aerial Vehicle

UCI. *See* Underwater Crime Investigators

Ultraviolet focus adjustments, 153–156

Ultraviolet light (UV), 153–156, 386, 391–398
  blood and, 396, 396f
  crime scene photography and, 389
  fiber/fingerprint fluorescence by, 397f
  filters blocking/transmitting, 394f

  filters for, 103–104, 105f, 394f
  fluorescence photography, 397–398
  photography and, 393–396
  semen stains with, 390f

Underexposures, 84f, 86f, 268f
  with digital cameras, 59
  digital processing of, 545, 545f
  "push processing" for, 542–545

Underwater Crime Investigators (UCI), 562, 562f

Underwater photography, 553–565, 554f, 563f
  camera for, 558f
  direct flash in, 561f
  distances in, 564
  flash used in, 557f, 559f
  lenses for, 564
  lighting in, 555–564
  of pistol, 554f, 555f, 559f
  problems in, 563–564
  safety concerns for, 564–565

Unmanned Aerial Vehicle (UAV), 549

Urine stains, 388f

UV. *See* Ultraviolet light

V

Vanishing point, 430

Vehicle identification number (VIN), 522, 524f

Vehicle photographs, 521f–523f, 530f, 573f

Vehicular damage, 522

Vertical parallel lines, 448f

Vertical walls, evidence, 627–629, 628f

View compression, 179–181

Viewfinder, 4–6, 113f

VIN. *See* Vehicle identification number

Visible light, 385, 386f, 406–416

Vivitar 285 HV flash, 227f, 228f, 229f

W

Washington, George, 534f

Wave theory of light, 199, 199f

Wavelength, 384f

Weather, 166, 515, 517

White balance, 56, 56f
White light, 386f
Wide angle flash diffuser, 240f, 244
Wide apertures, 18f, 39f, 93, 140f, 302
Wide-angle lenses, 62, 173, 181–186
  distortion from, 183–184, 350
  edges tilting in with, 183f
  field of view from, 181–183
  interior overall photographs from, 341f
  PWL using, 308f
Windshield damage, 530f
Witness viewpoints, 518–519
Wounds, 368f, 369f, 373f, 374f

Y
Y-point hash marks, 431f

Z
Zeiss, Carl, 131
Zone focusing, 147–150, 148f, 149f, 152f, 153f, 155f, 160f
Zoom lenses. *See also* Telephoto lenses
  camera with, 191f
  focusing with, 136–137
  at various focal lengths, 170f, 171f